Mechanisms of Mitotic Chromosome Segregation

Special Issue Editor

J. Richard McIntosh

MDPI

Special Issue Editor
J. Richard McIntosh
University of Colorado
USA

Editorial Office
MDPI AG
St. Alban-Anlage 66
Basel, Switzerland

This edition is a reprint of the Special Issue published online in the open access journal *Biology* (ISSN 2079-7737) from 2016–2017 (available at: http://www.mdpi.com/journal/biology/special_issues/mitosis).

For citation purposes, cite each article independently as indicated on the article page online and as indicated below:

Author 1; Author 2; Author 3 etc. Article title. *Journal Name*. **Year**. Article number/page range.

ISBN 978-3-03842-402-4 (Pbk)
ISBN 978-3-03842-403-1 (PDF)

The cover image shows a PtK_1 cell in anaphase. The picture was recorded on the Boulder high voltage electron microscope by Mary Morphew and Richard McIntosh. They prepared the sample by culturing the cells on gold grids coated with Formvar and carbon, lysing them in an equilibrium mixture of tubulin and microtubules in 100 mM PIPES buffer, pH 6.9, supplemented with 1 mM GTP, $MgSO_4$, and EGTA at 37° C, followed by fixation in 2% glutaraldehyde, post-fixation in 1% OsO_4, then dehydration in an ethanol series, followed by drying with the critical point method. Reprinted, courtesy of Cold Spring Harbor Press, doi: 0.1101/cshperspect.a023218.

Table of Contents

About the Guest Editor....................v

Preface to "Mechanisms of Mitotic Chromosome Segregation" vii

J. Richard McIntosh and Thomas Hays
A Brief History of Research on Mitotic Mechanisms
Reprinted from: *Biology* **2016**, *5*(4), 55; doi: 10.3390/biology5040055
http://www.mdpi.com/2079-7737/5/4/55 1

Tarun M. Kapoor
Metaphase Spindle Assembly
Reprinted from: *Biology* **2017**, *6*(1), 8; doi: 10.3390/biology6010008
http://www.mdpi.com/2079-7737/6/1/8 39

Andrea Musacchio and Arshad Desai
A Molecular View of Kinetochore Assembly and Function
Reprinted from: *Biology* **2017**, *6*(1), 5; doi: 10.3390/biology6010005
http://www.mdpi.com/2079-7737/6/1/5 75

Helder Maiato, Ana Margarida Gomes, Filipe Sousa and Marin Barisic
Mechanisms of Chromosome Congression during Mitosis
Reprinted from: *Biology* **2017**, *6*(1), 13; doi: 10.3390/biology6010013
http://www.mdpi.com/2079-7737/6/1/13 122

Ajit P. Joglekar
A Cell Biological Perspective on Past, Present and Future Investigations of the Spindle Assembly Checkpoint
Reprinted from: *Biology* **2016**, *5*(4), 44; doi: 10.3390/biology5040044
http://www.mdpi.com/2079-7737/5/4/44 178

Michael A. Lampson, M.A.; Grishchuk, E.L.
Mechanisms to Avoid and Correct Erroneous Kinetochore-Microtubule Attachments.
Reprinted from: *Biology* **2017**, *6*(1), 1; doi: 10.3390/biology6010001
http://www.mdpi.com/2079-7737/6/1/1 197

Moé Yamada and Gohta Goshima
Mitotic Spindle Assembly in Land Plants: Molecules and Mechanisms
Reprinted from: *Biology* **2017**, *6*(1), 6; doi: 10.3390/biology6010006
http://www.mdpi.com/2079-7737/6/1/6 215

Charles L. Asbury
Anaphase A: Disassembling Microtubules Move Chromosomes toward Spindle Poles
Reprinted from: *Biology* **2017**, *6*(1), 15; doi: 10.3390/biology6010015
http://www.mdpi.com/2079-7737/6/1/15 235

Jonathan M. Scholey, Gul Civelekoglu-Scholey and Ingrid Brust-Mascher
Anaphase B
Reprinted from: *Biology* **2016**, *5*(4), 51; doi: 10.3390/biology5040051
http://www.mdpi.com/2079-7737/5/4/51 267

Tamara Potapova and Gary J. Gorbsky
The Consequences of Chromosome Segregation Errors in Mitosis and Meiosis
Reprinted from: *Biology* **2017**, *6*(1), 12; doi: 10.3390/biology6010012
http://www.mdpi.com/2079-7737/6/1/12 297

About the Guest Editor

J. Richard McIntosh has been a student of mitosis since he finished graduate school in 1968. He has studied aspects of chromosome segregation in plants, animals, slime molds, and fungi, using methods of microscopy, biochemistry, genetics, and molecular biology. He has also been interested in conceptual models that might cast light on the hidden complexities of spindle function. His research initially focused on microtubule-dependent motor enzymes that might play a role in chromosome movement. Later he probed the ways in which microtubule depolymerization might generate forces to contribute to the process. His 50 years of working on this fascinating process have allowed him to meet and work with many important scholars of mitosis. The expertise of all the authors in this book reflects his wide-ranging friendships and acquaintances in the field.

Preface to "Mechanisms of Mitotic Chromosome Segregation"

Mitosis attracts the interest of many biologists because it is fundamental to the livelihood of all eukaryotes. It is also an esthetic pleasure to watch, thanks to the power of time-laps imaging and the elegance of chromosome segregation. Most biologists learned the basics of mitotic phenomenology in the early years of their career, but they have not kept up with the complexities that have emerged from more recent research. This book provides a convenient way in which to learn the remarkable advances that the field has achieved. The progress described in these chapters reflects both the skill and insight of many investigators and the power brought to biological research by the many technological advances that have occurred in recent decades.

Mitotic spindles have long been difficult to study because they are small, ephemeral, essential, and complex. Recent studies of mitosis have been revolutionized by progress in many fields of science and technology including optics, electronics, genetics, biochemistry, and molecular biology. Advances in all these fields have had significant impact on the way students of mitosis can answer questions about mechanisms. For example, improvements in cameras have allowed scholars to follow the behavior of specific molecules within mitotic structures. Techniques for labeling and purifying molecules have helped scholars understand the functions of particular spindle components. Genetics, in combination with molecular biology, has allowed scientists to relate processes in vivo to molecular functions in vitro. In short, research on mitotic mechanisms reflects the progress in all fields of cell biology. Mitosis is, in fact, a showcase for the power of methods and approaches that have revolutionized our understanding of cells in general.

This book is a compilation of reviews by experts in several aspects of mitosis: the formation of the mitotic spindle, the specializations that attach chromosomes to the spindle, the signaling processes that regulate mitotic progression to maximize accuracy, the mechanical processes that drive chromosome motion, and the consequences of mitotic mistakes. The authors of these chapters work on a range of organisms, so aspects of mitosis in fungi, plants, insects, lower vertebrates, and mammals are all considered. One goal of this book is to bridge the gaps that sometimes form because scholars of the same process in different organisms do not always communicate as closely as they should for the advancement of knowledge.

Students of mitoses must deal with the fact that they work in the context of more than a century of work by an international army of scientists. This situation can lead to review articles that either skip the earlier work and cite only the most recent papers, or get bogged down in the morass of early studies and fail to deal adequately with the most recent research. This book includes one chapter that addresses some of the history of mitosis research from its origins in the nineteenth century up to approximately 1980. This strategy has allowed other authors to focus on more recent findings and their significance for understanding mechanisms.

Given these attributes, the editor hopes that you will find this book interesting and useful. Its content represents an informed account of a complex and important field of biology; its style is directed towards a pleasant reading experience; and you can't beat the price for a scholarly volume that could readily be used for teaching an advanced course.

J. Richard McIntosh
Guest Editor

biology

MDPI

Review

A Brief History of Research on Mitotic Mechanisms

J. Richard McIntosh [1,*] and Thomas Hays [2]

1 Department of Molecular, Cellular and Developmental Biology, University of Colorado, Boulder, CO 80309, USA
2 Department of Genetics, Cell Biology and Development, Medical School and College of Biological Sciences, University of Minnesota, Saint Paul, MN 55455, USA; haysx001@umn.edu
* Correspondence: richard.mcintosh@colorado.edu; Tel.: +1-303-492-8533; Fax: +1-303-492-7744

Academic Editor: Chris O'Callaghan
Received: 1 October 2016; Accepted: 25 November 2016; Published: 21 December 2016

Abstract: This chapter describes in summary form some of the most important research on chromosome segregation, from the discovery and naming of mitosis in the nineteenth century until around 1990. It gives both historical and scientific background for the nine chapters that follow, each of which provides an up-to-date review of a specific aspect of mitotic mechanism. Here, we trace the fruits of each new technology that allowed a deeper understanding of mitosis and its underlying mechanisms. We describe how light microscopy, including phase, polarization, and fluorescence optics, provided descriptive information about mitotic events and also enabled important experimentation on mitotic functions, such as the dynamics of spindle fibers and the forces generated for chromosome movement. We describe studies by electron microscopy, including quantitative work with serial section reconstructions. We review early results from spindle biochemistry and genetics, coupled to molecular biology, as these methods allowed scholars to identify key molecular components of mitotic mechanisms. We also review hypotheses about mitotic mechanisms whose testing led to a deeper understanding of this fundamental biological event. Our goal is to provide modern scientists with an appreciation of the work that has laid the foundations for their current work and interests.

Keywords: mitosis; mitotic spindle; chromosome; kinetochore; microtubule; motor enzyme; centrosome; tubulin dynamics; force; accuracy

1. Discoveries about Mitosis from Early Descriptions of Mitotic Structures

The history of research on mitosis is intertwined with the development of the relevant technologies, particularly microscopy. This linkage derives from the sizes of spindles and their activities; it also reflects a need for significant signal amplification to study mitotic components and processes. Moreover, the isolation of dividing nuclei as a simplified system for biochemical studies has proven technically difficult, so in the early days of mitosis research the majority of information came from work on whole cells. Indeed, research on mitosis has motivated the development of several microscope technologies, including more effective modes of live cell imaging. We have therefore organized this presentation around the emergence of relevant technologies and the physical sciences that enabled them.

Initial work on mitosis took place in several laboratories, beginning around 1870. Pioneering studies by Friedrich Schneider [1] (Figure 1), Eduard Strasburger [2] and others independently described the structures and positions of chromosomes in fixed, dividing cells, while Eduard Van Beneden [3] identified objects at the spindle poles that we would now call centrosomes (Figure 2). It was, however, Walther Flemming [4] (translated into English and republished [5]) who named the "mitotic" process and first described a plausible chronology of chromosome behavior in anticipation of cell division (Figure 3). Much of his work is assembled in an elegant book, published in 1882 [6].

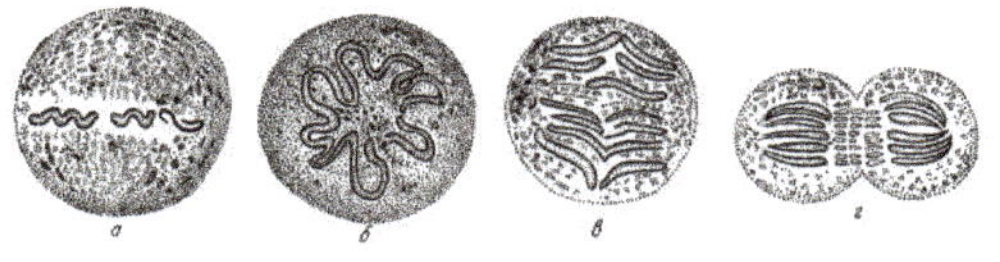

Figure 1. Drawing of dividing nuclei by Schneider, 1873 [1].

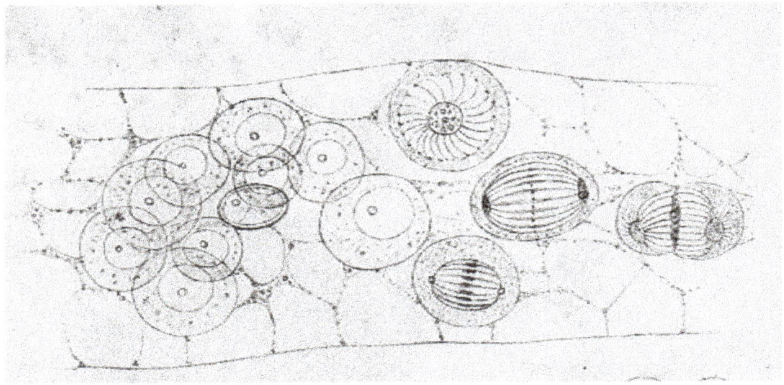

Figure 2. Drawing of mitotic figures that indicate structures at the spindle poles. van Benedin, 1876 [3] Image courtesy of Biodiversity Heritage Library. http://www.biodiversitylibrary.org. Drawing of mitotic figures that indicate structures at the spindle poles. van Benedin, 1876.

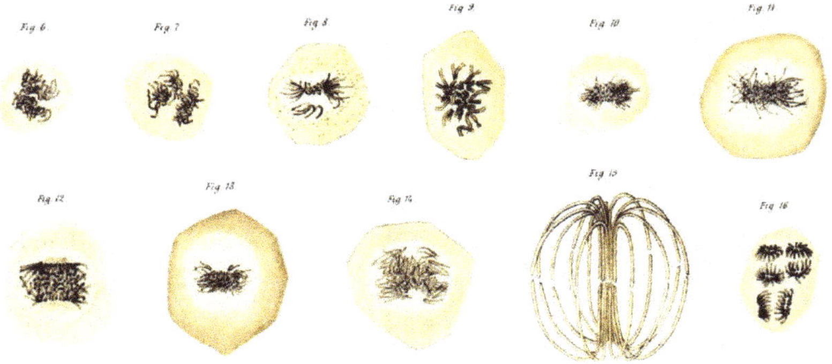

Figure 3. Drawings of mitotic figures by Flemming, 1878 [4,5]. This image is displayed under the terms of a Creative Commons License (Attribution-Noncommercial-Share Alike 3.0 Unported license, as described at http://creativecommons.org/licenses/by-nc-sa/3.0/.

Such work was possible because the imaging capabilities of the compound microscopes available at that time greatly exceeded those of the microscopes with which Hooke [7] first described cells in the 1660s, and with which van Leeuwenhoek [8] characterized the structures and behaviors of many single-celled organisms. Indeed, it was the invention of achromatic lenses (1823) that brought the resolution of light microscopy to ~1 μm, producing instruments that empowered Schleiden, a botanist, and Schwann, a zoologist, to demonstrate the ubiquity of cells, and Virchow [9] to realize that "all cells come from cells" (a powerful and important statement, despite its limited evolutionary perspective). They also provided both Pasteur and Koch with the tools they needed to recognize the importance of microorganisms in the propagation and progression of disease. Abbe's invention of a high numerical aperture condenser in 1875 [10] and the subsequent introduction of oil immersion lenses (1878) finally brought microscopes to a space resolution of ~0.2 μm, enabling the remarkably accurate drawings found in the best of the early descriptions of mitosis.

These early studies explored a range of mitotic cells in tissues of both animals and plants, mostly in specimens that were fixed and stained prior to examination. In this situation, what one saw depended quite strongly on the method of sample preparation. Moreover, without a camera to record the observations, structures were represented by hand drawings. These factors, and the variations in structure across the range of specimens examined, led to disagreement about the validity of any given set of observations. The first verbal description of mitosis in living cells was given by Mayzel in 1875 [11]. With Mayzel's permission, Flemming published a drawing of such a cell division [4] (Figure 4). Several other workers, such as Schleicher and Peremeschko, published images of chromosomes in live cells, but again it was Flemming who drew multiple stages in the division of a single cell type: epidermal cells from a salamander larva. The result was virtually a hand-drawn, time-laps movie (Figure 5) [4]. With time, additional images of mitosis in living cells were presented by scholars studying a range of organisms, and yet these descriptions generally involved only the chromosomes. At this point, there was no knowledge about the relationship between the "thick" fibers (chromosomes) and the "thin" ones (the spindle) seen in fixed material; both were thought to be manifestations of nuclear structure as this organelle prepared to divide.

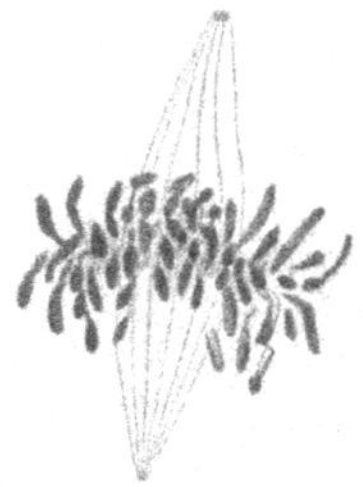

Figure 4. Drawing of a mitotic figure in a live cell prepared by Mayzel and published by Flemming, 1878. [4] This image is displayed under the terms of a Creative Commons License (Attribution-Noncommercial-Share Alike 3.0 Unported license, as described at http://creativecommons.org/licenses/by-nc-sa/3.0/.

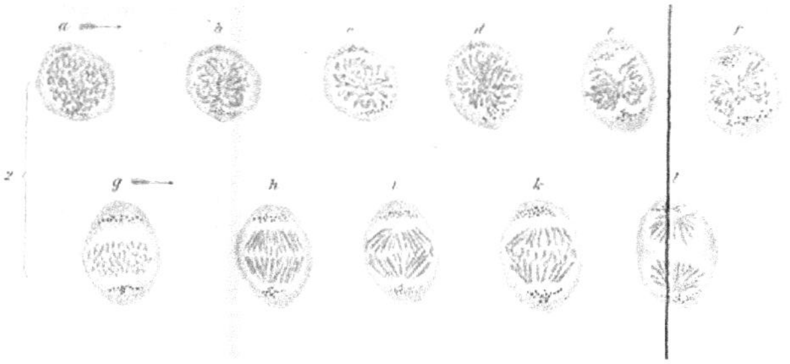

Figure 5. Drawings of chromosome segregation in living epidermal cells of a salamander larva. Flemming, 1878 [4]. This image is displayed under the terms of a Creative Commons License (Attribution-Noncommercial-Share Alike 3.0 Unported license, as described at http://creativecommons.org/licenses/by-nc-sa/3.0/.

The role of chromosomes as sites for the storage of a cell's genetic information was first proposed by Weismann in 1885 [12]; in 1903–1904 Boveri [13] and Sutton [14] published studies on the behavior of chromosomes during both normal cell division and the generation of gametes. They realized independently that chromosome motions and patterns of segregation were consistent with the then controversial idea that chromosomes carried a cell's genetic information. Most of these studies are well reviewed and summarized in the 1925 edition of E.B. Wilson's monograph on "The Cell in Development and Heredity" [15], an important resource for students of mitosis. From all this descriptive groundwork, the essential features of chromosome segregation were established, but the underlying mechanism for mitosis was still mysterious.

One limitation in this early work was the impact of the fixatives and stains used to visualize cellular infrastructure. Different fixation solutions were used by different investigators, but all such mixtures employed acids of various strengths and organic solvents, such as alcohols. Spindle fibers that might push and pull on chromosomes were seen by many, but only in fixed material, raising controversy about the legitimacy of these structures. While microscopists also saw mitotic events in living cells, in these specimens only the chromosomes were apparent. The very lack of visible spindle fibers in living cells cast doubt on the validity of the fibers seen in fixed cells, particularly since fixatives were known to induce the formation of aster-like structures in egg white and solutions of gelatin [16,17]. This observation led to the alternative concept that cytoplasm in living cells was colloidal, comprised of invisible particles and/or vesicles. In this view, fibers were artifacts of exposure to chemical fixatives, which triggered the condensation of invisible particles and/or vesicles into fibrous structures (reviewed in Wilson) [15].

The case for the reality of spindle fibers was supported by mitotic fibers that were evident in certain live cells, including diatoms as seen by Lauterborn in 1896 [18] (For an English translation, see [19]). Somewhat later, the case was enhanced by mechanical experiments in which Chambers used a microneedle to probe intracellular structures [20] (reviewed in [15]). These micromanipulation experiments showed that a spindle behaved as a coherent structure when twisted, rotated, displaced, or moved. The invention of phase optics by Zernike in the early 1930s (reviewed [21]), made spindle fibers more readily visible in some living cells, e.g., the flagellates living in the hind gut of the wood-feeding roach, *Cryptocercus* [22]. This work was particularly valuable, because the centrosomes in these unicellular organisms were much bigger than in most cells, allowing the first characterization of centrosome duplication and segregation during the cell cycle. Many workers in the field, however, viewed these results from "unusual cells" as unconvincing anomalies. Where were the spindle fibers in the mitotic cells of sea urchins, nematodes, amphibians, and mammals that had been the focus of so many studies?

Another imaging breakthrough came from the work of W.J. Schmidt [23] and F.O. Schmidt [24], each of whom employed polarized light microscopy to visualize the birefringence (BR), i.e., the two refractive indices that are visible in optically anisotropic materials. Viewed between crossed polarizers, the apparently homogenous material surrounding mitotic chromosomes was clearly if weakly birefringent, evidence for the presence of fibrous material in the living spindle. Spindle BR was also seen in mitotic cells from vertebrates in 1948 by Hughes and Swann [25]. Shinya Inoue pioneered several advances in the optics used to detect and measure BR, enhancing the value of polarized light microscopy for detailed observations on mitosis in living cells. He invented a way to compensate for the position-dependent optical activity of high numerical aperture lenses, allowing him to visualize spindle BR with high sensitivity (which depends on the extinction of the polarizing system) and at comparatively high space resolution (which depends on numerical aperture) (Figure 6) [26,27]. This invention also allowed Inoue to experiment with the factors that increased and decreased the amount of spindle BR [26], as described in more detail below. With this technology, Inoue saw time-dependent fluctuations in spindle birefringence and was able to use cinematography to capture the entire process of mitosis in livings cells from both plants and animals. These innovations led to an essentially universal acceptance of spindle fibers as a reality.

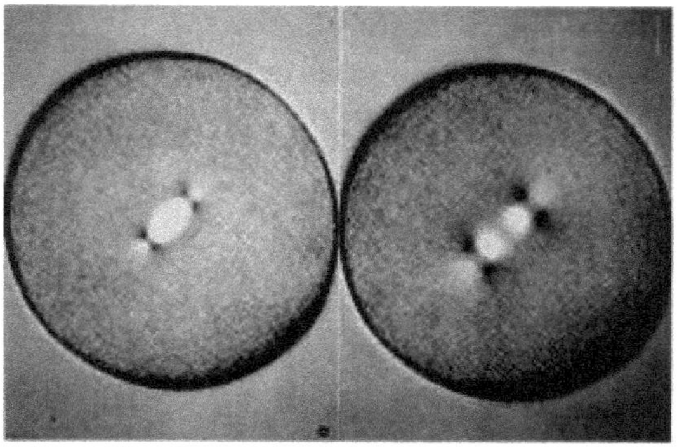

Figure 6. Mitotic spindles in living sea urchin eggs: metaphase (**left**) and mid-anaphase (**right**) viewed with polarization microscopy, similar to Inoue and Dan, 1951 [26]. Image from Salmon, E.D., 1982, Meth. Cell Biol. 25: 69–105. With permission from the author and the Copyright Clearance Center.

2. New Technologies for Structural Studies Advanced Our Understanding of Spindle Organization

The visualization of spindle fibers took another step forward when mitotic cells were successfully studied by electron microscopy. Initial work used the same harsh fixations that had produced fibers for view in the light microscope; now seen at higher resolution, the fibers appeared as bundles of much finer fibrils [28,29]. The "fine structure" of these fibrils was later seen with greater clarity by Harris in sea urchin blastomeres [30] (Figure 7) and by Roth and Daniels in amebae [31] that had been fixed with osmium tetroxide, either at low pH or in the presence of divalent cations. In this work, spindle fibers corresponded to bundles of 15 nm filaments that appeared tubular. With the subsequent discovery of glutaraldehyde as a fixative [32], similar and better-preserved tubular fibers, now 25 nm in diameter, were found in all spindles studied (Figure 8). Some of these spindle "microtubules" (MTs) were seen by Brinkley and Stubblefield to attach to specializations on each chromatid of a metaphase chromosome. These specializations appeared as paired structures at the chromosome's primary constriction or "centromere" (Figure 9) [33]. The attachment sites were identified as loci of MT binding and called "kinetochores", a term given earlier to the chromosomal regions responsible for chromosome motion. The spindle thereby became visible as an organized assembly of MTs that must somehow exert forces on chromosomes. This idea has served ever since as the framework for most work on mitosis ever since.

Considerable effort has gone into the structural characterization of the spindle's MT component. Most of the early work used electron microscopy of serial sections cut from fixed and plastic-embedded samples; this approach provided the resolution in 3-dimensions (3-D) necessary to distinguish the individual but tightly bunched MTs and to reveal the overall architecture of spindle fibers (Figure 10). Initial quantitative work on spindle structure was based on counts of the numbers of MTs in spindle cross-sections, presented as a function of position along the spindle axis, which was assessed by the number of sections cut since the one that included a spindle pole (Figure 11A–C) [34–36]. As techniques improved, investigators were able to track each MT from section to section, allowing displays of some aspects of spindle geometry in 3-D, e.g., the interdigitation of MTs associated with each half of the spindle in the region near the midplane of an anaphase spindle. This arrangement formed a robust interpolar bundle, the structure seen by light microscopy as the "continuous" or "pole-to-pole" spindle [34,37] (Figure 12).

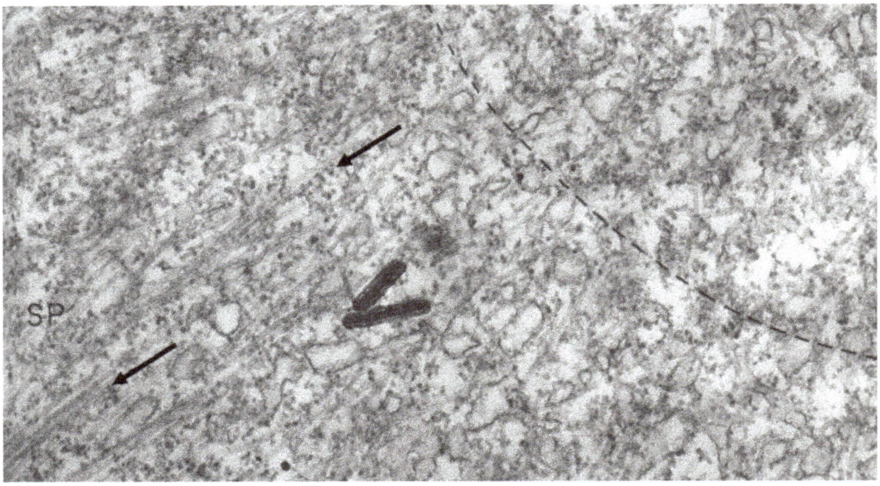

Figure 7. A portion of a sea urchin mitotic spindle (SP) imaged in an electron microscope, showing the MTs (arrows) that make up the spindle fibers that had been seen by light microscopy. The curved dashed line marks the polar end of the spindle and the beginning of a specialized region that surrounds the spindle pole in these cells. (Dark rods are contamination.) Harris, 1965 [30]. This image is displayed under the terms of a Creative Commons License (Attribution-Noncommercial-Share Alike 3.0 Unported license, as described at http://creativecommons.org/licenses/by-nc-sa/3.0/.

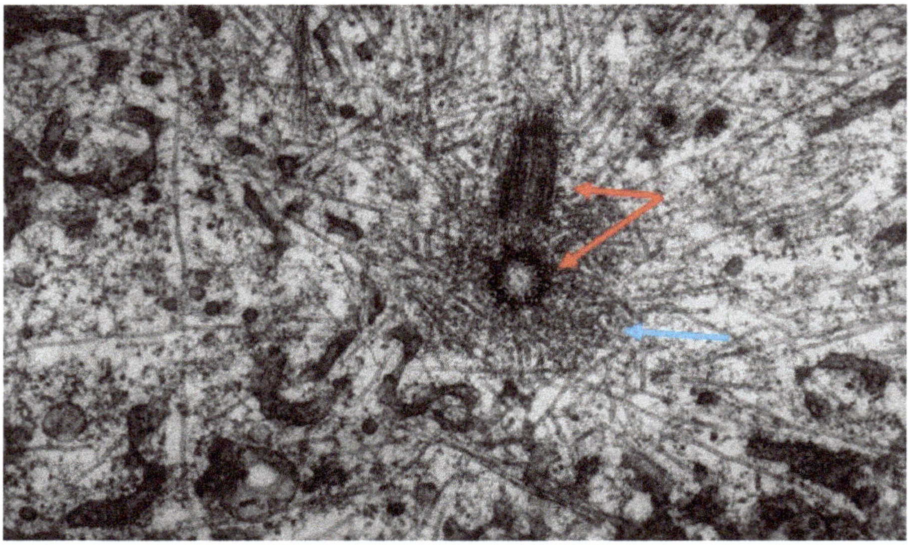

Figure 8. Electron micrograph of a mitotic spindle pole in a cultured mammalian cell, fixed with glutaraldehyde. Red arrows mark centrioles, the blue arrow indicates pericentriolar material where MTs are nucleated. Image kindness of Kent McDonald, Univ. California, Berkeley.

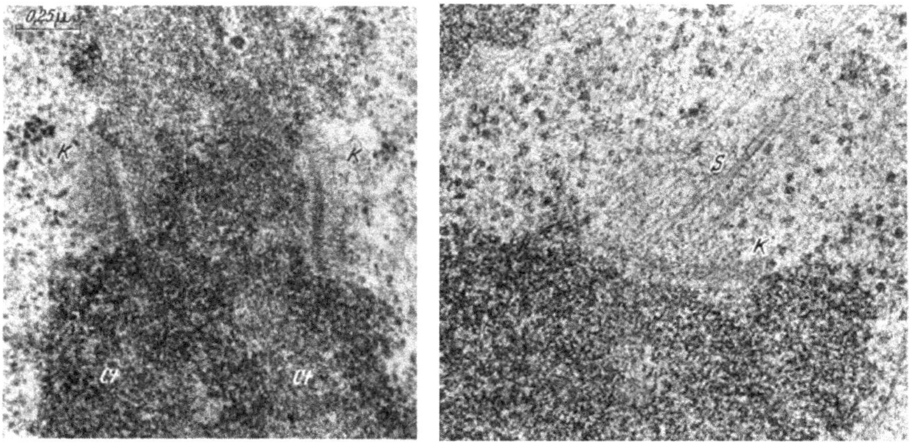

(A) (B)

Figure 9. Kinetochores (*K*) are the specializations on mitotic chromosomes (*Ct*) that bind MTs. (**A**) = sister kinetochores in a mammalian cell, strain CHO, treated with colcemid to block MT formation; (**B**) = a kinetochore after removal of the drug and regrowth of spindle MTs (*S*). From Brinkley and Stubblefield, 1966 [33]. With permission from Elsevier Publishing.

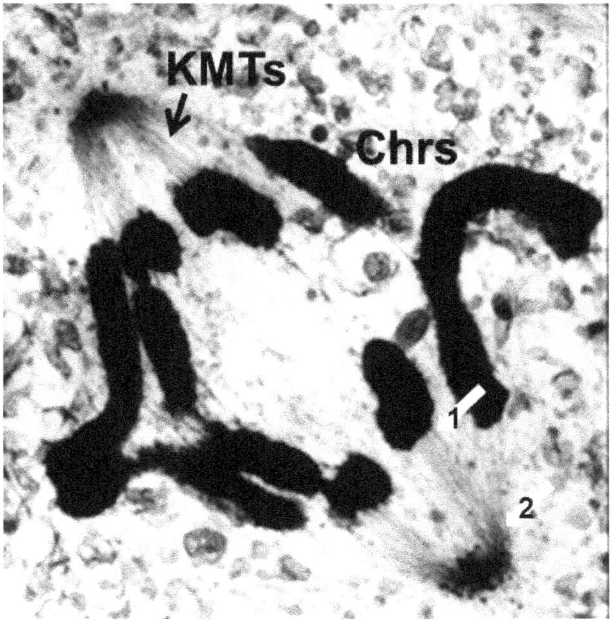

Figure 10. Thick section of a mammalian cell in anaphase, lysed before fixation to reduce the complexity of background staining. KMTs = kinetochore microtubules; Chrs = chromosomes. White arrows indicate sites of apparent attachment between MTs and a chromosome (1) and a pole (2). From McIntosh et al., 1975b [37]. By permission of the author.

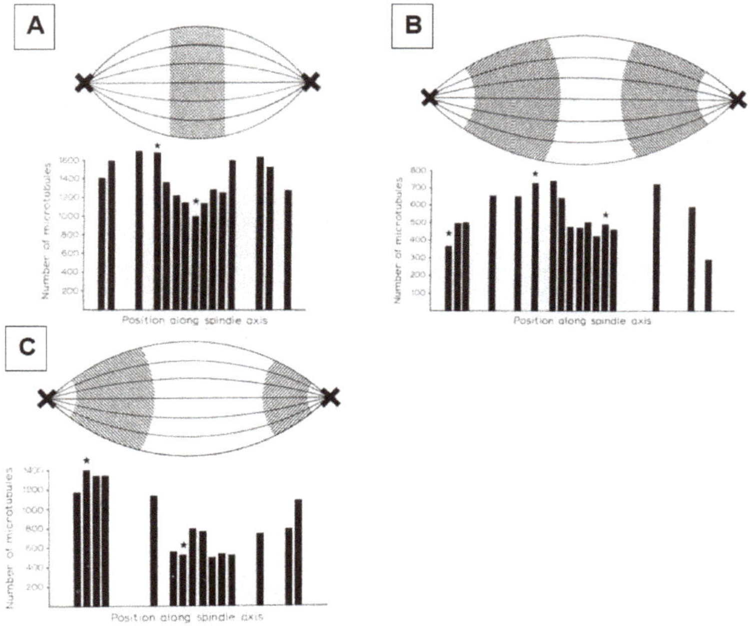

Figure 11. Counts of total numbers of MTs seen on successive spindle cross-sections from pole to pole at three stages of mitosis: (**A**) = metaphase; (**B**) = early anaphase; (**C**) = mid anaphase. From McIntosh and Landis, 1971 [34]. This image is displayed under the terms of a Creative Commons License (Attribution-Noncommercial-Share Alike 3.0 Unported license, as described at http://creativecommons.org/licenses/by-nc-sa/3.0/.

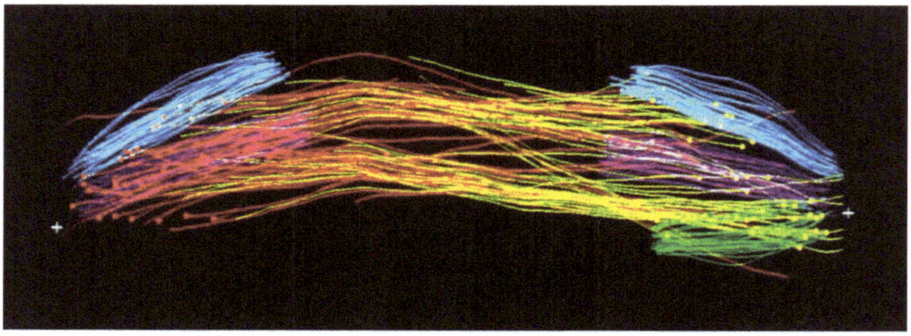

Figure 12. The paths of many spindle MTs, traced through serial sections of a mammalian cell (strain PtK2) in anaphase. White crosses mark the positions of the poindle poles. The shorter bundles of colored lines represent the kinetochore MTs that cluster to form the kinetochore fibers visible in the light microscope. (Colors are used simply to make these clusters stand out.) The red and yellow lines represent non-kinetochore MTs, which are associated with one pole or the other and interdigitate at the spindle's midplane to make the "interpolar" spindle. These MTs slide and elongate during anaphase B. Image kindness of D. Mastronarde, Univ. Colorado.

Some students of spindle structure used thin sections cut parallel to the axis of the spindle and traced each MT as it appeared on a single section; they then super-imposed these traces to make a representation of spindle structure that served useful comparative purposes. Although these views were drawings, not full reconstructions of spindle organization in a sub-volume of the overall structure, they provided informative views of the spindle after an experimental treatment [38].

The small spindles found in micro-organisms provided a particularly fruitful field for study by electron microscopy. The first group to capitalize on these cells used high voltage electrons to image comparatively thick sections cut parallel to the spindle axes in cells that had been lysed during fixation, removing much of the cytoplasmic density that is characteristic of small cells [39]. With stereo views at distinct stages of spindle formation and function, one could get a good overview of spindle organization and its changes with time as the MTs grew from the centrosomes, formed a bi-polar array, then organized the chromosomes and segregated them, largely through spindle elongation. A more detailed view of spindles in small cells emerged from the use of larger numbers of serial thin sections cut perpendicular to the spindle axis. With these sections one could track MTs in 3-D to characterize their distribution. The well-ordered interpolar spindles of diatoms were the first to yield information about changes in MT arrangement as a function of spindle elongation in anaphase B [40]. Subsequent work extended these discoveries to other diatoms, then to a cellular slime mold [41] and budding yeast [42]. This work, in sum, revealed a consistent pattern of structure in which one or a few MTs associated end-on with each chromosome, and a bundle of MTs formed between the two spindle poles, setting up an interdigitating framework of anti-parallel MTs whose interactions near the spindle midplane could drive spindle elongation through the sliding apart of two MT families, commonly accompanied by MT growth (Figure 13).

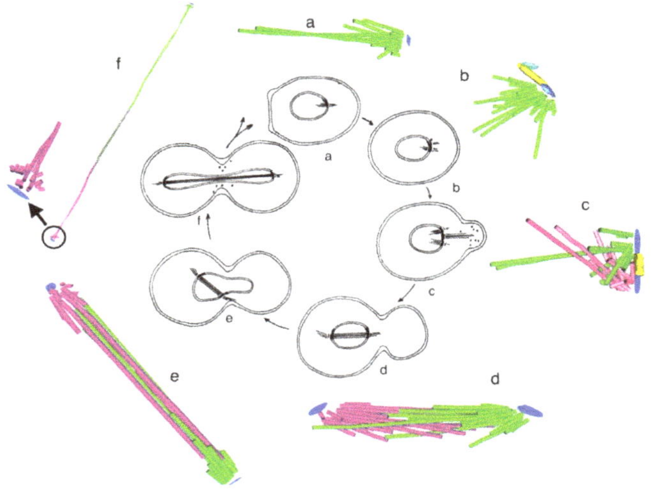

Figure 13. The spindle cycle in budding yeast. In the center of the figure, drawings represent the structure of budding yeast cells as they traverse the cell cycle. Around the edges are models made from tomographic reconstructions of the MT component of yeast spindles at each stage of mitosis. (**a**) There is only one centrosome but MTs grow from it into the cell's nucleus; (**b**) The centrosome is duplicated and more shorter MTs project into the nucleus; (**c**) There are now two functional centrosomes, sitting side by side, each projecting MTs into the nucleus. At this stage, the spindle is in the process of attaching sister kinetochores to sister spindle poles; (**d**) A bi-polar spindle has formed; (**e**) The cell is advanced in anaphase B and the sister chromosomes are well separated; (**f**) A long, slender spindle runs from pole to pole (green and magenta MTs), and the chromosomes are drawn tightly around each pole. This spindle severs as the cell divides at cytokinesis, and the cell returns to state A. Redrawn from [43] by Eileen O'Toole. This image is displayed under the terms of a Creative Commons License (Attribution-Noncommercial-Share Alike 3.0 Unported license, as described at http://creativecommons.org/licenses/by-nc-sa/3.0/.

An interesting reflection on the value of technological improvements in the progress of mitosis research is seen in a comparison of the results cited above with work done after the power of fluorescence microscopy became widely appreciated. Studies of budding yeast spindles by indirect immunofluorescence produced images quite like those that were laboriously prepared by serial section electron microscopy [43]. In the early work the resolving power of the electron microscope was used largely for fiber classification (MTs vs. microfilaments, etc.) and to resolve closely spaced fibers; such detail was not necessary for the study of gross fiber motions. Later chapters of this book will show how the clever use of light microscopy, together with various techniques for image contrast generation, have contributed tremendously to our current understanding of spindle mechanics.

Serial cross-sectioning and electron microscopy also provided important early information about the MT components of bigger spindles. The structure of the cold-stable bundle of MTs that associates with each kinetochore in a mammalian cells was elucidated in this way [44], and the structure of both kinetochore-associated MTs [45] and other spindle MTs that contribute to mammalian spindle structure [46] were similarly studied (Figure 14A–D). Unlike the small spindles, in which all MTs had one end on a spindle pole, these larger structures included many MTs, both of whose ends appeared free in the body of the spindle. Moreover, not all of the MTs with one end on a kinetochore were long enough to reach the area around the pole. In addition, the kinetochore-associated fibers visible in the light microscope were seen to contain MTs that did not end on a kinetochore. This structural complexity challenged the perspective that chromosome segregation was accomplished in all cells by a common mitotic mechanism. Instead, the structural variation suggested that spindles did not simply scale up in size; big spindles might involve different structural and functional principles than small ones. Perhaps bigger cells with more and bigger chromosomes placed different demands on spindle mechanics, so different solutions for chromosome segregation were required.

Figure 14. Two-dimensional projections of all the MTs in a volume that includes ~one-half of an early anaphase spindle from a PtK1 cell. (**A**) = all MTs traced; (**B**) = all kinetochore-associated MTs seen; (**C,D**) = all non-KMTs associated with the two spindle poles. From Mastronarde et al., 1993 [47]. This image is displayed under the terms of a Creative Commons License (Attribution-Noncommercial-Share Alike 3.0 Unported license, as described at http://creativecommons.org/licenses/by-nc-sa/3.0/.

An important limitation with all the early structural studies of spindle MTs was their inability to detect the polar orientation of these polymers. MTs were demonstrated to be polar, i.e., to be vectors, by Amos and Klug [47], although the concept of MT polarity had been identified as important for mitosis somewhat earlier [48]. There were indications from experiments, both in vivo [49,50] and in vitro [51,52], that MTs could grow from either the centrosomes at the spindle pole or the kinetochores of metaphase chromosomes, suggesting that the MTs in any half-spindle pointed in opposite directions. The issue of spindle MT polarity was settled in several steps: (1) Experiments in vitro revealed a kinetic polarity in MT growth; the polymers had a fast and a slow growing end [53]. Work from the Borisy lab showed that MTs growing from centrosomes were oriented with their fast-growing "plus" ends distal to the centrosome [54]; (2) A method was discovered by which the protein subunit of MTs would add to the walls of existing MTs, forming hooks whose direction of curvature revealed the polar orientation of the original MT lattice [55]; (3) The application of hooks to spindles showed that both the MTs emanating from the spindle poles and those associated with kinetochores were oriented in the same direction: their fast-growing ends were distal to the spindle pole [56,57]; (4) The flagellar ATPase, dynein was identified as an additional polarity marker, binding along the MT lattice in a polarized fashion and confirming the underlying MT orientation in spindles [58]. Thus, the polar orientation of spindle MTs turned out to be strikingly simple (Figure 15). The ability of kinetochores to promote the nucleation of MTs then posed a mystery: are these MTs initiated upside down or does the spindle contain some MTs that are oppositely oriented? Intriguingly, the structural evidence argued strongly against the latter possibility, but exactly how kinetochores can initiate MTs of the right orientation is still an unsolved problem.

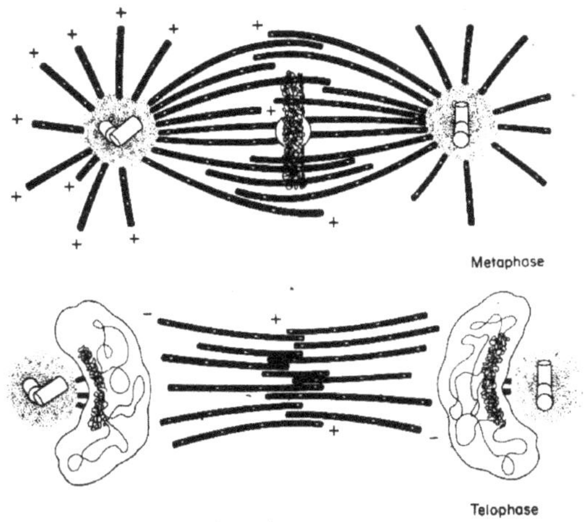

Figure 15. Diagrams showing the polar orientation of Spindle MTs, as assessed by the tubulin-containing hooks. Euteneuer and McIntosh 1981, 1982 [56,57]. This image is displayed under the terms of a Creative Commons License (Attribution-Noncommercial-Share Alike 3.0 Unported license, as described at http://creativecommons.org/licenses/by-nc-sa/3.0/.

3. Comparisons of Spindles across Phyla

The fact that mitosis occurs in all eukaryotic cells has meant that students of diverse organisms have contributed to the literature of the field. Although the earliest studies were focused on organisms whose cells were comparatively large and whose chromosomes were particularly visible, e.g., the amphibians, insects, nematodes, oocytes of marine invertebrates, and certain plants, subsequent investigations reached out more widely. Cells from the endosperm in plant seeds have been particular useful because they make almost no cell walls, which improves both the clarity of images obtained by light microscopy and the ease with which cells can be flattened to reduce their 3-D image complexity. One of the most productive scholars of plant mitosis was Andrew Bajer, who initially used phase microscope with cine-recording, then polarization optics, then electron microscopy (reviewed in [59]) and subsequently immune-staining of spindle components to characterize mitosis in these large and beautiful cells [60]. Bajer and his students described large bundles of MTs associated with each chromosome and many additional spindle MTs, assembled into a structure that resembled the interpolar spindle of animal cells, except for its lack of a focus on defined polar structures.

Micro-organisms were also subjects for detailed and informative study. Hans Ris and Donna Kubai used electron microscopy to describe spindles in several dinoflagellates, which were thought at the time to represent a particularly primitive kind of eukaryote whose spindles might, therefore, give insight into the evolutionary origins of mitosis [61]. These organisms carried out mitosis within their nucleus, and the images then available suggested that cytoplasmic MTs were simply a guide for the direction that chromosomes might move. It was proposed that motive force might come from the membranes of the nuclear envelope, by analogy with a then popular model for the mechanism of chromosome segregation in bacteria [62]. In retrospect, the apparent absence of connections between chromosomes and MTs may have reflected procedural difficulties in preserving these cells for structural studies. This view is supported by the observation that in at least some dinoflagellates MTs terminate on a knob-like differentiation at the site of contact between kinetochores and the nuclear envelope [63] (Figure 16). Some of the difficulties of preparing micro-organisms for electron microscopy have been solved by the introduction of rapid freezing, followed by fixation through the substitution of cytoplasmic water with organic solvents at low temperature (freeze-substitution) [64]. In the study of cellular fine structure, these methods provided a significant improvement in cell preservation [65].

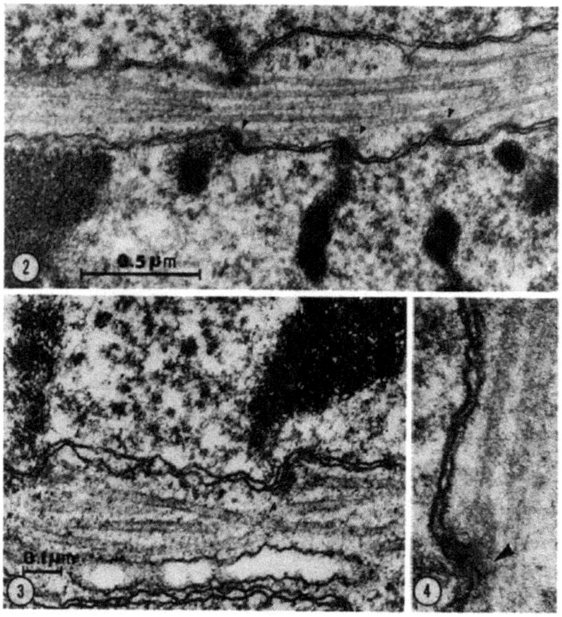

Figure 16. MTs, chromosomes and their interactions in the dinoflagellate, *Amphidium*. MTs run in a cytoplasmic channel, but some of them are connected to chromosomes through the nuclear envelope. From Oakley and Dodge [63]. This image is displayed under the terms of a Creative Commons License (Attribution-Noncommercial-Share Alike 3.0 Unported license, as described at http://creativecommons.org/licenses/by-nc-sa/3.0/.

The biological diversity of micro-organisms has been informative about different ways in which organisms solve the problem of reliable chromosome segregation [66,67]. However, this knowledge has been a two-edged sword; it has provided a legitimate sense of variation but also a bewildering sense of complexity. The wide range of mitotic and meiotic structures and processes has nurtured the belief that there could not possibly be a "universal" mechanism for chromosome movement. For clarity in addressing this issue, a brief summary of mitotic and meiotic events is presented here, followed by some examples of diversity in the patterns of chromosome motion.

4. A Summary Description of Mitotic and Meiotic Events

Despite diversity in the patterns of chromosome segregation, some common features have been identified. Small spindles, as found in most fungi and algae, build a distinct interpolar spindle (sometimes called the central spindle, the continuous spindle, or the core bundle, referring to its tightly bunched MTs). One or more MTs that are not in this bundle run directly from one of the spindle poles to each kinetochore, making a mechanical connection that is essential for normal chromosome motion. Commonly, the chromosomes congress to a metaphase plate, but the connection between sister chromatids is not rigid, allowing the metaphase chromosomes to separate transiently, showing a "breathing" as they move back and forth about the spindle equator [68]. In some diatoms, this breathing is so extreme that sister kinetochores are pulled to opposite spindle poles before anaphase begins [69]. The onset of anaphase is usually abrupt [70], and most chromosomes separate due to two kinds of motions: an approach to the spindle poles (Anaphase A) and an elongation of the distance between the poles (Anaphase B).

In medium sized spindles, such as those in the cells of mammals, nematodes, and fruit flies, there are multiple MTs (5–40) attached to each kinetochore, and some but not all of them run from kinetochore to pole [71]. Again, there is an interpolar spindle, but not all of the MTs in this structure have one end associated with a pole; many of them start and end in the body of the spindle, and some of them commingle with the kinetochore-attached MTs, making the "kinetochore fiber" that is visible by light microscopy a mixture of MTs that are kinetochore associated and those that are not [46]. While centrosomes are present and contribute to spindle organization, they have been shown by several experiments to be dispensable, so long as slower mitotic progression can be tolerated. In large spindles, such as those that form in early frog embryos or in the endosperm of higher plants, there are many thousands of MTs. In frog eggs, most of these form in association with chromosomes, rather than centrosomes, and they form a bipolar spindle through a combination of MT rearrangements and dynamics, as will be discussed in later chapters of this book. Connections of MTs directly to the poles are rare. Thus, there appear to be several structures to understand more fully in our efforts to elucidate mitotic mechanisms.

Meiotic spindles resemble mitotic spindles in many ways, but the essential function of reducing the chromosome complement from diploid to haploid requires a few important differences. Foremost, there are two rounds of chromosome segregation with only one round of DNA replication. Secondly, in preparation for the first meiotic division, homologous chromosomes pair and become connected by the crossings-over that occur during meiotic prophase; a set of four chromatids (two identical and two homologous) form a single meiotic chromosome, called either a "bivalent" or a "tetrad" (Figure 17). These congress to a metaphase plate and segregate by the separation of "half-bivalents", which serve as anaphase chromosomes for meiosis I. Following telophase, a short interphase that lacks DNA replication, the chromosomes enter meiosis II, which resembles a mitotic division, except that the number of chromosomes at the metaphase plate is half the number found in a diploid mitosis (Compare Figure 17, left and right).

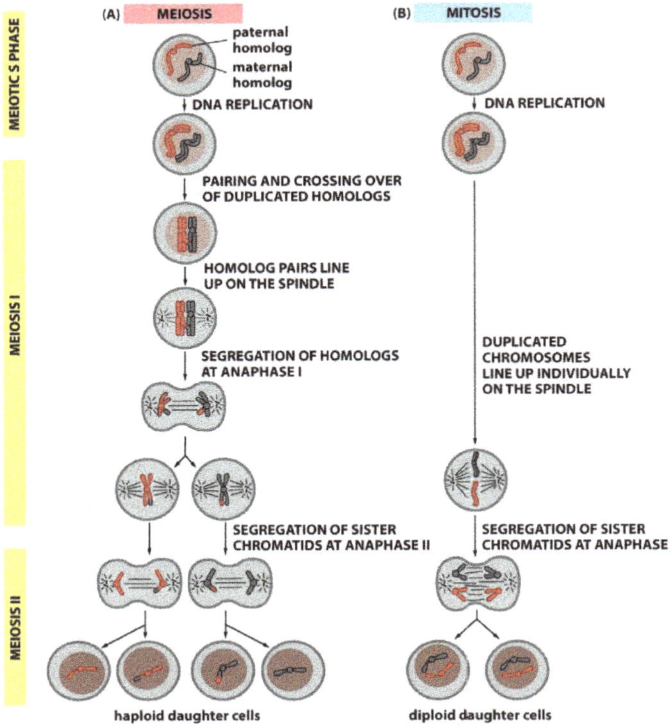

Figure 17. Diagrams of meiotic (**A**) and mitotic (**B**) cell divisions. With permission from the Taylor Francis Group, publishers.

Diversity in the mechanisms of chromosome segregation is found in several published works. For example, the first meiotic division of spermatocytes in crane flies includes a normal anaphase separation of paired autosomal bivalents, but a delayed separation of the two unpaired sex chromosomes [72]. The replicated but unpaired sex chromosomes, exhibit amphitelic attachments (i.e., sister kinetochores are attached to sister spindle poles). After autosomal segregation is complete, the sex chromosomes segregate almost simultaneously to opposite poles, retaining their amphitelic attachments as they move in opposite directions by a mechanism that remains unexplained.

Another anomaly is found in the fly, *Sciara copraphilia*, where meiosis I in males includes segregation of the male-contributed chromosomes from the female chromosomes on a monopolar spindle, which discards the former and draws in the latter to a single spindle pole [73,74]. Again, the mechanism underlying this chromosome behavior is still mysterious. At various times, students of mitosis have focused on these unusual behaviors and demanded that any successful hypothesis for mitotic mechanism must explain all these unusual phenomena. With the advent of molecular biology and a subsequent focus on chromosome segregation in a relatively few "model organisms", this view has faded. Regardless of one's opinion on this matter, there are certainly things to be learned from biological diversity, including a better understanding of conserved regulatory elements that may bridge diverse mitotic mechanisms.

As an example of informative mitotic diversity, one can look at mitotic behavior of the nuclear envelope. During mitosis in higher plants and animals the nuclear envelope disperses, allowing commingling of the nucleo- and cytoplasms. In many unicellular organisms, on the other hand, the envelope remains intact throughout mitosis. This appears to be a sharp distinction and a fundamental difference, but it turns out that there are examples of intermediate conditions that argue otherwise. In the green alga, *Chlamydomonas*, the nuclear envelope is largely intact, but near the spindle poles there is a window that allows centrosomes in the cytoplasm to extend MTs into the nucleoplasm and thereby affect chromosome position [75]. A similar situation is found in the nuclei of *Drosophila* embryos at the syncytial blastoderm stage [76]. In addition, there are cases where the spindle forms in the cytoplasm and the nuclear envelope remains intact. The mitotic chromosomes associate with the inner surface of the nuclear envelope and somehow form MT attachment sites on the outer surface of the envelope. As a result, the cytoplasmic spindle can organize chromosome segregation, even while the spindle MTs are excluded from the nucleoplasm [63,77,78]. These examples are all consistent with the simple view that mitosis requires a bi-polar array of MTs that interacts with the already duplicated chromosomes, linking sister kinetochores to sister poles. Exactly how this arrangement is achieved and regulated can vary, and may be of secondary importance.

With the advent of molecular biology, both to help identify the protein products of genes and to go from biochemical analyses to a set of genetic loci, mitosis research focused down on a relatively few organisms whose combination of genetic manipulability and susceptibility to molecular transformation and microscopic imaging made them useful experimental systems. Thus, as will become apparent in the chapters that follow, there is now a huge amount of information about the spindles of budding and fission yeasts, of the nematode, *Caenorhabditis elegans*, of the fly *Drosophila melanogaster*, and of mammalian cells grown in culture. Likewise, a few plant systems are under intense scrutiny. Yet, as the sequencing of DNA has become relatively inexpensive, and as molecular tools and capabilities have expanded, there is now a trend away from this biological myopia. We can look forward to a greater interrogation of biological diversity and a resulting improvement in understanding of this complex and important biological process.

5. Biochemical Work to Characterize the Mitotic Machinery

The first successful efforts to purify mitotic spindles took advantage of the eggs of marine invertebrates, which could be obtained in large numbers and induced to enter an almost synchronous mitosis by the addition of the corresponding sperm [79,80]. Cell numbers, and therefore the amounts of isolated spindles were sufficient for biochemical study, but the first spindles isolated were non-physiologically stable and inert with respect to mitotic activity. Robert Kane improved this isolation protocol in 1962, leading to some characterization of the factors important for spindle stability [81]: pH around 6.0 and a not well understood aspect of lowering the solution's dielectric constant. However, these spindles too were inactive, and even when the method was further improved by using glycerol and dimethylsulfoxide to preserve spindle stability in a reversible way [82], the isolates were still too inactive and complex to be very informative. The latter spindles showed some of the dynamic properties of spindles in living cells, e.g., sensitivity to hydrostatic pressure, but they did not move chromosomes, so it was hard to assess the functional significance of either their dynamic properties or any specific components. Still other media [83] were used to isolate labile spindles that yielded information about a few proteins that associated with spindle MTs in a polymerization-dependent way [84]. The discovery of Taxol as an agent to promote MT stability [85] spawned additional efforts to isolate useful mitotic spindles [86], but again, the lack of functional assays for the many components of the isolates frustrated attempts to use these preparations for a molecular analysis of mitotic mechanism.

One of the most significant advances in spindle chemistry came from the lab of Edwin Taylor. It had been known for years that colchicine disrupted mitotic spindle structure and function but had little effect on a cell's progression through interphase. Based on this specificity, Taylor surmised that the drug bound to the protein subunit of spindle fibers, i.e., the subunits of MTs. He made a radioactive form of this spindle poison and used then-standard methods of protein biochemistry to purify the colchicine-binding component of spindles [87,88]. Ironically, the material that provided the best yield of colchicine-binding protein was mammalian brain, a tissue known for its dearth of cell divisions. This observation provided important evidence that the colchicine-binding protein was used in multiple cellular settings, just as MTs had been found to be essentially ubiquitous. Subsequent work by Richard Weisenberg, identified conditions in which the isolated protein, now called "tubulin", would assemble into MTs in vitro [89]. These advances opened up multiple approaches to the experimental study of mitosis.

Several labs used version of the conditions discovered by Weisenberg to make cell-free models of spindles that preserved aspects of spindle function and allow limited biochemical study, if not the analysis of components. This approach was pioneered by Hoffmann-Berling in the 1950s; he used glycerol to preserve aspects of the cytoskeleton as the cell membrane dissolved, then Mg^{2+}-ATP to activate cellular contractility. He applied the same approach to mitotic cells [90], but with only limited success, perhaps because the conditions that preserved microfilaments and their interactions with myosin did not work for spindles. Maintaining a labile spindle became practical as soon as one knew methods to purify tubulin and promote its polymerization [89]. By lysing cells with non-ionic detergents in an equilibrium mixture of tubulin and MTs, investigators were able to support anaphase motions after membrane permeability was disrupted [91]. However, without a way to initiate anaphase in these lysed cells, one could only look for conditions that supported chromosome motion after it had started. For this purpose, solutions containing Mg-ATP and a high molecular weight polyethylene glycol were sufficient; soluble tubulin was unnecessary for anaphase A and limited anaphase B. Subsequent work showed that the presence of tubulin could increase the extent of spindle elongation [92], but mechanisms still remained elusive.

An alternative biochemical approach emerged from work on extracts of frog eggs. These huge cells can be broken and fractionated to yield undiluted meiotic cytoplasm in sufficient quantities to facilitate experimentation on factors that induce a mitotic state [93]. It was this system that allowed the first purification of "maturation promoting factor" [94], which later came to be known as cyclin-dependent kinase 1. This discovery attracted legions of workers to the frog egg system for biochemical analysis of cell cycle control. Upon the addition of frog sperm, this system will produce cycling nuclei and mitotic spindles whose action in a cell-free environment is remarkably life-like. These spindles have served as fruitful experimental material for many laboratories, as will be described in later chapters of this book.

Another productive approach to spindle biochemistry was based on the use of labeled antibodies, which could identify and localize specific spindle components. Some investigators raised antibodies to proteins of interest, such as MT-associated proteins (MAPs) or motor enzymes and looked by indirect immunofluorescence to see whether these molecules were present in spindles [95–97]. This approach did, however, lead the field on some false trails, because the spindle contains a very large number of proteins, not all of which are functionally significant for mitosis. For example, the muscle protein, actin, appears to be a major spindle component [97], but later work showed that most spindle actin is monomeric; fibrous actin, which is more likely to be of mechanical importance, lies largely outside the spindle [98]. Nonetheless, the discovery through immuno-fluorescence that gamma-tubulin is localized at spindle poles [99] and that the minus end-directed motor, dynein, is localized to mitotic kinetochores [100,101] are examples of discoveries that opened a wealth of opportunities for experimentation, many of which will be discussed in other chapters. In summary, however, this rather targeted approach to component identification did not have the power necessary to reveal important but previously unknown spindle parts.

Substantial progress in the identification of spindle components was made serendipitously through the discovery that certain auto-immune syndromes in humans led to the production of antibodies that bound to unknown spindle components localized at intriguing sites, such as kinetochores [102]. The potential value of these antisera was recognized by several students of spindle structure, but it was the chromatin structure group at Johns Hopkins, led by Bill Earnshaw and Don Cleveland, that first capitalized on these sera as tools to identify centromere and kinetochore proteins [103]. This work produced a series of important papers in which centromere proteins (CENPs A–F) were identified as specific polypeptides. The significance of some of these proteins for mitosis was implied by the failure of mitosis following injection of these antibodies into mitotic cells [104]. CENP-A was purified and identified as a histone-like protein by the clever use of spermatozoa as a cell type that retains this protein, even as histones are removed to permit chromatin transport in a compact sperm head [105]. CENP-A and other kinetochore components identified in this way have become central players in mitotic function, as will be discussed in later chapters.

6. Spindle Genetics as a Route to Understanding Mitotic Mechanism

Students of mitosis have long hoped that genetics would provide deep insights into the functionally significant components of the mitotic spindle. Hints supporting this hope came occasionally from the communities that studied organisms suitable for classical genetics, e.g., fruit flies. Mutants with interesting mitotic phenotypes were occasionally reported, e.g., some alleles of the *claret* locus, which is important for *Drosophila* eye color. These mutants included the phenotypes of abnormal mitoses in early embryonic cells [106] and of meiotic chromosome loss [107]. The reason for the surprising coupling of eye color with mitosis remained obscure for many years. It was only with the cloning of both the DNA around the *claret* locus and the gene for *Drosophila* kinesin that it became clear that this claret allele included a deletion of DNA from an adjacent gene, which encoded a motor enzyme important for mitosis [108].

Some workers sought mitotic mutants in normally diploid organisms by taking advantage of their biology. Ostergren used radiation mutagenesis of lilies and a screen for cells with aberrant mitosis during the brief haplophase that follows meiosis, but although he collected many such strains, the molecular tools need to analyze them were not available, so little progress resulted [109]. Likewise, students of fly biology sought male sterile meiotic mutants, maternal effect embryonic lethals, and late larval lethals in an effort to identify mutations affecting chromosome behavior and cell division. These mutants identified genes whose wild-type functions are important for chromosome condensation and integrity, for the progression of nuclear and centrosomal cycles, and completion of chromosome segregation. [110,111], but they did not cast light on mitotic mechanism. With only a genetic locus and a phenotype as guides, it was impossible to see mechanistic connections between mutant and wild type mitosis.

A full scale genetic assault on mitosis, and on cell cycle regulation more generally, did not occur until Hartwell began his screen for temperature sensitive (ts) mutants of the budding yeast, *Saccharomyces cerevisiae* [112] and Nurse pursued the same strategy with the fission yeast, *Schizosaccharomyces pombe* [113]. Meanwhile, Morris undertook an analogous screen for ts mutants in *Aspergillus* with a specific focus on those with an aberrant mitosis [114], and Yanagida sought mutants of *S. pombe* that would pertain specifically to mitosis. His lab achieved singular success through the study of strains that drove a septum inappropriately through an undivided nucleus, the "cut" mutants [115]. Other students of yeast biology looked for mutations that led to an increased frequency of chromosome loss. These "chromosome instability" (CIN) mutants again identified genes important for high fidelity mitosis [116]. With the advent of DNA cloning for fungal genes by reversion of mutants to the wild phenotype through transformation with a library of wild type DNA, it became possible to learn the DNA sequences of genes whose mutation had led to mitotic failure. In this way both Morris's lab and those of Rose and Hoyt discovered "pioneer" sequences of genes important for successful chromosome behavior in *A. nidulans* and *S. cerevisiae*. The mechanistic significance of several of these mitotic mutants became evident upon the cloning and sequencing of the gene for kinesin from *Drosophila melanogaster* [117]. Numerous fungal mutants with mitotic phenotypes could now be recognized as mutations in a family of MT-dependent motors that played important roles in the formation and function of mitotic spindles and related processes, such as karyogamy (the coming together of gamete nuclei to form a diploid nucleus). These genes and their analysis will be described in later chapters of this book.

The ability to clone mutant genes in fungi with comparative ease led to an amazing spate of progress in identifying key molecular functions important for mitosis. Kinases that regulate the activity of both centrosomes and kinetochores were found in each of the organisms studied. Genes whose products were essential for chromosome attachment to the spindle and for centrosome duplication were discovered and analyzed. Among the most important discoveries made through a genetic approach was the identification of proteins that provide key functions in the spindle's system of quality control, the "spindle assembly checkpoint". As discussed in Chapter 5, clever genetic screens identified several proteins that contribute to a system that delays anaphase onset until the chromosomes are properly attached to the spindle [118,119]. Additional progress was made through a marriage of biochemistry with molecular and genetic approaches, as in the work by the Kilmartin, lab, who workers isolated both spindles and the "spindle pole bodies" (SPBs), which are the centrosomes of budding yeast, followed by a combination of raising antibodies and screening libraries to identify genes and gene products that were components of these important mitotic structures [120]. With the more recent rise of genome sequencing and mass spectroscopy researchers in mitosis have come to possess what is probably a complete "parts list" for the mitotic machinery of small cells. Again, this work will be reviewed in later chapters. In sum, the emergence of modern molecular technology has made a dramatic difference to the information about mitosis that can be gleaned from genetics.

7. Insights into Mitotic Mechanism from Studies of Mitotic Physiology In Vivo

Early cytologists described the structures and organelles involved in spindle assembly and function based on correlations between fixed and living cells. As they achieved greater clarity in the descriptions of mitotic events, they realized that an understanding of underlying mechanisms would require experimentation. Their investigations were supported by the development of companion technologies such as micromanipulation, microinjection, and fluorescent labelling of spindle molecules, all employed with a new mindset of reaching into the cell to perturb the mitotic process.

a. **Experiments on Kinetochores**. Improved microscopy enabled descriptions of both localized and diffuse kinetochores, as seen in certain insects [121]. Experimental work using X-rays to fragment chromosome [122,123] provided direct evidence for the role of kinetochores in spindle attachment and chromosome segregation. X-ray induced fragmentation of chromosomes with localized kinetochore produced multiple fragments, only one of which contained a kinetochore. This fragments attached to the spindle and moved normally, but the fragments that lacked a kinetochore failed to attach and were lost at subsequent divisions [124]. By contrast, chromosomes with diffuse kinetochores were connected to spindle fibers by the entire poleward surface of each chromosome. X-ray-induced fragments of these chromosomes retained kinetic capacity; regardless of how small the pieces became, each chromosome fragment was pulled to opposite spindle poles by its associated kinetochore fibers [123]. This work established the importance of kinetochores in chromosome motion.

b. **Observations and Experiments on Chromosome Movements.** Some diversity in anaphase was discovered by quantitative descriptions of chromosome movements in living cells. Changes in spindle length and kinetochore separation revealed two phases of chromosome movement [25,125]. One involved the shortening of chromosomal fibers, the other an elongation of the spindle with consequent chromosome movement. Early descriptive work revealed variation in the extent to which organisms relied on Anaphase A or B. Some used only Anaphase A (Tradescantia) and some only Anaphase B (Primary spermatocytes of the Aphid, Tamalia [125]). Some cells used both though separated in time (Secondary spermatocytes and embryonic cells of the Aphid, Tamalia; Hemiptera and Homoptera; [125]), and others used both anaphase mechanisms overlapping in time and therefore difficult to distinguish (grasshoppers; chick tissue culture cells [25]).

The two phases of anaphase were experimentally distinguished by their responses to chemical inhibition; chloral hydrate blocked anaphase B but not A [125]. In a different approach, Brinkley and colleagues [126] noted that low concentrations of colcemid had little effect on kinetochore fibers, but non-kinetochore spindle fibers were lost, resulting in a monopolar spindle. Other investigators exposed cells to hypertonic media, X-rays, or elevated temperatures and studied the abnormal anaphase configurations that arose as a result of cross-linked chromosomes (e.g., [127,128]). Following these perturbations, the forces that would normally have pulled chromosomes to the poles now pulled the spindle poles together, suggesting that chromosome-to-pole forces were stronger than the forces driving spindle elongation. The failure of these spindles to elongate correlated with a delay or absence of cleavage furrow formation, providing some of the first evidence that furrow formation is linked to the spindle. These and similar observations promoted the view that spindles comprised two components: the chromosomal fibers that pulled chromosomes poleward, and the spindle body that could elongate, pushing the spindle poles apart, indirectly moving the pole-attached chromosomes.

Early studies of spindle dynamics using polarized light were paralleled by detailed studies of chromosome behaviors in living cells using improved phase contrast optics and later, differential interference microscopy. The major features of chromosome movements during cell division were distilled from observations of plant endosperm divisions (Reviewed in [59]), as well as spermatocyte divisions in crane fly [128], grasshoppers [129,130], mantids [131], and phasmids [132]. In general, chromosomes moved along the curvature of the spindle, parallel to the spindle long axis. These movements are irregular, with occasional pauses and reversals of direction, but they are always directed towards one or the other spindle poles, and are always progressively becoming more centered on the equator of the spindle at metaphase. At anaphase, chromosomes separate, marking the end of metaphase, and move poleward, continuing on paths parallel to the interpolar spindle. These motions are independent but are far more coordinated than prometaphase motions. In both prometaphase and anaphase, localized kinetochores were seen to direct the independent movements of chromosomes.

c. **Chromosome Pulling Forces Were Detected in Prometaphase**. Descriptions of chromosome behavior in a variety of meiotic cells revealed a "pre-metaphase stretch" in which the distance between kinetochores on homologous chromosomes was greatly extended by action of the spindle [131,132], e.g., mantids and phasmids (Figure 18). This pre-metaphase stretch was followed by the gradual re-contraction of chromosomes and a resumption of normal prometaphase congression. Notably, the timing of the stretch and the resumption of congression were asynchronous on adjacent chromosomes, implying that mitotic forces acted independently on individual chromosomes. Taken together, these studies suggested the autonomy of chromosome movements during mitosis, eliminating models of collective transport by the spindle. These observations contributed to the rise of traction fiber models, in which kinetochores and their associated chromosomes were pulled individually towards the spindle poles to which they were attached.

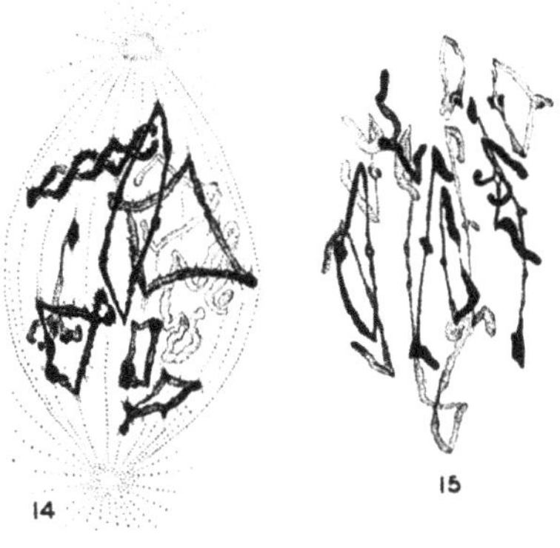

Figure 18. Typical pre-metaphase stretch of bivalent chromosomes in *Stagmomantis carolina,* a mantid. The kinetochores of homologous chromosomes are pulled far apart during Meiosis I in this species. 14 = early prometaphase, 15 is later. From Hughes-Schrader, 1943 [131]. With permission from the University of Chicago Press.

d. **Physical Perturbations of the Spindle as a Whole**. Some of the most convincing evidence for the existence of chromosomal fibers in vivo came from experimental manipulations of dividing cells. The centrifugation of mitotic cells distorted the spindle [121], stretching it [133] and/or severing chromosome attachment sites [134]. These perturbations also separated anaphase spindles into two half spindles, showing that chromosome-spindle pole attachments were strong enough to resist the centrifugal forces that distorted bivalent chromosomes. Centrifugation experiments further suggested a gel-like mechanical texture of the spindle.

The nature of this gel and of chromosomal fibers was further studied by the reversible application of hydrostatic pressure [49]. Pease found that progressive increases in hydrostatic pressure (e.g., 2000 to 6000 psi) progressively reduced the rate of anaphase chromosome movement, culminating in its complete cessation. Indeed, Pease provided early evidence for a functional linkage between the presence of a spindle gel structure and chromosome movement. When the rigidity of the fibrous spindle was reduced at elevated hydrostatic pressures, chromosome movement ceases. Remarkably, however, after a release of hydrostatic pressure, spindle fibers reassembled and chromosome movement restarted. Pease proposed that forces are imparted to the chromosomes by two phase transitions: a sol to gel transition, which added spindle fiber material and a solation of chromosomal fibers, which occurred at the spindle poles.

Quantification of spindle birefringence under various physical and pharmacological conditions provided key insights into the reversible assembly of the mitotic spindle (reviewed in [135–137]). A sharp dependency of spindle birefringence on temperature suggested a low energy bonding between aligned subunits of the birefringent spindle fibers and provided compelling evidence for the self-assembly of fibers at physiological temperatures [135]. Inoue and Sato [135] proposed the dynamic equilibrium model of spindle assembly in which aligned subunits forming birefringent spindle fibers were in equilibrium with a pool of unaligned, non-birefringent subunits. Consistent with Inoue's equilibrium model, Edwin Taylor [138,139] showed that the protein synthesis required for spindle assembly is completed well before the onset of prophase. However, neither an appreciable loss of kinetochore fiber birefringence nor the predicted transient acceleration of poleward chromosome velocity in anaphase was observed upon the rapid cooling of crane fly or grasshopper spermatocytes [140].

In eggs of the marine worm, *Chaetopterus*, the metaphase-arrested spindle is attached to the egg cortex. When spindle MTs were depolymerized by exposure to cooling, colchicine or high hydrostatic pressure, the spindle shortened, transmitting forces along the chromosomal fibers to pull the metaphase chromosomes closer to the cortex [141]. Moreover, reversal of treatment, which allowed the repolymerization of spindle fibers, pushed chromosomes away from the cortex. Taken together these studies provided the first physiochemical evidence for a link between the reversible assembly of spindle fibers and force production for chromosome movements. The dynamic equilibrium model was later revised, based on the understanding that spindle birefringence is due largely to MTs and that spindle MTs self-assemble from a pool of tubulin subunits [142,143].

e. **Local Perturbations of Spindle Structure and Function**. Experiments with microbeams of ultraviolet (UV) light raised provocative questions about the nature of MT dynamics and the roles of MTs in chromosome movements. Forer [144] reported that UV microbeam irradiation of chromosomal fibers in crane fly spermatocytes at metaphase produced localized "areas of reduced birefringence" (ARBs). ARBs subsequently moved poleward, even while the chromosomes remained aligned at the metaphase plate. Such observations indicated that chromosomal fibers were not static and suggested a continuous poleward flux of materials within the chromosomal fiber (Figure 19). Moreover, in anaphase, Forer reported variability in the impact of ARBs on chromosome movement [144]. Two thirds of the time, chromosome to pole movements ceased when irradiated fibers contained ARBs. One third of the ARBs, however, had no impact on chromosome movement. Moreover, two thirds of irradiated fibers did not develop ARBs, yet chromosome to pole movements were blocked as frequently as when an ARB formed. Forer interpreted these results to mean that chromosomal fibers contained two components: birefringent MTs, which neither produce nor transmit the forces required for chromosome movements, and a non-birefringent element that was required for traction the forces that pulled chromosomes poleward in anaphase A. Electron microscopy later confirmed that in at least some experiments, MTs within the ARBs were severed and/or depolymerized [145,146]. Forer's complicated results from anaphase spindles have never really been explained.

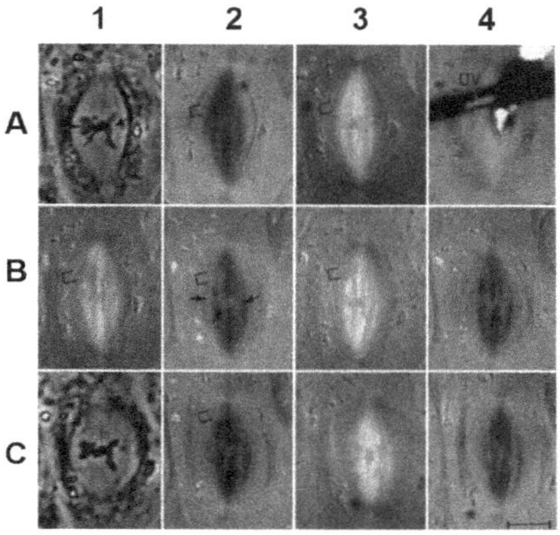

Figure 19. A crane fly spermatocyte irradiated during metaphase. (**A1**) autosomes labeled with arrow; (**A2**,**A3**) the position to be irradiated is indicated by a bracket; (**A4**) UV = the ultraviolet irradiation; (**B1**,**B2**,**B3**,**C2**) The position of the area of reduced birefringence (on the chromosomal fiber of the left bivalent) is indicated by a bracket; (**B2**) the autosomes labeled with arrows. The times of the photographs in minutes relative to the time of irradiation. A1, −11; A2, −7; A3, −5; A4, −0.5; B1, +2; B2, +2.5; B3, +6; B4, +7; C1, +10; C2, +14.5; C3, +18.5; C4, +19.5. The area of reduced birefringence moved to the pole, and did not displace the pole when it arrived there. From Forer, 1966 [144]. This image is displayed under the terms of a Creative Commons License (Attribution-Noncommercial-Share Alike 3.0 Unported license, as described at http://creativecommons.org/licenses/by-nc-sa/3.0/.

With regard to spindle MTs that did not attach to kinetochores, a subset of these MTs from opposing poles were found to interdigitate at the spindle midzone (see above); these fibers exhibit even greater stability than was found with the fibers connected to chromosomes [147]. Experiments showed that this interpolar bundle was under compression from the forces pulling chromosomes poleward; when the bundle was severed by a UV microbeam, the spindle of a diatom collapsed [145]. Much progress has recently been made in understanding the organization and force production associated with the zone of overlap at the spindle equator, as discussed by Scholey et al., this volume [148].

f. **Studies on MT Dynamics in Vivo**. Salmon and coworkers [149] microinjected high concentrations of colchicine into dividing sea urchin eggs, blocking spindle MT polymerization, and allowing the rate of depolymerization to be measured. A rapid decrease in birefringence (BR) reflected the disappearance of non-kinetochore MTs, while kinetochore MTs were again differentially stable (Figure 20). Surprisingly, the calculated rate of depolymerization (180–992 dimers per second) was significantly faster than predicted from the in vitro parameters of tubulin dynamics. It was, however, consistent with a significantly different, "dynamic instability" pathway concurrently proposed for MT behavior [150]. At about the same time, several investigators capitalized on photobleaching, as well as on the incorporation of injected tubulin and a specially labeled tubulin, using fully functional tubulin "analogues" to interrogate the pathways of tubulin assembly and disassembly during mitosis and the mechanistic basis of chromosome movements [151–153].

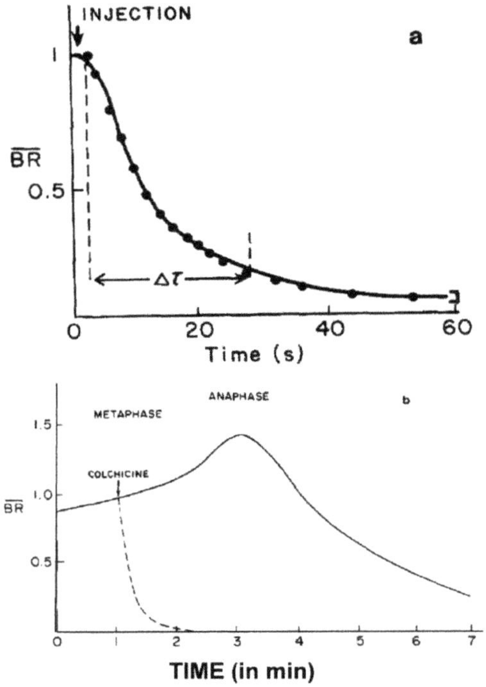

Figure 20. Changes in spindle birefringence (BR) in the half spindle as measured by a video spot meter. (**a**) Tracing of the chart record of the video voltage following injection of colchicine (0.2 mM) into a first division metaphase cell. The characteristic time, Δτ, of nonkinetochore MT depolymerization is measured by the time between the onset of BR decay and the time where the video voltage decreases to 10% of the initial value. The line is a first-order decay curve for k = 0.092. Spindle BR is normalized by the initial value at the time of injection; (**b**) Comparison of the rate of disappearance of normalized spindle BR after 1.5 mM intracellular colchicine injection with the normal rate of half-spindle disassembly at late anaphase. From Salmon et al., 1984 [149]. This image is displayed under the terms of a Creative Commons License (Attribution-Noncommercial-Share Alike 3.0 Unported license, as described at http://creativecommons.org/licenses/by-nc-sa/3.0/.

In particular, fluorescence recovery after photobleaching (FRAP) experiments in sea urchin spindles in vivo revealed two phases of fluorescence recovery: a fast phase, reflecting a quick turnover of tubulin within the labile polar MTs ($t_{1/2}$ = 19 s), and a slow phase associated with the comparatively stable kinetochore MTs ($t_{1/2}$ = 60–90 s [153]). Moreover, photobleached bar patterns positioned within the half spindle showed no evidence for directional, poleward movement during the rapid recovery phase, indicating that poleward flux of unbleached subunits incorporated at the dynamic MT plus ends was not the mechanism by which fluorescence was recovering [154,155]. Instead, the fast recovery was explained by the dynamic instability of polar MTs.

In complete agreement with FRAP analyses, the incorporation of labeled tubulin injected into mammalian cells also show that non-kinetochore MTs quickly become fully labeled (<1 min) following the injection of fluorescent or biotinylated tubulin, while kinetochore MTs were not completely labeled a full 10 min after injection [156]. Further analysis of tubulin incorporation determined the sites of insertion, and removal, of tubulin subunits in the more stable kinetochore MTs, and identified a poleward flux of tubulin within kinetochore MTs [156,157]. Of particular note, Mitchison [157] developed a caged fluorescent tubulin that could be activated by a focused beam of UV light to turn on fluorescence in a narrow bar across mitotic spindles in vertebrate cells. Over time, the fluorescent bar moved toward the spindle pole, demonstrating a poleward flux of kinetochore MTs and their disassembly at spindle poles (Figure 21). These results added yet another layer of dynamics to kinetochore MT assembly with implications for the polar linkage of kinetochore MTs, as well as a potential mechanisms of force production that could power chromosome movement or the spindle elongation.

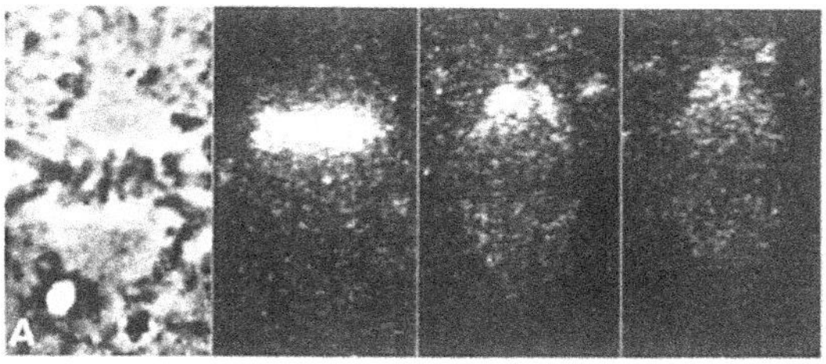

Figure 21. Phase-contrast and three fluorescence images of the cell to be analyzed. Times relative to photoactivation are (from left to right): 418, 5, 429, 674 s. Pole to pole distance = 18.8 μm. From Mitchison, 1989 [157]. This image is displayed under the terms of a Creative Commons License (Attribution-Noncommercial-Share Alike 3.0 Unported license, as described at http://creativecommons.org/licenses/by-nc-sa/3.0/.

g. **Investigating Spindle-Chromosome Interactions by Micro-Manipulation**. In a classic series of studies, Nicklas and coworkers used the tips of fine microneedles to tug on chromosomes in grasshopper spermatocytes and characterize their attachment to the meiotic spindle. They showed that chromosome-attached spindle fibers were essentially inextensible, that they were bound to the rest of the spindle near the pole, and that they were readily displaced, either laterally or toward the pole, without losing their chromosome attachment; moreover, each chromosome was quite independent of its neighbors [158]. Each chromosome behaved like a pendulum suspended by a thin wire and free to swing from a pivot point at the spindle pole. The chromosomal fiber was stiff under tension but flexible to compression and bending. Nicklas further demonstrated that during prometaphase and metaphase chromosomes could be detached from their fibers by repeated tugging with the needle. Once released from the needle, they would reattach to the spindle (See Supplementary Movie 1). In contrast, chromosomes in anaphase could not be detached from their fibers, despite great effort. The segregation of chromosomes following these micromanipulations was indistinguishable from that in control, unmanipulated cells.

Begg and Ellis [159] extended these observations, relating the mechanical connection between spindle and chromosome to the associated birefringent fiber. By selectively manipulating the chromosomal fiber they confirmed that this structure was attached to the chromosome. They showed that colchicine treatment abolished both the visible birefringent fiber and the mechanical connection between the chromosome and the spindle. During anaphase, if an already segregating chromosome was pushed poleward in advance of its neighbors, the birefringence of the associated kinetochore fiber was greatly reduced, presumably due to distortion of the fiber and the splaying of its MTs. Intriguingly, the splayed fiber remained intact and continued to shorten, as inferred from the eventual resumption of poleward movement by this chromosome in concert with its neighbors. These observations provided early evidence that the regulations of spindle-generated force and the mechanical linkage between spindle and chromosome changed at anaphase onset. During prometaphase, a kinetochore fiber that was not subjected to tension was unstable, and the linkage to the spindle could be lost as the chromosome reoriented. The onset of anaphase stabilized the chromosome's spindle attachment, even though the chromosomal fiber was shortening. Indeed, the coupling between force production and mechanical linkage appeared to be fundamental to proper chromosome orientation and segregation [160,161].

Nicklas next interrogated the process of kinetochore orientation and reorientation during prometaphase [161]. Chromosomes were detached and monitored during reattachment. (For a visual description of this process, see Movie 1 in Supplementary Material. This video was made from an original film recording made by Nicklas and kindly donated for use in this review.) The results showed that kinetochore orientation was the primary determinant of the spindle pole to which that kinetochore would reattach. Regardless of where a detached chromosome was placed, the kinetochore interacted preferentially with the pole it most nearly faced. Second, once one kinetochore oriented towards a pole, the sister kinetochore was constrained to face the opposite spindle pole, as a consequence of chromosome structure. The bipolar configuration of each kinetochore favored the bipolar attachment of chromosomes.

Early in prometaphase, as chromosomes begin to engage with the spindle, kinetochore position and orientation are disorganized; almost any arrangement of kinetochore-MT interactions can be found (reviewed in [162]). Some of these initial arrangements result in inappropriate attachments between chromosomes and the spindle, e.g., sister kinetochores attached to a single pole. These attachments must be aborted and corrected for the spindle to accomplish proper chromosome segregation [158,160,161]. To study such reorientations, Nicklas and Koch [160] capitalized on their ability to induce chromosomal mal-orientations by micromanipulation. Chromosomes were detached, then bent and reoriented, so both of their kinetochores pointed towards and then attached to the same pole. This unipolar attachment resulted in an initial poleward movement of the mal-oriented chromosome, followed by a reorientation of one or the other kinetochore, so it faced the opposite pole, leading to a stable bipolar orientation (all visible in Movie 1). Reorientation of the manipulated chromosome was indistinguishable from the reorientations of spontaneously occurring mal-orientations. The manipulated chromosome, together with neighboring, unmanipulated chromosomes, entered anaphase and segregated normally during cell division.

Since the spindle is applying tension to a chromosome when it is in a bipolar orientation, with sister kinetochores being pulled in opposite directions, Nicklas and Koch tested whether tension was the stabilizing factor in chromosome orientation. Tension was applied by using a microneedle to pull against the two kinetochore fibers of a mono-oriented chromosome, imitating the tension that would normally have been provided by each kinetochore's attaching to the opposite spindle pole. Remarkably, applied tension induced stability in monopolar orientations, whereas control, monopolar chromosomes reoriented in the absence of tension. The importance of tension was further supported by experiments in which two chromosomes in monopolar orientations to opposite spindle poles where interlocked [163]. The poleward forces on one chromosome pulled against the poleward forces on the second chromosome, providing tension and stabilizing the interlocking configuration. In other work, Nicklas and coworkers, correlated the reorientation behavior of mal-oriented chromosomes with the distribution of kinetochore MTs [38,163–165]. Most MTs associated with a reorienting kinetochore appeared to have disconnected from the pole, though they were only slightly rotated from their initial orientations. Significantly, a few MTs did extend from the reorienting kinetochore towards the opposite pole. The origin of this minor population of MTs has not yet been determined. It is not known whether these MTs were captured from the opposite pole or initiated anew from the reorienting kinetochore.

This experimental work provided important evidence for a suggestion previously made by Dietz [128] that chromosome orientation and reorientation is based on selection; chromosomes explore multiple connections to the spindle, attaching and reattaching until a stable attachment is achieved. The difference between stable and unstable orientations is a result of the tension generated by forces acting to pull sister kinetochores toward opposite spindle poles. The stability of kinetochore linkage to the spindle is dependent on tension. The molecular mechanisms by which tension contributes to stability remains a key unsolved piece of the mitotic puzzle.

h. **Assessment of Spindle-Generated Forces.** The first studies of mitotic force magnitudes considered the relationship between chromosome size and velocity. These studies showed that velocity was independent of size and thus of viscous load over a limited range. Chromosomes varying more than 2-fold in size exhibit the same speeds in both prometaphase and anaphase [130]. Similarly, McNeil and Berns [166] showed in metaphase PtK_2 cells that when a single kinetochore is irradiated with a high powered laser, the unirradiated sister kinetochore transports twice the normal amount of chromatin at the same velocity as a normal anaphase chromosome.

Nicklas [130] and Taylor [167] estimated the force needed to move a chromosome at the speeds of mitosis through cytoplasm, whose viscosity they could estimate; both these calculations concluded that only ~1×10^{-8} dynes (0.1 pN) was required for anaphase motions in either grasshopper or newt cells. Subsequently, Nicklas measured the force an anaphase meiotic spindle could exert on a chromosome whose movement was impeded by a calibrated microneedle [168]. The force required to stop chromosome motion was 10,000 times greater than the force needed to overcome viscous drag. Forces less than ~10^{-6} dynes (10 pN) had no effect on chromosome speed; greater forces slowed the chromosome, with speed falling to zero as the opposing force approached ~7×10^{-5} dynes or 700 pN. Moreover, Nicklas estimated that the tension forces on a prometaphase chromosome (0.5×10^{-5} dynes; 50 pN) are significantly higher than the forces required for normal anaphase (10^{-8} dynes; 0.1 pN). The clear implication of these calculations and measurements was that spindles can generate far more force than is necessary simply to drag a chromosome through mitotic cytoplasm.

One significant implication of these large forces is the existence of a mechanism to regulate force production under normal conditions. Such a regulatory mechanism was also suggested by the load independence of chromosome velocity. Thus, regardless of how hard the mitotic motor can pull on a chromosome, the rate of chromosome movements is regulated, probably by kinetochore MT assembly/disassembly. Meanwhile, changing the length of kinetochore MTs can determine chromosome position [169,170]. Whether chromosome movements are powered by molecular motors or by MT dynamics remains under active investigation.

i. **Experiments to Investigate Chromosome Congression to the Metaphase Plate.** Students of mitosis have long recognized the importance of metaphase in establishing a uniform initial condition for subsequent chromosome segregation. Numerous hypotheses were advanced for how chromosomes are brought to the spindle midplane. A simple and important idea emerged from the studies by Rashevsky [171], Hughes-Schrader [131], and Ostergren [172,173]: the poleward force on a chromosome might increase with distance from the pole. Paired chromosomes would then congress to the spindle equator because that position allowed the opposing forces to be balanced. Considerable evidence supports this force-balance theory, including observations on the consequences of upsetting the force balance. If sister chromatids are disconnected, either naturally at anaphase or artificially at metaphase [174], the opposing forces are uncoupled and each chromatid moves poleward. Similarly, a laser micro-beam can be used to destroy one metaphase kinetochore, and the chromosome then moves towards the pole to which the undamaged kinetochore is attached [166,175]. Thus, prometaphase and metaphase chromosomes are clearly being pulled in two directions at once.

One way to unify the fact of opposing tensions and Ostergren's force-distance hypothesis is to suppose that the force applied to a kinetochore is proportional to the length of the kinetochore fiber. Evidence for this proposition was found in naturally occurring multivalent chromosomes, which were asymmetrically positioned on the metaphase spindle [176]. To test the force-length relationship more rigorously, the metaphase positions of experimentally generated multivalent chromosomes were analyzed in living grasshopper spermatocytes [177]. At metaphase, asymmetrically oriented, multivalent chromosomes lay closer to the pole to which the greater number of kinetochores attached. The quantification of the fiber bundle lengths for the grasshopper multivalents supported the hypothesis that the pole-directed force acting on a chromatid is linearly proportional to the kinetochore-to-pole fiber length. In a follow up study, Hays and Salmon [178] used a microbeam to ablate portions of a single kinetochore on metaphase bivalent grasshopper chromosomes. Irradiations of a single kinetochore caused the chromosome to shift to a new equilibrium metaphase position closer to the pole to which the unirradiated kinetochore was attached. After each experiment, the cells were fixed and examined by electron microscopy to determine the MT numbers on both the damage and the intact kinetochores. The greater the kinetochore damage, the fewer MTs remained, and the farther the chromosome position shifted. Assuming a balance of forces on the chromosome at metaphase, there was a direct correlation between the poleward force at the kinetochore and the number of kinetochore MTs. These results are consistent with models of chromosome congression in which the metaphase equilibrium position reflects a balance of forces. The results can be interpreted in light of "traction fiber" models in which poleward force producers are distributed along the length of kinetochore fiber MTs, or in the context of newer models in which kinetochore motors pull the chromosome poleward along the kinetochore MTs and against a "polar ejection force" that increases in strength towards the pole [179] (See also discussion below).

Polar-ejection, or "elimination" forces were originally discussed by Darlington [180] and Ostergren [173]. They have subsequently been characterized by the movements of mono-oriented chromosomes on monopolar spindles in which a single kinetochore fiber connects each chromosome to the pole. The sister kinetochore lacks a spindle fiber and is inactive. If the only mitotic forces on a chromosome were the poleward directed forces acting on kinetochores, then each chromosome should be pulled all the way to the pole. This is not observed. Rieder and colleagues [179] found that the positions of monopolar chromosomes and the oscillations they display are a result of both the pole-directed forces on kinetochores and forces associated with the polar array of MTs that push the chromosome arms away from the spindle pole. If the chromosomal arms are cut off by a laser microbeam, then the free arms are expelled radially from the polar region, while the smaller kinetochore fragment moves closer to the pole. Similarly, after cold or nocodozole treatment to depolymerize the polar MTs, the polar ejection forces are reduced and the mono-oriented chromosomes are pulled closer to the pole.

The molecular basis of the polar ejection forces and their potential regulation in vivo is a subject of great current interest. Regardless of mechanism, the significant point is that the magnitude of force is maximal near the spindle poles and decreases with distance from the pole.

8. Models of Mitotic Mechanisms

The significant and intriguing events of mitosis have inspired many generations of scientists to speculate about the mechanisms that underlie chromosome segregation. At every stage in this segment of intellectual history, theoreticians have used the information available to propose ideas about the forces needed to cause chromosome motions. The earliest of these ideas were formulated before anyone knew the physical nature of proteins and nucleic acids, so the hypotheses put forward have in general not been helpful. For example, when looking at the spindles of marine eggs and embryos, physicists were impressed by their similarity to the lines of magnetic force visualized by iron filings or by probes of the electric field around a pair of fixed charges. Several models for mitosis were proposed, based on these similarities, but none of them led to experiments that clarified mitotic mechanism (reviewed in Wilson [15]). Other non-specific models based on colloids and exclusions between "phases" were similarly unproductive.

Cytologists who watched mitosis in the cells of diverse organisms recognized at least two sites of force generation: a "traction force" between the chromosomes and the poles to achieve anaphase A and an extensive force between the poles to produce anaphase B. Ideas about the origins of these forces depended on whether that scientist believed that the spindle fibers seen in fixed material were present in living cells. As Schrader pointed out in his essay "On the reality of spindle fibers" [121], one had to believe that the fibers in fixed cells represented either a structural feature that was cryptic in live material or lines of force produced by the spindle mechanism. The first ideas about mitotic mechanism that channeled thought in useful ways were ones that assigned contractile properties to the apparent connections between chromosomes and poles (summarized in [181]) and expansive properties to the interpolar fibers, first discussed by Bělař [127]. Actomyosin was commonly invoked for the former and an unknown fiber system for the latter. It was not until the quantitative study of spindle birefringence by polarization microscopy and the discovery of spindle MTs by electron microscopy that spindle modeling became specific.

Inoue and Sato [135] described a model for both chromosome to pole and pole-from-pole motions based on a labile association of the subunits that formed spindle fibers. They posited a dynamic equilibrium between soluble and assembled states of spindle subunits, using the formulation:

$$A_0 - B \rightleftharpoons B \tag{1}$$

where A_0 represented the total amount of polymerizable material and B the amount in polymer at any given time and temperature. A quantity proportional to B could be measured by polarization microscopy, allowing a reformulation of the equation simply in terms of the observable quantity, birefringent retardation (Γ), where Γ was assumed to be proportional to the amount of polymerized material, B. Thus, Equation (1) was rewritten

$$\frac{G_0}{k} - \frac{G}{k} \rightleftharpoons \frac{G}{k} \tag{2}$$

where k was the constant of proportionality.

An equilibrium constant could then be computed directly from measured Γ, since

$$K_{eq} = \frac{G}{G_{\max} - G} \tag{3}$$

By measuring Γ as a function of temperature and assuming that $\Gamma_{\max}$ was a good estimate of Γ_0, Inoue was able to estimate the Gibbs free energy, as well as ΔH and ΔS for the polymerization reaction. This analysis showed that spindle formation was entropy driven and allowed estimates of the force such a system could generate. However, this model for protein polymerization was hard to relate to then current ideas about polymer assembly [182]. A reformulation of the data in the context of normal polymerization reactions brought the initial interpretation into concurrence with ideas about protein polymerization and was important advance [141]. Such in vivo experiments are, however, subject to the concern that changing the temperature of a cell may affect processes other than spindle birefringence, such as the concentration of free calcium ions or various important nucleotides, as well as the properties of membranes. Nonetheless, the idea that MT assembly and disassembly might be the source of forces for chromosome movement gained wide-spread attention and favor, dominating the field for several years.

Alternative views were provided from several sources. Some scholars were staunch supporters of the idea that actin and myosin were likely to be the sources of all intracellular motions, mitosis included. Arthur Forer supported this view by experiments that implied a lack of correlation between spindle birefringence and the ability of spindles to exert forces on chromosomes [183]. He went on to find evidence for actin in spindles, a view supported by later work with immunofluorescence. Subirana developed an hypothesis for how spindle fibers, e.g., actin, might interact with matrix components through a mechanochemical enzyme, such as myosin, to push these fibers through cytoplasm in directions defined by their intrinsic polarity [184]. However, a substantial body of work with both actin inhibitors and probes for actin localization has largely discredited this point of view. Actin and related proteins may play roles at some sites of spindle force generation, such as some kinetochores during spindle fiber attachment [185]. Moreover, in very large spindles there is now compelling evidence that actin contributes to the gathering in of chromosomes during prometaphase, facilitating subsequent attachment of kinetochores to spindle fibers [186]. Other than that, most thinking about mitotic forces has focused on MTs.

Following the discovery of axonemal dynein [187], investigators realized that MT-dependent ATPases could be the major source of mitotic forces. One model explored the possibility that forces for chromosome motion were generated between spindle MTs by a dynein-like motor enzyme [48]. This proposal considered chromosomes simply as passengers attached to a subset of spindle MTs. The model was sufficiently specific that the authors found only one set of MT orientations that would allow a single motor to accomplish the then known functions of mitosis. While the idea was attractive and widely considered, the model's assumptions about spindle MT polarity were subsequently shown to be wrong [56,57,188]. The idea of motor-driven interactions between MTs as a mechanism for anaphase B has persisted in a useful way, as discussed in Scholey et al., this volume [148], but as a way to achieve anaphase A and other aspects of chromosome motion, such as congression to the metaphase plate, this model must be dismissed.

With the discovery of kinesin and its role in vesicle motion, motor enzymes re-entered mitotic modeling with the postulate that kinetochores might be sites of motor localization, allowing chromosomes to move over spindle MTs in an active way. This important innovation was a departure from the longstanding idea that chromosomes were simply passengers for whatever mitotic force generators might exist. Mazia expressed this view most lucidly with the adage that "the role in mitosis of the chromosome arms, which carry most of the genetic material, may be compared with that of a corpse at a funeral: they provide the reason for the proceedings but do not take an active part in them" [189]. With advances in the genetics and molecular biology of yeasts and fruit flies, it became clear that cells contain a large repertoire of kinesins and dyneins, so there was no dearth of motors to help chromosomes move. As the kinesin super-family grew in size and as it was realized that even a simple yeast cell made five or more kinesins, as well as a cytoplasmic dynein, the ideal of a simple model for mitosis faded rapidly. Data on motor localization demonstrated that there were force generators at kinetochores, on chromosome arms and poles, on spindle fibers, in asters, and at the cell cortex. The question became one of learning how much each of these motors contributed to each kind of chromosome motion. These issues are discussed in later chapters of this book.

As the polymerization behavior of tubulin was elucidated by experiments on purified components in vitro, the idea of MT dynamics as an origin for mitotic forces resurfaced in a new and exciting form. One of the most innovative models was based on the fact that MTs at apparent polymerization equilibrium are actually at steady state; they can polymerize at one end and depolymerize at the other, making individual polymers into treadmills [190]. While this property might appear to violate the second law of thermodynamics, no such problem is encountered, because the system is driven by hydrolysis of the GTP bound to tubulin during its polymerization. This model envisioned kinetochores as sites of MT attachment that were able to allow tubulin addition during metaphase, so MTs could treadmill poleward, losing subunits at the spindle poles. The kinetochore was endowed with the capabilities of either allowing tubulin addition or not, so at anaphase onset it ceased to permit polymerization and then followed the MT plus end as it was pulled poleward by tubulin loss at the MT minus ends near the spindle poles. It is noteworthy that many complex aspects of MT dynamics that have recently been observed with markers for MT movements are completely consistent with the predictions of this model.

The discovery of dynamic instability [150] added further interest and complexity to the properties of MT dynamics. One of the most important tasks of the mitotic spindle is to form stable attachments with kinetochores such that sister chromatids are linked to sister poles. The behavior of dynamically instable MTs led to an insightful model for the attachment process [191], which has subsequently been developed in quantitative form by several groups [192,193]. Recent work on this important issue will be discussed elsewhere in this book.

Theoreticians of the 1980s were also interested in MT dynamics as a potential force generator, so in spite of the lure of motor enzymes, several papers were published, developing this idea. One of these explored MT (or actin) dynamics as a source of mechanical work, using principles of thermodynamics [194]. Another analyzed a structural model for the attachment between spindle MTs and the kinetochore, based on a "sleeve", as suggested earlier by Margolis and Wilson [190], that would surround the polymer end and provide a movable attachment between a kinetochore and MTs as the polymers shortened [195]. Shortly thereafter, the Kirschner group obtained evidence that depolymerizing MTs could retain their attachment to kinetochores in vitro [195]. These authors suggested that this retention was sufficiently strong to allow MT depolymerization to generate the forces necessary for anaphase A, either by a sleeve of the kind proposed by Hill or with a ring-like structure that could surrounded the MT and ride a "conformational wave" that might be generated by MT depolymerization [196].

These ideas received considerable support from subsequent observations, both in vitro and in vivo. Members of the McIntosh lab demonstrated minus end-directed chromosome motion in vitro; isolated chromosomes could follow the depolymerizing end of shortening MTs at physiological speeds in the absence of soluble nucleotide triphosphates [197]. This and subsequent work by this and other groups on the generation of force by MT depolymerization will be reviewed in a later chapter. It is noteworthy, however, that the observations cited above made the experimental landscape of spindle physiology remarkably complex. There is force generation by MT depolymerization, by motors of the kinesin super-family, and by dynein, all present and active in the spindle. Modeling chromosome motion is now a significant challenge. Future progress will require a painstakingly careful union of a large body of facts about each cell type under consideration and a deep knowledge of the relevant chemical physics.

Anaphase B, one the other hand, is enough simpler that several useful models for spindle elongation have been put forward. Early work, formulated in the context of the idea that those spindle MTs not connected to chromosomes were "continuous", led to the straightforward proposal that spindles grew longer simply by the elongation of their component MTs [135]. Following the discovery that continuous MTs are rare, if they exist at all, and that the interpolar spindle is made from two interdigitating MTs families, one emanating from each pole [34,198], the concept of spindle elongation by motor-driven sliding of antiparallel MTs has been favored [199,200]. More recent and complete work on spindle elongation in fruit flies has greatly extended and formalized those ideas with a quantitative model that is a rigorous accounting of a considerable amount of structural, biochemical, and physiological data, as reviewed in Scholey et al., this volume [148].

Knowledge about small spindles, such as those that form in yeast cells, is now sufficiently complete that several groups are trying to formulate quantitative models with predictive power [201,202]. These models too will be addressed in subsequent chapters, but in summary we can say that our knowledge of mitotic phenomenology is now sufficiently complete that the dream of a single "model for mitosis", cherished by many earlier investigators, is simply that: a dream. The biological significance of mitotic events is sufficient that cells have invoked multiple mechanisms to achieve the goal of accurate chromosome segregation. The book that follows is a testimony to the skill and hard work of a small army of investigators who have contributed to our current understanding of the process. While no simple model for mitosis is currently available, or perhaps ever will be, our thinking about the process is becoming ever more precise, providing important new understanding about key molecules and targets that may well advance our clinical treatments for human disease. We can plausibly hope that the mechanisms now understood will comprise the framework for a deep understanding of mitosis in the not too far distant future.

9. Conclusions

This chapter describes a remarkable series of advances in our understanding of a complex and important biological problem: the mechanisms by which eukaryotic cells transmit a complete, undamaged genome to their daughters. The progress made reflects not only the skill, diligence, and cleverness of many investigators, it shows how each generation of scientists has depended on both previous progress in descriptions of natural phenomena and on the development of new technologies. The small size and delicate structure of mitotic spindles has made them a "poster child" for the importance of technological advances, first in microscopy, then in the intricate ways that genetics, biochemistry, and molecular biology can cooperate to identify important functional components of a complex process. In addition, however, the research on mitosis described here has demonstrated the importance of mechanical processes in cell organization and physiology. We hope that future generations of biologists will take note of this important aspect of cell biology.

Supplementary Materials: The following are available online at www.mdpi.com/2079-7737/5/55/s1.

Acknowledgments: The structural work from the lab of J. Richard McIntosh was supported by numerous grants from the NIH and the ACS. Thomas Hays also acknowledges his NIH, ACS, AHA and Pew Foundation support for related research on mitosis and motor proteins. The authors thank Edward Salmon, Robert Margolis, and the reviewers for their critical readings of the manuscript and many helpful suggestions. We also thank Manfred Schliwa for help in understanding the contributions of early scholars of mitosis who wrote in German.

Author Contributions: The authors planned this paper together. J. Richard McIntosh wrote the sections on early history, spindle structure and diversity, biochemistry, genetics, and modeling. Thomas Hays wrote the sections about experiments on living cells that probed mitotic physiology. Both authors edited and revised the work.

Conflicts of Interest: The authors state that they have no conflict of interests.

References

1. Schneider, A. Untersuchungen über Plathelminthen. In *Bericht der Oberhessischen Gesellschaft für Natur- und Heilkunde*; Upper-Hessian Society for Natural and Medical Science: Giessen, Hesse, Germany, 1873; Volume 14. (In German)
2. Strasburger, E. *Zellbildung und Zelltheilung "Cell Formation and Cell Division"*; Gustav Fischer: Jena, Germany, 1880.
3. Van Beneden, E. Recherches sur les Dicyemides. *Bull. Acad. R.* **1876**, *41*, 1–111.
4. Flemming, W. Zur Kenntniss der Zelle und ihre Lebenserscheinungen. *Arch. Mikr. Anat.* **1878**, *16*, 302–436. [CrossRef]
5. Flemming, W. Contributions to the knowledge of the cell and its vital processes. *J. Cell Biol.* **1965**, *25*, 3–69. [PubMed]
6. Flemming, W. *Zellsubstanz, Kern und Zelltheilung*; Vogel: Leipzig, Germany, 1882. (In German)
7. Hooke, R. *Micrographia: Or Some Physiological Descriptions of Minute Bodies Made by Magnifying Glasses with Observations and Inquiries Thereupon*; James Allestry: London, UK, 1667.
8. Van Leeuwenhoek, A. *Arcana Naturae Detacta*; Apud Henricum a Krooneveld: Delft, the Netherlands, 1695.
9. Virchow, R. *Die Cellularpathologie in Ihrer Begründung auf Physiologische und Pathologische Gewebelehre*; Verlag von August Hirschwald: Berlin, Germany, 1858.
10. Abbe, E. A New Illuminating Apparatus for the Microscope. *Mon. Microsc. J.* **1875**, *13*, 77–82. [CrossRef]
11. Mayzel, W. Ueber eigenthümiche Vorgänge bei der Theilung der Kerne in Epithelialzellen" (On peculiar events during the division of nuclei of epithelial cells). *Zentralbl. Med. Wiss.* **1885**, *13*, 849–852.
12. Weissmann, A. *Die Continuität des Keimplasma's als Grundlage einer Theorie der Vererbung*; Fischer: Jena, Germany, 1885.
13. Boveri, T.H. *Ergebnisse über die Konstitution der Chromatischen Substanz des Zelkerns*; Fisher: Jena, Germany, 1904.
14. Sutton, W.S. The chromosomes in heredity. *Biol. Bull.* **1903**, *4*, 231–251. [CrossRef]
15. Wilson, E.B. *The Cell in Development and Heredity*; MacMillan, Inc.: New York, NY, USA, 1925.
16. Fischer, A. *Fixirung, Färbung und Bau des Protoplasmas: Kritische Untersuchungen über Technik und Theorie in der Neueren Zellforschung*; Fisher: Jena, Germany, 1899.
17. Hardy, W.B. On Spindles. *J. Physiol.* **1899**, *24*, 158–210. [CrossRef] [PubMed]
18. Lauterborn, R. *Untersuchengen ueber Bau, Kernteilung und Bewegung der Diatomeen*; Wilhelm Engelmann: Leipzig, Germany, 1896.

19. Lauterborn, R. Cell Division in Diatoms. *Protoplasma* **1984**, *120*, 132–154.
20. Chambers, R. Microdissection studies. II. The cell aster, a reversible gelation phenomenon. *J. Exp. Zool.* **1917**, *23*, 483–504. [CrossRef]
21. Zernike, F. How I discovered Phase optics. *Science* **1955**, *121*, 345–349. [CrossRef]
22. Cleveland, L.R.; Hall, S.R.; Sanders, E.P. *The Wood-Feeding Roach Cryptocercus, Its Protozoa, and the Symbiosis between Protozoa and Roach*, 1st ed.; American Academy of Arts and Sciences: Cambridge, MA, USA, 1934; Volume 17, p. 406.
23. Schmidt, W.J. *Doppelbrechung von Karyoplasma, Metaplasma und Zytoplasma*; Gebrueder Borntraeger: Berlin, Germany, 1937. (In German)
24. Schmidt, F.O. The ultrastructure of protoplasmic constituents. *Physiol. Rev.* **1939**, *19*, 270–302.
25. Hughes, A.F.; Swann, M.M. Anaphase Movements in the Living Cell. *J. Exp. Biol.* **1948**, *25*, 45–72.
26. Inoue, S.; Dan, K. Birefringence of the dividing cell. *J. Morph.* **1951**, *89*, 423–456. [CrossRef]
27. Inoue, S.; Hyde, W.L. Studies on depolarization of light at microscope lens surfaces, II. The simultaneous realization of high resolution and high sensitivity with the polarizing microscope. *J. Cell Biol.* **1957**, *3*, 831–838. [CrossRef]
28. Rozsa, G.; Wyckoff, R.W.G. The electron microscopy of dividing cells. *Biochim. Biophys. Acta* **1950**, *6*, 334–339. [CrossRef]
29. Bernhard, W.; De Harven, E. Electron microscopic study of the ultrastructure of centrioles in vertebra. *Z. Zellforsch. Mikrosk. Anat.* **1956**, *45*, 378–398. [PubMed]
30. Harris, P. Some Observations Concerning Metakinesis in Sea Urchin Eggs. *J. Cell Biol.* **1965**, *25*, 73–77. [CrossRef]
31. Roth, L.E.; Daniels, E.W. Electron microscopic studies of mitosis in amebae: II The Giant Ameba Pelomyxa carolinensis. *J. Cell Biol.* **1962**, 57–78. [CrossRef]
32. Sabatini, D.D.; Bensch, K.; Barrnett, R.J. Cytochemistry and electron microscopy. The preservation of cellular ultrastructure and enzymatic activity by aldehyde fixation. *J. Cell Biol.* **1963**, *17*, 19–58. [CrossRef] [PubMed]
33. Brinkley, B.R.; Stubblefield, E. The fine structure of the kinetochore of a mammalian cell in vitro. *Chromosoma* **1966**, *19*, 28–43. [CrossRef] [PubMed]
34. McIntosh, J.R.; Landis, S.C. The Distribution of Spindle Microtubules during Mitosis in Cultured Human Cells. *J. Cell Biol.* **1971**, *49*, 468–497. [CrossRef] [PubMed]
35. Fuge, H. Microtubule distribution in metaphase and anaphase spindles of the spermatocytes of Pales ferruginea. A quantitative analysis of serial cross-sections (author's transl). *Chromosoma* **1973**, *43*, 109–143. [CrossRef] [PubMed]
36. McIntosh, J.R.; Cande, W.Z.; Snyder, J.A.; Vanderslice, K. Studies on the mechanism of mitosis. *Ann. N.Y. Acad. Sci.* **1975**, *253*, 383–406. [CrossRef]
37. McIntosh, J.R.; Cande, W.Z.; Snyder, J.A. Structure and physiology of the mammalian mitotic spindle. *Soc. Gen. Physiol. Ser.* **1975**, *30*, 31–76. [PubMed]
38. Nicklas, R.B.; Kubai, D.F.; Hays, T.S. Spindle microtubules and their mechanical associations after micromanipulation in anaphase. *J. Cell Biol.* **1982**, *95*, 91–104. [CrossRef] [PubMed]
39. Peterson, J.B.; Ris, H. Electron-microscopic study of the spindle and chromosome movement in the yeast Saccharomyces cerevisiae. *J. Cell Sci.* **1976**, *22*, 219–242. [PubMed]
40. McDonald, K.; Pickett-Heaps, J.D.; McIntosh, J.R.; Tippit, D.H. On the mechanism of anaphase spindle elongation in Diatoma vulgare. *J. Cell Biol.* **1977**, *74*, 377–388. [CrossRef] [PubMed]
41. McIntosh, J.R.; Roos, U.P.; Neighbors, B.; McDonald, K.L. Architecture of the microtubule component of mitotic spindles from *Dictyostelium discoideum*. *J. Cell Sci.* **1985**, *75*, 93–129. [PubMed]
42. Winey, M.; Mamay, C.L.; O'Toole, E.T.; Mastronarde, D.N.; Giddings, T.H.; McDonald, K.L.; McIntosh, J.R. Three-dimensional ultrastructural analysis of the *Saccharomyces cerevisiae* mitotic spindle. *J. Cell Biol.* **1995**, *129*, 1601–1615. [CrossRef] [PubMed]
43. Kilmartin, J.V.; Adams, A.E. Structural rearrangements of tubulin and actin during the cell cycle of the yeast *Saccharomyces*. *J. Cell Biol.* **1984**, *98*, 922–933. [CrossRef] [PubMed]
44. Rieder, C.L.; Borisy, G.G. The attachment of kinetochores to the pro-metaphase spindle in PtK1 cells. Recovery from low temperature treatment. *Chromosoma* **1981**, *82*, 693–716. [CrossRef] [PubMed]
45. McDonald, K.L.; O'Toole, E.T.; Mastronarde, D.N.; McIntosh, J.R. Kinetochore microtubules in PTK cells. *J. Cell Biol.* **1992**, *118*, 369–383. [CrossRef] [PubMed]

46. Mastronarde, D.N.; McDonald, K.L.; Ding, R.; McIntosh, J.R. Interpolar spindle microtubules in PTK cells. *J. Cell Biol.* **1993**, *123*, 1475–1489. [CrossRef] [PubMed]
47. Amos, L.; Klug, A. Arrangement of subunits in flagellar microtubules. *J. Cell Sci.* **1974**, *14*, 523–549. [PubMed]
48. McIntosh, J.R.; Hepler, P.K.; Van Wie, D.G. Model for Mitosis. *Nature* **1969**, *224*, 659–663. [CrossRef]
49. Pease, D.C. Hydrostatic pressure effects upon the spindle figure and chromosome movements. *Biol. Bull.* **1946**, *91*, 145–169. [CrossRef] [PubMed]
50. Witt, P.L.; Ris, H.; Borisy, G.G. Origin of kinetochore microtubules in Chinese hamster ovary cells. *Chromosoma* **1980**, *81*, 483–505. [CrossRef] [PubMed]
51. Weisenberg, R.; Rosenfeld, A. Role of intermediates in microtubule assembly in vivo and in vitro. *Ann. N.Y. Acad. Sci.* **1975**, *253*, 78–89. [CrossRef] [PubMed]
52. Telzer, B.R.; Moses, M.J.; Rosenbaum, J.L. Assembly of microtubules onto kinetochores of isolated mitotic chromosomes of HeLa cells. *Proc. Natl. Acad. Sci. USA* **1975**, *72*, 4023–4027. [CrossRef] [PubMed]
53. Bergen, L.G.; Borisy, G.G. Head-to-tail polymerization of microtubules in vitro. Electron microscope analysis of seeded assembly. *J. Cell Biol.* **1980**, *84*, 141–150. [CrossRef] [PubMed]
54. Bergen, L.G.; Kuriyama, R.; Borisy, G.G. Polarity of microtubules nucleated by centrosomes and chromosomes of Chinese hamster ovary cells in vitro. *J. Cell Biol.* **1980**, *84*, 151–159. [CrossRef] [PubMed]
55. Heidemann, S.R.; McIntosh, J.R. Visualization of the structural polarity of microtubules. *Nature* **1980**, *286*, 517–519. [CrossRef] [PubMed]
56. Euteneuer, U.; McIntosh, J.R. Polarity of midbody and phragmoplast microtubules. *J. Cell Biol.* **1980**, *87*, 509–515. [CrossRef] [PubMed]
57. McIntosh, J.R.; Euteneuer, U. Tubulin hooks as probes for microtubule polarity: An analysis of the method and an evaluation of data on microtubule polarity in the mitotic spindle. *J. Cell Biol.* **1984**, *98*, 525–533. [CrossRef] [PubMed]
58. Telzer, B.R.; Haimo, L.T. Decoration of spindle microtubules with Dynein: Evidence for uniform polarity. *J. Cell Biol.* **1981**, *89*, 373–378. [CrossRef] [PubMed]
59. Bajer, A.S.; Mole-Bajer, J. Spindle Dynamics and Chromosome Movement. *Int. Rev. Cytol. Suppl.* **1972**, *34*, 1–271.
60. De Mey, J.; Lambert, A.M.; Bajer, A.S.; Moeremans, M.; De Brabander, M. Visualization of microtubules in interphase and mitotic plant cells of Haemanthus endosperm with the immuno-gold staining method. *Proc. Natl. Acad. Sci. USA* **1982**, *79*, 1898–1902. [CrossRef] [PubMed]
61. Kubai, D.F.; Ris, H. Division in the dinoflagellate *Gyrodinium cohnii* (Schiller). A new type of nuclear reproduction. *J. Cell Biol.* **1969**, *40*, 508–528. [CrossRef] [PubMed]
62. Jacob, F.; Brenner, S.; Cuzin, F. On the regulation of DNA replication in bacteria. *Cold Spring Harb. Symp. Quant. Biol.* **1963**, *28*, 329–348. [CrossRef]
63. Oakley, B.R.; Dodge, J.D. Mitosis in the Cryptophyceae. *Nature* **1973**, *244*, 521–522. [CrossRef] [PubMed]
64. Feder, N.; Sidman, R.L. Methods and principles of fixation by freeze-substitution. *J. Cell Biol.* **1958**, *4*, 593–600. [CrossRef]
65. Heath, I.B. Mitosis in the fungus *Thraustotheca clavata*. *J. Cell Biol.* **1974**, *60*, 204–220. [CrossRef] [PubMed]
66. Ris, H.; Kubai, D.F. An unusual mitotic mechanism in the parasitic protozoan *Syndinium* sp. *J. Cell Biol.* **1974**, *60*, 702–720. [CrossRef] [PubMed]
67. Kubai, D.F. The evolution of the mitotic spindle. *Int. Rev. Cytol.* **1975**, *43*, 167–227. [PubMed]
68. He, X.; Asthana, S.; Sorger, P.K. Transient sister chromatid separation and elastic deformation of chromosomes during mitosis in budding yeast. *Cell* **2000**, *101*, 763–775. [CrossRef]
69. Pickett-Heaps, J.D.; Tippit, D.H.; Leslie, R. Light and electron microscopic observations on cell division in two large pennate diatoms, Hantzschia and Nitzschia. I. Mitosis in vivo. *Eur. J. Cell Biol.* **1980**, *21*, 1–11. [PubMed]
70. Nabeshima, K.; Nakagawa, T.; Straight, A.F.; Murray, A.; Chikashige, Y.; Yamashita, Y.M.; Hiraoka, Y.; Yanagida, M. Dynamics of centromeres during metaphase-anaphase transition in fission yeast: Dis1 is implicated in force balance in metaphase bipolar spindle. *Mol. Biol. Cell* **1998**, *9*, 3211–3225. [CrossRef] [PubMed]
71. Rieder, C.L. The structure of the cold-stable kinetochore fiber in metaphase PtK1 cells. *Chromosoma* **1981**, *84*, 145–158. [CrossRef] [PubMed]

72. Fuge, H. The arrangement of microtubules and the attachment of chromosomes to the spindle during anaphase in tipulid spermatocytes. *Chromosoma* **1974**, *45*, 245–260. [CrossRef] [PubMed]
73. Metz, C.W. An apparent case of monocentric mitosis in Sciara (Diptera). *Science* **1926**, *63*, 190–191. [CrossRef] [PubMed]
74. Kubai, D.F. Meiosis in *Sciara coprophila*: Structure of the spindle and chromosome behavior during the first meiotic division. *J. Cell Biol.* **1982**, *93*, 655–669. [CrossRef] [PubMed]
75. Johnson, U.G.; Porter, K.R. Fine structure of cell division in *Chlamydomonas reinhardi*. Basal bodies and microtubules. *J. Cell Biol.* **1968**, *38*, 405–425. [CrossRef]
76. Stafstrom, J.P.; Staehelin, L.A. Dynamics of the nuclear envelope and of nuclear pore complexes during mitosis in the Drosophila embryo. *Eur. J. Cell Biol.* **1984**, *34*, 179–189. [PubMed]
77. Oakley, B.R.; Dodge, J.D. Kinetochores associated with the nuclear envelope in the mitosis of a dinoflagellate. *J. Cell Biol.* **1974**, *63*, 322–325. [CrossRef] [PubMed]
78. Ritter, H.; Inoue, S.; Kubai, D. Mitosis in Barbulanympha. I. Spindle structure, formation, and kinetochore engagement. *J. Cell Biol.* **1978**, *77*, 638–654. [CrossRef] [PubMed]
79. Mazia, D.; Dan, K. The Isolation and Biochemical Characterization of the Mitotic Apparatus of Dividing Cells. *Proc. Natl. Acad. Sci. USA* **1952**, *38*, 825–835. [CrossRef]
80. Mazia, D.; Mitchison, J.M.; Medina, H.; Harris, P. The direct isolation of the mitotic apparatus. *J. Cell Biol.* **1961**, *10*, 467–474. [CrossRef]
81. Kane, R.E. The mitotic apparatus. Physical-chemical factors controlling stability. *J. Cell Biol.* **1965**, *25*, 137–144. [CrossRef]
82. Forer, A. Characteristics of sea-urchin mitotic apparatus isolated using a dimethyl sulphoxide/glycerol medium. *J. Cell Sci.* **1974**, *16*, 481–497. [PubMed]
83. Sakai, H.S.; Shimoda, S.; Hiramoto, Y. Mass isolation of mitotic apparatus using a glycerol/Mg^{2+}/Triton X-100 medium. *Exp. Cell Res.* **1977**, *104*, 457–461. [CrossRef]
84. Murphy, D.B. Identification of microtubule-associated proteins in the meiotic spindle of surf clam oocytes. *J. Cell Biol.* **1980**, *84*, 235–245. [CrossRef] [PubMed]
85. Schiff, P.B.; Fant, J.; Horwitz, S.B. Promotion of microtubule assembly in vitro by taxol. *Nature* **1979**, *277*, 665–667. [CrossRef] [PubMed]
86. Kuriyama, R.; Keryer, G.; Borisy, G.G. The mitotic spindle of Chinese hamster ovary cells isolated in taxol-containing medium. *J. Cell Sci.* **1984**, *66*, 265–275. [PubMed]
87. Taylor, E.W. The Mechanism of Colchicine Inhibition of Mitosis. I. Kinetics of Inhibition and the Binding of H3-Colchicine. *J. Cell Biol.* **1965**, *25*, 145–160. [CrossRef]
88. Borisy, G.G.; Taylor, E.W. The mechanism of action of colchicine. Binding of colchincine-3H to cellular protein. *J. Cell Biol.* **1967**, *34*, 525–533. [CrossRef] [PubMed]
89. Weisenberg, R.C. Microtubule formation in vitro in solutions containing low calcium concentrations. *Science* **1972**, *177*, 1104–1105. [CrossRef] [PubMed]
90. Hoffmann-Berling, H. Adenosinetriphosphate as the energy substance for cell movement. *Biochim. Biophys. Acta* **1954**, *14*, 182–194. [PubMed]
91. Cande, W.Z.; Snyder, J.; Smith, D.; Summers, K.; McIntosh, J.R. A functional mitotic spindle prepared from mammalian cells in culture. *Proc. Natl. Acad. Sci. USA* **1974**, *71*, 1559–1563. [CrossRef] [PubMed]
92. Rebhun, L.I.; Palazzo, R.E. In vitro reactivation of anaphase B in isolated spindles of the sea urchin egg. *Cell Motil. Cytoskelet.* **1988**, *10*, 197–209. [CrossRef] [PubMed]
93. Lohka, M.J.; Maller, J.L. Induction of nuclear envelope breakdown, chromosome condensation, and spindle formation in cell-free extracts. *J. Cell Biol.* **1988**, *101*, 518–523. [CrossRef]
94. Lohka, M.J.; Hayes, M.K.; Maller, J.K. Purification of maturation-promoting factor, an intracellular regulator of early mitotic events. *Proc. Natl. Acad. Sci. USA* **1988**, *85*, 3009–3013. [CrossRef] [PubMed]
95. Fujiwara, K.; Pollard, T.D. Fluorescent antibody localization of myosin in the cytoplasm, cleavage furrow, and mitotic spindle of human cells. *J. Cell Biol.* **1976**, *71*, 848–875. [CrossRef] [PubMed]
96. Izant, J.G.; Weatherbee, J.A.; McIntosh, J.R. A microtubule-associated protein in the mitotic spindle and the interphase nucleus. *Nature* **1982**, *295*, 248–250. [CrossRef] [PubMed]
97. Cande, W.Z.; Lazarides, E.; McIntosh, J.R. A comparison of the distribution of actin and tubulin in the mammalian mitotic spindle as seen by indirect immunofluorescence. *J. Cell Biol.* **1977**, *72*, 552–567. [CrossRef] [PubMed]

98. Barak, L.S.; Nothnagel, E.A.; DeMarco, E.F.; Webb, W.W. Differential staining of actin in metaphase spindles with 7-nitrobenz-2-oxa-1,3-diazole-phallacidin and fluorescent DNase: Is actin involved in chromosomal movement? *Proc. Natl. Acad. Sci. USA* **1981**, *78*, 3034–3038. [CrossRef] [PubMed]
99. Oakley, C.E.; Oakley, B.R. Identification of gamma-tubulin, a new member of the tubulin superfamily encoded by mipA gene of *Aspergillus nidulans*. *Nature* **1989**, *338*, 662–664. [CrossRef] [PubMed]
100. Pfarr, C.M.; Coue, M.; Grissom, P.M.; Hays, T.S.; Porter, M.E.; McIntosh, J.R. Cytoplasmic dynein is localized to kinetochores during mitosis. *Nature* **1990**, *345*, 263–265. [CrossRef] [PubMed]
101. Steuer, E.R.; Wordeman, L.; Schroer, T.A.; Sheetz, M.P. Localization of cytoplasmic dynein to mitotic spindles and kinetochores. *Nature* **1990**, *345*, 266–268. [CrossRef] [PubMed]
102. Moroi, Y.; Peebles, C.; Fritzler, M.J.; Steigerwald, J.; Tan, E.M. Autoantibody to centromere (kinetochore) in scleroderma sera. *Proc. Natl. Acad. Sci. USA* **1980**, *77*, 1627–1631. [CrossRef] [PubMed]
103. Earnshaw, W.C.; Halligan, N.; Cooke, C.; Rothfield, N. The kinetochore is part of the metaphase chromosome scaffold. *J. Cell Biol.* **1984**, *98*, 3352–3357. [CrossRef]
104. Bernat, R.L.; Borisy, G.G.; Rothfield, N.F.; Earnshaw, W.C. Injection of anticentromere antibodies in interphase disrupts events required for chromosome movement at mitosis. *J. Cell Biol.* **1990**, *111*, 1519–1533. [CrossRef] [PubMed]
105. Palmer, D.K.; O'Day, K.; Trong, H.L.; Charbonneau, H.; Margolis, R.L. Purification of the centromere-specific protein CENP-A and demonstration that it is a distinctive histone. *Proc. Natl. Acad. Sci. USA* **1991**, *88*, 3734–3738. [CrossRef] [PubMed]
106. Wald, H. Cytologic studies of the abnormal development of the eggs of the Claret mutant type of *Drosophila simulans*. *Genetics* **1935**, *21*, 264–281.
107. Davis, D.G. Chromosome Behavior under the Influence of Claret-Nondisjunctional in *Drosophila melanogaster*. *Genetics* **1969**, *61*, 577–594. [PubMed]
108. Walker, R.A.; Salmon, E.D.; Endow, S.A. The *Drosophila claret* segregation protein is a minus-end directed motor molecule. *Nature* **1990**, *347*, 780–782. [CrossRef] [PubMed]
109. Ostergren, G. Description of unpublished work on mitotic mutants in Lily, given to the author during a meeting at the Swedish. University of Agricultural Sciences in Uppsala: Sweden, 1976.
110. Baker, B.S. Paternal loss (pal): A meiotic mutant in *Drosophila melanogaster* causing loss of paternal chromosomes. *Genetics* **1975**, *80*, 267–296. [PubMed]
111. Gatti, M.; Baker, B.S. Genes controlling essential cell-cycle functions in *Drosophila melanogaster*. *Genes Dev.* **1989**, *3*, 438–453. [CrossRef] [PubMed]
112. Hartwell, L.; Culotti, J.; Ried, B. Genetic control of the cell-division cycle in yeast. I. Detection of mutants. *Proc. Natl. Acad. Sci. USA* **1962**, *66*, 352–359. [CrossRef]
113. Nurse, P. Genetic control of cell size at cell division in yeast. *Nature* **1975**, *256*, 547–551. [CrossRef] [PubMed]
114. Morris, N.R. Mitotic mutants of *Aspergillus nidulans*. *Genet. Res.* **1975**, *26*, 237–254. [CrossRef] [PubMed]
115. Hirano, T.; Funahashi, S.; Uemura, T.; Yanagida, M. Isolation and characterization of *Schizosaccharomyces pombe* cutmutants that block nuclear division but not cytokinesis. *EMBO J.* **1986**, *5*, 2973–2979. [PubMed]
116. Hoyt, M.A.; Stearns, T.; Botstein, D. Chromosome instability mutants of *Saccharomyces cerevisiae* that are defective in microtubule-mediated processes. *Mol. Cell. Biol.* **1990**, *10*, 223–234. [CrossRef] [PubMed]
117. Yang, J.T.; Saxton, W.M.; Goldstein, L.S. Isolation and characterization of the gene encoding the heavy chain of *Drosophila kinesin*. *Proc. Natl. Acad. Sci. USA* **1988**, *85*, 1864–1868. [CrossRef] [PubMed]
118. Li, R.; Murray, A.W. Feedback control of mitosis in budding yeast. *Cell* **1991**, *66*, 519–531. [CrossRef]
119. Hoyt, M.A.; He, L.; Loo, K.K.; Saunders, W.S. Two *Saccharomyces cerevisiae* kinesin-related gene products required for mitotic spindle assembly. *J. Cell Biol.* **1992**, *118*, 109–120. [CrossRef] [PubMed]
120. Rout, M.P.; Kilmartin, J.V. Components of the yeast spindle and spindle pole body. *J. Cell Biol.* **1990**, *111*, 1913–1927. [CrossRef] [PubMed]
121. Schrader, F. On the reality of spindle fibers. *Biol. Bull.* **1934**, *67*, 519–533. [CrossRef]
122. Carlson, J.G. Mitotic Behavior of Induced Chromosomal Fragments Lacking Spindle Attachments in the Neuroblasts of the Grasshopper. *Proc. Natl. Acad. Sci. USA* **1938**, *24*, 500–507. [CrossRef] [PubMed]
123. Hughes-Schrader, S.; Ris, H. The diffuse spindle attachment of coccids, verified by the mitotic behavior of induced chromosome fragments. *J. Exp. Zool.* **1941**, *87*, 429–456. [CrossRef]
124. White, M.J.D. The Effect of X-Rays on the First Meiotic Division in Three Species of Orthoptera. *Proc. R. Soc. Lond. Ser. B* **1937**, *124*, 183–196. [CrossRef]

125. Ris, H. A quantitative study of anaphase movement in the aphid Tamalia. *Biol. Bull.* **1943**, *85*, 164–178. [CrossRef]
126. Brinkley, B.R.; Stubblefield, E.; Hsu, T.C. The effects of colcemid inhibition and reversal on the fine structure of the mitotic apparatus of Chinese hamster cells in vitro. *J. Ultrastruct. Res.* **1967**, *19*, 1–18. [CrossRef]
127. Belar, K. Beiträge zur Kausalanalyse der Mitose. *Roux Arch Entw Mech Org* **1929**, *118*, 359–484. [CrossRef]
128. Dietz, R. Multiple Geschlechchromosomen bei den cypriden Ostracoden, ihre Evolution and ihr Teilungsverhalten. *Chromosoma* **1958**, *9*, 359–440. [CrossRef] [PubMed]
129. Nicklas, R.B. Recurrent pole-to-pole movements of the sex chromosome during prometaphase I in Melanoplus differentialis spermatocytes. *Chromosoma* **1961**, *12*, 97–115. [CrossRef] [PubMed]
130. Nicklas, R.B. Chromosome Velocity during Mitosis as a Function of Chromosome Size and Position. *J. Cell Biol.* **1965**, *25*, 119–135. [CrossRef]
131. Hughes-Schrader, S. Polarization, kinetochore movements, and bivalent structure in the meiosis of male mantids. *Biol. Bull.* **1943**, *85*, 265–300. [CrossRef]
132. Hughes-Schrader, S. The "Pre-Metaphase Stretch" and kinetochore orientation in Phasmids. *Chromosoma* **1946**, *3*, 1–21. [CrossRef]
133. Shimamura, T. On the mechanism of nuclear division and chromosome arrangement. VI. Studies on the effect of the centrifugal force upon nuclear division. *Cytologia* **1940**, *11*, 186–216.
134. Beams, H.W.; King, R.L. The effect of ultracentrifuging upon chick embryonic cells, with special reference to the "resting" nucleus and the mitotic spindle. *Biol. Bull.* **1936**, *71*, 188–198. [CrossRef]
135. Inoue, S.; Sato, H. Cell motility by labile association of molecules. The nature of mitotic spindle fibers and their role in chromosome movement. *J. Gen. Physiol.* **1967**, *50*, 259–292. [CrossRef]
136. Rebhun, L.I.; Sawada, N. Augmentation and dispersion of the in vivo mitotic apparatus of living marine eggs. *Protoplasma* **1969**, *68*, 1–22. [CrossRef] [PubMed]
137. Inoue, S. The effect of colchicine on the microscopic and submicroscopic structure of the mitotic spindle. *Exp. Cell Res. Suppl.* **1952**, *2*, 305–318.
138. Taylor, E.W. Dynamics of Spindle Formation and its Inhibition by Chemicals. *J. Cell Biol.* **1959**, *6*, 193–196. [CrossRef]
139. Taylor, E.W. Relation of Protein Synthesis to the Division Cycle in Mammalian Cell Cultures. *J. Cell Biol.* **1963**, *19*, 1–18. [CrossRef] [PubMed]
140. Salmon, E.D.; Begg, D.A. Functional implications of cold-stable microtubules in kinetochore fibers of insect spermatocytes during anaphase. *J. Cell Biol.* **1980**, *85*, 853–865. [CrossRef] [PubMed]
141. Salmon, E.D. Spindle microtubules: Thermodynamics of in vivo assembly and role in chromosome movement. *Ann. N. Y. Acad. Sci.* **1975**, *253*, 383–406. [CrossRef] [PubMed]
142. Sato, H.; Ellis, G.W.; Inoue, S. Microtubular origin of mitotic spindle form birefringence. Demonstration of the applicability of Wiener's equation. *J. Cell Biol.* **1975**, *67*, 501–517. [CrossRef] [PubMed]
143. Inoue, S. *Microtubule dynamics and chromosome motion. In Cell Motility; Cold Spring Harbor Conferences on Cell Proliferation*; Cold Spring Harbor Press: Cold Spring Harbor, NY, USA, 1976; Volume 3, pp. 1317–1328.
144. Forer, A. Characterization of the mitotic traction system, and evidence that birefringent spindle fibers neither produce nor transmit force for chromosome movement. *Chromosoma* **1966**, *19*, 44–98. [CrossRef] [PubMed]
145. Leslie, R.J.; Pickett-Heaps, J.D. Ultraviolet microbeam irradiations of mitotic diatoms: Investigation of spindle elongation. *J. Cell Biol.* **1983**, *96*, 548–561. [CrossRef] [PubMed]
146. Wilson, P.J.; Forer, A. Ultraviolet microbeam irradiation of chromosomal spindle fibres shears microtubules and permits study of the new free ends in vivo. *J. Cell Sci.* **1988**, *91*, 455–468. [PubMed]
147. McDonald, K.L.; Edwards, M.K.; McIntosh, J.R. Cross-sectional structure of the central mitotic spindle of Diatoma vulgare. Evidence for specific interactions between antiparallel microtubules. *J. Cell Biol.* **1979**, *83*, 443–461. [CrossRef] [PubMed]
148. Scholey, J.M.; Civelekoglu-Scholey, G.; Brust-Mascher, I.; Anaphase, B. *Biology* 2017. [CrossRef]
149. Salmon, E.D.; McKeel, M.; Hays, T. Rapid rate of tubulin dissociation from microtubules in the mitotic spindle in vivo measured by blocking polymerization with colchicine. *J. Cell Biol.* **1984**, *99*, 1066–1075. [CrossRef] [PubMed]
150. Mitchison, T.; Kirschner, M. Dynamic instability of microtubule growth. *Nature* **1984**, *312*, 237–242. [CrossRef] [PubMed]

151. Leslie, R.J.; Saxton, W.M.; Mitchison, T.J.; Neighbors, B.; Salmon, E.D.; McIntosh, J.R. Assembly properties of fluorescein-labeled tubulin in vitro before and after fluorescence bleaching. *J. Cell Biol.* **1984**, *99*, 2146–2156. [CrossRef] [PubMed]
152. Gorbsky, G.J.; Sammak, P.J.; Borisy, G.G. Chromosomes move poleward in anaphase along stationary microtubules that coordinately disassemble from their kinetochore ends. *J. Cell Biol.* **1987**, *104*, 9–18. [CrossRef] [PubMed]
153. Salmon, E.D.; Leslie, R.J.; Saxton, W.M.; Karow, M.L.; McIntosh, J.R. Spindle microtubule dynamics in sea urchin embryos: analysis using a fluorescein-labeled tubulin and measurements of fluorescence redistribution after laser photobleaching. *J. Cell Biol.* **1984**, *99*, 2165–2174. [CrossRef] [PubMed]
154. Wadsworth, P.; Salmon, E.D. Analysis of the treadmilling model during metaphase of mitosis using fluorescence redistribution after photobleaching. *J. Cell Biol.* **1986**, *102*, 1032–1038. [CrossRef] [PubMed]
155. Cassimeris, L.; Inoue, S.; Salmon, E.D. Microtubule dynamics in the chromosomal spindle fiber: Analysis by fluorescence and high-resolution polarization microscopy. *Cell Motil. Cytoskelet.* **1988**, *10*, 185–196. [CrossRef] [PubMed]
156. Mitchison, T.; Evans, L.; Schulze, E.; Kirschner, M. Sites of microtubule assembly and disassembly in the mitotic spindle. *Cell* **1986**, *45*, 515–527. [CrossRef]
157. Mitchison, T.J. Polewards microtubule flux in the mitotic spindle: Evidence from photoactivation of fluorescence. *J. Cell Biol.* **1989**, *109*, 637–652. [CrossRef] [PubMed]
158. Nicklas, R.B.; Staehly, C.A. Chromosome micromanipulation. I. The mechanics of chromosome attachment to the spindle. *Chromosoma* **1967**, *21*, 1–16. [CrossRef] [PubMed]
159. Begg, D.A.; Ellis, G.W. Micromanipulation studies of chromosome movement. II. Birefringent chromosomal fibers and the mechanical attachment of chromosomes to the spindle. *J. Cell Biol.* **1979**, *82*, 542–554. [CrossRef] [PubMed]
160. Nicklas, R.B.; Koch, C.A. Chromosome micromanipulation. 3. Spindle fiber tension and the reorientation of mal-oriented chromosomes. *J. Cell Biol.* **1969**, *43*, 40–50. [CrossRef] [PubMed]
161. Nicklas, R.B. Chromosome micromanipulation. II. Induced reorientation and the experimental control of segregation in meiosis. *Chromosoma* **1967**, *21*, 17–50. [CrossRef] [PubMed]
162. Rieder, C.L. The formation, structure, and composition of the mammalian kinetochore and kinetochore fiber. *Int. Rev. Cytol.* **1982**, *79*, 1–58. [PubMed]
163. Henderson, S.A.; Koch, C.A. Co-orientation stability by physical tension: A demonstration with experimentally interlocked bivalents. *Chromosoma* **1970**, *29*, 207–216. [CrossRef] [PubMed]
164. Nicklas, R.B.; Kubai, D.F. Microtubules, chromosome movement, and reorientation after chromosomes are detached from the spindle by micromanipulation. *Chromosoma* **1985**, *92*, 313–324. [CrossRef] [PubMed]
165. Ault, J.G.; Nicklas, R.B. Tension, microtubule rearrangements, and the proper distribution of chromosomes in mitosis. *Chromosoma* **1989**, *98*, 33–39. [CrossRef] [PubMed]
166. McNeill, P.A.; Berns, M.W. Chromosome behavior after laser microirradiation of a single kinetochore in mitotic PtK2 cells. *J. Cell Biol.* **1981**, *88*, 543–553. [CrossRef] [PubMed]
167. Taylor, E.W. Brownian and saltatory movements of cytoplasmic granules and the movement of anaphase chromosomes. In *Proceedings of the Fourth International Congress on Rheology*; Copley, A.L., Ed.; NASA: New York, NY, USA, 1965; Part 4; pp. 175–191.
168. Nicklas, R.B. Measurements of the force produced by the mitotic spindle in anaphase. *J. Cell Biol.* **1983**, *97*, 542–548. [CrossRef] [PubMed]
169. Nicklas, R.B. Chromosome movement: Current models and experiments on living cells. In *"Molecules and Cell Movement"*; Inoué, S., Stephens, R.E., Eds.; Raven Press: Raven, New York, NY, USA, 1975; pp. 97–118.
170. Forer, A. Possible roles of microtubules and actin-like filaments during cell-division. In *Cell Cycle Controls*; Academic Press: New York, NY, USA, 1974; pp. 319–336.
171. Rashevsky, N. Some remarks on the movement of chromosomes during cell division. *Bull. Math. Biophys.* **1941**, *3*, 1–3. [CrossRef]
172. Ostergren, G. Equilibrium of trivalents and the mechanism of chromosome movement. *Hereditas* **1945**, *31*, 498–511.
173. Ostergren, G. Considerations on some elementary features of mitosis. *Hereditas* **1950**, *36*, 1–19. [CrossRef]
174. Wise, D. On the mechanism of prometaphase congression: Chromosome velocity as a function of position on the spindle. *Chromosoma* **1978**, *69*, 231–241. [CrossRef] [PubMed]

175. Wada, B.; Izutsu, K. Effects of ultraviolet microbeam irradiations on mitosis studied in Tradescantia cells in vivo. *Cytologia* **1961**, *26*, 480–491. [CrossRef]
176. Ostergren, G. The mechanism of co-ordination of bivalents and multivalents. *Hereditas* **1951**, *37*, 85–156. [CrossRef]
177. Hays, T.S.; Wise, D.; Salmon, E.D. Traction force on a kinetochore at metaphase acts as a linear function of kinetochore fiber length. *J. Cell Biol.* **1982**, *93*, 374–389. [CrossRef] [PubMed]
178. Hays, T.S.; Salmon, E.D. Poleward force at the kinetochore in metaphase depends on the number of kinetochore microtubules. *J. Cell Biol.* **1990**, *110*, 391–404. [CrossRef] [PubMed]
179. Rieder, C.L.; Davison, E.A.; Jensen, L.C.; Cassimeris, L.; Salmon, E.D. Oscillatory movements of monooriented chromosomes and their position relative to the spindle pole result from the ejection properties of the aster and half-spindle. *J. Cell Biol.* **1986**, *103*, 581–591. [CrossRef] [PubMed]
180. Darlington, C.D. *Recent Advances in Cytology*; Blankiston's Son & Co. Inc: Philadelphia, PA, USA, 1937.
181. Cornman, I. A summary of evidence in favor of the traction fiber in mitosis. *Am. Nat.* **1944**, *78*, 410–422. [CrossRef]
182. Oosawa, F.; Kasai, M. A theory of linear and helical aggregations of macromolecules. *J. Mol. Biol.* **1962**, *4*, 10–21. [CrossRef]
183. Forer, A. Local Reduction of Spindle Fiber Birefringence in Living Nephrotoma Suturalis (Loew) Spermatocytes Induced by Ultraviolet Microbeam Irradiation. *J. Cell Biol.* **1965**, *25*, 95–117. [CrossRef]
184. Subirana, J.A. Role of spindle microtubules in mitosis. *J. Theor. Biol.* **1968**, *20*, 117–123. [CrossRef]
185. Schibler, M.J.; Pickett-Heaps, J.D. Mitosis in Oedogonium: Spindle microfilaments and the origin of the kinetochore fiber. *Eur. J. Cell Biol.* **1980**, *22*, 687–698. [PubMed]
186. Field, C.M.; Lénárt, P. Bulk cytoplasmic actin and its functions in meiosis and mitosis. *Curr. Biol.* **2011**, *21*, R825–R830. [CrossRef] [PubMed]
187. Gibbons, I.R.; Rowe, A.J. Dynein: A Protein with Adenosine Triphosphatase Activity from Cilia. *Science* **1965**, *149*, 424–426. [CrossRef] [PubMed]
188. Euteneuer, U.; McIntosh, J.R. Structural polarity of kinetochore microtubules in PtK1 cells. *J. Cell Biol.* **1981**, *89*, 338–345. [CrossRef] [PubMed]
189. Mazia, D. The Cell in Mitosis. In *The Cell*; Academic Press: New York, NY, USA, 1961; Volume 3.
190. Margolis, R.L.; Wilson, L.; Keifer, B.I. Mitotic mechanism based on intrinsic microtubule behaviour. *Nature* **1978**, *272*, 450–452. [CrossRef] [PubMed]
191. Kirschner, M.; Mitchison, T. Beyond self-assembly: From microtubules to morphogenesis. *Cell* **1986**, *45*, 329–342. [CrossRef]
192. Holy, T.E.; Leibler, S. Dynamic instability of microtubules as an efficient way to search in space. *Proc. Natl. Acad. Sci. USA* **1994**, *91*, 5682–5685. [CrossRef] [PubMed]
193. Paul, R.; Wollman, R.; Silkworth, W.T.; Nardi, I.K.; Cimini, D.; Mogilner, A. Computer simulations predict that chromosome movements and rotations accelerate mitotic spindle assembly without compromising accuracy. *Proc. Natl. Acad. Sci. USA* **2009**, *106*, 15708–15713. [CrossRef] [PubMed]
194. Hill, T.L.; Kirschner, M.W. Bioenergetics and kinetics of microtubule and actin filament assembly-disassembly. *Int. Rev. Cytol.* **1982**, *78*, 1–125. [PubMed]
195. Hill, T.L. Theoretical problems related to the attachment of microtubules to kinetochores. *Proc. Natl. Acad. Sci. USA* **1985**, *82*, 4404–4408. [CrossRef] [PubMed]
196. Koshland, D.E.; Mitchison, T.J.; Kirschner, M.W. Polewards chromosome movement driven by microtubule depolymerization in vitro. *Nature* **1988**, *331*, 499–504. [CrossRef] [PubMed]
197. Coue, M.; Lombillo, V.A.; McIntosh, J.R. Microtubule depolymerization promotes particle and chromosome movement in vitro. *J. Cell Biol.* **1991**, *112*, 1165–1175. [CrossRef] [PubMed]
198. Brinkley, B.R.; Cartwright, J. Ultrastructural analysis of mitotic spindle elongation in mammalian cells in vitro. Direct microtubule counts. *J. Cell Biol.* **1971**, *50*, 416–431. [CrossRef] [PubMed]
199. Tippit, D.H.; Fields, C.T.; O'Donnell, K.L.; Pickett-Heaps, J.D.; McLaughlin, D.J. The organization of microtubules during anaphase and telophase spindle elongation in the rust fungus *Puccinia*. *Eur. J. Cell Biol.* **1984**, *34*, 34–44. [PubMed]
200. Masuda, H.; Hirano, T.; Yanagida, M.; Cande, W.Z. In vitro reactivation of spindle elongation in fission yeast nuc2 mutant cells. *J. Cell Biol.* **1990**, *110*, 417–425. [CrossRef] [PubMed]

201. Gardner, M.K.; Odde, D.J. Modeling of chromosome motility during mitosis. *Curr. Opin. Cell Biol.* **2006**, *18*, 639–647. [CrossRef] [PubMed]
202. Nedelec, F.J.; Surrey, T.; Maggs, A.C.; Leibler, S. Self-organization of microtubules and motors. *Nature* **1997**, *389*, 305–308. [PubMed]

biology

MDPI

Review

Metaphase Spindle Assembly

Tarun M. Kapoor

Laboratory of Chemistry and Cell Biology, the Rockefeller University, New York, NY 10065, USA; kapoor@mail.rockefeller.edu

Academic Editor: J. Richard McIntosh
Received: 3 October 2016; Accepted: 19 January 2017; Published: 3 February 2017

Abstract: A microtubule-based bipolar spindle is required for error-free chromosome segregation during cell division. In this review I discuss the molecular mechanisms required for the assembly of this dynamic micrometer-scale structure in animal cells.

Keywords: mitosis; cell division; microtubules; kinesin; dynein; chromosome; metaphase

1. Introduction

The equal partitioning of replicated genomes into two daughter cells depends on the assembly and function of a microtubule-based bipolar spindle, which can be several micrometers in size. The assembly of this cellular structure involves multiple steps, including the breakdown of the nuclear envelope, separation of centrosomes, organization of microtubules into a bipolar spindle, and attachment of sister chromatids to microtubules from opposite spindle poles. Many of these steps can be completed within minutes and may occur in parallel. The need for accurate chromosome segregation is likely balanced against the requirement for the rapid completion of the process as many cellular functions—including those that safeguard against damage of the genome—are largely suppressed during cell division [1].

The idea that the assembly dynamics of filaments plays a key role in spindle function came from the rapid and reversible responses to perturbation [2]. Following the identification of tubulin and the characterization of its very rapid polymerization and depolymerization rates (which are on the order of 10–50 μm/min), it became clear that tubulin's properties are crucial for the fast timescales of spindle assembly and chromosome segregation (reviewed in [3]). Additional studies analyzing signal recovery after photo-bleaching of fluorescent tubulin incorporated into spindles revealed rapid turnover ($t_{1/2}$: ~20 s) of tubulin at steady state [4,5]. In the following years, other imaging methods—including photoactivation of fluorescence and fluorescent speckle microscopy—confirmed these fast dynamics of tubulin and also showed that microtubules flux poleward in many animal spindles, a persistent motion at ~1–2 μm/min of the microtubule lattice towards each spindle pole [6,7]. These complex dynamics are likely to be the convolution of microtubule nucleation, directional transport, and dynamic instability, which involves the co-existence of growing or shrinking filaments and stochastic transitions between these two states [8].

At the start of cell division, the disruption of the interphase microtubule array can be abrupt, and is associated with an increase in microtubule polymer levels, as well as an increase in turnover [9]. The interphase microtubule array is typically arranged around one organizing center. In dividing cells, the separating centrosomes help organize two interacting filament arrays. Direct measurement of the dynamics of individual filaments has revealed the specific changes in dynamic instability parameters—including an increase in catastrophe frequency and the duration of depolymerization events—as cells enter mitosis [10]. A recent study which tracked microtubule dynamics in 3-D showed that the filament growth velocities in metaphase can be twice the interphase growth rates [11]. Together, centrosome separation and these changes in tubulin polymerization dynamics help to assemble the bipolar spindle, which is comprised of three types of microtubules (Figure 1). First, kinetochore microtubules link kinetochores to spindle poles, and directly contribute to chromosome motion. Second, interpolar microtubules interact with microtubules from the opposite spindle pole, but do not directly interact with kinetochores. These filaments help establish the spindle's shape and mechanical framework. Third, astral microtubules, which extend from each spindle pole to the cell cortex, help orient and position the cell division apparatus.

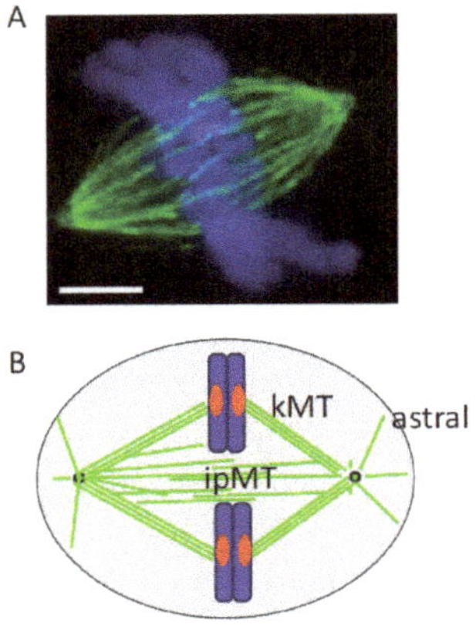

Figure 1. The metaphase spindle. (**A**) Overlay shows tubulin (green) and DNA (blue) in a mammalian cell. The cell was fixed and processed for immunofluorescence. Scale bar, 5 µm; (**B**) Schematic highlights kinetochore (kMT), interpolar (ipMT), and astral microtubules. DNA: blue; kinetochore: red; tubulin: green; centrosome: black circle.

The persistent and fast turnover of spindle microtubules suggests that this micrometer-scale structure continuously rebuilds itself. The spindle not only maintains its shape, but can correct defects. For example, when a microneedle is used to "cut" the spindle, it recovers its bipolar shape [12]. What is even more remarkable is that when two spindles are close enough to interact, they can fuse to form a single spindle of the same size as the original individual spindles [13–15]. One might assume that spindle size and shape are somehow intrinsically set by the biochemical components, and therefore such repair and fusion are possible. However, this is not likely to be the case as spindle size can change at different stages of development, when cell size reduces ~100-fold due to rapid cell divisions without cell growth [16]. Variation in the size of spindles independent of changes in the composition of the cytoplasm has been elegantly dissected in recent studies that combine the use of the *Xenopus* egg extract system and microfluidics technology [17,18]. These studies show that spindle size changes, comparable to those that occur during development, can be achieved by simply changing the volume of the cytoplasm in which the spindle assembles. Together, these findings suggest that the metaphase spindle is a dynamically self-assembling cellular structure that can autonomously maintain its organization, but can also respond to external cues.

Genome sequencing, large-scale loss-of-function studies, and proteomics have essentially identified all the proteins needed for spindle assembly in human cells. It is becoming clear that spindle assembly depends on multiple—at least partly redundant—molecular mechanisms that can act in parallel. A major challenge now is to unravel how this structure, which can be several micrometers in size, is assembled by nanometer-sized proteins. For example, we need to understand how simple geometric features, which can be 1000 times the size of the proteins required for microtubule organization, are measured in dividing cells to regulate distinct functional outputs. In this review, I discuss how metaphase spindles assemble, highlighting recent findings in the context of earlier work, and focus mainly on cell division in animal cells.

2. The Dynamic Architecture of the Metaphase Spindle

The metaphase spindle in animal cells is comprised of thousands of microtubules, whose densities are so high that we cannot resolve individual filaments by standard light microscopy. Therefore, insights into the architecture of the animal metaphase spindle have come from careful electron microscopy studies, which have helped establish the polarity, spacing, and overlap of the different spindle microtubule subtypes [19–21].

These electron microscopy studies revealed that kinetochore microtubules are organized in bundles of ~25 filaments [20,22]. The minus-ends of these filaments are located close to the spindle poles (within ~1 μm of the centriole), and the plus-ends interact with kinetochores [20]. While the number of microtubules in a bundle can vary, it does not appear to be correlated with the direction of chromosome motion [22].

The interpolar microtubules have minus-ends distributed away from the spindle pole (1–2 μm) and have mean lengths of ~4.5 μm in cells with half-spindle lengths of ~5 μm, resulting in many filaments extending past the spindle mid-plane [21]. Several bundles of two to six microtubules with close spacing (~40 nm) can be observed during metaphase, and are likely to be precursors of the microtubule bundles that persist during anaphase and become part of the central spindle. Interestingly, antiparallel microtubules are more strongly associated than parallel ones [21]. These early studies also revealed that interpolar microtubule minus-ends interact with kinetochore microtubule bundles, forming a branched "fir tree"-type arrangement. Similar microtubule branching has been described in other systems, including higher plants [23]. These early studies suggest that some of the interpolar microtubules could be nucleated at sites distal to the centrosomes [21]—an idea supported by more recent findings (see below).

Light microscopy-based analyses have revealed that the dynamics of kinetochore and non-kinetochore microtubules can differ in two ways. First, the interpolar microtubule turnover rate ($t_{1/2}$: ~20 s) is more rapid than that of kinetochore microtubules ($t_{1/2}$: ~420 s) [4,24,25]. Second, the rate of poleward flux for kinetochore microtubules can be ~10% slower than that for interpolar microtubules [26]. The biochemical basis of these differences is poorly understood.

The fast turnover of interpolar microtubules has raised the possibility that the lengths and positions of individual microtubule filaments may not be accurately revealed by imaging methods that require sample fixation. This may be a more significant issue in cases where these non-kinetochore microtubules comprise ~95% of the total filaments, such as the large vertebrate meiotic spindles [27]. The EB (end-binding) proteins allow growing plus-ends of single filaments to be tracked in dense networks, and have served as valuable probes to analyze microtubule organization in dividing cells [28]. However, we lack reliable reporters to track single filament minus-ends in dividing cells. The recently described CAMSAP/patronin proteins only selectively label microtubule minus-ends in interphase cells, and other proteins (e.g., ASP) have only been shown to help locate the minus-ends of microtubule bundles [29–31]. Therefore, analyses of microtubule distributions have relied on indirect approaches, with many studies focusing on the metaphase spindle assembled in *Xenopus* egg extracts. This cell-free system is particularly well-suited for these analyses, as it allows the addition of reagents (e.g., fluorescent proteins) at selected concentrations as well as microsurgery (needle and laser-based) [32,33].

Burbank and colleagues used fluorescent speckle microscopy to determine microtubule orientation and fluorescent tubulin incorporation to localize plus-ends in metaphase spindles assembled in *Xenopus* egg extracts [34]. These data indicated that the minus-ends of microtubules are distributed throughout the spindle, with highest concentrations at spindle poles. A study from my laboratory—in collaboration with the Danuser laboratory—analyzed the motion of single fluorescent tubulin molecules in the metaphase spindle to examine microtubule organization [35]. Briefly, the poleward motion of microtubules was found to be locally heterogeneous, with standard deviations in instantaneous velocity of ~1 μm/min (~30% of the mean instantaneous velocity). Therefore, we hypothesized that the correlated motion of two single fluorophores, aligned along the spindle's long axis, would indicate that both fluorophores likely reside on a single filament. The distance between these two fluorophores would be the minimum length of that filament. A mathematical model based on the measurements of hundreds of such fluorophore pairs indicated that the most of the non-kinetochore microtubules are shorter than the spindles' half-length. In addition, our data indicated that these relatively short filaments are distributed throughout the spindle, consistent with a tiled organization of these microtubules in the metaphase spindle (Figure 2) [35]. Evidence for a tiled array of spindle microtubules was also obtained from elegant electron tomography and 3-D modeling studies of *Caenorhabditis elegans* oocyte meiotic spindles [36].

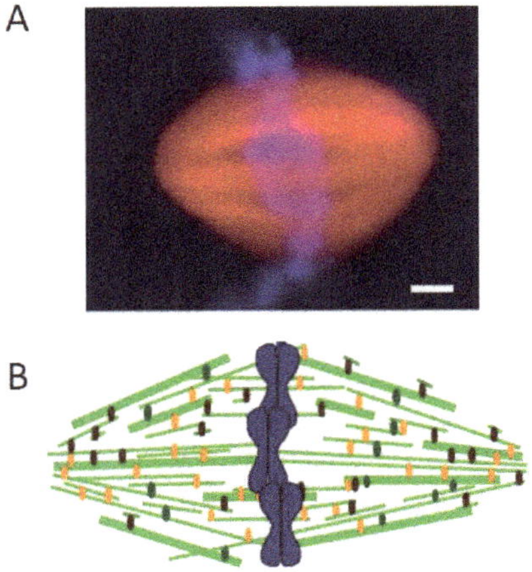

Figure 2. Metaphase spindle assembled in *Xenopus* egg extracts. (**A**) Overlay shows tubulin (red) and DNA (blue) in a metaphase spindle assembled around demembraned sperm DNA. Rhodamine-labeled tubulin was added to visualize microtubules, and Hoescht was used to stain DNA. Scale bar, 5 µm; (**B**) Schematic for the spindle assembled in *Xenopus* egg extracts. Tubulin: green, thicker lines indicate filament bundles; DNA: blue).

A more recent study combined microsurgery and quantitative fluorescence microscopy to analyze spindle microtubule length and position in metaphase spindles assembled in *Xenopus* egg extracts [37]. In this study, a laser was used to rapidly cut thin (~0.1 µm) rectangular regions perpendicular to the spindle's long axis. The microtubule plus-ends generated by the cuts rapidly depolymerize, while the new minus-ends persist. As there is antiparallel microtubule overlap, the fluorescence intensity reduction due to filament depolymerization propagates towards each spindle pole. The relative ratios of these reduced intensity regions can be used to determine the relative orientations of the microtubules at the site of the cut. Using two laser cuts and a computational model, the authors estimated the plus- and minus-end densities and the lengths of microtubules at different locations in the metaphase spindle. These analyses revealed that microtubule lengths are exponentially distributed at all spindle locations, with mean lengths being shortest near the poles (2 µm) and longest in the middle (13.7 µm). Suppression of microtubule poleward flux removes this spatial variation in microtubule length distributions, resulting in a mean microtubule length of ~7 µm. Remarkably, this mean length is similar for microtubules nucleated in the cytoplasm, away from the spindle, by *Tetrahymena* pellicles [37], or centrosomes [38]. Based on these data and other findings, the authors propose that microtubule stability does not vary across the spindle, consistent with earlier single-fluorophore-based analyses [39]. Instead, they propose that spindle microtubule organization depends on the spatial variation of nucleation—which is highest at the center of the spindle—and directional transport-dependent sorting of microtubules [37]. Additional studies are needed to further test this model in different cellular contexts and dissect where and how microtubules are nucleated during cell division.

Together, these findings shed light on the dynamic architecture of the metaphase spindle. The more stable kinetochore microtubule bundles extend from kinetochore to spindle pole [20]. The more dynamic interpolar microtubules are likely distributed across the spindle in a tiled-array [34–37].

3. Micromechanics of the Metaphase Spindle

The earliest observations of cell division suggested that forces acted on chromosomes during segregation [40]. The studies by Nicklas were the first to provide a direct measurement of these forces [41]. He used force-calibrated glass microneedles to oppose the forces generated by the spindle to move chromosomes during anaphase and found that in grasshopper spermatocytes nanonewton-scale forces were needed to stall anaphase chromosome motion. Remarkably, these forces were 10,000-fold greater than what was needed to move chromosomes, not attached to the spindle, in the cytoplasm of the same cells.

Active forces which involve the conversion of chemical energy to mechanical work are generated in the spindle by motor proteins and microtubule polymerization dynamics. Individual motor proteins walking to the plus- or minus-end of the microtubules can generate forces on the order of ~5 pN [42]. Microtubule assembly and disassembly can also generate forces of comparable magnitude [43]. These active forces are balanced against each other, and against elasticity and friction, the passive forces in the bipolar spindle. Here, I focus on some recent advances in our understanding of the metaphase spindle's micro-mechanical properties.

In principle, elasticity of the spindle can be related to that of microtubules, whose flexural rigidity has been directly measured [44]. However, establishing a precise relationship between these measurements of individual microtubules and those in the spindle requires an understanding of the number of filaments in bundles, the number and type of crosslinks (e.g., do they resist relative filament motion), and the properties of the surrounding medium. The major source of friction in the spindle is likely due to the breaking of non-covalent bonds (e.g., between microtubules and associated motor or non-motor proteins during motion). The magnitude of this resistive force increases with the rate of motion. Valuable insights in to the viscoelastic properties have been gained through analyses of cytoskeleton networks reconstituted with purified proteins [45–47]. However, unlike these well-studied polymer networks, the spindle is anisotropic (e.g., microtubule orientation, microtubule types, distribution of binding proteins) and the polymers are dynamic.

To directly probe the micro-mechanical properties of the metaphase spindle, my laboratory in collaboration with the Ishiwata laboratory focused on bipolar spindles assembled in *Xenopus* egg extracts [48–52]. There are no cell membranes in this system, and force probes can directly contact spindles that "float" in the cytoplasm and are stable for several minutes. We first employed cantilever-based probes, and found that the spindle's response to small deformations was viscoelastic, and larger compression resulted in more plastic responses [48]. This study, along with work from other laboratories [53,54], indicates that spindles' deformation response depends on the orientation of the applied force. In particular, we found that ~4 nN force was needed to shorten the metaphase spindle by ~1 μm along its pole-to-pole axis. Less force was needed to compress the spindle across its width [48]. These differences are likely linked to the orientation of microtubules that mainly align with the spindle's long axis.

The cantilever-based set-up was not well-suited for fluorescence imaging, and therefore we switched to force-calibrated glass microneedle [50]. We devised a two-needle system with one stiff needle (stiffness >50 nN/μm) that could be used to apply force and another flexible needle (stiffness 0.2–0.5 nN/μm) whose bending could be used to measure force. Both needles were passivated to reduce non-specific associations with spindle components. These needles were inserted into selected sites within the spindle, and forces were applied in different orientations. This set-up also allowed us to apply forces across a wide range of timescales. This is important, as dynamics in the spindle occur across a similarly wide range of timescales, with motor proteins stepping quickly (~10–100 ms), turnover of interpolar microtubules occurring at intermediate timescales (~10 s), and kinetochore microtubule turnover being much slower (~5 min).

Our analyses revealed that the spindle's response to deformations along the long axis is mainly viscous [50]. Based on these measurements, we can estimate that a microtubule moving at the rate of poleward flux would experience a frictional force of 10–20 pN/μm, suggesting that the active force—likely generated by a few motor proteins—would be of this magnitude.

Along the spindle's short axis, the response to deformation is a more complex timescale-dependent combination of viscosity and elasticity [50]. The viscous response is highest on the timescale of tens of seconds. Importantly, this timescale matches that of chromosome motion and suggests that the deformation associated with the motion of a chromosome—which is large compared to the average mesh size of spindle microtubules—would be dissipated locally with limited effect of the spindle's overall stability. The spindle's response to deformations is more elastic at slower or faster timescales. The elastic response to short-acting forces can be linked to interpolar microtubule mechanics, while that to more persistent forces can be linked to kinetochore microtubule dynamics.

In a more recent study using force-calibrated microneedles that were not passivated (i.e., coated to block non-specific interactions), we were able to stretch spindles by applying forces at each pole [52]. The spindle's response to these forces can be described by a Zener-type model—the model that also describes responses to forces along the spindle's short axis [50,52]. The elastic stiffness and frictional coefficients were 5–7-fold greater along the spindle's long axis compared to the short axis. The next major steps are to combine these measurements with biochemical perturbations to link the mechanical responses to specific protein activities and dynamics.

A tensile element that is not comprised of microtubules referred to as the "spindle matrix" has been hypothesized to play an important role in spindle assembly [55–57]. A variety of proteins (e.g., NuMA (Nuclear Mitotic Apparatus protein) [58], Skeletor [59], and nuclear lamins [60]), poly(ADPribose) [61], or endoplasmic reticulum (ER) membranes [62]) have been proposed to be components of such a "spindle matrix". A recent study has also revealed interesting biophysical properties of a protein that may be associated with the spindle matrix, involving phase-transitions to form liquid droplets [63]. Based on all these studies, it appears that many of these proteins, other bio-polymers, or membranes may contribute to spindle organization in different systems. However, a direct contribution to spindle mechanics has not been firmly established. Our studies directly probing spindle mechanics [48,50], along with another study by Gatlin and colleagues [54], do not support the hypothesis that a non-microtubules-based structure in the spindle makes a substantial contribution to its overall mechanics.

4. Overlapping Mechanisms of Microtubule Formation

There are three major mechanisms for microtubule formation during spindle assembly.

4.1. Centrosomes as Sites of Microtubule Nucleation

The earliest models for spindle formation considered the centrosome—an organelle occupying a central position in the cell—to be the organizing center for microtubules [40]. The centrosome is comprised of two centrioles that organize hundreds of proteins to form the pericentriolar material, which surrounds the centrioles. Evidence for the function of centrosomes as microtubule nucleating sites came from studies with permeabilized mitotic cells and isolated centrosomes from mitotic cells [64–67]. These studies also revealed that the centrosome matures (or "ripens") upon mitotic entry [66]. Consistent with cell cycle-dependent changes, centrosomes from mitotic cells were found to generate ~five-fold more microtubules than those isolated from interphase cells [68]. This study also showed that the capacity of the centrosome to nucleate microtubules does not depend on centriole number [68]. It is now clear that the centrosomes are the major sites of microtubule nucleation in many dividing cells (discussed in [19]).

The maturation of the centrosome involves the recruitment of additional pericentriolar material, and depends on at least two kinases: Polo-like kinase-1 and Aurora A kinase. For example, it has been proposed that Polo-like kinase-1 can help recruit pericentriolar proteins such as pericentrin via phosphorylation [69]. This kinase can also activate another kinase, NEK9, which in turn phosphorylates NEDD1 to help recruit γ-tubulin to centrosomes [70]. It is generally accepted that the main microtubule nucleator in cells is γ-tubulin, which functions with several associated proteins to form large multiprotein complexes [71]. The functions and regulation of γ-tubulin complexes are discussed in more detail below (see Section 8).

Early evidence indicated that microtubules—with plus-ends extending outwards—grow from centrosomes with spherical symmetry [72]. This process would allow the microtubule plus-ends to effectively "search and capture" kinetochores in the cytosol [72]. An interaction with the kinetochore could stabilize the microtubule and over time, lead to the polarization of the microtubule array. Direct evidence for this "search and capture" mechanism was obtained in vertebrate cells (newt lung cells) [73]. This "search and capture" model for spindle formation has strongly influenced research in the field. These studies revealed that additional—possibly redundant and overlapping—mechanisms must also contribute to bipolar spindle formation and proper chromosome attachment.

Many lines of evidence have revealed that centrosomes alone are not sufficient for spindle formation [74]. For example, experiments in unfertilized *Xenopus* oocytes showed that injection of centrosomes alone did not promote microtubule aster formation [75]. Centrosomes induced microtubule formation only when nuclei were also injected. In one key experiment, the nuclear envelope was prematurely ruptured during prophase, and spindle assembly was examined in living grasshopper cells [76]. Spindles formed rapidly under these conditions, but failed to form if the nuclei or the centrosomes were microsurgically removed. Along with additional tests, these experiments indicated that chromosomes and centrosomes are needed for spindle assembly in this system.

The importance of centrosome-independent spindle assembly is clear, as many cell types (e.g., plant cells and oocytes from several species) divide successfully without centrosomes [74]. Multiple lines of evidence from different experimental systems indicate that bipolar spindles can assemble without centrosomes in cells that normally have these organelles. For example, in *Drosophila*, functional spindles lacking astral microtubules (i.e., anastral spindles) assemble in the presence of mutations in proteins (asterless (asl) or centrosomin (cnn)) that disrupt centrosome function [77,78]. Remarkably, adult flies—albeit with some altered phenotypes (e.g., male sterility)—develop in the presence of centrosomin mutations [78]. It should also be noted that studies in other animal models (such as *C. elegans*) indicate that centrosomes are needed for bipolar spindle formation [79,80].

Probably the best evidence that functional spindles can assemble in somatic cells without centrosomes was obtained from two different microsurgery-based experiments. In one study, a laser was used to ablate the centrosome at the start of mitosis [81]. Bipolar spindles assembled with normal morphologies and recruited spindle pole proteins (e.g., NuMA), but not centrosome-associated proteins (e.g., γ-tubulin or pericentrin). Importantly, the kinetics of spindle assembly were similar to that in cells with centrosomes. In another study, microneedles were used to cut an interphase cell to generate a fragment that contained the nucleus, but lacked centrosomes [82]. These cell fragments entered mitosis and assembled a morphologically-normal bipolar spindle that lacked centrosomes. An important feature of these experiments is that centrosome-independent microtubule formation was revealed without artificially raising tubulin concentration (e.g., by treating cells with a microtubule depolymerizing drug or injecting additional tubulin), which may favor pathways that may not contribute significantly at normal physiological tubulin concentrations.

4.2. The Roles of Chromosomes and Kinetochores in Microtubule Formation

Several lines of evidence suggest that the chromosomes are not merely passive cargoes, but play a key role in assembling the bipolar spindle that will eventually segregate them. Among the first direct tests of contributions of chromosomes to spindle assembly were the micromanipulation experiments reported by Marek [83]. Chromosomes in spermatocytes from two different grasshopper species were removed from the dividing cell or detached from the spindle and reintroduced at a later point. The "volume birefringence" measured using polarized light microcopy provided an estimate of the total microtubule content before and after micromanipulation. These experiments revealed that the number of microtubules in the spindle was proportional to the number of chromosomes. Nicklas and Gordon confirmed these conclusions with electron microscopy-based measurements. They found that the total length of spindle microtubules scaled with the number of chromosomes in the spindle [84]. In addition, studies of microtubule nucleation by isolated kinetochores in vitro [85] and microtubule formation in cells after recovery from treatments with chemical inhibitors of microtubule assembly [86,87] indicated that kinetochores can promote the formation of microtubules. Together, these studies also led to an important new hypothesis, that microtubule formation could be promoted by a diffusible signal generated by kinetochores (Figure 3A) [87].

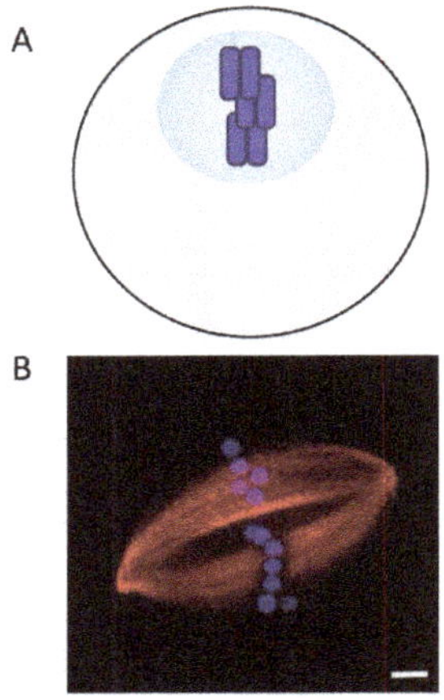

Figure 3. (**A**) Chromosomes (blue) generate signals to promote the formation of microtubules in their vicinity; (**B**) A spindle assembled around chromatinized DNA-beads added to *Xenopus* egg extracts. DNA: blue; tubulin: red. Scale bar, 5 μm.

A key finding was that DNA from various sources (including bacteriophage lambda) injected into an unfertilized egg and assembled into chromatin could promote local microtubule polymerization [75]. These experiments indicated that chromatin—even in the absence of a kinetochore—was sufficient to generate a signal that can promote microtubule formation. In parallel, micromanipulation studies in grasshopper spermatocytes revealed that a single chromosome can induce the formation of a mini-spindle [15]. Electron microscopy analyses indicated that only a small fraction (~4%) of the total microtubules were kinetochore-associated, leading to the proposal that the chromosome—and not just the kinetochore—contributes to microtubule formation. Evidence that chromosomes promote microtubule formation in their vicinity also came from studies in *Drosophila* oocytes [88].

The *Xenopus* egg extract system allowed additional tests of the roles of chromosomes in microtubule formation. Addition of demembraned sperm nuclei to egg extracts induced the formation of microtubule arrays that were polarized towards chromatin [89]. Compelling evidence that chromatin can induce spindle assembly in the absence of kinetochores or centrosomes came from a study using plasmid DNA attached to beads (Figure 3B) [90]. DNA-coated micrometer-sized beads were added to interphase extracts to induce chromatin formation. Upon transfer to M-phase extracts, these beads induced the formation of microtubules that self-organized into bipolar spindles within minutes.

Evidence for chromosome-dependent microtubule formation in somatic cells, in the presence of centrosomes, has come from studies in which proteins (i.e., HSET and NuMA) required for spindle pole formation were inhibited [91]. Under these conditions, centrosomes dissociated from the assembling spindle, and kinetochore microtubule bundles were still observed. These data suggest that kinetochore microtubules assemble via mechanisms independent of attachments to centrosomes in somatic cells.

Directly observing the formation of microtubules around chromosomes (or kinetochores) during mitosis is challenging due to the high density of spindle microtubules. Monastrol—a cell permeable chemical inhibitor of kinesin-5—provided a simple assay to observe chromosome- and kinetochore-associated microtubule formation in mammalian cells [92,93]. In the presence of monastrol, cells arrested with monopolar spindles as centrosome separation was inhibited. In treated cells, most chromosomes were positioned at the periphery of a radial microtubule array and oriented such that one kinetochore pointed towards the centrosome, while its sister kinetochore pointed away. Microtubules were observed forming from the kinetochore pointing away from the centrosome. As this kinetochore was shielded by chromosome arms from the dynamic centrosome-associated microtubule plus-ends, it was unlikely that these kinetochore-associated microtubules were derived from the centrosomes in the dividing cell.

There are at least two possible mechanisms for the formation of these kinetochore microtubules. First, microtubules may nucleate near the chromosome, and filament plus-ends that interact with the kinetochore get stabilized and organized into a bundle, with minus-ends pointing away from the kinetochore. Second, the kinetochore may directly nucleate microtubules such that minus-ends point away and growth occurs by the addition of tubulin at the filament plus-ends, as is the case during the polewards flux of kinetochore fibers. An elegant study in *Drosophila* S2 cells combined laser-based microsurgery and live-cell fluorescence microscopy and showed that tubulin subunits are continuously incorporated at kinetochores, even for kinetochore fibers that are severed and do not directly interact with spindle poles [94]. These data suggest that once formed, these kinetochore microtubules can grow by a poleward flux-type mechanism with minus-ends pointing away from the kinetochore.

4.3. Microtubule-Dependent Microtubule Formation

Studies in plants—which lack centrosomes or a readily apparent single microtubule organizing center (MTOC)—have provided valuable insights into non-centrosomal pathways of microtubule formation. An early proposal was that plants may have a "diffuse centrosome" [95]. Analyses of microtubule formation that involved tracking EB-proteins at the *Arabadopsis* cell cortex supported this hypothesis [96]. A competing hypothesis based on studies in the green alga *Nitella* suggested that microtubules could themselves recruit nucleation sites to promote the formation of new microtubules (Figure 4) [97]. In this study, microtubule formation after relief from chemical inhibitor treatments revealed highly-branched filament clusters. The microtubule-dependent microtubule formation hypothesis is also supported by analyses of γ-tubulin localization. In particular, γ-tubulin is found along microtubules within the mitotic spindle [98], and along filaments in asters assembled in vitro [99].

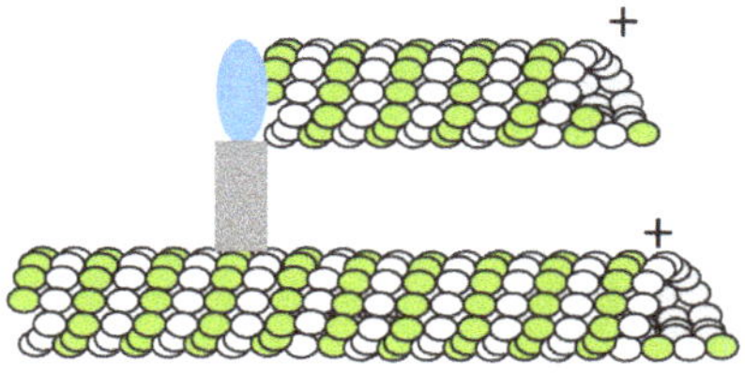

Figure 4. Schematic for microtubule-dependent microtubule formation. The microtubule (tubulin dimer: white, green) can recruit and activate γ-tubulin ring complex (γ-TURC, blue), possibly via the augmin complex (grey).

More direct evidence for microtubule-dependent microtubule formation came from studies analyzing MAP65 function in *Arabadopsis* [100]. In this study, microtubule formation within a microtubule bundle was directly observed. Even stronger evidence for this mechanism was obtained in an elegant study analyzing interphase microtubule organization in fission yeast [101]. Importantly, this study also provided data supporting the functional significance of recruiting γ-tubulin to the sides of existing microtubules to nucleate new ones. Additional studies tracking microtubule growth using EB-proteins in cultured insect cells suggested that a similar microtubule formation mechanism likely contributes to mitotic spindle assembly [102]. Analysis of microtubule formation in *Xenopus* egg extracts also suggested that microtubule-stimulated microtubule formation may contribute to spindle assembly [103].

There has been significant progress in our understanding of the molecular mechanisms underlying microtubule formation in dividing cells. These advances are discussed in more detail below (see Sections 8 and 9).

5. The Influence of Centrosomes on Spindle Shape

The number of centrosomes in a cell is tightly controlled, and a typical vertebrate somatic cell divides with two centrosomes at opposite ends of a bipolar spindle. The importance of centrosomes in building bipolar spindles was first revealed by observations that multi-polar spindles (i.e., with more than two poles) assembled in cells with more than two centrosomes [104]. In a dividing cell with only one centrosome, bipolar spindles do not form, and a monopolar spindle—a single radial array of microtubules surrounded by chromosomes at their periphery—is observed. This has been revealed through the analysis of mutants in different model organisms, and also by specific manipulations. For example, centriole disjoining can be induced in sea urchin embryos using chemical reducing agents [105]. Under these conditions, daughter cells are generated with only one centrosome, and assemble monopolar spindles.

In many cell types, the two duplicated centrosomes separate during prophase. If this process fails, monopolar spindles can assemble. For example, when cells are treated with a chemical inhibitor of kinesin-5 (a microtubule-based motor protein), monopolar spindles accumulate [93]. Relief from chemical inhibition results in centrosome separation and bipolar spindle assembly [106]. Briefly, in addition to kinesin-5, at least three other activities can contribute to centrosome separation (recently reviewed in [107]). First, microtubule polymerization itself can generate forces to push centrosomes apart. Second, astral microtubules can interact with dynein—another microtubule-based motor protein—at the cell cortex. Dynein walking towards the minus-ends of these microtubules, or maintaining attachment to depolymerizing filament ends can pull centrosomes towards the cell cortex. Third, cortical flows generated by actomyosin at the cortex may contribute to centrosome separation via interactions with astral microtubules.

It is noteworthy that in some mutant backgrounds (e.g., *urchin*, allelic to KLP61F/kinesin-5), a bipolar spindle can form even when centrosome separation fails, indicating that bipolar spindle formation and centrosome separation can be uncoupled [108]. In this study, monoastral bipolar spindles (i.e., one pole in bipolar spindle had an associated microtubule aster while the other pole did not) were observed. More recently, a genome-wide RNAi screen revealed that mono-astral bipolar spindles can form after knockdown of a number of proteins, including the transcription factor Myb and the E2 ubiquitin conjugating enzyme UbcD [109].

Multiple mechanisms contribute to assembling bipolar spindles even when dividing cells have more than two centrosomes [110,111]. These mechanisms, which can directly cluster the extra centrosomes at two poles or act more indirectly, include the spindle assembly checkpoint, cell adhesion, and microtubule-associated proteins that help organize spindle poles (e.g., dynein, NuMA, and HSET/kinesin-14) [112,113]. It is noteworthy that multipolar spindles are not observed in the absence of centrosomes [74].

It is now generally accepted that while the centrosome may not be essential for building a functional bipolar spindle, these organelles do have important roles in dividing cells. When present, the centrosomes are the major sites of microtubule formation. In addition, centrosome-nucleated microtubules have at least two key functions. First, the centrosome-nucleated astral microtubules—which can interact with the cell cortex—play a crucial role in positioning the bipolar spindle in the dividing cell (reviewed in [114]). Second, the centrosome-associated astral microtubules can not only "search and capture" kinetochores, but also capture other microtubules. In particular, astral microtubules can interact with the minus-ends of microtubule bundles that are associated with kinetochores, but not anchored at poles [92]. This "minus-end capture" process effectively increases the kinetochore target size. In particular, the plus-ends of astral microtubules need not only find the relatively small kinetochores in the dividing cell, but can establish productive contacts with kinetochore-associated filaments that can extend micrometers beyond each kinetochore. In addition, the astral microtubules also capture interphase microtubules that remain when a cell enters mitosis. Direct imaging of green fluorescent protein (GFP)–tubulin-expressing animal cells in prophase revealed microtubules that not directly associated with the centrosome and are present at the periphery of the cell can form bundles that get transported towards the centrosome [115]. The transport of these "pre-existing" microtubules and kinetochore-associated microtubules towards centrosomes is likely mediated by the minus-end-directed motor protein cytoplasmic dynein [92,115].

6. The Influence of Chromosomes on Spindle Shape

Analysis of microtubule dynamics in asters formed in the presence of centrosomes and sperm nuclei added to egg extracts revealed that chromosomes polarize these filament arrays via a short-range effect on dynamic instability parameters [116]. Growth velocity and catastrophe frequencies were reduced, while rescue frequencies were increased for filaments close to or in contact with chromosomes. This study also found evidence for a weaker but longer-range effect which could guide microtubules towards chromosomes without direct interactions. For these analyses, the geometry of chromatin structures was controlled by attaching DNA to micro-patterned gold stripes on glass coverslips. A subsequent study in which chromatinized DNA-beads were used suggested that the long-range (~10 micrometer) microtubule aster polarization effect is possibly stronger than previously considered, and also induces the directional migration of arrays towards chromatin [117]. Tracking growing microtubule plus-ends in somatic cells also revealed asymmetry in microtubule growth from centrosomes [118]. During prophase and prometaphase, microtubules proximal to the nucleus/chromosomes are longer than those oriented away. Together, these studies suggest that chromosomes can influence the spatial organization of microtubule arrays that form in dividing cells.

To examine the influence of chromosomes on the spatial organization of metaphase spindles, the chromatinized DNA-bead assay has served as a powerful experimental system [90]. My laboratory, along with our collaborators, designed a setup that employed magnetic fields to align chromatinized DNA-coated paramagnetic beads into linear arrays that resembled beads on a string. These arrays ranged from ~10 to ~90 micrometers in length [119]. When these bead arrays were added to *Xenopus* egg extracts, they moved freely in the cytosol and promoted microtubule formation along their lengths. The arrays also changed shape by bending and forming kinks, likely due to interactions with microtubules that were being organized into bipolar spindles. Remarkably, the length or width of the spindles did not scale with the length of the DNA-bead string, but was similar to that of spindles that formed around unaligned DNA-bead clusters or other forms of chromatin in these extracts. In another study, Nedelec and co-workers examined the influence of chromatin on spindle assembly using chromatinized DNA-beads immobilized on surfaces with a lithographic micro-pattern [120]. Again, microtubules assembled along and proximal to the DNA-based structures. Findings from both studies indicate that the organization (e.g., aspect ratio or length) of individual spindles was largely independent of the shape of the DNA-based arrays. One key difference between the observations reported in these two studies was that multiple spindle structures were observed on the surface immobilized DNA-beads, while on the more flexible free-floating DNA bead arrays, only one spindle formed. I favor the possibility that the flexibility of the DNA-strings that my laboratory used allowed more efficient fusion of spindle poles, while in the other study, the glass surface to which the beads were fixed likely inhibited spindle pole fusion and favored spindle pole splitting. This hypothesis is supported by several observations, including the finding that when two bipolar spindles are brought close together in egg extracts, these spindles fuse to form one spindle [14].

Together, these findings indicate that while chromosomes regulate microtubule formation and polarize filament arrays, the activities of microtubule-associated proteins (e.g., motor proteins) establish spindle bipolarity and overall shape.

7. Dissecting the Chromosome-Based Signal for Spindle Assembly

7.1. Ran-GTP

Ran (Ras-like nuclear G protein) is an evolutionarily-conserved GTPase involved in diverse aspects of nuclear function. Characteristic properties of these Ras-related GTPases are that they slowly hydrolyze or exchange GTP. Nucleotide hydrolysis is promoted by GAPs (GTPase activating proteins), and nucleotide exchange depends on GEFs (guanine nucleotide exchange factors). There are advanced models for how Ran establishes the direction of nuclear transport via import and export receptor proteins during interphase [121–123].

The first clues that Ran regulates microtubule organization came from studies in budding yeast. A screen for genes whose overexpression suppressed the phenotype due to conditional α-tubulin mutations led to the identification of RCC1, the budding yeast homolog of RanGEF [124]. A few years later, a mutation in the budding yeast homolog of RanBP1—a protein that binds Ran-GTP and functions as an accessory factor that promotes RanGAP-mediated GTP hydrolysis—was characterized and shown to have no observable defect in nucleocytoplasmic transport, but blocked cell growth [125]. The phenotypes associated with the mutation included improper spindle positioning, likely due to failure in the formation of astral microtubules. In the same year, Nishimoto and co-workers reported the characterization of RanBPM, a Ran-binding protein [126] they had previously identified in a yeast two-hybrid screen using Ran as bait [127]. RanBPM preferentially interacted with GTP-bound Ran, associated with centrosomes in cultured cells, and its over-expression induced the formation of ectopic microtubule asters. The assembly of microtubule asters nucleated by isolated centrosomes could be suppressed by antibodies to RanBPM, and also by the addition of Ran-GTP (e.g., Ran loaded with a non-hydrolyzable GTP analog) [126]. Together, these findings indicate that Ran can regulate microtubule organization independent of its role in nucleocytoplasmic transport.

Evidence that Ran was coopted for the regulation of microtubule formation during M-phase came the following year, from studies by multiple independent research groups [128–132]. All of these groups used *Xenopus* egg extracts, an experimental system that allowed analyses of Ran's contribution to microtubule organization in M-phase without concerns about its functions in other parts of the cell cycle. Reduction of Ran-GTP levels, by immunodepleting RCC1, adding mutant forms of Ran that mainly bind GDP (T24N), or adding RanBP1, suppressed the formation of microtubule asters from centrioles added to these extracts. By contrast, increasing Ran-GTP concentration via addition of RCC1 or mutant forms of Ran that are "locked" in the GTP-bound state promoted centriole-dependent aster formation. Remarkably, increasing GTP-Ran levels induced aster formation in the absence of added chromatin or centrioles. In two of these studies, the addition of GTP-locked Ran mutants (G19V bound to GTPγS [129]; or Ran L45E [131]) led to the formation of bipolar spindle-like microtubule-based structures. It was also shown that other reagents (e.g., DMSO or taxol) that promoted microtubule aster formation in these extracts did not lead to bipolar structures. Together, these findings suggest a role for Ran-GTP in microtubule formation during cell division.

The requirement of Ran's GTP-hydrolysis cycle in regulating microtubule organization is supported by two lines of evidence. First, asters formed by Ran-GTPγS—which cannot convert to RanGDP—were smaller in size compared to those formed in the presence of RanGTP. Second, microtubule aster assembly in egg extracts depleted of the RanGEF RCC1 can be rescued by the addition of RanGTP but not RanGDP [130]. These studies, along with the fact that RCC1 is mainly bound to mitotic chromatin, led to the proposal that Ran-GTP could be the sought-after chromatin-generated "enzyme factor" that promotes spatially-restricted microtubule stabilization [87].

Ran itself does not target microtubule asters or spindles assembled in *Xenopus* egg extracts [133], suggesting that it functions through effector proteins, such as the transport receptors importin-α and β. Studies on nucleocytoplasmic transport had already established that nuclear import receptors bind their cargo in the cytosol where Ran-GTP levels are low, while in the nucleus they bind Ran-GTP and release cargoes. In particular, the transport of proteins bearing the nuclear localization signal (NLS) depends on a complex formed by importin-β that binds Ran-GTP, and importin-α, an adaptor that recognizes the NLS-bearing proteins. Importin-β can also directly recognize and transport cargoes independent of importin-α. Export receptors bind their cargoes—along with Ran-GTP—in the nucleus, and release cargoes in the cytosol after GTP hydrolysis. Guided by these models, Gruss and co-workers designed experiments to test if transport receptors contribute to Ran-dependent microtubule formation [121,123–135]. Specifically, they showed that the addition of importin-α, but not a mutant form that cannot bind NLS-containing cargoes, inhibited Ran-GTP-induced microtubule aster assembly in *Xenopus* egg extracts [134]. Immunodepletion of importin-α from these extracts also suppressed the formation of Ran-induced microtubule-based structures [134]. In another study, Nachury and co-workers reported that the depletion of proteins that bind a GTP-locked Ran mutant (Q69L) from egg extracts induces the formation of microtubule-based structures, while the addition of importin-β inhibits the formation of these structures [133]. Together, these findings, along with additional data, support a model in which Ran-GTP is generated proximal to chromosomes by chromosome-bound RCC1 and promotes the assembly of microtubule-based structures by locally releasing "cargoes" from transport receptors. Away from chromosomes, the concentration of Ran-GTP is lower (likely due to RanBP1 and RanGAP1 promoting GTP-hydrolysis), and the transport receptors inhibit the proteins that promote microtubule formation.

The microtubule-associated proteins TPX2 and NuMA were the first "cargoes" of the transport receptor shown to be involved in Ran-dependent microtubule formation [133,134,136]. Subsequent studies have identified several additional proteins involved in spindle assembly that can be regulated by Ran. These include non-motor MAPs (e.g., NUSAP and HURP) [137–139], motor proteins (e.g., XKid and HSET) [140,141], and the RNA-binding protein Rae1 [142], Cdk11 [143], and nuclear lamins [60]. Other targets, possibly via TPX2, include kinesin-5 (or Eg5/Kif11/KSP) and Aurora kinase [131,144].

Consistent with the hypothesis that Ran-GTP can function as a diffusible signal regulating microtubule organization proximal to chromosomes, a spatial Ran-GTP gradient can be detected during M-phase. The first evidence for this gradient came from studies using FRET (Forster Resonance Energy Transfer)-based sensors that have the donor and acceptor fluorescent proteins linked by a peptide [145]. One sensor was designed to detect Ran-GTP, and incorporated a peptide corresponding to RanBP1's Ran-GTP binding region. The other sensor was engineered to detect the release of importin-α from importin-β, and employed a peptide corresponding to importin-α's importin-β-binding domain. These sensors indicated that Ran-GTP concentration was high near chromosomes and importin-α/importin-β binding, and therefore "cargo" inhibition, were low near chromosomes in metaphase spindles assembled in *Xenopus* egg extracts.

The overall size and shape of the Ran-GTP gradients revealed by the two sensors were similar, extending over micrometers but not reaching the spindle poles, which in these spindles can be ~30 micrometers apart. Subsequent studies employed different sensors and fluorescence lifetime imaging (FLIM) rather than the measurement of donor/acceptor signal ratios alone, which can be sensitive to fluorophore concentration and bleed-through of fluorescence signal [146,147]. These measurements were consistent with a chromosome-centered Ran-GTP-dependent signal, which can release spindle assembly factors from importin-β, covering distances that extend all the way across the spindle [146,147]. A possible explanation for how this gradient could induce asymmetry in microtubule aster organization came from modeling and experimental data that indicate that the Ran-gradient may be combined with the activities of a Ran-regulated kinase (CDK11) and phosphatases [148]. FRET-based sensors also revealed the presence of a Ran-GTP gradient in somatic cells. This spatial gradient was much steeper, and extended across a shorter distance (3–4 μm) compared to what was detected in spindles assembled in *Xenopus* egg extracts [147]. A more recent study reported an even more localized spatial gradient, extending only ~2 μm in dividing somatic cells [149].

7.2. Chromosomal Passenger Complex (CPC)

Early ideas for how chromosomes generate a microtubule-formation signal focused on kinase- and phosphatase-based protein phosphorylation, rather than on the Ran-pathway. A specific proposal suggested that a phosphatase could be chromosomally localized, and could counteract a kinase that freely diffuses and phosphorylates microtubule-associated proteins that control filament stabilization [150]. Evidence supporting this model came from a study examining Stathmin/Op18, a 17 kDa protein that may bind tubulin subunits to suppress microtubule assembly [151,152]. Op18 can have multiple phosphorylations, many of which are present during interphase. In the presence of chromatin, additional residues in Op18 get phosphorylated. The addition of wildtype Op18 or an Op18 mutant lacking these phosphorylation sites rapidly (within ~3 min) disrupted bipolar spindles assembled in egg extracts, reducing microtubule density and overall spindle size. In egg extracts depleted of Op18, the formation of microtubules around chromatinized DNA-beads was accelerated. Other studies suggest that phosphorylation reduces Op18's binding to tubulin [153]. Together, these data are consistent with phosphorylation suppressing Op18's inhibitory effect on microtubule formation, and support a model in which chromosomes control the activity of microtubule assembly factors.

A FRET-based sensor whose signal changes when Op18 binds tubulin revealed a spatial gradient that is centered at chromosomes and extends towards the cell periphery [154]. These measurements and a simple calculation suggest that a phosphorylation gradient may extend 4–8 micrometers from chromosomes. However, immunodepletion of Op18—which altered early stages of spindle assembly—did not affect the shape and size of bipolar spindles assembled in egg extracts [152]. These data suggest that Op18 may not be the key effector of chromosome signals that promote microtubule assembly in dividing cells. While functional redundancy due to Op18-related proteins is difficult to exclude, the observation that a mouse knock-out of Op18 is viable is also consistent with this hypothesis [155,156].

Depletion of Polo-like kinase disrupted spindle assembly around chromatinized DNA-beads, indicating that this cell cycle kinase, which can associate with chromosomes, plays an important role in microtubule formation during M-phase [153]. Interestingly, the chromatin-induced phosphorylation of Op18 was suppressed in the absence of Polo-like kinase [153]. However, Op18 was not shown to be a direct substrate of Polo-like kinase, and subsequent studies revealed that the relevant kinase is Aurora B—a protein in the "chromosomal passenger complex" (CPC) [157,158].

Aurora B, along with Incenp, Survivin/BIR, and Dasra/Borealin form the CPC, which is enriched at the inner centromere during metaphase and associates with microtubules in the central spindle after anaphase in dividing cells [157]. CPC function is needed for multiple different aspects of cell division, including chromosome–microtubule attachment, the spindle assembly checkpoint, and cytokinesis. Compelling evidence that the CPC is needed for spindle assembly came from a study by Funabiki and co-workers [159]. Immunodepletion of the CPC via Dasra/Borealin or Incenp antibodies disrupted the assembly of spindles in *Xenopus* egg extracts. By contrast, centrosome-dependent microtubule aster formation was not affected by CPC depletion. However, these asters did not associate with chromosomes, indicating that chromosome-associated signaling was disrupted.

Kinesin-13/MCAK had been characterized as a CPC substrate whose microtubule depolymerization activity can be suppressed by phosphorylation in vitro [160–162]. Funabiki and co-workers showed that depletion of MCAK, along with the CPC, resulted in the rescue of microtubule formation around chromatinized DNA beads in egg extracts.

Systematic analyses of CPC function using depletion/add-back-type approaches in *Xenopus* egg extracts revealed that the Dasra subunit promoted CPC's chromosome binding, and this interaction was needed for spindle assembly [163]. The major CPC-dependent phosphorylation site (ser-16) in Op18 was identified and used, along with a canonical CPC substrate (histone H3 ser-10), as a reporter of CPC activity in egg extracts. The phosphorylation of both substrates can be induced by the addition of chromatin to egg extracts or by the addition of stabilized microtubules [163,164]. In fact, the addition of antibodies to cluster together multiple CPC complexes resulted in kinase activation in egg extracts. This antibody-based activation could—independent of Dasra binding—promote chromosome-associated or centrosome-associated microtubules in egg extracts [163]. Follow-up studies uncovered a microtubule-binding site in the Incenp subunit, and showed that the CPC—which mainly localizes to metaphase chromosomes—can be detected within the spindle [165]. Analyses using a microtubule-targeted FRET-based sensor for CPC activity [166] suggested that the CPC can phosphorylate spindle microtubule-associated substrates [165].

Together, these data have led to a model in which the CPC, initially activated by chromosomes, must be targeted to microtubules to promote spindle formation. Near chromosomes, CPC-dependent phosphorylation likely promotes microtubule formation by suppressing the activities that increase microtubule catastrophe. Once present, microtubules can bind and activate the CPC, and thereby promote additional microtubule assembly, effectively establishing a positive feed-back loop triggered by chromosomes. This "dual detection" of chromosomes and microtubules provides a plausible explanation for how CPC-dependent microtubule formation is spatially restricted around chromosomes in egg extracts [165].

7.3. Interplay between Ran-GTP and the CPC

To dissect the relative contributions of the Ran- and CPC-signals to spindle formation, Funabiki and co-workers used the *Xenopus* egg extract system. In particular, they took advantage of their findings that chromatinized DNA-beads can induce microtubule assembly in egg extracts co-depleted of the CPC and kinesin-13/MCAK [159]. The addition of RanT24N, which mimics the nucleotide-free state and binds RCC1 with high affinity, suppressed microtubule formation. In addition, they showed that the addition of a GTP-locked Ran mutant to egg extracts depleted of the CPC promoted microtubule aster formation [159]. Further, the addition of RanT24N did not inhibit Op18 hyper-phosphorylation or substantially alter microtubule assembly by the antibody-mediated activation of the CPC [163]. Together, these data indicate that the CPC and the Ran pathways can act independently to promote microtubule formation in egg extracts.

The interplay between Ran and CPC-signaling was further examined by Maresca and co-workers [167]. The authors used combinations of two Ran mutants—RanT24N, which mimics the nucleotide free state, and RanQ69L, which is deficient in GTP-hydrolysis activity and mimics the GTP-bound state—to "flatten" the Ran-GTP spatial gradient during spindle assembly in egg extracts. They found that spindles did not assemble around chromatinized DNA-beads under these conditions. Importantly, mixing DNA-beads with CPC-beads (CPC linked to beads via antibodies to Incenp) promoted spindle formation, even when the Ran-GTP spatial gradient was "flattened". Under similar conditions, the CPC-beads alone promoted microtubule formation, but not the organization of these filaments into bipolar spindles. Together, these data suggest that two independent signals are generated by chromatin to promote microtubule organization during M-phase [167].

Interestingly, the addition of EB1—a microtubule +TIP (microtubule plus-end tracking protein) that can promote microtubule formation in egg extracts—along with RanT24N also rescues spindle assembly around sperm nuclei [28,167]. As EB1 is not known to be regulated by Ran, these data suggest that Ran-GTP signals and its down-stream effectors are not required for spindle assembly when microtubule formation is at sufficiently high levels.

The contributions of both the RanGTP and the CPC pathways to chromosome-mediated spindle assembly in somatic cells is supported by a study by Wadsworth and colleagues [168]. These researchers combined the recovery of microtubule formation after treatment with nocodazole—a modification of the assay developed by De Brabander [87]—with fluorescence microscopy and inhibition of selected proteins. In this assay, the rate of microtubule formation near chromosomes was slower than that from centrosomes, but the amount of polymer generated near chromosomes was greater than that near centrosomes. Injection of importin-β suppressed chromosome-associated microtubule formation, but not centrosome-associated microtubule formation. In fact, centrosome-associated microtubule formation was slightly enhanced. These data, along with RNAi-mediated knockdown of TPX2 alone, survivin alone, or the knockdown of both survivin and kinesin-13/MCAK, indicate that the Ran-GTP and the CPC pathways contribute to the formation of microtubules proximal to chromosomes in porcine cells [168]. It is noteworthy that knockdown of kinetochore proteins did not block microtubule formation, suggesting that microtubule stabilization promoted by chromosomes does not require kinetochore–microtubule attachment [168].

Several other studies support a role for Ran-GTP in somatic cell division, including the following three. First, microinjection of importin-β's cargo binding domain (aa 71-876) into mammalian cells (Ptk1) severely disrupted spindle assembly [133]. Second, injection of importin-β's cargo-binding domain (aa 71-876) into other cell types (e.g., HeLa), disrupted early stages of spindle organization and caused a delay in the prometaphase-to-metaphase transition [147]. This study also showed that micro-injection of a GTP-locked Ran mutant (Q69L) led to ectopic microtubule nucleation, and aster formation and injection of full-length recombinant importin-β resulted in spindle pole "splitting" in dividing cells [147]. The authors suggest that Ran-GTP signals contribute to early stages of spindle assembly, but can be dispensable once bipolar spindles are assembled in these somatic cells. Third, studies using RNAi to knockdown Ran-GTP "effector" proteins support a role for this pathway in somatic cell division [169–171]. However, the findings from these knockdown studies can be more difficult to interpret, as these "effector" proteins are regulated by multiple inputs and may have functions during stages of the cell cycle other than mitosis.

The possibility that the Ran pathway has only a relatively minor role in somatic cell division is supported by different lines of evidence, including the following four. First, studies of tsBN2 cells (which lack normal RCC1 function) revealed that morphologically normal appearing spindles can assemble around chromosomes [172]. Second, siRNA-mediated knockdown of RanGAP significantly altered the shape of the Ran-GTP gradient, but did not impact bipolar spindle formation in cultured human cells [149]. Third, a study showed that in *Drosophila* S2 cells (in which a relatively steep Ran-GTP gradient can be detected during mitosis), depletion of RCC1 did not disrupt spindle assembly in the presence or absence of centrosomes [173]. Remarkably, this study also revealed that RCC1 depletion did not impact microtubule assembly during recovery from depolymerization. Fourth, a systematic dissection of the role of Ran-pathway was carried out by Khodjakov and co-workers by modifying an assay first developed by Brinkley and co-worker [174,175]. In this assay, mitosis with unreplicated genomes (MUG) is induced in cultured cells. The bulk of the chromosomes separate from small kinetochore fragments in these dividing cells. Khodjakov and co-workers used FRET-based sensors to show that the chromosomes generate spatial gradients of Ran-GTP during MUG. In addition, these chromosomes can induce asymmetry in the growth of astral microtubules, consistent with the presence of chromosome-derived signals. Remarkably, this MUG assay revealed that bipolar spindles can assemble largely independently of where the bulk of the chromosomes are positioned and where the Ran-GTP concentration is likely to be highest. The authors also showed that spindle assembly under their assay conditions did not require centrosomes, but did depend on kinetochores. Interestingly, these authors also found that inhibition of the CPC using chemical inhibitors of Aurora kinase did not suppress spindle formation during MUG.

A recent study by Needleman and co-workers may help explain these conflicting data on the role of Ran-GTP in dividing somatic cells [149]. These authors apply an approach called TIMMA (time-integrated multipoint moment analysis, a multipoint fluorescence fluctuation spectroscopy method) to determine protein concentration and measure diffusion constants at several locations in a single living cell. They find that Ran exists in fast- and slow-diffusing forms in dividing cells. The slow-diffusing form—likely bound to importins—is enriched proximal to chromosomes, while the fast-diffusing form is uniformly distributed across the cell. This method also revealed that the Ran-regulated microtubule associated protein TPX2 is also present in fast- and slow-diffusing forms. However, unlike Ran, both species of TPX2 are strongly enriched proximal to chromosomes. Microtubule depolymerization disrupts the soluble TPX2, but not the Ran, spatial gradients. These data, along with analyses of two other Ran-regulated proteins, suggest that the spatial distribution of spindle assembly factors is not only influenced by Ran, but also by interactions with microtubules. The spindle assembly factors activated near chromosomes by Ran can promote local microtubule formation, and interactions with these newly formed filaments can lead to local feed-back influencing spatial organization and function. This model can help explain how spindle size can be uncoupled from the shape of the Ran gradient. It also provides plausible explanations for findings from the MUG assays and how spindle size scale may vary with cell volume [17,18,175].

8. Targeting and Activating γ-Tubulin

Genetic studies of microtubule organization and function in *Aspergillus nidulans*—an important model organism in which many key mitosis genes have been discovered [176]—led to the identification of γ-tubulin as a suppressor of a conditional lethal mutation in β-tubulin [177]. Subsequent studies localized γ-tubulin to centrosomes in different cell types, and revealed that it is part of a large multi-protein complex called γ-TURC (or γ-tubulin ring complex, reviewed in [71]). It has been established that γ-tubulin has an essential role in microtubule formation in a variety of cell types, including those that do not depend on the centrosome for cell division (e.g., plants [178] and *Drophophila* oocytes [179]). Consistent with these data, γ-tubulin is also implicated in assembling microtubules from non-centrosomal sites (such as kinetochores [180]), and can be found located within the spindle [179,181]. In addition to promoting microtubule nucleation, γ-tubulin may also function as a microtubule minus-end cap [182].

Reconstituted γ-TURC complexes from *S. cerevisiae* have been characterized and found to be much less efficient in nucleating microtubules when compared to centrosomes [183]. Specific structure-guided crosslinks of the γ-TURC complex into a "closed" complex only led to modest increases in nucleation activity [71]. A possible explanation for these observations has come from a recent study by Brouhard and co-workers that shows that microtubule nucleation from templates such as γ-TURC is kinetically unfavorable in vitro [184]. They suggest that this is due to a structural mismatch between the ring-shaped templates and growing microtubule plus-ends, which may exist as sheets [185]. Brouhard and colleagues also show that microtubule-associated proteins that promote catastrophe (e.g., MCAK) inhibit nucleation and suppressors of catastrophe (e.g., TPX2, also see below) promote nucleation. These data, along with other findings, suggest different models for how microtubule-associated proteins can promote γ-tubulin-dependent microtubule nucleation. First, microtubule-associated proteins can bind and activate γ-TURC complexes. Second, an indirect mechanism would involve microtubule-associated proteins promoting nucleation by inhibiting catastrophe events to prevent the loss of newly-nucleated filaments. Third, suggested by a recent study from Surrey and co-workers (discussed in more detail below), microtubule-associated proteins can directly promote microtubule nucleation [186,187]. In this case, γ-TURC complexes may stabilize or "cap" polymers with a specific organization (e.g., a particular protofilament number).

8.1. Microtubule Targeting of γ-Tubulin

Clues for how γ-tubulin could be targeted to microtubules in dividing cells came from a study by Stearns and co-workers characterizing NEDD1 (or GCP-WD), a subunit of the human γ-TuRC [188]. These authors showed that NEDD1 is required for the localization of γ-tubulin at centrosomes and within mitotic spindles. Assays analyzing microtubule regrowth after drug-induced depolymerization revealed that NEDD1 is needed for centrosome-dependent and -independent microtubule formation in dividing cells. In particular, a specific phosphorylation of NEDD1 (at Ser418, likely by CDK1) helps recruit γ-tubulin to the mitotic spindle, but not the centrosome. Consistent with this localization, this mitotic NEDD1 phosphorylation was shown to be required only for centrosome-independent microtubule formation. Based on these findings, the authors proposed that NEDD1 could recruit γ-TURC to the sides of microtubules to promote microtubule-dependent microtubule formation, similar to the mechanism suggested by Tran and co-workers examining interphase microtubule organization in fission yeast [101]. They also suggest that their observations could also be explained by an indirect mechanism in which NEDD1 contributes to the proper distributions of filament minus-ends in the spindle, and γ-tubulin "caps" these filament ends. This latter hypothesis is supported by findings that inhibition of the microtubule-severing protein katanin reduces the amount of γ-tubulin in the spindle [189].

Other proteins that recruit γ-tubulin to spindle microtubules were discovered by Goshima and co-workers [109]. These authors carried out a genome-wide screen that employed high-throughput microscopy to analyze mitotic phenotypes in cultured *Drosophila* S2 cells. They first confirmed that γ-tubulin recruitment to spindle poles depends on centriolar proteins (e.g., Sas-6) and polo kinase. They found that bipolar spindles assembled after RNAi-mediated knockdown of these centriolar proteins, indicating that the recruitment of γ-tubulin to the spindle pole is not essential for cell division in these cells. They also discovered that the recruitment of γ-tubulin to the spindle depends on a set of previously uncharacterized proteins that they named Dgt2–6 (for dim γ-tubulin). The knockdown of these Dgt proteins—which also localize to spindle microtubules—reduced spindle microtubule density and caused defects in spindle organization, chromosome alignment, and cell cycle progression. Importantly, knockdown-associated phenotypes became more severe when centrosomal and Dgt proteins were co-depleted, indicating that the Dgt proteins contribute to centrosome-independent microtubule formation. Additional work revealed that Dgt proteins form a heteroctameric-protein complex that was named "augmin" [190], and is conserved across metazoans [191,192]. To recruit γ-tubulin to spindles, augmin's Dgt6 (also named HAUS6) subunit's C-terminal domain likely binds NEDD1 [191], the protein previously shown to be involved in recruiting γ-TURC to spindle microtubules [188].

It is tempting to speculate that augmin functions in a manner similar to the Arp2/3 complex, which can bind along the side of an actin filament and promote the nucleation of a daughter actin filament [190,193]. In such a model augmin would bind to the side of a microtubule and recruit γ-TURC to promote the formation of a daughter filament oriented parallel to the mother filament (Figure 4). Consistent with this model, an elegant electron tomography study by Kamasaki and co-workers detected a ~29 nm rod-shaped structure at microtubule minus-ends that could serve as a link to the side of another microtubule in the mitotic spindle [194]. However, it is unclear if augmin is indeed this rod-shaped structure, and additional work (e.g., immuno-electron microscopy analysis) is needed.

Currently, evidence that augmin is involved in microtubule-dependent microtubule formation comes from a study by Petry and co-workers [195]. In this study, microtubule aster formation in *Xenopus* egg extracts was induced by the addition of a GTP-locked Ran mutant (RanQ69L) and the microtubule-associated protein TPX2. The formation of branched microtubule networks could be directly observed under these conditions. Immunodepletion of augmin suppressed the organization of microtubules into asters in this assay, consistent with its role in microtubule-dependent microtubule formation. Interpreting these results is not entirely straightforward, as TPX2 itself can directly promote microtubule formation (discussed below in more detail).

The loss of augmin function has been studied in different organisms. In zebrafish, mutation in augmin leads to defects in hematopoiesis [196]. In flies, augmin mutants are viable but female sterile [197,198]. In filamentous fungus, disruption of augmin genes does not affect mitosis [199]. By contrast, the mouse knockout of the augmin subunit Dgt6/HAUS6, generated by Watanabe and co-workers, indicates that augmin is needed for mouse embryonic development [200]. Centrioles are absent for the first divisions during mouse development, and spindle assembly involves the clustering of multiple MTOCs. Interestingly, MTOC clustering fails without augmin. This phenotype is similar to what has been reported for augmin RNAi in cultured cells [192,201,202].

To dissect the role of augmin in MTOC clustering, Watanabe and co-workers overexpressed Polo-like kinase 4 (PLK4) in HeLa cells, which can lead to spindles with multiple poles [200]. Their findings using this assay—along with augmin knockdown and disruption of NEDD1-dependent γ-tubulin targeting to the spindle—suggest that the γ-tubulin associated with spindle microtubules contributes to centrosome clustering. It is noteworthy that the electron microscopy studies by Kamasaki and co-workers had found defects in centriolar microtubules after knockdown of augmin, suggesting a more direct role for augmin in centrosome organization [194]. Therefore, additional studies are needed to properly dissect how augmin contributes to centrosome (or MTOC) clustering during cell division.

In an effort to dissect augmin function, my laboratory reconstituted this hetero-octameric complex with recombinant proteins expressed in insect cells [203]. Our biochemical and electron microscopy-based studies of the "holo-complex" and different stable sub-complexes revealed how the eight proteins may interact to form a Y-shaped structure. In assays with purified proteins, augmin bound the sides of stabilized microtubules with micromolar affinity and diffused in 1-D with short association times (seconds), but did not reveal any preference for microtubule ends. It is noteworthy that these microtubule binding lifetimes are similar to what has been measured for augmin turnover in dividing cells (e.g., GFP-Dgt5/HAUS5; $t_{1/2}$ = 4 s [190]). Further, the electron microscopy study by Kamasaki and co-workers shows that there is typically only one daughter microtubule associated with a mother filament, suggesting short-lived association of the new and the pre-existing filaments [194]. If these associations were long-lived, multiple daughter filaments would be associated with one mother filament, as daughter filaments would have nucleated additional filaments.

We also showed that the addition of augmin holo-complex and sub-complexes to *Xenopus* egg extracts promoted the formation of microtubule asters [203]. This activity is increased in the presence of RanQ69L (GTP-locked mutant) and depended on the HAUS8/Hice1-subunit's microtubule-binding site, but did not require the domain in HAUS6/Dgt6 needed to bind NEDD1 and recruit γ-TURC. Comparisons of asters induced by octameric and sub-complexes with similar in vitro microtubule binding properties indicated that proper asymmetry and microtubule bundling in these asters required all eight subunits in the augmin complex. In these assays, we were unable to detect augmin at branch points between filaments. Consistent with our studies with purified microtubules, the augmin complexes associated along the lengths of microtubules. Therefore, additional studies are needed to determine if augmin does indeed work in a manner similar to the Arp2/3 complex, or if it promotes aster formation by directly stabilizing microtubules and promoting their bundling. These models for augmin function need not be mutually exclusive, and together may help explain augmin's function during cell division.

8.2. A Direct Role for TPX2 (Targeting Protein for Xklp2) in Microtubule Formation

This microtubule associated protein was identified as a factor needed for the recruitment of the motor protein XKLP2 (human kif15, kinesin-12) to spindle poles [204]. While the potential role of TPX2 in directly regulating XKLP2 function is not yet fully understood, it is clear that TPX2 may be a protein with several distinct functions, including the spindle targeting of kinesin-5 and Aurora A kinase [205]. Here, I highlight recent progress in our understanding of its function in promoting microtubule formation and its regulation by Ran-GTP.

An important finding was that recombinant TPX2 added to *Xenopus* egg extracts can induce the formation of microtubule asters [206]. Interestingly, in another study, TPX2 was found in HeLa cell nuclear extract fractions that can promote microtubule aster formation when added to *Xenopus* egg extracts [134]. Additional analyses using assays with importin-α depleted egg extracts, GTP-locked Ran mutants, and recombinant TPX2 led to the proposal that TPX2 may be the only importin-α binding protein required for Ran-GTP-dependent microtubule formation [134]. The involvement of TPX2 and several other effectors of Ran-dependent microtubule formation has raised the question of whether it is possible that the importins can quantitatively sequester and inhibit the functions of all these proteins. This is relevant, as the concentration of NLS-containing proteins can be high, and may compete for transport receptor binding. One solution to this potential issue has been uncovered by recent structural and biochemical analyses of TPX2-importin-α binding [207]. Via its central domain, TPX2 binds at a site on importin-α that is distinct from the site bound by many NLS-containing proteins, thereby effectively reducing direct competition. However, it is noteworthy that TPX2-immunoprecipitations do not efficiently co-deplete importin-α from egg extracts, and additional analyses are needed to properly dissect this interaction in cellular contexts [134]. Importantly, additional in vitro studies have shown that purified recombinant TPX2 (albeit at high concentrations) can promote the formation of tubulin aggregates and filament bundles [208].

More convincing evidence for a direct role of TPX2 in microtubule formation and its regulation by importin-α has come from a recent study by Surrey and co-workers [186]. Characterization of full-length TPX2 revealed that it is a monomer in solution and has some preference for the GMPCPP-bound lattice and the growing tip of a microtubule, but not the end of a shrinking filament [186]. These findings, along with other data, suggest that TPX2 recognizes a specific tubulin conformation at the growing filament end, but this feature is likely distinct from that recognized by the EB proteins [209]. TPX2 also suppresses microtubule catastrophes and slows depolymerization, thereby increasing microtubule lifetimes. Surrey and co-workers also find that while TPX2 does not strongly bind soluble tubulin dimers, it can promote the formation of "stubs" that are likely multimers of tubulin. The growth of these "stubs"—which may be microtubule nucleation intermediates—is blocked by TPX2.

Human chTOG (or XMAP215/Stu2p/Dis1/Alp14 homolog) was also characterized in this study, and was shown to be a microtubule polymerase similar to other proteins in the XMAP215 family [28,186,210]. Additionally, consistent with an earlier study of other XMAP215 proteins, chTOG only weakly promoted the nucleation of microtubules. Remarkably, chTOG and TPX2 together strongly promoted microtubule nucleation [186].

Interestingly, in controlled in vitro experiments, purified importin-α/β could inhibit the nucleation of microtubules by TPX2 and chTOG, and blocked the formation of tubulin "stubs" by TPX2 [186]. These studies provide straightforward explanations for why TPX2 and XMAP215 are needed for Ran-GTP-dependent microtubule formation in *Xenopus* egg extracts, and the observation that TPX2 promotes microtubule nucleation from XMAP215 immobilized on beads [211]. Future work characterizing the TPX2-nucleated "stubs" will help strengthen this model and advance our understanding for how TPX2, along with γ-TURC and augmin, contributes to microtubule formation during cell division.

9. Sliding and Sorting Microtubules

The importance of the relative sliding and sorting of microtubules for spindle assembly and function was appreciated long before many of the microtubule-based motor proteins needed for cell division had been identified [212]. The identification of several kinesins required for cell division resulted from studies in the genetically-tractable fungi *Aspergillus nidulans* and *Saccharomyces cerevisiae* [176,213]. Studies in *S. cerevisiae* also provided the first compelling evidence supporting a model in which counter-acting forces generated by motor proteins help assemble spindles [213,214]. In particular, the phenotypes due to the loss of kinesin-5 (Cin8 and Kip1) function could be partially suppressed by the deletion of the kinesin-14 (Kar3) gene in *S. cerevisiae* [214]. Subsequent work from several labs in different model organisms has helped to further develop this model for how motor proteins can push or pull microtubules to assemble spindles. These mainly cell biological studies suggest how spindle length may be controlled and how microtubules may be focused at spindle poles. These findings have been extensively reviewed [53,74,107,215–217]. Advances in our understanding of the regulation of microtubule length—another key parameter for proper spindle assembly and motor protein-based sorting—has also been recently reviewed [210]. Here I focus mainly on recent biochemical and biophysical studies of mitotic motor proteins and how they help explain spindle assembly.

9.1. Kinesin-5 (or Eg5/KSP/Klp61F)

Kinesin-5 is a homo-tetrameric microtubule plus-end-directed motor protein that can crosslink filaments [218–220]. A recent study has revealed how a long four-helix bundle orients pairs of motor domains at opposite ends of this extended ~80 nm dumbbell-shaped molecule [221]. Kinesin-5 also has a non-motor microtubule binding domain at its C-terminus [222]. My lab, in collaboration with the Schmidt lab, showed that full-length *Xenopus laevis* kinesin-5 can slide apart microtubules it crosslinks [223]. Briefly, we established a "microtubule sandwich" assay in which we first immobilized axonemes (bundles of microtubules) on a glass coverslip in a flow-cell. Kinesin-5 constructs, along with stabilized microtubules and ATP, were then added, and the microtubules were imaged. Full-length kinesin-5, but not truncated dimeric constructs, captured and crosslinked microtubules from the solution and moved them relative to the surface-attached filaments. Analyses of polarity-marked filaments indicated that while kinesin-5 can slide apart antiparallel microtubules, kinesin-5-crosslinked parallel filaments do not move relative to each other. Analyses of the relative sliding of microtubules that were crosslinked at approximately right angles indicated that the motor protein walked along each filament it crosslinked. As a result, the relative sliding of two antiparallel filaments was twice the velocity of the kinesin walking along a single microtubule (Figure 5A). Subsequent studies confirmed these findings and demonstrated that kinesin-5 from other model organisms can also slide antiparallel microtubules apart [224,225].

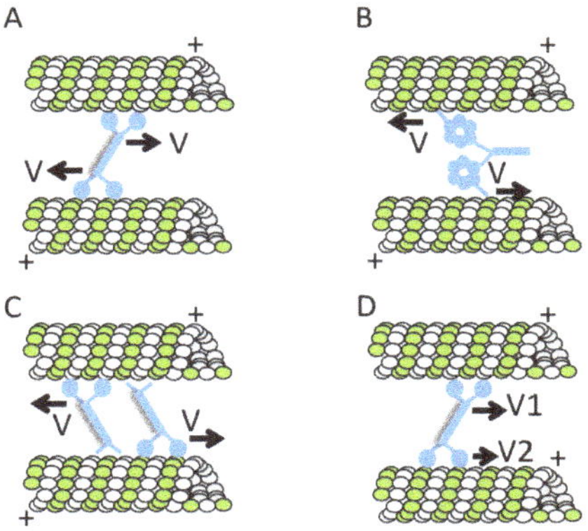

Figure 5. Schematics for how different motor proteins can crosslink and slide two microtubules apart. (**A**) A kinesin-5 homotetramer can walk towards the plus-end of each filament it crosslinks; (**B**) Kinesin-14 dimers crosslink microtubules via motor (circles) and non-motor (lines) domains. The motor domains can bind either filament and walk towards its minus-end; (**C**) Dynein dimers can walk towards the minus-ends of the microtubules. Each motor domain in the dimer may interact with a different filament; (**D**) Kinesin-12 homotetramers may slide parallel microtubules relative to each other by walking faster on one filament in the pair. Tubulin dimer: green, white; motor proteins: blue; V = velocity; in D, V1 is greater than V2; plus-end of the microtubule is also indicated (+).

Single molecule studies with GFP-tagged full-length *Xenopus laevis* kinesin-5 revealed that the motion of this motor protein along a single microtubule includes ATP-hydrolysis independent 1-D diffusion, in addition to its ATP-dependent directional motion [226]. We also found that crosslinking two microtubules stimulates kinesin-5's directional motility [227]. Analyses of a homotetrameric kinesin-5 construct lacking the C-terminal non-motor microtubule binding region revealed that tetramerization of the relatively low processivity motor domains is not sufficient for relative microtubule sliding [228]. Four motor domains along with four non-motor domains in the homotetramer are needed to tune kinesin-5's microtubule interactions for its filament sliding function.

Our findings suggest a model in which kinesin-5 molecules make long (~30 s) associations with single microtubules and explore their lengths via 1-D diffusion, which together can increase the probability that kinesin-5 will crosslink another filament. If the second filament is antiparallel, kinesin-5 motility is triggered, and the motor protein walks towards each filament's plus-end to slide them apart (Figure 5A). Interaction with a parallel filament would lead to crosslinking, but not persistent relative motion.

Recently, we have further modified the relative filament sliding assays and now combined fluorescence imaging (TIRF (total internal reflection fluorescence) -based) with optical trapping [229] (Figure 6). This assay, which we named "mini-spindle" assay, allows imaging of microtubule and kinesin motion, as well as measurement of piconewton forces generated within this minimal structural unit of the spindle. The relative orientation and motion of the microtubules can also be directly controlled. This "mini-spindle" assay revealed that ensembles of kinesin-5 crosslinking two antiparallel microtubules can generate pushing forces that are proportional to filament overlap length.

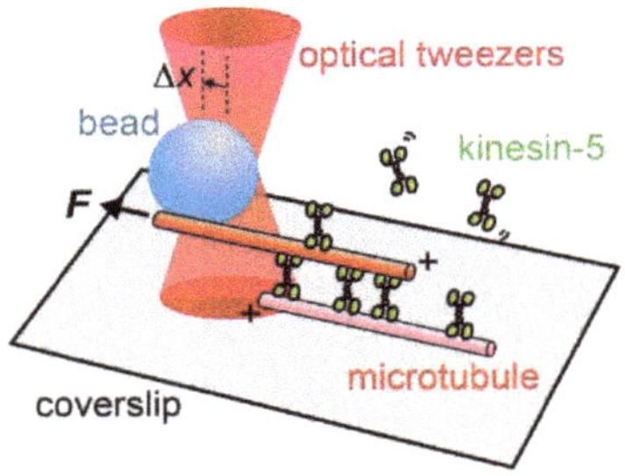

Figure 6. Schematic for the "mini-spindle" assay. Optical trapping and TIRF (total internal reflection fluorescence) microscopy are combined to examine forces generated by kinesin-5 sliding two microtubules apart.

Within the spindle, the microtubules flux poleward [6,7]. As a result, antiparallel interpolar microtubules continuously slide apart at constant relative velocities (2–3 μm/min). Our "mini-spindle" assay can mimic this relative filament motion, and also allow measurements of forces generated by kinesin-5 [229]. We find that kinesin-5 does not generate a strong pushing force when the antiparallel filament sliding velocity matches that of the motor protein's unloaded filament sliding velocity. At slower velocities, kinesin-5 generates a pushing force to assist relative motion, and at faster microtubule sliding velocities, kinesin-5 generates a braking force. Importantly, in all these cases with moving filaments, the forces generated by kinesin-5 ensembles scale with microtubule overlap lengths.

Kinesin-5 crosslinking two parallel microtubules does not generate forces to push these filaments apart [229]. However, it does generate a substantial force to resist relative filament motion, and this force increases with relative sliding velocity. Importantly, this braking force also scales with filament overlap length.

Long-standing models for spindle assembly have predicted a force in the spindle that scales with a micrometer-scale geometric feature (e.g., filament or overlap length). Our studies with purified proteins in vitro reveal that kinesin-5 could be this activity. These findings help to interpret several observations made in the context of "whole" spindles [229]. For example, our data indicate that kinesin-5 in the spindle could help coordinate polewards flux across the dynamic micrometer-scale structure. Microtubules that slide faster relative to other filaments would experience a kinesin-5-dependent braking force, and slower filaments would experience a force that can accelerate their motion. Small microtubule "seeds" added to spindles move polewards at velocities consistent with faster dynein-dependent transport [90]. This faster polewards transport is likely possible as kinesin-5 may generate a smaller braking force for these filaments that can only achieve short overlap lengths.

9.2. Kinesin-12s (or hKif15/Xklp2)

These are also microtubule plus-end-directed motor proteins involved in cell division [230,231]. The addition of a dominant negative construct of the *Xenopus laevis* kinesin-12 (Xklp2) to spindle assembly reactions in egg extracts resulted in monopolar spindles [230]. Subsequent studies showed that immunodepletion of Xklp2 did not disrupt spindle assembly in the same egg extract system [206]. These findings, along with RNAi-mediated knockdown studies [232], suggest that this kinesin is not likely to be essential for spindle assembly in animal cells. However, studies in *C. elegans* have shown that kinesin-12 (KLP18) is needed for meiotic spindle formation, but not mitotic spindle formation, and may contribute to the bundling of parallel microtubules [233].

The field was relatively stuck until the Medema and Vernos laboratories re-examined kinesin-12 function in somatic cell division [234,235]. These studies devised assays based on my earlier finding that while chemical inhibition of kinesin-5 blocked spindle assembly, acute inhibitor treatments did not collapse assembled bipolar spindles in somatic cells [106]. Knockdown of kinesin-12 by RNAi in human cells revealed this motor protein was needed to maintain spindle bipolarity when kinesin-5 function was blocked using chemical inhibitors [234,235]. Knockdown of kinesin-12 alone had a modest effect on spindle assembly, but increased the fraction of monopolar spindles formed upon kinesin-5 inhibition.

Interestingly, studying somatic cell resistance to kinesin-5 inhibitors has also shed light on how kinesin-12 contributes to bipolar spindle assembly [231,236]. Ohi and co-workers have shown that kinesin-12 preferentially localizes to kinetochore microtubules, but its overexpression can also target this kinesin to interpolar microtubules [231]. This altered localization may allow kinesin-12 to function redundantly with kinesin-5 to push or keep antiparallel interpolar microtubules apart, and thereby confer resistance to kinesin-5 inhibition. Live imaging studies suggest that kinesin-12's function in dividing cells likely involves bundling parallel microtubules. In another study, Ohi and co-workers found that resistance to kinesin-5 inhibitors can arise without over-expression of kinesin-12, but involves its function [237]. Their findings suggest that increased microtubule bundling—which can be caused by a mutation in kinesin-5—can promote kinesin-12-dependent microtubule organization and help assemble spindles in the absence of kinesin-5 motility.

Like the cell biological studies, different biochemical studies of recombinant kinesin-12 have not been entirely consistent with each other. One study from the Ohi laboratory shows that this kinesin is a homodimer with motor and non-motor microtubule binding domains [238]. Its binding to single filaments is transient, and involves the formation of an inactive "closed" conformation. By contrast, kinesin-12's interactions with two microtubules in a bundle are long-lived. This study also showed that kinesin-12 can slide two microtubules apart [238]. However, another study from the McAinsh laboratory reports that kinesin-12 is a homotetramer that can crosslink microtubules, but cannot slide antiparallel filaments apart [239]. Both studies show that microtubule crosslinking in vitro does not require TPX2, the putative kinesin-12 spindle targeting protein. It is possible that TPX2 may promote the formation of microtubule bundles that would be the preferred substrate for kinesin-12.

Another more recent study from the McAinsh laboratory examined kinesin-12 crosslinking and sliding of dynamic microtubules in vitro [240]. The authors found that kinesin-12 can track the plus-ends of growing microtubules and suppress catastrophe when multiple motor protein molecules accumulate at filament ends. They also showed that kinesin-12 molecules can slide parallel microtubule filaments relative to each other, and suggest that this motility is achieved by differences in the velocities at which the motor walks on each filament in crosslinks (Figure 5D). Further work is needed to dissect this motor protein's functions, and sort through some of the discrepancies in the literature.

9.3. Kinesin-14s (or XCTK2/HSET/Ncd)

These microtubule minus-end-directed kinesins have their motor domains at their C-terminus and a non-motor microtubule binding site at their N-terminus [241–243]. TIRF-based single molecule studies of the *Drosophila melanogaster* kinesin-14 (Ncd, a homodimer in solution) show that it diffuses in 1-D along single microtubules in the presence of ATP [244]. Kinesin-14 can also crosslink two microtubules in parallel or antiparallel orientations. Robust relative filament sliding (~100 nm/s) with plus-ends leading was observed only for antiparallel microtubules. Parallel filaments moved apart for a few seconds after the initial encounter, and then stopped moving. These observations can be explained by considering that when kinesin-14 molecules crosslink two filaments, the dimeric motor domains can interact with either filament (e.g., top or bottom) (Figure 5C). In the antiparallel case, motor domains walking towards the minus-end of each filament would assist each other to slide the filaments apart. In the parallel case, motor proteins walking towards the minus-ends of each filament would oppose each other in a molecular tug-of-war, and the filaments would not move relative to each other. Analyses of the *S. pombe* kinesin-14 (klp2) revealed a similar activity [245], suggesting that potentially all kinesin-14s slide antiparallel filaments apart and generate static crosslinks between parallel filaments.

These elegant in vitro studies indicate that kinesin-14s could oppose the relative filament sliding of antiparallel microtubules driven by kinesin-5, as these motor proteins would push filaments apart in opposite directions. It is noteworthy that while kinesin-5 molecules would be stationary between two filaments that slide apart, kinesin-14 molecules would translocate with the moving filaments, diffusing and switching orientations between the two crosslinked filaments.

9.4. Cytoplasmic Dynein

This microtubule-based motor protein is a member of the AAA+ (ATPases associated with diverse cellular activities) family [246]. It functions as an ~1.2 MDa multi-protein complex with two heavy chains, each of which contains a motor domain at the C-terminus and a "tail" domain that mediates interactions with accessory factors at its N-terminus [247–249]. Cytoplasmic dynein is the major microtubule minus-end-directed motor protein that transports a wide range of cargoes (e.g., organelles, proteins, and mRNA) in eukaryotic cells. There have been several advances in our understanding of cytoplasmic dynein's structure, biochemistry, and motility [247–249]. Here I highlight dynein's function in spindle organization, focusing on its microtubule sliding and sorting functions. This activity is needed to organize the prophase microtubule array [115], assemble spindle poles [250–252], and to generate forces that counteract the activities of kinesin-5 and kinesin-12 [234, 235,253–255]. These functions have been linked to the transport of other microtubule-associated proteins (e.g., NuMA [250]), and to the relative sliding of microtubules [14,251].

Early biochemical studies demonstrated the crosslinking and bundling of microtubules by dynein in vitro [256,257]. A recent study by Tannenbaum and co-workers has shown that cytoplasmic dynein can also slide antiparallel microtubules apart [236]. The use of well-characterized truncated constructs revealed that this relative microtubule sliding activity is independent of "tail"-domain-mediated interactions with accessory proteins. Single molecule studies revealed that dynein, which walks towards the minus-ends of single filaments, has a much more complex behavior when interacting with two antiparallel microtubules. Dynein molecules can pause for seconds before making directional runs along either filament they crosslink. These data, together with additional findings, suggest a model in which dynein can crosslink and slide microtubules in a manner distinct from kinesin-5 and kinesin-14. Each AAA domain in the dimeric molecule can bind a different filament, and walking on each filament would lead to relative motion, pulling microtubule minus-ends together (Figure 5B).

9.5. In Vitro Studies of Motor Proteins Generating Opposing Forces

It is tempting to speculate that motor proteins with opposing activities should be able to establish stable antiparallel microtubule overlap—a recurring filament configuration within the metaphase spindle. However, theoretical work has suggested that motor protein mixtures would lead to unstable states [258–260]. Experiments are consistent with the theory. Directional instability (i.e., with frequent back and forth motion of the microtubules) has been observed when mixtures of dynein and kinesin-1 [261], or kinesin-5 and kinesin-14, are immobilized on surfaces [222]. Another study that examined the relative sliding of two antiparallel microtubules by mixtures of kinesin-5 and kinesin-14 also indicated that static overlap cannot be achieved [224]. In fact, efforts to balance the motor proteins' activities resulted in directional instability, with a broad distribution of instantaneous velocities [224]. Interestingly, this study also reveals that a much smaller number of kinesin-5 molecules can compete against a large number of kinesin-14 molecules crosslinking antiparallel microtubules. Importantly, this study shows that stable antiparallel microtubule overlap can be achieved by incorporating an engineered microtubule crosslinking protein along with the motor proteins. There are several filament crosslinking proteins that are known to be involved in cell division, and it is likely that these proteins may contribute to stabilizing filament overlap. Constant antiparallel microtubule overlap in animal spindles must be maintained in the presence of persistent motion due to poleward flux [6], adding another layer of complexity to the efforts in reconstituting spindle assembly with purified proteins.

10. Outlook

As outlined in this review, there have been several major advances in our understanding of how the cell division apparatus assembles. We are now poised to unravel the basic biochemical principles underlying many of the processes, such as microtubule nucleation and force generation, required for successful cell division. New cell-based experiments will help test specific predictions from biochemical studies. The multiple regulatory inputs controlling each of these essential processes will have to be carefully teased apart. I believe that now is a very exciting time to study cell division, as it provides a unique opportunity to establish the basic principles for how micrometer-scale cellular structures can be assembled by nanometer-sized proteins, and also explore how mechanical and biochemical inputs intersect to control distinct protein outputs. Powerful new methodologies (e.g., to image cell division in different tissues in vivo) and tools (e.g., cell-permeable chemical inhibitors of key proteins) will likely be developed and used to answer these questions.

Acknowledgments: I am grateful to the NIH (National Institutes of Health) (GM65933) for funding and to members of laboratory for helpful comments. I apologize for any work that I may have unintentionally omitted.

Conflicts of Interest: The author declares no conflict of interest.

References

1. Foley, E.A.; Kapoor, T.M. Microtubule attachment and spindle assembly checkpoint signalling at the kinetochore. *Nat. Rev. Mol. Cell Biol.* **2013**, *14*, 25–37. [CrossRef] [PubMed]
2. Inoue, S.; Sato, H. Cell motility by labile association of molecules. The nature of mitotic spindle fibers and their role in chromosome movement. *J. Gen. Physiol.* **1967**, *50*, 259–292. [CrossRef]
3. Desai, A.; Mitchison, T.J. Microtubule polymerization dynamics. *Annu. Rev. Cell Dev. Biol.* **1997**, *13*, 83–117. [CrossRef] [PubMed]
4. Saxton, W.M.; Stemple, D.L.; Leslie, R.J.; Salmon, E.D.; Zavortink, M.; McIntosh, J.R. Tubulin dynamics in cultured mammalian cells. *J. Cell Biol.* **1984**, *99*, 2175–2186. [CrossRef] [PubMed]
5. Salmon, E.D.; Leslie, R.J.; Saxton, W.M.; Karow, M.L.; McIntosh, J.R. Spindle microtubule dynamics in sea urchin embryos: Analysis using a fluorescein-labeled tubulin and measurements of fluorescence redistribution after laser photobleaching. *J. Cell Biol.* **1984**, *99*, 2165–2174. [CrossRef] [PubMed]
6. Mitchison, T.J. Polewards microtubule flux in the mitotic spindle: Evidence from photoactivation of fluorescence. *J. Cell Biol.* **1989**, *109*, 637–652. [CrossRef] [PubMed]
7. Kwok, B.H.; Kapoor, T.M. Microtubule flux: Drivers wanted. *Curr. Opin. Cell Biol.* **2007**, *19*, 36–42. [CrossRef] [PubMed]
8. Mitchison, T.; Kirschner, M. Dynamic instability of microtubule growth. *Nature* **1984**, *312*, 237–242. [CrossRef] [PubMed]
9. Zhai, Y.; Kronebusch, P.J.; Simon, P.M.; Borisy, G.G. Microtubule dynamics at the G2/M transition: Abrupt breakdown of cytoplasmic microtubules at nuclear envelope breakdown and implications for spindle morphogenesis. *J. Cell Biol.* **1996**, *135*, 201–214. [CrossRef] [PubMed]

10. Rusan, N.M.; Fagerstrom, C.J.; Yvon, A.M.; Wadsworth, P. Cell cycle-dependent changes in microtubule dynamics in living cells expressing green fluorescent protein-alpha tubulin. *Mol. Biol. Cell* **2001**, *12*, 971–980. [CrossRef] [PubMed]
11. Chen, B.C.; Legant, W.R.; Wang, K.; Shao, L.; Milkie, D.E.; Davidson, M.W.; Janetopoulos, C.; Wu, X.S.; Hammer, J.A., 3rd; Liu, Z.; et al. Lattice light-sheet microscopy: Imaging molecules to embryos at high spatiotemporal resolution. *Science* **2014**. [CrossRef] [PubMed]
12. Tirnauer, J.S.; Salmon, E.D.; Mitchison, T.J. Microtubule plus-end dynamics in *Xenopus* egg extract spindles. *Mol. Biol. Cell* **2004**, *15*, 1776–1784. [CrossRef] [PubMed]
13. Savoian, M.S.; Rieder, C.L. Mitosis in primary cultures of *Drosophila melanogaster* larval neuroblasts. *J. Cell Sci.* **2002**, *115*, 3061–3072. [PubMed]
14. Gatlin, J.C.; Matov, A.; Groen, A.C.; Needleman, D.J.; Maresca, T.J.; Danuser, G.; Mitchison, T.J.; Salmon, E.D. Spindle fusion requires dynein-mediated sliding of oppositely oriented microtubules. *Curr. Biol.* **2009**, *19*, 287–296. [CrossRef] [PubMed]
15. Church, K.; Nicklas, R.B.; Lin, H.P. Micromanipulated bivalents can trigger mini-spindle formation in *Drosophila melanogaster* spermatocyte cytoplasm. *J. Cell Biol.* **1986**, *103*, 2765–2773. [CrossRef] [PubMed]
16. Wuhr, M.; Chen, Y.; Dumont, S.; Groen, A.C.; Needleman, D.J.; Salic, A.; Mitchison, T.J. Evidence for an upper limit to mitotic spindle length. *Curr. Biol.* **2008**, *18*, 1256–1261. [CrossRef] [PubMed]
17. Hazel, J.; Krutkramelis, K.; Mooney, P.; Tomschik, M.; Gerow, K.; Oakey, J.; Gatlin, J.C. Changes in cytoplasmic volume are sufficient to drive spindle scaling. *Science* **2013**, *342*, 853–856. [CrossRef] [PubMed]
18. Good, M.C.; Vahey, M.D.; Skandarajah, A.; Fletcher, D.A.; Heald, R. Cytoplasmic volume modulates spindle size during embryogenesis. *Science* **2013**, *342*, 856–860. [CrossRef] [PubMed]
19. Rieder, C.L. The formation, structure, and composition of the mammalian kinetochore and kinetochore fiber. *Int. Rev. Cytol.* **1982**, *79*, 1–58. [PubMed]
20. McDonald, K.L.; O'Toole, E.T.; Mastronarde, D.N.; McIntosh, J.R. Kinetochore microtubules in PTK cells. *J. Cell Biol.* **1992**, *118*, 369–383. [CrossRef] [PubMed]
21. Mastronarde, D.N.; McDonald, K.L.; Ding, R.; McIntosh, J.R. Interpolar spindle microtubules in PTK cells. *J. Cell Biol.* **1993**, *123*, 1475–1489. [CrossRef] [PubMed]
22. McEwen, B.F.; Heagle, A.B.; Cassels, G.O.; Buttle, K.F.; Rieder, C.L. Kinetochore fiber maturation in PtK1 cells and its implications for the mechanisms of chromosome congression and anaphase onset. *J. Cell Biol.* **1997**, *137*, 1567–1580. [CrossRef] [PubMed]
23. Bajer, A.S.; Mole-Bajer, J. Reorganization of microtubules in endosperm cells and cell fragments of the higher plant Haemanthus in vivo. *J. Cell Biol.* **1986**, *102*, 263–281. [CrossRef] [PubMed]
24. Zhai, Y.; Kronebusch, P.J.; Borisy, G.G. Kinetochore microtubule dynamics and the metaphase-anaphase transition. *J. Cell Biol.* **1995**, *131*, 721–734. [CrossRef] [PubMed]
25. Cimini, D.; Wan, X.; Hirel, C.B.; Salmon, E.D. Aurora kinase promotes turnover of kinetochore microtubules to reduce chromosome segregation errors. *Curr. Biol.* **2006**, *16*, 1711–1718. [CrossRef] [PubMed]
26. Maddox, P.; Straight, A.; Coughlin, P.; Mitchison, T.J.; Salmon, E.D. Direct observation of microtubule dynamics at kinetochores in *Xenopus* extract spindles: Implications for spindle mechanics. *J. Cell Biol.* **2003**, *162*, 377–382. [CrossRef] [PubMed]
27. Ohi, R.; Burbank, K.; Liu, Q.; Mitchison, T.J. Nonredundant functions of Kinesin-13s during meiotic spindle assembly. *Curr. Biol.* **2007**, *17*, 953–959. [CrossRef] [PubMed]
28. Akhmanova, A.; Steinmetz, M.O. Control of microtubule organization and dynamics: Two ends in the limelight. *Nat. Rev. Mol. Cell Biol.* **2015**, *16*, 711–726. [CrossRef] [PubMed]
29. Akhmanova, A.; Hoogenraad, C.C. Microtubule minus-end-targeting proteins. *Curr. Biol.* **2015**, *25*, R162–R171. [CrossRef] [PubMed]
30. Goodwin, S.S.; Vale, R.D. Patronin regulates the microtubule network by protecting microtubule minus ends. *Cell* **2010**, *143*, 263–274. [CrossRef] [PubMed]
31. Ito, A.; Goshima, G. Microcephaly protein Asp focuses the minus ends of spindle microtubules at the pole and within the spindle. *J. Cell Biol.* **2015**, *211*, 999–1009. [CrossRef] [PubMed]
32. Budde, P.P.; Desai, A.; Heald, R. Analysis of microtubule polymerization in vitro and during the cell cycle in *Xenopus* egg extracts. *Methods* **2006**, *38*, 29–34. [CrossRef] [PubMed]
33. Maresca, T.J.; Heald, R. Methods for studying spindle assembly and chromosome condensation in *Xenopus* egg extracts. *Methods Mol. Biol.* **2006**, *322*, 459–474. [PubMed]

34. Burbank, K.S.; Mitchison, T.J.; Fisher, D.S. Slide-and-cluster models for spindle assembly. *Curr. Biol.* **2007**, *17*, 1373–1383. [CrossRef] [PubMed]
35. Yang, G.; Houghtaling, B.R.; Gaetz, J.; Liu, J.Z.; Danuser, G.; Kapoor, T.M. Architectural dynamics of the meiotic spindle revealed by single-fluorophore imaging. *Nat. Cell Biol.* **2007**, *9*, 1233–1242. [CrossRef] [PubMed]
36. Srayko, M.; O'Toole E, T.; Hyman, A.A.; Muller-Reichert, T. Katanin disrupts the microtubule lattice and increases polymer number in *C. elegans* meiosis. *Curr. Biol.* **2006**, *16*, 1944–1949. [CrossRef] [PubMed]
37. Brugues, J.; Nuzzo, V.; Mazur, E.; Needleman, D.J. Nucleation and transport organize microtubules in metaphase spindles. *Cell* **2012**, *149*, 554–564. [CrossRef] [PubMed]
38. Tournebize, R.; Popov, A.; Kinoshita, K.; Ashford, A.J.; Rybina, S.; Pozniakovsky, A.; Mayer, T.U.; Walczak, C.E.; Karsenti, E.; Hyman, A.A. Control of microtubule dynamics by the antagonistic activities of XMAP215 and XKCM1 in *Xenopus* egg extracts. *Nat. Cell Biol.* **2000**, *2*, 13–19. [PubMed]
39. Needleman, D.J.; Groen, A.; Ohi, R.; Maresca, T.; Mirny, L.; Mitchison, T. Fast microtubule dynamics in meiotic spindles measured by single molecule imaging: Evidence that the spindle environment does not stabilize microtubules. *Mol. Biol. Cell* **2010**, *21*, 323–333. [CrossRef] [PubMed]
40. Mitchison, T.J.; Salmon, E.D. Mitosis: A history of division. *Nat. Cell Biol.* **2001**, *3*, E17–E21. [CrossRef] [PubMed]
41. Nicklas, R.B. Measurements of the force produced by the mitotic spindle in anaphase. *J. Cell Biol.* **1983**, *97*, 542–548. [CrossRef] [PubMed]
42. Svoboda, K.; Block, S.M. Force and velocity measured for single kinesin molecules. *Cell* **1994**, *77*, 773–784. [CrossRef]
43. Dogterom, M.; Kerssemakers, J.W.; Romet-Lemonne, G.; Janson, M.E. Force generation by dynamic microtubules. *Curr. Opin. Cell Biol.* **2005**, *17*, 67–74. [CrossRef] [PubMed]
44. Gittes, F.; Mickey, B.; Nettleton, J.; Howard, J. Flexural rigidity of microtubules and actin filaments measured from thermal fluctuations in shape. *J. Cell Biol.* **1993**, *120*, 923–934. [CrossRef] [PubMed]
45. Gardel, M.L.; Kasza, K.E.; Brangwynne, C.P.; Liu, J.; Weitz, D.A. Chapter 19: Mechanical response of cytoskeletal networks. *Methods Cell Biol.* **2008**, *89*, 487–519. [PubMed]
46. Lin, Y.; Koenderink, G.H.; MacKintosh, F.C.; Weitz, D.A. Viscoelastic properties of microtubule networks. *Macromolecules* **2007**, *40*, 7714–7720. [CrossRef]
47. Sato, M.; Schwartz, W.H.; Selden, S.C.; Pollard, T.D. Mechanical properties of brain tubulin and microtubules. *J. Cell Biol.* **1988**, *106*, 1205–1211. [CrossRef] [PubMed]
48. Itabashi, T.; Takagi, J.; Shimamoto, Y.; Onoe, H.; Kuwana, K.; Shimoyama, I.; Gaetz, J.; Kapoor, T.M.; Ishiwata, S. Probing the mechanical architecture of the vertebrate meiotic spindle. *Nat. Methods* **2009**, *6*, 167–172. [CrossRef] [PubMed]
49. Shimamoto, Y.; Kapoor, T.M. Microneedle-based analysis of the micromechanics of the metaphase spindle assembled in *Xenopus laevis* egg extracts. *Nat. Protoc.* **2012**, *7*, 959–969. [CrossRef] [PubMed]
50. Shimamoto, Y.; Maeda, Y.T.; Ishiwata, S.; Libchaber, A.J.; Kapoor, T.M. Insights into the micromechanical properties of the metaphase spindle. *Cell* **2011**, *145*, 1062–1074. [CrossRef] [PubMed]
51. Takagi, J.; Itabashi, T.; Suzuki, K.; Kapoor, T.M.; Shimamoto, Y.; Ishiwata, S. Using micromanipulation to analyze control of vertebrate meiotic spindle size. *Cell Rep.* **2013**, *5*, 44–50. [CrossRef] [PubMed]
52. Takagi, J.; Itabashi, T.; Suzuki, K.; Shimamoto, Y.; Kapoor, T.M.; Ishiwata, S. Micromechanics of the vertebrate meiotic spindle examined by stretching along the pole-to-pole axis. *Biophys. J.* **2014**, *106*, 735–740. [CrossRef] [PubMed]
53. Dumont, S.; Mitchison, T.J. Compression regulates mitotic spindle length by a mechanochemical switch at the poles. *Curr. Biol.* **2009**, *19*, 1086–1095. [CrossRef] [PubMed]
54. Gatlin, J.C.; Matov, A.; Danuser, G.; Mitchison, T.J.; Salmon, E.D. Directly probing the mechanical properties of the spindle and its matrix. *J. Cell Biol.* **2010**, *188*, 481–489. [CrossRef] [PubMed]
55. Pickett-Heaps, J.D.; Tippit, D.H.; Porter, K.R. Rethinking mitosis. *Cell* **1982**, *29*, 729–744. [CrossRef]
56. Scholey, J.M.; Rogers, G.C.; Sharp, D.J. Mitosis, microtubules, and the matrix. *J. Cell Biol.* **2001**, *154*, 261–266. [CrossRef] [PubMed]
57. Zheng, Y. A membranous spindle matrix orchestrates cell division. *Nat. Rev. Mol. Cell Biol.* **2010**, *11*, 529–535. [CrossRef] [PubMed]
58. Dionne, M.A.; Howard, L.; Compton, D.A. NuMA is a component of an insoluble matrix at mitotic spindle poles. *Cell Motil. Cytoskelet.* **1999**, *42*, 189–203. [CrossRef]

59. Walker, D.L.; Wang, D.; Jin, Y.; Rath, U.; Wang, Y.; Johansen, J.; Johansen, K.M. Skeletor, a novel chromosomal protein that redistributes during mitosis provides evidence for the formation of a spindle matrix. *J. Cell Biol.* **2000**, *151*, 1401–1412. [CrossRef] [PubMed]
60. Tsai, M.Y.; Wang, S.; Heidinger, J.M.; Shumaker, D.K.; Adam, S.A.; Goldman, R.D.; Zheng, Y. A mitotic lamin B matrix induced by RanGTP required for spindle assembly. *Science* **2006**, *311*, 1887–1893. [CrossRef] [PubMed]
61. Chang, P.; Jacobson, M.K.; Mitchison, T.J. Poly(ADP-ribose) is required for spindle assembly and structure. *Nature* **2004**, *432*, 645–649. [CrossRef] [PubMed]
62. Lu, L.; Ladinsky, M.S.; Kirchhausen, T. Cisternal organization of the endoplasmic reticulum during mitosis. *Mol. Biol. Cell* **2009**, *20*, 3471–3480. [CrossRef] [PubMed]
63. Jiang, H.; Wang, S.; Huang, Y.; He, X.; Cui, H.; Zhu, X.; Zheng, Y. Phase transition of spindle-associated protein regulate spindle apparatus assembly. *Cell* **2015**, *163*, 108–122. [CrossRef] [PubMed]
64. Gould, R.R.; Borisy, G.G. The pericentriolar material in Chinese hamster ovary cells nucleates microtubule formation. *J. Cell Biol.* **1977**, *73*, 601–615. [CrossRef] [PubMed]
65. McGill, M.; Brinkley, B.R. Human chromosomes and centrioles as nucleating sites for the in vitro assembly of microtubules from bovine brain tubulin. *J. Cell Biol.* **1975**, *67*, 189–199. [CrossRef] [PubMed]
66. Snyder, J.A.; McIntosh, J.R. Initiation and growth of microtubules from mitotic centers in lysed mammalian cells. *J. Cell Biol.* **1975**, *67*, 744–760. [CrossRef] [PubMed]
67. Spiegelman, B.M.; Lopata, M.A.; Kirschner, M.W. Multiple sites for the initiation of microtubule assembly in mammalian cells. *Cell* **1979**, *16*, 239–252. [CrossRef]
68. Kuriyama, R.; Borisy, G.G. Microtubule-nucleating activity of centrosomes in Chinese hamster ovary cells is independent of the centriole cycle but coupled to the mitotic cycle. *J. Cell Biol.* **1981**, *91*, 822–826. [CrossRef] [PubMed]
69. Lee, K.; Rhee, K. PLK1 phosphorylation of pericentrin initiates centrosome maturation at the onset of mitosis. *J. Cell Biol.* **2011**, *195*, 1093–1101. [CrossRef] [PubMed]
70. Sdelci, S.; Schutz, M.; Pinyol, R.; Bertran, M.T.; Regue, L.; Caelles, C.; Vernos, I.; Roig, J. Nek9 phosphorylation of NEDD1/GCP-WD contributes to Plk1 control of gamma-tubulin recruitment to the mitotic centrosome. *Curr. Biol.* **2012**, *22*, 1516–1523. [CrossRef] [PubMed]
71. Kollman, J.M.; Merdes, A.; Mourey, L.; Agard, D.A. Microtubule nucleation by gamma-tubulin complexes. *Nat. Rev. Mol. Cell Biol.* **2011**, *12*, 709–721. [CrossRef] [PubMed]
72. Kirschner, M.; Mitchison, T. Beyond self-assembly: From microtubules to morphogenesis. *Cell* **1986**, *45*, 329–342. [CrossRef]
73. Rieder, C.L.; Alexander, S.P. Kinetochores are transported poleward along a single astral microtubule during chromosome attachment to the spindle in newt lung cells. *J. Cell Biol.* **1990**, *110*, 81–95. [CrossRef] [PubMed]
74. Wadsworth, P.; Khodjakov, A. E pluribus unum: Towards a universal mechanism for spindle assembly. *Trends Cell Biol.* **2004**, *14*, 413–419. [CrossRef] [PubMed]
75. Karsenti, E.; Newport, J.; Kirschner, M. Respective roles of centrosomes and chromatin in the conversion of microtubule arrays from interphase to metaphase. *J. Cell Biol.* **1984**, *99*, 47s–54s. [CrossRef] [PubMed]
76. Zhang, D.; Nicklas, R.B. Chromosomes initiate spindle assembly upon experimental dissolution of the nuclear envelope in grasshopper spermatocytes. *J. Cell Biol.* **1995**, *131*, 1125–1131. [CrossRef] [PubMed]
77. Bonaccorsi, S.; Giansanti, M.G.; Gatti, M. Spindle self-organization and cytokinesis during male meiosis in asterless mutants of *Drosophila melanogaster*. *J. Cell Biol.* **1998**, *142*, 751–761. [CrossRef] [PubMed]
78. Megraw, T.L.; Kao, L.R.; Kaufman, T.C. Zygotic development without functional mitotic centrosomes. *Curr. Biol.* **2001**, *11*, 116–120. [CrossRef]
79. Kemp, C.A.; Kopish, K.R.; Zipperlen, P.; Ahringer, J.; O'Connell, K.F. Centrosome maturation and duplication in *C. elegans* require the coiled-coil protein SPD-2. *Dev. Cell* **2004**, *6*, 511–523. [CrossRef]
80. Hamill, D.R.; Severson, A.F.; Carter, J.C.; Bowerman, B. Centrosome maturation and mitotic spindle assembly in *C. elegans* require SPD-5, a protein with multiple coiled-coil domains. *Dev. Cell* **2002**, *3*, 673–684. [CrossRef]
81. Khodjakov, A.; Cole, R.W.; Oakley, B.R.; Rieder, C.L. Centrosome-independent mitotic spindle formation in vertebrates. *Curr. Biol.* **2000**, *10*, 59–67. [CrossRef]
82. Hinchcliffe, E.H.; Miller, F.J.; Cham, M.; Khodjakov, A.; Sluder, G. Requirement of a centrosomal activity for cell cycle progression through G1 into S phase. *Science* **2001**, *291*, 1547–1550. [CrossRef] [PubMed]

83. Marek, L.F. Control of spindle form and function in grasshopper spermatocytes. *Chromosoma* **1978**, *68*, 367–398. [CrossRef]
84. Nicklas, R.B.; Gordon, G.W. The total length of spindle microtubules depends on the number of chromosomes present. *J. Cell Biol.* **1985**, *100*, 1–7. [CrossRef] [PubMed]
85. Borisy, G.G. Polarity of microtubules of the mitotic spindle. *J. Mol. Biol.* **1978**, *124*, 565–570. [CrossRef]
86. Witt, P.L.; Ris, H.; Borisy, G.G. Origin of kinetochore microtubules in Chinese hamster ovary cells. *Chromosoma* **1980**, *81*, 483–505. [CrossRef] [PubMed]
87. De Brabander, M.; Geuens, G.; De Mey, J.; Joniau, M. Nucleated assembly of mitotic microtubules in living PTK2 cells after release from nocodazole treatment. *Cell Motil.* **1981**, *1*, 469–483. [CrossRef] [PubMed]
88. Theurkauf, W.E.; Hawley, R.S. Meiotic spindle assembly in Drosophila females: Behavior of nonexchange chromosomes and the effects of mutations in the nod kinesin-like protein. *J. Cell Biol.* **1992**, *116*, 1167–1180. [CrossRef] [PubMed]
89. Sawin, K.E.; Mitchison, T.J. Mitotic spindle assembly by two different pathways in vitro. *J. Cell Biol.* **1991**, *112*, 925–940. [CrossRef] [PubMed]
90. Heald, R.; Tournebize, R.; Blank, T.; Sandaltzopoulos, R.; Becker, P.; Hyman, A.; Karsenti, E. Self-organization of microtubules into bipolar spindles around artificial chromosomes in *Xenopus* egg extracts. *Nature* **1996**, *382*, 420–425. [CrossRef] [PubMed]
91. Gordon, M.B.; Howard, L.; Compton, D.A. Chromosome movement in mitosis requires microtubule anchorage at spindle poles. *J. Cell Biol.* **2001**, *152*, 425–434. [CrossRef] [PubMed]
92. Khodjakov, A.; Copenagle, L.; Gordon, M.B.; Compton, D.A.; Kapoor, T.M. Minus-end capture of preformed kinetochore fibers contributes to spindle morphogenesis. *J. Cell Biol.* **2003**, *160*, 671–683. [CrossRef] [PubMed]
93. Mayer, T.U.; Kapoor, T.M.; Haggarty, S.J.; King, R.W.; Schreiber, S.L.; Mitchison, T.J. Small molecule inhibitor of mitotic spindle bipolarity identified in a phenotype-based screen. *Science* **1999**, *286*, 971–974. [CrossRef] [PubMed]
94. Maiato, H.; Khodjakov, A.; Rieder, C.L. *Drosophila* CLASP is required for the incorporation of microtubule subunits into fluxing kinetochore fibres. *Nat. Cell Biol.* **2005**, *7*, 42–47. [CrossRef] [PubMed]
95. Mazia, D. Centrosomes and mitotic poles. *Exp. Cell Res.* **1984**, *153*, 1–15. [CrossRef]
96. Chan, J.; Calder, G.M.; Doonan, J.H.; Lloyd, C.W. EB1 reveals mobile microtubule nucleation sites in Arabidopsis. *Nat. Cell Biol.* **2003**, *5*, 967–971. [CrossRef] [PubMed]
97. Wasteneys, G.O.; Williamson, R.E. Reassembly of microtubules in Nitella tasmanica: Assembly of coritcal microtubules in branching clusters and its relevance to steady-state microtubule assembly. *J. Cell Sci.* **1989**, *93*, 704–714.
98. Lajoie-Mazenc, I.; Tollon, Y.; Detraves, C.; Julian, M.; Moisand, A.; Gueth-Hallonet, C.; Debec, A.; Salles-Passador, I.; Puget, A.; Mazarguil, H.; et al. Recruitment of antigenic gamma-tubulin during mitosis in animal cells: Presence of gamma-tubulin in the mitotic spindle. *J. Cell Sci.* **1994**, *107 Pt 10*, 2825–2837. [PubMed]
99. Stearns, T.; Kirschner, M. In vitro reconstitution of centrosome assembly and function: The central role of gamma-tubulin. *Cell* **1994**, *76*, 623–637. [CrossRef]
100. Van Damme, D.; Van Poucke, K.; Boutant, E.; Ritzenthaler, C.; Inze, D.; Geelen, D. In vivo dynamics and differential microtubule-binding activities of MAP65 proteins. *Plant Physiol.* **2004**, *136*, 3956–3967. [CrossRef] [PubMed]
101. Janson, M.E.; Setty, T.G.; Paoletti, A.; Tran, P.T. Efficient formation of bipolar microtubule bundles requires microtubule-bound gamma-tubulin complexes. *J. Cell Biol.* **2005**, *169*, 297–308. [CrossRef] [PubMed]
102. Mahoney, N.M.; Goshima, G.; Douglass, A.D.; Vale, R.D. Making microtubules and mitotic spindles in cells without functional centrosomes. *Curr. Biol.* **2006**, *16*, 564–569. [CrossRef] [PubMed]
103. Clausen, T.; Ribbeck, K. Self-organization of anastral spindles by synergy of dynamic instability, autocatalytic microtubule production, and a spatial signaling gradient. *PLoS ONE* **2007**, *2*, e244. [CrossRef] [PubMed]
104. Boveri, T. Concerning the origin of malignant tumours by Theodor Boveri. Translated and annotated by Henry Harris. *J. Cell Sci.* **2008**, *121*, 1–84. [CrossRef] [PubMed]
105. Sluder, G.; Rieder, C.L. Centriole number and the reproductive capacity of spindle poles. *J. Cell Biol.* **1985**, *100*, 887–896. [CrossRef] [PubMed]

106. Kapoor, T.M.; Mayer, T.U.; Coughlin, M.L.; Mitchison, T.J. Probing spindle assembly mechanisms with monastrol, a small molecule inhibitor of the mitotic kinesin, Eg5. *J. Cell Biol.* **2000**, *150*, 975–988. [CrossRef] [PubMed]
107. Tanenbaum, M.E.; Medema, R.H. Mechanisms of centrosome separation and bipolar spindle assembly. *Dev. Cell* **2010**, *19*, 797–806. [CrossRef] [PubMed]
108. Wilson, P.G.; Fuller, M.T.; Borisy, G.G. Monastral bipolar spindles: Implications for dynamic centrosome organization. *J. Cell Sci.* **1997**, *110*, 451–464. [PubMed]
109. Goshima, G.; Wollman, R.; Goodwin, S.S.; Zhang, N.; Scholey, J.M.; Vale, R.D.; Stuurman, N. Genes required for mitotic spindle assembly in *Drosophila* S2 cells. *Science* **2007**, *316*, 417–421. [CrossRef] [PubMed]
110. Brinkley, B.R. Managing the centrosome numbers game: From chaos to stability in cancer cell division. *Trends Cell Biol.* **2001**, *11*, 18–21. [CrossRef]
111. Ring, D.; Hubble, R.; Kirschner, M. Mitosis in a cell with multiple centrioles. *J. Cell Biol.* **1982**, *94*, 549–556. [CrossRef] [PubMed]
112. Kwon, M.; Godinho, S.A.; Chandhok, N.S.; Ganem, N.J.; Azioune, A.; Thery, M.; Pellman, D. Mechanisms to suppress multipolar divisions in cancer cells with extra centrosomes. *Genes Dev.* **2008**, *22*, 2189–2203. [CrossRef] [PubMed]
113. Quintyne, N.J.; Reing, J.E.; Hoffelder, D.R.; Gollin, S.M.; Saunders, W.S. Spindle multipolarity is prevented by centrosomal clustering. *Science* **2005**, *307*, 127–129. [CrossRef] [PubMed]
114. Grill, S.W.; Hyman, A.A. Spindle positioning by cortical pulling forces. *Dev. Cell* **2005**, *8*, 461–465. [CrossRef] [PubMed]
115. Rusan, N.M.; Tulu, U.S.; Fagerstrom, C.; Wadsworth, P. Reorganization of the microtubule array in prophase/prometaphase requires cytoplasmic dynein-dependent microtubule transport. *J. Cell Biol.* **2002**, *158*, 997–1003. [CrossRef] [PubMed]
116. Dogterom, M.; Felix, M.A.; Guet, C.C.; Leibler, S. Influence of M-phase chromatin on the anisotropy of microtubule asters. *J. Cell Biol.* **1996**, *133*, 125–140. [CrossRef] [PubMed]
117. Carazo-Salas, R.E.; Karsenti, E. Long-range communication between chromatin and microtubules in *Xenopus* egg extracts. *Curr. Biol.* **2003**, *13*, 1728–1733. [CrossRef] [PubMed]
118. Piehl, M.; Cassimeris, L. Organization and dynamics of growing microtubule plus ends during early mitosis. *Mol. Biol. Cell* **2003**, *14*, 916–925. [CrossRef] [PubMed]
119. Gaetz, J.; Gueroui, Z.; Libchaber, A.; Kapoor, T.M. Examining how the spatial organization of chromatin signals influences metaphase spindle assembly. *Nat. Cell Biol.* **2006**, *8*, 924–932. [CrossRef] [PubMed]
120. Dinarina, A.; Pugieux, C.; Corral, M.M.; Loose, M.; Spatz, J.; Karsenti, E.; Nedelec, F. Chromatin shapes the mitotic spindle. *Cell* **2009**, *138*, 502–513. [CrossRef] [PubMed]
121. Gorlich, D.; Mattaj, I.W. Nucleocytoplasmic transport. *Science* **1996**, *271*, 1513–1518. [CrossRef] [PubMed]
122. Melchior, F.; Gerace, L. Two-way trafficking with Ran. *Trends Cell Biol.* **1998**, *8*, 175–179. [CrossRef]
123. Moore, M.S.; Blobel, G. A G protein involved in nucleocytoplasmic transport: The role of Ran. *Trends Biochem. Sci.* **1994**, *19*, 211–216. [CrossRef]
124. Kirkpatrick, D.; Solomon, F. Overexpression of yeast homologs of the mammalian checkpoint gene RCC1 suppresses the class of alpha-tubulin mutations that arrest with excess microtubules. *Genetics* **1994**, *137*, 381–392. [PubMed]
125. Ouspenski, II A RanBP1 mutation which does not visibly affect nuclear import may reveal additional functions of the ran GTPase system. *Exp. Cell Res.* **1998**, *244*, 171–183.
126. Nakamura, M.; Masuda, H.; Horii, J.; Kuma, K.; Yokoyama, N.; Ohba, T.; Nishitani, H.; Miyata, T.; Tanaka, M.; Nishimoto, T. When overexpressed, a novel centrosomal protein, RanBPM, causes ectopic microtubule nucleation similar to gamma-tubulin. *J. Cell Biol.* **1998**, *143*, 1041–1052. [CrossRef] [PubMed]
127. Yokoyama, N.; Hayashi, N.; Seki, T.; Pante, N.; Ohba, T.; Nishii, K.; Kuma, K.; Hayashida, T.; Miyata, T.; Aebi, U.; et al. A giant nucleopore protein that binds Ran/TC4. *Nature* **1995**, *376*, 184–188. [CrossRef] [PubMed]
128. Carazo-Salas, R.E.; Guarguaglini, G.; Gruss, O.J.; Segref, A.; Karsenti, E.; Mattaj, I.W. Generation of GTP-bound Ran by RCC1 is required for chromatin-induced mitotic spindle formation. *Nature* **1999**, *400*, 178–181. [PubMed]
129. Kalab, P.; Pu, R.T.; Dasso, M. The ran GTPase regulates mitotic spindle assembly. *Curr. Biol.* **1999**, *9*, 481–484. [CrossRef]

130. Ohba, T.; Nakamura, M.; Nishitani, H.; Nishimoto, T. Self-organization of microtubule asters induced in *Xenopus* egg extracts by GTP-bound Ran. *Science* **1999**, *284*, 1356–1358. [CrossRef] [PubMed]
131. Wilde, A.; Zheng, Y. Stimulation of microtubule aster formation and spindle assembly by the small GTPase Ran. *Science* **1999**, *284*, 1359–1362. [CrossRef] [PubMed]
132. Zhang, C.; Hughes, M.; Clarke, P.R. Ran-GTP stabilises microtubule asters and inhibits nuclear assembly in *Xenopus* egg extracts. *J. Cell Sci.* **1999**, *112*, 2453–2461. [PubMed]
133. Nachury, M.V.; Maresca, T.J.; Salmon, W.C.; Waterman-Storer, C.M.; Heald, R.; Weis, K. Importin beta is a mitotic target of the small GTPase Ran in spindle assembly. *Cell* **2001**, *104*, 95–106. [CrossRef]
134. Gruss, O.J.; Carazo-Salas, R.E.; Schatz, C.A.; Guarguaglini, G.; Kast, J.; Wilm, M.; Le Bot, N.; Vernos, I.; Karsenti, E.; Mattaj, I.W. Ran induces spindle assembly by reversing the inhibitory effect of importin alpha on TPX2 activity. *Cell* **2001**, *104*, 83–93. [CrossRef]
135. Melchior, F.; Gerace, L. Mechanisms of nuclear protein import. *Curr. Opin. Cell Biol.* **1995**, *7*, 310–318. [CrossRef]
136. Wiese, C.; Wilde, A.; Moore, M.S.; Adam, S.A.; Merdes, A.; Zheng, Y. Role of importin-beta in coupling Ran to downstream targets in microtubule assembly. *Science* **2001**, *291*, 653–656. [CrossRef] [PubMed]
137. Sillje, H.H.; Nagel, S.; Korner, R.; Nigg, E.A. HURP is a Ran-importin beta-regulated protein that stabilizes kinetochore microtubules in the vicinity of chromosomes. *Curr. Biol.* **2006**, *16*, 731–742. [CrossRef] [PubMed]
138. Ribbeck, K.; Groen, A.C.; Santarella, R.; Bohnsack, M.T.; Raemaekers, T.; Kocher, T.; Gentzel, M.; Gorlich, D.; Wilm, M.; Carmeliet, G.; et al. NuSAP, a mitotic RanGTP target that stabilizes and cross-links microtubules. *Mol. Biol. Cell* **2006**, *17*, 2646–2660. [CrossRef] [PubMed]
139. Koffa, M.D.; Casanova, C.M.; Santarella, R.; Kocher, T.; Wilm, M.; Mattaj, I.W. HURP is part of a Ran-dependent complex involved in spindle formation. *Curr. Biol.* **2006**, *16*, 743–754. [CrossRef] [PubMed]
140. Ems-McClung, S.C.; Zheng, Y.; Walczak, C.E. Importin alpha/beta and Ran-GTP regulate XCTK2 microtubule binding through a bipartite nuclear localization signal. *Mol. Biol. Cell* **2004**, *15*, 46–57. [CrossRef] [PubMed]
141. Trieselmann, N.; Armstrong, S.; Rauw, J.; Wilde, A. Ran modulates spindle assembly by regulating a subset of TPX2 and Kid activities including Aurora A activation. *J. Cell Sci.* **2003**, *116*, 4791–4798. [CrossRef] [PubMed]
142. Blower, M.D.; Nachury, M.; Heald, R.; Weis, K. A Rae1-containing ribonucleoprotein complex is required for mitotic spindle assembly. *Cell* **2005**, *121*, 223–234. [CrossRef] [PubMed]
143. Yokoyama, H.; Gruss, O.J.; Rybina, S.; Caudron, M.; Schelder, M.; Wilm, M.; Mattaj, I.W.; Karsenti, E. Cdk11 is a RanGTP-dependent microtubule stabilization factor that regulates spindle assembly rate. *J. Cell Biol.* **2008**, *180*, 867–875. [CrossRef] [PubMed]
144. Tsai, M.Y.; Wiese, C.; Cao, K.; Martin, O.; Donovan, P.; Ruderman, J.; Prigent, C.; Zheng, Y. A Ran signalling pathway mediated by the mitotic kinase Aurora A in spindle assembly. *Nat. Cell Biol.* **2003**, *5*, 242–248. [CrossRef] [PubMed]
145. Kalab, P.; Weis, K.; Heald, R. Visualization of a Ran-GTP gradient in interphase and mitotic *Xenopus* egg extracts. *Science* **2002**, *295*, 2452–2456. [CrossRef] [PubMed]
146. Caudron, M.; Bunt, G.; Bastiaens, P.; Karsenti, E. Spatial coordination of spindle assembly by chromosome-mediated signaling gradients. *Science* **2005**, *309*, 1373–1376. [CrossRef] [PubMed]
147. Kalab, P.; Pralle, A.; Isacoff, E.Y.; Heald, R.; Weis, K. Analysis of a RanGTP-regulated gradient in mitotic somatic cells. *Nature* **2006**, *440*, 697–701. [CrossRef] [PubMed]
148. Athale, C.A.; Dinarina, A.; Mora-Coral, M.; Pugieux, C.; Nedelec, F.; Karsenti, E. Regulation of microtubule dynamics by reaction cascades around chromosomes. *Science* **2008**, *322*, 1243–1247. [CrossRef] [PubMed]
149. Oh, D.; Yu, C.H.; Needleman, D.J. Spatial organization of the Ran pathway by microtubules in mitosis. *Proc. Natl. Acad. Sci. USA* **2016**, *113*, 8729–8734. [CrossRef] [PubMed]
150. Karsenti, E.; Verde, F.; Felix, M.A. Role of type 1 and type 2A protein phosphatases in the cell cycle. *Adv. Protein Phosphorus* **1991**, *6*, 453–482.
151. Andersen, S.S. Balanced regulation of microtubule dynamics during the cell cycle: A contemporary view. *Bioessays* **1999**, *21*, 53–60. [CrossRef]
152. Andersen, S.S.; Ashford, A.J.; Tournebize, R.; Gavet, O.; Sobel, A.; Hyman, A.A.; Karsenti, E. Mitotic chromatin regulates phosphorylation of Stathmin/Op18. *Nature* **1997**, *389*, 640–643. [CrossRef] [PubMed]
153. Budde, P.P.; Kumagai, A.; Dunphy, W.G.; Heald, R. Regulation of Op18 during spindle assembly in *Xenopus* egg extracts. *J. Cell Biol.* **2001**, *153*, 149–158. [CrossRef] [PubMed]
154. Niethammer, P.; Bastiaens, P.; Karsenti, E. Stathmin-tubulin interaction gradients in motile and mitotic cells. *Science* **2004**, *303*, 1862–1866. [CrossRef] [PubMed]

155. Schubart, U.K.; Yu, J.; Amat, J.A.; Wang, Z.; Hoffmann, M.K.; Edelmann, W. Normal development of mice lacking metablastin (P19), a phosphoprotein implicated in cell cycle regulation. *J. Biol. Chem.* **1996**, *271*, 14062–14066. [PubMed]
156. Cassimeris, L. The oncoprotein 18/stathmin family of microtubule destabilizers. *Curr. Opin. Cell Biol.* **2002**, *14*, 18–24. [CrossRef]
157. Carmena, M.; Wheelock, M.; Funabiki, H.; Earnshaw, W.C. The chromosomal passenger complex (CPC): From easy rider to the godfather of mitosis. *Nat. Rev. Mol. Cell Biol.* **2012**, *13*, 789–803. [CrossRef] [PubMed]
158. Gadea, B.B.; Ruderman, J.V. Aurora B is required for mitotic chromatin-induced phosphorylation of Op18/Stathmin. *Proc. Natl. Acad. Sci. USA* **2006**, *103*, 4493–4498. [CrossRef] [PubMed]
159. Sampath, S.C.; Ohi, R.; Leismann, O.; Salic, A.; Pozniakovski, A.; Funabiki, H. The chromosomal passenger complex is required for chromatin-induced microtubule stabilization and spindle assembly. *Cell* **2004**, *118*, 187–202. [CrossRef] [PubMed]
160. Andrews, P.D.; Ovechkina, Y.; Morrice, N.; Wagenbach, M.; Duncan, K.; Wordeman, L.; Swedlow, J.R. Aurora B regulates MCAK at the mitotic centromere. *Dev. Cell* **2004**, *6*, 253–268. [CrossRef]
161. Lan, W.; Zhang, X.; Kline-Smith, S.L.; Rosasco, S.E.; Barrett-Wilt, G.A.; Shabanowitz, J.; Hunt, D.F.; Walczak, C.E.; Stukenberg, P.T. Aurora B phosphorylates centromeric MCAK and regulates its localization and microtubule depolymerization activity. *Curr. Biol.* **2004**, *14*, 273–286. [CrossRef] [PubMed]
162. Ohi, R.; Sapra, T.; Howard, J.; Mitchison, T.J. Differentiation of cytoplasmic and meiotic spindle assembly MCAK functions by Aurora B-dependent phosphorylation. *Mol. Biol. Cell* **2004**, *15*, 2895–2906. [CrossRef] [PubMed]
163. Kelly, A.E.; Sampath, S.C.; Maniar, T.A.; Woo, E.M.; Chait, B.T.; Funabiki, H. Chromosomal enrichment and activation of the aurora B pathway are coupled to spatially regulate spindle assembly. *Dev. Cell* **2007**, *12*, 31–43. [CrossRef] [PubMed]
164. Kuntziger, T.; Gavet, O.; Sobel, A.; Bornens, M. Differential effect of two stathmin/Op18 phosphorylation mutants on *Xenopus* embryo development. *J. Biol. Chem.* **2001**, *276*, 22979–22984. [CrossRef] [PubMed]
165. Tseng, B.S.; Tan, L.; Kapoor, T.M.; Funabiki, H. Dual detection of chromosomes and microtubules by the chromosomal passenger complex drives spindle assembly. *Dev. Cell* **2010**, *18*, 903–912. [CrossRef] [PubMed]
166. Tan, L.; Kapoor, T.M. Examining the dynamics of chromosomal passenger complex (CPC)-dependent phosphorylation during cell division. *Proc. Natl. Acad. Sci. USA* **2011**, *108*, 16675–16680. [CrossRef] [PubMed]
167. Maresca, T.J.; Groen, A.C.; Gatlin, J.C.; Ohi, R.; Mitchison, T.J.; Salmon, E.D. Spindle assembly in the absence of a RanGTP gradient requires localized CPC activity. *Curr. Biol.* **2009**, *19*, 1210–1215. [CrossRef] [PubMed]
168. Tulu, U.S.; Fagerstrom, C.; Ferenz, N.P.; Wadsworth, P. Molecular requirements for kinetochore-associated microtubule formation in mammalian cells. *Curr. Biol.* **2006**, *16*, 536–541. [CrossRef] [PubMed]
169. Di Fiore, B.; Ciciarello, M.; Mangiacasale, R.; Palena, A.; Tassin, A.M.; Cundari, E.; Lavia, P. Mammalian RanBP1 regulates centrosome cohesion during mitosis. *J. Cell Sci.* **2003**, *116*, 3399–3411. [CrossRef] [PubMed]
170. Gruss, O.J.; Wittmann, M.; Yokoyama, H.; Pepperkok, R.; Kufer, T.; Sillje, H.; Karsenti, E.; Mattaj, I.W.; Vernos, I. Chromosome-induced microtubule assembly mediated by TPX2 is required for spindle formation in HeLa cells. *Nat. Cell Biol.* **2002**, *4*, 871–879. [CrossRef] [PubMed]
171. Guarguaglini, G.; Renzi, L.; D'Ottavio, F.; Di Fiore, B.; Casenghi, M.; Cundari, E.; Lavia, P. Regulated Ran-binding protein 1 activity is required for organization and function of the mitotic spindle in mammalian cells in vivo. *Cell Growth Differ.* **2000**, *11*, 455–465. [PubMed]
172. Nishitani, H.; Ohtsubo, M.; Yamashita, K.; Iida, H.; Pines, J.; Yasudo, H.; Shibata, Y.; Hunter, T.; Nishimoto, T. Loss of RCC1, a nuclear DNA-binding protein, uncouples the completion of DNA replication from the activation of cdc2 protein kinase and mitosis. *EMBO J.* **1991**, *10*, 1555–1564. [PubMed]
173. Moutinho-Pereira, S.; Stuurman, N.; Afonso, O.; Hornsveld, M.; Aguiar, P.; Goshima, G.; Vale, R.D.; Maiato, H. Genes involved in centrosome-independent mitotic spindle assembly in *Drosophila* S2 cells. *Proc. Natl. Acad. Sci. USA* **2013**, *110*, 19808–19813. [CrossRef] [PubMed]
174. Wise, D.A.; Brinkley, B.R. Mitosis in cells with unreplicated genomes (MUGs): Spindle assembly and behavior of centromere fragments. *Cell Motil. Cytoskelet.* **1997**, *36*, 291–302. [CrossRef]
175. O'Connell, C.B.; Loncarek, J.; Kalab, P.; Khodjakov, A. Relative contributions of chromatin and kinetochores to mitotic spindle assembly. *J. Cell Biol.* **2009**, *187*, 43–51. [CrossRef] [PubMed]
176. Morris, N.R.; Osmani, S.A.; Engle, D.B.; Doonan, J.H. The genetic analysis of mitosis in Aspergillus nidulans. *Bioessays* **1989**, *10*, 196–201. [CrossRef] [PubMed]

177. Weil, C.F.; Oakley, C.E.; Oakley, B.R. Isolation of mip (microtubule-interacting protein) mutations of Aspergillus nidulans. *Mol. Cell Biol.* **1986**, *6*, 2963–2968. [CrossRef] [PubMed]
178. Murata, T.; Sonobe, S.; Baskin, T.I.; Hyodo, S.; Hasezawa, S.; Nagata, T.; Horio, T.; Hasebe, M. Microtubule-dependent microtubule nucleation based on recruitment of gamma-tubulin in higher plants. *Nat. Cell Biol.* **2005**, *7*, 961–968. [CrossRef] [PubMed]
179. Hughes, S.E.; Beeler, J.S.; Seat, A.; Slaughter, B.D.; Unruh, J.R.; Bauerly, E.; Matthies, H.J.; Hawley, R.S. Gamma-tubulin is required for bipolar spindle assembly and for proper kinetochore microtubule attachments during prometaphase I in Drosophila oocytes. *PLoS Genet.* **2011**, *7*, e1002209. [CrossRef] [PubMed]
180. Mishra, R.K.; Chakraborty, P.; Arnaoutov, A.; Fontoura, B.M.; Dasso, M. The Nup107-160 complex and gamma-TuRC regulate microtubule polymerization at kinetochores. *Nat. Cell Biol.* **2010**, *12*, 164–169. [CrossRef] [PubMed]
181. Endow, S.A.; Hallen, M.A. Anastral spindle assembly and gamma-tubulin in *Drosophila* oocytes. *BMC Cell Biol.* **2011**. [CrossRef] [PubMed]
182. Wiese, C.; Zheng, Y. A new function for the gamma-tubulin ring complex as a microtubule minus-end cap. *Nat. Cell Biol.* **2000**, *2*, 358–364. [CrossRef] [PubMed]
183. Vinh, D.B.; Kern, J.W.; Hancock, W.O.; Howard, J.; Davis, T.N. Reconstitution and characterization of budding yeast gamma-tubulin complex. *Mol. Biol. Cell* **2002**, *13*, 1144–1157. [CrossRef] [PubMed]
184. Wieczorek, M.; Bechstedt, S.; Chaaban, S.; Brouhard, G.J. Microtubule-associated proteins control the kinetics of microtubule nucleation. *Nat. Cell Biol.* **2015**, *17*, 907–916. [CrossRef] [PubMed]
185. Chretien, D.; Fuller, S.D.; Karsenti, E. Structure of growing microtubule ends: Two-dimensional sheets close into tubes at variable rates. *J. Cell Biol.* **1995**, *129*, 1311–1328. [CrossRef] [PubMed]
186. Roostalu, J.; Cade, N.I.; Surrey, T. Complementary activities of TPX2 and chTOG constitute an efficient importin-regulated microtubule nucleation module. *Nat. Cell Biol.* **2015**, *17*, 1422–1434. [CrossRef] [PubMed]
187. Groen, A.C.; Maresca, T.J.; Gatlin, J.C.; Salmon, E.D.; Mitchison, T.J. Functional overlap of microtubule assembly factors in chromatin-promoted spindle assembly. *Mol. Biol. Cell* **2009**, *20*, 2766–2773. [CrossRef] [PubMed]
188. Luders, J.; Patel, U.K.; Stearns, T. GCP-WD is a gamma-tubulin targeting factor required for centrosomal and chromatin-mediated microtubule nucleation. *Nat. Cell Biol.* **2006**, *8*, 137–147. [CrossRef] [PubMed]
189. Buster, D.; McNally, K.; McNally, F.J. Katanin inhibition prevents the redistribution of gamma-tubulin at mitosis. *J. Cell Sci.* **2002**, *115*, 1083–1092. [PubMed]
190. Goshima, G.; Mayer, M.; Zhang, N.; Stuurman, N.; Vale, R.D. Augmin: A protein complex required for centrosome-independent microtubule generation within the spindle. *J. Cell Biol.* **2008**, *181*, 421–429. [CrossRef] [PubMed]
191. Uehara, R.; Nozawa, R.S.; Tomioka, A.; Petry, S.; Vale, R.D.; Obuse, C.; Goshima, G. The augmin complex plays a critical role in spindle microtubule generation for mitotic progression and cytokinesis in human cells. *Proc. Natl. Acad. Sci. USA* **2009**, *106*, 6998–7003. [CrossRef] [PubMed]
192. Lawo, S.; Bashkurov, M.; Mullin, M.; Ferreria, M.G.; Kittler, R.; Habermann, B.; Tagliaferro, A.; Poser, I.; Hutchins, J.R.; Hegemann, B.; et al. HAUS, the 8-subunit human Augmin complex, regulates centrosome and spindle integrity. *Curr. Biol.* **2009**, *19*, 816–826. [CrossRef] [PubMed]
193. Pollard, T.D. Regulation of actin filament assembly by Arp2/3 complex and formins. *Annu. Rev. Biophys. Biomol. Struct.* **2007**, *36*, 451–477. [CrossRef] [PubMed]
194. Kamasaki, T.; O'Toole, E.; Kita, S.; Osumi, M.; Usukura, J.; McIntosh, J.R.; Goshima, G. Augmin-dependent microtubule nucleation at microtubule walls in the spindle. *J. Cell Biol.* **2013**, *202*, 25–33. [CrossRef] [PubMed]
195. Petry, S.; Groen, A.C.; Ishihara, K.; Mitchison, T.J.; Vale, R.D. Branching microtubule nucleation in *Xenopus* egg extracts mediated by augmin and TPX2. *Cell* **2013**, *152*, 768–777. [CrossRef] [PubMed]
196. Du, L.; Xu, J.; Li, X.; Ma, N.; Liu, Y.; Peng, J.; Osato, M.; Zhang, W.; Wen, Z. Rumba and Haus3 are essential factors for the maintenance of hematopoietic stem/progenitor cells during zebrafish hematopoiesis. *Development* **2011**, *138*, 619–629. [CrossRef] [PubMed]
197. Wainman, A.; Buster, D.W.; Duncan, T.; Metz, J.; Ma, A.; Sharp, D.; Wakefield, J.G. A new Augmin subunit, Msd1, demonstrates the importance of mitotic spindle-templated microtubule nucleation in the absence of functioning centrosomes. *Genes Dev.* **2009**, *23*, 1876–1881. [CrossRef] [PubMed]

198. Meireles, A.M.; Fisher, K.H.; Colombie, N.; Wakefield, J.G.; Ohkura, H. Wac: A new Augmin subunit required for chromosome alignment but not for acentrosomal microtubule assembly in female meiosis. *J. Cell Biol.* **2009**, *184*, 777–784. [CrossRef] [PubMed]
199. Edzuka, T.; Yamada, L.; Kanamaru, K.; Sawada, H.; Goshima, G. Identification of the augmin complex in the filamentous fungus Aspergillus nidulans. *PLoS ONE* **2014**, *9*, e101471. [CrossRef] [PubMed]
200. Watanabe, S.; Shioi, G.; Furuta, Y.; Goshima, G. Intra-spindle Microtubule Assembly Regulates Clustering of Microtubule-Organizing Centers during Early Mouse Development. *Cell Rep.* **2016**, *15*, 54–60. [CrossRef] [PubMed]
201. Leber, B.; Maier, B.; Fuchs, F.; Chi, J.; Riffel, P.; Anderhub, S.; Wagner, L.; Ho, A.D.; Salisbury, J.L.; Boutros, M.; Kramer, A. Proteins required for centrosome clustering in cancer cells. *Sci. Transl. Med.* **2010**. [CrossRef] [PubMed]
202. Kleylein-Sohn, J.; Pollinger, B.; Ohmer, M.; Hofmann, F.; Nigg, E.A.; Hemmings, B.A.; Wartmann, M. Acentrosomal spindle organization renders cancer cells dependent on the kinesin HSET. *J. Cell Sci.* **2012**, *125*, 5391–5402. [CrossRef] [PubMed]
203. Hsia, K.C.; Wilson-Kubalek, E.M.; Dottore, A.; Hao, Q.; Tsai, K.L.; Forth, S.; Shimamoto, Y.; Milligan, R.A.; Kapoor, T.M. Reconstitution of the augmin complex provides insights into its architecture and function. *Nat. Cell Biol.* **2014**, *16*, 852–863. [CrossRef] [PubMed]
204. Wittmann, T.; Boleti, H.; Antony, C.; Karsenti, E.; Vernos, I. Localization of the kinesin-like protein Xklp2 to spindle poles requires a leucine zipper, a microtubule-associated protein, and dynein. *J. Cell Biol.* **1998**, *143*, 673–685. [CrossRef] [PubMed]
205. Wadsworth, P. Tpx2. *Curr. Biol.* **2015**, *25*, R1156–R1158. [CrossRef] [PubMed]
206. Wittmann, T.; Wilm, M.; Karsenti, E.; Vernos, I. TPX2, A novel *Xenopus* MAP involved in spindle pole organization. *J. Cell Biol.* **2000**, *149*, 1405–1418. [CrossRef] [PubMed]
207. Giesecke, A.; Stewart, M. Novel binding of the mitotic regulator TPX2 (target protein for *Xenopus* kinesin-like protein 2) to importin-alpha. *J. Biol. Chem.* **2010**, *285*, 17628–17635. [CrossRef] [PubMed]
208. Schatz, C.A.; Santarella, R.; Hoenger, A.; Karsenti, E.; Mattaj, I.W.; Gruss, O.J.; Carazo-Salas, R.E. Importin alpha-regulated nucleation of microtubules by TPX2. *EMBO J.* **2003**, *22*, 2060–2070. [CrossRef] [PubMed]
209. Maurer, S.P.; Fourniol, F.J.; Bohner, G.; Moores, C.A.; Surrey, T. EBs recognize a nucleotide-dependent structural cap at growing microtubule ends. *Cell* **2012**, *149*, 371–382. [CrossRef] [PubMed]
210. Howard, J.; Hyman, A.A. Microtubule polymerases and depolymerases. *Curr. Opin. Cell Biol.* **2007**, *19*, 31–35. [CrossRef] [PubMed]
211. Tsai, M.Y.; Zheng, Y. Aurora A kinase-coated beads function as microtubule-organizing centers and enhance RanGTP-induced spindle assembly. *Curr. Biol.* **2005**, *15*, 2156–2163. [CrossRef] [PubMed]
212. McIntosh, J.R.; Helper, P.K.; Van Wie, D.G. Model for Mitosis. *Nature* **1969**, *224*, 659–663. [CrossRef]
213. Hoyt, M.A.; Geiser, J.R. Genetic analysis of the mitotic spindle. *Annu. Rev. Genet.* **1996**, *30*, 7–33. [CrossRef] [PubMed]
214. Saunders, W.S.; Hoyt, M.A. Kinesin-related proteins required for structural integrity of the mitotic spindle. *Cell* **1992**, *70*, 451–458. [CrossRef]
215. Walczak, C.E.; Heald, R. Mechanisms of mitotic spindle assembly and function. *Int. Rev. Cytol.* **2008**, *265*, 111–158. [PubMed]
216. Karsenti, E.; Vernos, I. The mitotic spindle: A self-made machine. *Science* **2001**, *294*, 543–547. [CrossRef] [PubMed]
217. Goshima, G.; Scholey, J.M. Control of mitotic spindle length. *Annu. Rev. Cell Dev. Biol.* **2010**, *26*, 21–57. [CrossRef] [PubMed]
218. Cole, D.G.; Saxton, W.M.; Sheehan, K.B.; Scholey, J.M. A "slow" homotetrameric kinesin-related motor protein purified from *Drosophila* embryos. *J. Biol. Chem.* **1994**, *269*, 22913–22916. [PubMed]
219. Gordon, D.M.; Roof, D.M. The kinesin-related protein Kip1p of *Saccharomyces cerevisiae* is bipolar. *J. Biol. Chem.* **1999**, *274*, 28779–28786. [CrossRef] [PubMed]
220. Kashina, A.S.; Baskin, R.J.; Cole, D.G.; Wedaman, K.P.; Saxton, W.M.; Scholey, J.M. A bipolar kinesin. *Nature* **1996**, *379*, 270–272. [CrossRef] [PubMed]
221. Scholey, J.E.; Nithianantham, S.; Scholey, J.M.; Al-Bassam, J. Structural basis for the assembly of the mitotic motor Kinesin-5 into bipolar tetramers. *Elife* **2014**. [CrossRef] [PubMed]

222. Tao, L.; Mogilner, A.; Civelekoglu-Scholey, G.; Wollman, R.; Evans, J.; Stahlberg, H.; Scholey, J.M. A homotetrameric kinesin-5, KLP61F, bundles microtubules and antagonizes Ncd in motility assays. *Curr. Biol.* **2006**, *16*, 2293–2302. [CrossRef] [PubMed]
223. Kapitein, L.C.; Peterman, E.J.; Kwok, B.H.; Kim, J.H.; Kapoor, T.M.; Schmidt, C.F. The bipolar mitotic kinesin Eg5 moves on both microtubules that it crosslinks. *Nature* **2005**, *435*, 114–118. [CrossRef] [PubMed]
224. Hentrich, C.; Surrey, T. Microtubule organization by the antagonistic mitotic motors kinesin-5 and kinesin-14. *J. Cell Biol.* **2010**, *189*, 465–480. [CrossRef] [PubMed]
225. Van den Wildenberg, S.M.; Tao, L.; Kapitein, L.C.; Schmidt, C.F.; Scholey, J.M.; Peterman, E.J. The homotetrameric kinesin-5 KLP61F preferentially crosslinks microtubules into antiparallel orientations. *Curr. Biol.* **2008**, *18*, 1860–1864. [CrossRef] [PubMed]
226. Kwok, B.H.; Kapitein, L.C.; Kim, J.H.; Peterman, E.J.; Schmidt, C.F.; Kapoor, T.M. Allosteric inhibition of kinesin-5 modulates its processive directional motility. *Nat. Chem. Biol.* **2006**, *2*, 480–485. [CrossRef] [PubMed]
227. Kapitein, L.C.; Kwok, B.H.; Weinger, J.S.; Schmidt, C.F.; Kapoor, T.M.; Peterman, E.J. Microtubule cross-linking triggers the directional motility of kinesin-5. *J. Cell Biol.* **2008**, *182*, 421–428. [CrossRef] [PubMed]
228. Weinger, J.S.; Qiu, M.; Yang, G.; Kapoor, T.M. A nonmotor microtubule binding site in kinesin-5 is required for filament crosslinking and sliding. *Curr. Biol.* **2011**, *21*, 154–160. [CrossRef] [PubMed]
229. Shimamoto, Y.; Forth, S.; Kapoor, T.M. Measuring Pushing and Braking Forces Generated by Ensembles of Kinesin-5 Crosslinking Two Microtubules. *Dev. Cell* **2015**, *34*, 669–681. [CrossRef] [PubMed]
230. Boleti, H.; Karsenti, E.; Vernos, I. Xklp2, a novel *Xenopus* centrosomal kinesin-like protein required for centrosome separation during mitosis. *Cell* **1996**, *84*, 49–59. [CrossRef]
231. Sturgill, E.G.; Ohi, R. Kinesin-12 differentially affects spindle assembly depending on its microtubule substrate. *Curr. Biol.* **2013**, *23*, 1280–1290. [CrossRef] [PubMed]
232. Zhu, C.; Zhao, J.; Bibikova, M.; Leverson, J.D.; Bossy-Wetzel, E.; Fan, J.B.; Abraham, R.T.; Jiang, W. Functional analysis of human microtubule-based motor proteins, the kinesins and dyneins, in mitosis/cytokinesis using RNA interference. *Mol. Biol. Cell* **2005**, *16*, 3187–3199. [CrossRef] [PubMed]
233. Segbert, C.; Barkus, R.; Powers, J.; Strome, S.; Saxton, W.M.; Bossinger, O. KLP-18, a Klp2 kinesin, is required for assembly of acentrosomal meiotic spindles in *Caenorhabditis elegans*. *Mol. Biol. Cell* **2003**, *14*, 4458–4469. [CrossRef] [PubMed]
234. Tanenbaum, M.E.; Macurek, L.; Janssen, A.; Geers, E.F.; Alvarez-Fernandez, M.; Medema, R.H. Kif15 cooperates with eg5 to promote bipolar spindle assembly. *Curr. Biol.* **2009**, *19*, 1703–1711. [CrossRef] [PubMed]
235. Vanneste, D.; Takagi, M.; Imamoto, N.; Vernos, I. The role of Hklp2 in the stabilization and maintenance of spindle bipolarity. *Curr. Biol.* **2009**, *19*, 1712–1717. [CrossRef] [PubMed]
236. Tanenbaum, M.E.; Vale, R.D.; McKenney, R.J. Cytoplasmic dynein crosslinks and slides anti-parallel microtubules using its two motor domains. *Elife* **2013**. [CrossRef] [PubMed]
237. Sturgill, E.G.; Norris, S.R.; Guo, Y.; Ohi, R. Kinesin-5 inhibitor resistance is driven by kinesin-12. *J. Cell Biol.* **2016**, *213*, 213–227. [CrossRef] [PubMed]
238. Sturgill, E.G.; Das, D.K.; Takizawa, Y.; Shin, Y.; Collier, S.E.; Ohi, M.D.; Hwang, W.; Lang, M.J.; Ohi, R. Kinesin-12 Kif15 targets kinetochore fibers through an intrinsic two-step mechanism. *Curr. Biol.* **2014**, *24*, 2307–2313. [CrossRef] [PubMed]
239. Drechsler, H.; McHugh, T.; Singleton, M.R.; Carter, N.J.; McAinsh, A.D. The Kinesin-12 Kif15 is a processive track-switching tetramer. *Elife* **2014**. [CrossRef] [PubMed]
240. Drechsler, H.; McAinsh, A.D. Kinesin-12 motors cooperate to suppress microtubule catastrophes and drive the formation of parallel microtubule bundles. *Proc. Natl. Acad. Sci. USA* **2016**, *113*, E1635–E1644. [CrossRef] [PubMed]
241. Chandra, R.; Salmon, E.D.; Erickson, H.P.; Lockhart, A.; Endow, S.A. Structural and functional domains of the *Drosophila* Ncd microtubule motor protein. *J. Biol. Chem.* **1993**, *268*, 9005–9013. [PubMed]
242. Karabay, A.; Walker, R.A. Identification of microtubule binding sites in the Ncd tail domain. *Biochemistry* **1999**, *38*, 1838–1849. [CrossRef] [PubMed]
243. McDonald, H.B.; Stewart, R.J.; Goldstein, L.S. The kinesin-like ncd protein of *Drosophila* is a minus end-directed microtubule motor. *Cell* **1990**, *63*, 1159–1165. [CrossRef]
244. Fink, G.; Hajdo, L.; Skowronek, K.J.; Reuther, C.; Kasprzak, A.A.; Diez, S. The mitotic kinesin-14 Ncd drives directional microtubule-microtubule sliding. *Nat. Cell Biol.* **2009**, *11*, 717–723. [CrossRef] [PubMed]

245. Braun, M.; Drummond, D.R.; Cross, R.A.; McAinsh, A.D. The kinesin-14 Klp2 organizes microtubules into parallel bundles by an ATP-dependent sorting mechanism. *Nat. Cell Biol.* **2009**, *11*, 724–730. [CrossRef] [PubMed]
246. Erzberger, J.P.; Berger, J.M. Evolutionary relationships and structural mechanisms of AAA+ proteins. *Annu. Rev. Biophys. Biomol. Struct.* **2006**, *35*, 93–114. [CrossRef] [PubMed]
247. Bhabha, G.; Johnson, G.T.; Schroeder, C.M.; Vale, R.D. How Dynein Moves Along Microtubules. *Trends Biochem. Sci.* **2016**, *41*, 94–105. [CrossRef] [PubMed]
248. Hook, P.; Vallee, R.B. The dynein family at a glance. *J. Cell Sci.* **2006**, *119*, 4369–4371. [CrossRef] [PubMed]
249. Schmidt, H.; Carter, A.P. Review: Structure and mechanism of the dynein motor ATPase. *Biopolymers* **2016**, *105*, 557–567. [CrossRef] [PubMed]
250. Merdes, A.; Ramyar, K.; Vechio, J.D.; Cleveland, D.W. A complex of NuMA and cytoplasmic dynein is essential for mitotic spindle assembly. *Cell* **1996**, *87*, 447–458. [CrossRef]
251. Heald, R.; Tournebize, R.; Habermann, A.; Karsenti, E.; Hyman, A. Spindle assembly in *Xenopus* egg extracts: Respective roles of centrosomes and microtubule self-organization. *J. Cell Biol.* **1997**, *138*, 615–628. [CrossRef] [PubMed]
252. Compton, D.A. Focusing on spindle poles. *J. Cell Sci.* **1998**, *111*, 1477–1481. [PubMed]
253. Tanenbaum, M.E.; Macurek, L.; Galjart, N.; Medema, R.H. Dynein, Lis1 and CLIP-170 counteract Eg5-dependent centrosome separation during bipolar spindle assembly. *EMBO J.* **2008**, *27*, 3235–3245. [CrossRef] [PubMed]
254. Mitchison, T.J.; Maddox, P.; Gaetz, J.; Groen, A.; Shirasu, M.; Desai, A.; Salmon, E.D.; Kapoor, T.M. Roles of polymerization dynamics, opposed motors, and a tensile element in governing the length of *Xenopus* extract meiotic spindles. *Mol. Biol. Cell* **2005**, *16*, 3064–3076. [CrossRef] [PubMed]
255. Ferenz, N.P.; Paul, R.; Fagerstrom, C.; Mogilner, A.; Wadsworth, P. Dynein antagonizes eg5 by crosslinking and sliding antiparallel microtubules. *Curr. Biol.* **2009**, *19*, 1833–1838. [CrossRef] [PubMed]
256. Toba, S.; Toyoshima, Y.Y. Dissociation of double-headed cytoplasmic dynein into single-headed species and its motile properties. *Cell Motil. Cytoskelet.* **2004**, *58*, 281–289. [CrossRef] [PubMed]
257. Amos, L.A. Brain dynein crossbridges microtubules into bundles. *J. Cell Sci.* **1989**, *93*, 19–28. [PubMed]
258. Muller, M.J.; Klumpp, S.; Lipowsky, R. Tug-of-war as a cooperative mechanism for bidirectional cargo transport by molecular motors. *Proc. Natl. Acad. Sci. USA* **2008**, *105*, 4609–4614. [CrossRef] [PubMed]
259. Grill, S.W.; Kruse, K.; Julicher, F. Theory of mitotic spindle oscillations. *Phys. Rev. Lett.* **2005**. [CrossRef] [PubMed]
260. Badoual, M.; Julicher, F.; Prost, J. Bidirectional cooperative motion of molecular motors. *Proc. Natl. Acad. Sci. USA* **2002**, *99*, 6696–6701. [CrossRef] [PubMed]
261. Vale, R.D.; Malik, F.; Brown, D. Directional instability of microtubule transport in the presence of kinesin and dynein, two opposite polarity motor proteins. *J. Cell Biol.* **1992**, *119*, 1589–1596. [CrossRef] [PubMed]

biology

MDPI

Review

A Molecular View of Kinetochore Assembly and Function

Andrea Musacchio [1,2,*] and Arshad Desai [3,4,*]

1 Department of Mechanistic Cell Biology, Max Planck Institute of Molecular Physiology, Otto-Hahn Straße 11, Dortmund 44227, Germany
2 Centre for Medical Biotechnology, Faculty of Biology, University Duisburg-Essen, Essen 45117, Germany
3 Ludwig Institute for Cancer Research, La Jolla, CA 92093, USA
4 Department of Cellular & Molecular Medicine, 9500 Gilman Dr., La Jolla, CA 92093, USA
* Correspondence: andrea.musacchio@mpi-dortmund.mpg.de (A.M.); abdesai@ucsd.edu (A.D.); Tel.: +49-231-133-2101 (A.M.); +1-858-534-9698 (A.D.); Fax: +49-231-133-2199 (A.M.)

Academic Editor: J. Richard McIntosh
Received: 13 December 2016; Accepted: 17 January 2017; Published: 24 January 2017

Abstract: Kinetochores are large protein assemblies that connect chromosomes to microtubules of the mitotic and meiotic spindles in order to distribute the replicated genome from a mother cell to its daughters. Kinetochores also control feedback mechanisms responsible for the correction of incorrect microtubule attachments, and for the coordination of chromosome attachment with cell cycle progression. Finally, kinetochores contribute to their own preservation, across generations, at the specific chromosomal loci devoted to host them, the centromeres. They achieve this in most species by exploiting an epigenetic, DNA-sequence-independent mechanism; notable exceptions are budding yeasts where a specific sequence is associated with centromere function. In the last 15 years, extensive progress in the elucidation of the composition of the kinetochore and the identification of various physical and functional modules within its substructure has led to a much deeper molecular understanding of kinetochore organization and the origins of its functional output. Here, we provide a broad summary of this progress, focusing primarily on kinetochores of humans and budding yeast, while highlighting work from other models, and present important unresolved questions for future studies.

Keywords: centromere; kinetochore; cell division; mitosis; meiosis; KMN; CCAN; CENP-A

---

1. An Overview of Kinetochore Structure and Functions

In eukaryotes, the kinetochore is a proteinaceous multi-subunit assembly whose main function is to generate load-bearing attachments of sister chromatids (the replicated chromosomes held together by the protein complex cohesin) to spindle microtubules during cell division (mitosis or meiosis) (Figure 1A). Kinetochores couple sister chromatids to dynamic microtubules during congression and anaphase, allowing their separation and partition to the daughter cells [1–3].

Kinetochores assemble on a specialized chromatin locus named the centromere (which, when large enough to be observed, coincides with the primary constriction on chromosome spreads in karyotype analysis) [4–6]. Even if the name 'centromere' implies a position at the center of the chromosome, centromeres in different organisms can occupy very different positions, and are generally defined as metacentric (if they are in the middle of the chromosome), acrocentric (if they separate chromosome arms of different length), or telocentric (if they are positioned very close to a chromosome's end). In organisms, such as nematodes, several insects, and lower plants, centromeres extend along the entire length of the chromosome (so-called holocentric centromeres, in opposition to spatially delimited monocentric centromeres). The size of the chromosome segment required to assemble a functional kinetochore varies wildly from species to species, from ~125 base pairs (bps) in *Saccharomyces cerevisiae* to one or more million bps in humans. Most centromeres are defined by a specific chromatin signature rather than a specific DNA sequence (with notable exceptions discussed in Section 2). This property is generally referred to as epigenetic specification of centromeres [4–6]. Despite the considerable compositional and positional variety of centromeres, a common molecular architecture is clearly discernible in kinetochores across the eukaryotic world, with the significant known exception of kinetoplastids (see Section 8).

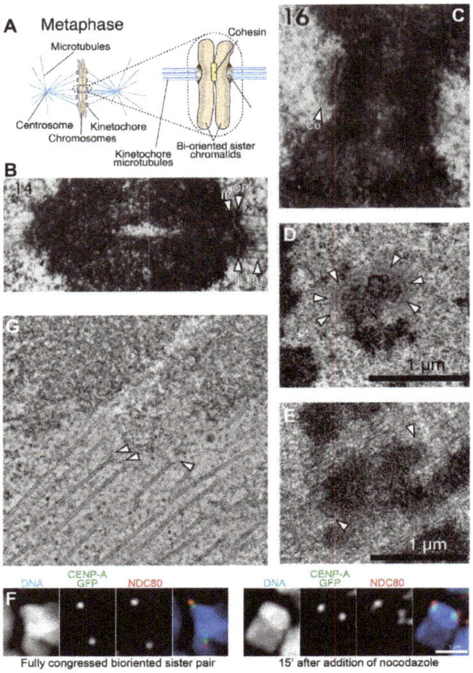

Figure 1. Kinetochore morphology in vertebrate cells (**A**) Schematic showing the attachment of chromosomes to spindle microtubules through kinetochores; (**B**) Early work on the kinetochore identified inner and outer plates, separated by a translucent layer. Microtubules terminate end-on on the kinetochore outer plate. Arrowheads indicate inner plate (IP), outer plate (OP), translucent layer (TL), and kinetochore microtubules (MT). Image reproduced with permission from reference [7]; (**C**) The corona (Co) is a fibrous structure that is more clearly visible on kinetochores prior to microtubule attachment. Image reproduced with permission from reference [7]; (**D**,**E**) Prior to microtubule attachment (**D**), vertebrate kinetochores adopt a crescent-like shape. The latter is not visible on fully congressed and bi-oriented kinetochores. Images courtesy of Alexey Khodjakov. See also reference [8]; (**F**) Left: at metaphase, the distributions of two proteins in the inner and outer kinetochores (NDC80 and CENP-A respectively), are similar; Right: After treatment with a microtubule-depolymerizing drug (nocodazole), proteins in the corona (not shown) and in the outer kinetochore undergo an expansion and form the crescent-like shape already shown in **D**. Image courtesy of Alexey Khodjakov. See also reference [8]; (**G**) A prometaphase PtK2 cell prepared for electron microscopy by high-pressure freezing and freeze-substitution in glutaraldehyde and Osmium tetroxide. The cell was then embedded in plastic, serial-sectioned with 300 nm sections, and imaged by serial tilting. A 3D reconstruction was computed by back-projection, using the IMOD software package. The slice shown here is about 5 nm thick, and represents the average of two consecutive tomographic planes. Arrowheads indicate slender fibrils connecting the end of microtubules to the kinetochore. Image courtesy of J. Richard McIntosh.

Besides mediating interactions with spindle microtubules, kinetochores are mechanosensors that control stability of microtubule attachment to favor the bi-orientation of sister chromatids (or of the bivalents during meiosis), instead of incomplete or incorrect configurations such as mono-orientation, syntelic attachment, or merotelic attachment [9]. This property of kinetochores is generally referred to as error correction, and its molecular basis remains rather poorly understood. The pioneering experiments of Nicklas and colleagues [10], as well as more recent functional analyses [11–13], suggest that the development of tension within the kinetochore or between sister kinetochores contributes to discerning correct attachment from incorrect ones (see chapter by Lampson and Grishchuk, reference [14]). Kinetochores also regulate the spindle assembly checkpoint (SAC, also named the metaphase checkpoint), a feedback mechanism required to couple the initiation of mitotic exit with the completion of sister chromatid bi-orientation [15,16]. The trigger of mitotic exit is the inactivation of mitotic Cyclin-dependent kinase (CDK) activity and the activation of the protease activity that eliminates sister chromatid cohesion. Both processes are regulated by Ubiquitin-dependent proteolysis, and the SAC inhibits this regulated proteolysis to prevent premature mitotic exit in presence of unattached or improperly attached kinetochores (see chapter by Ajit Joglekar, reference [17]). There is overlap between the functions of the SAC as a mechanism to gain time when chromosome attachment is incomplete or erroneous, and the function of the error correction apparatus that aims to favor bi-orientation. Indeed, common molecular machinery regulates these processes, at least at the apex of the pathway.

The ultrastructure of the vertebrate kinetochore is described based on early electron microscopy (EM) studies employing glutaraldehyde fixation that identified kinetochores as trilaminar structures, approximately 250 nm wide and 80 nm deep, with an electron-opaque inner plate juxtaposed to the centromeric chromatin, a translucent gap layer, and an electron-opaque, chromatin-distal outer plate apparently embedding the plus ends of spindle microtubules (defined as end-on attachment, Figure 1B) [7,18]. Furthermore, in the absence of microtubules, a fibrous structure named the corona becomes apparent externally to the outer plate [19–23] (Figure 1C). The corona, which is not morphologically discernable following microtubule binding, triggers a significant expansion of the kinetochore in a crescent-like shape [7,8,24–26] (Figure 1D,E). Studies with improved fixation (high pressure freezing followed by freeze substitution) failed to confirm the existence of a clearly defined trilaminar plate structure in the kinetochore, and have rather redefined the kinetochore as a disordered fibrous mesh in which the plus ends of microtubules are embedded [8,27,28]. Depolymerizing protofilaments of microtubules were shown to establish connections to the kinetochore through slender fibrils [29] (Figure 1F).

A significant limitation in our understanding of kinetochores until the early 2000s was that a molecular description of their architecture was largely missing. The advent of mass spectrometry-based proteomics and functional genomics has led to substantial progress in the identification of kinetochore subunits and sub-complexes, their reconstitution and purification, and their structural characterization at high-resolution by X-ray crystallography and EM [3]. In particular, the structure of most of the components of the outer kinetochore is now known, or can be inferred through cross-linking experiments, and parts of the inner kinetochore are also beginning to be characterized. Below, we first present a brief summary of studies on the centromere and its epigenetic definition. We then review progress toward defining the structural organization of kinetochores, with references to older foundational work and to accompanying chapters in this issue that focus on the functional output of kinetochores, including microtubule-dependent force generation, error correction, and the SAC (see references [14,17,30]). In our discussion of kinetochore structural organization, we will focus on human kinetochores and lessons from recent biochemical reconstitution efforts but will refer to work in other models to highlight important parallels/differences, and also to present questions that emerge from cross-model comparisons.

2. The Centromere

As discussed in the previous section, the centromere is the specialized chromatin region on which kinetochores assemble. The DNA sequence at the centromere, however, varies considerably from organism to organism. The short *cis*-acting DNA segments of *S. cerevisiae* centromeres (usually designated as CENs) have overall conserved sequence features among the 16 chromosomes and are sufficient for kinetochore assembly [31–35]. This type of centromere, found in *S. cerevisiae* and related fungi, is referred to as a point centromere [36] (Figure 2A–C). The complexity of centromeres in most other organisms, however, vastly exceeds that of the *S. cerevisiae* centromere. In the majority of model systems studied to date, centromeres consist of highly repetitive DNA elements, including retro-transposons or tandem repeat arrays, or combinations of both [4,6,37]. These centromeres span chromosome regions in a range from tens of thousands to millions bps, and have therefore been defined as regional centromeres [36]. For instance, human centromeres consist of a large number of tandem 171-bps repeats, called α-satellite repeats, which extend within domains of ~0.2–4.0 Mbps [38] (Figure 2D–F). Tandem repetitive sequences, unrelated to those in humans, are also identified at centromeres of mice, fission yeast, flies, and plants, among others [4,6,37]. These complex regional centromeres have a central portion where the kinetochore is assembled, flanked by pericentromeric regions that are often also repetitive, heterochromatic in nature and accumulate cohesin complexes.

In contrast to *S. cerevisiae*, it has not been possible to identify, in these larger and more complex centromeres, a univocal relation between the underlying sequence of the centromere and the ability to seed a kinetochore [4,6,37]. For instance, conversion of a non-functional centromere to a functional one on mini-chromosomes can occur in the absence of apparent sequence, structural, or chemical changes in *S. pombe* [39]. Stably inherited dicentric chromosomes (chromosomes with two distinct repeat arrays normally associated with centromere function) invariably show inactivation of one of the two predicted centromeres, indicating that the DNA sequence is insufficient to establish the kinetochore [40,41]. On the same line, functional neo-centromeres can form at euchromatic regions of human supernumerary marker chromosomes in the absence of alphoid DNA [42–44], showing that repetitive DNA is not necessary for a functional centromere. Similarly, acentric (i.e., centromere lacking) chromosome fragments produced by irradiation can be transmitted quite faithfully in *D. melanogaster* cells because they acquire neo-centromere activity at non-repetitive sequences [45]. That the presence of repetitive sequences is not an absolute requirement for centromere identity is also supported by the observation that centromeres of several organisms are devoid of them [46–50]. Repetitive sequences, however, are likely to contribute to the stabilization of centromere organization and function. Evolutionarily new centromeres (ENCs), which are initially generated by centromere repositioning in non-repetitive, "gene desert" regions of the genome without additional chromosomal changes, re-acquire repetitive DNA sequences in short evolutionary times [51–53].

Collectively, the observations discussed in the previous paragraph provided the foundation for the idea that centromeres in most organisms are defined epigenetically rather than through specific DNA sequences [54–56]. Over the years, the search for crucial determinants of epigenetic specification of centromere identity has narrowed to CENP-A (also called CenH3) [57,58]. CENP-A is a histone H3 variant [59–61] (Figure 3A). (Specific features of CENP-A that distinguish it from canonical H3 are discussed in Section 2.) With some exceptions [62], CENP-A is present at functional centromeres, from *S. cerevisiae* (where it is called Cse4) to humans [61,63,64]. CENP-A is required for kinetochore recruitment of all other kinetochore components [65–69], and is sufficient to promote kinetochore assembly when targeted artificially to ectopic locations [70–76].

While the association of CENP-A, a crucial epigenetic factor, with the sequence-specific yeast centromere may seem counterintuitive, it is important to note that CENP-A not only serves as an epigenetic mark but also functions as the foundation for kinetochore assembly (as discussed in Section 3), and this latter function is retained by *S. cerevisiae* CENP-A^{Cse4}. In *S. cerevisiae*, the CEN DNA-binding CBF3 complex helps target CENP-A^{Cse4}, making propagation of CENP-A nucleosomes genetically specified and restricted to a specific location. The concept of epigenetic specification in other species relates to the fact that the presence of CENP-A on a defined segment of DNA is (largely) sequence independent, yet extremely stable and self-propagating at that particular locus (the centromere) through multiple cell generations. The molecular basis for this phenomenon is discussed in Section 7.

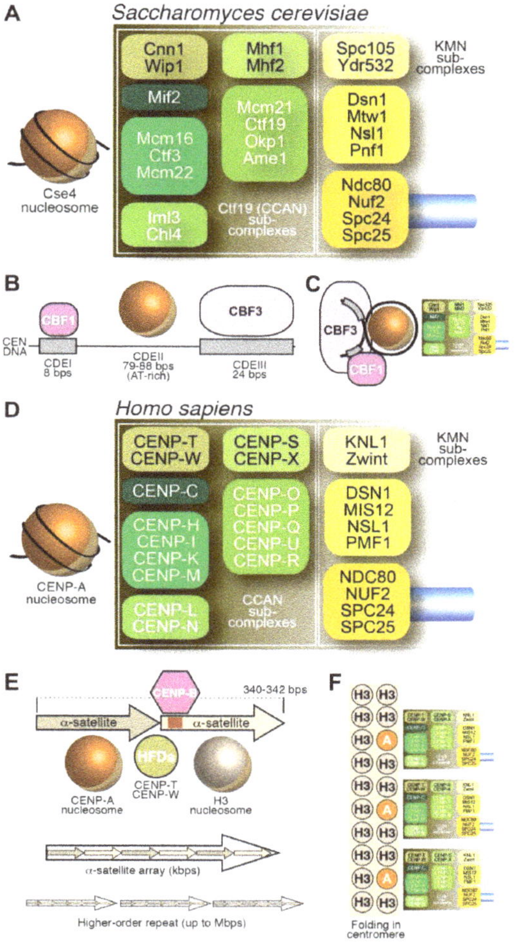

Figure 2. Schematic summary of the structural organization of budding yeast and human kinetochores. Related colors highlight conserved components/complexes. (**A**) Schematic of the *S. cerevisiae* kinetochore with subunit names; (**B**) The *S. cerevisiae* centromere (CEN) DNA is stereotyped and contains CDEI, CDEII, and CDEIII regions, which bind CBF1, Cse4$^{\text{CENP-A}}$, and CBF3, respectively; (**C**) Folding of CEN DNA around a Cse4 nucleosome brings CBF1 and CBF3 in close proximity; (**D**) Schematic of the *H. sapiens* kinetochore. Orthologous complexes are shown in the same order as in (**A**); (**E**) The unit of human centromere assembly may consist of a pair of α-satellite repeats, each precisely wrapping around a nucleosome. One of the two α-satellite repeats carries a CENP-B box. The CENP-TW complex may interact in the inter-nucleosomal region through its histone-fold domain (HFDs) [77,78]. Repeats of this unit give rise to α-satellite arrays, which in turn may organize themselves in higher order repeats (HORs); (**F**) The human centromere arises from folding of centromeric chromatin in three dimensions to facilitate the participation of several CENP-A nucleosomes in kinetochore assembly.

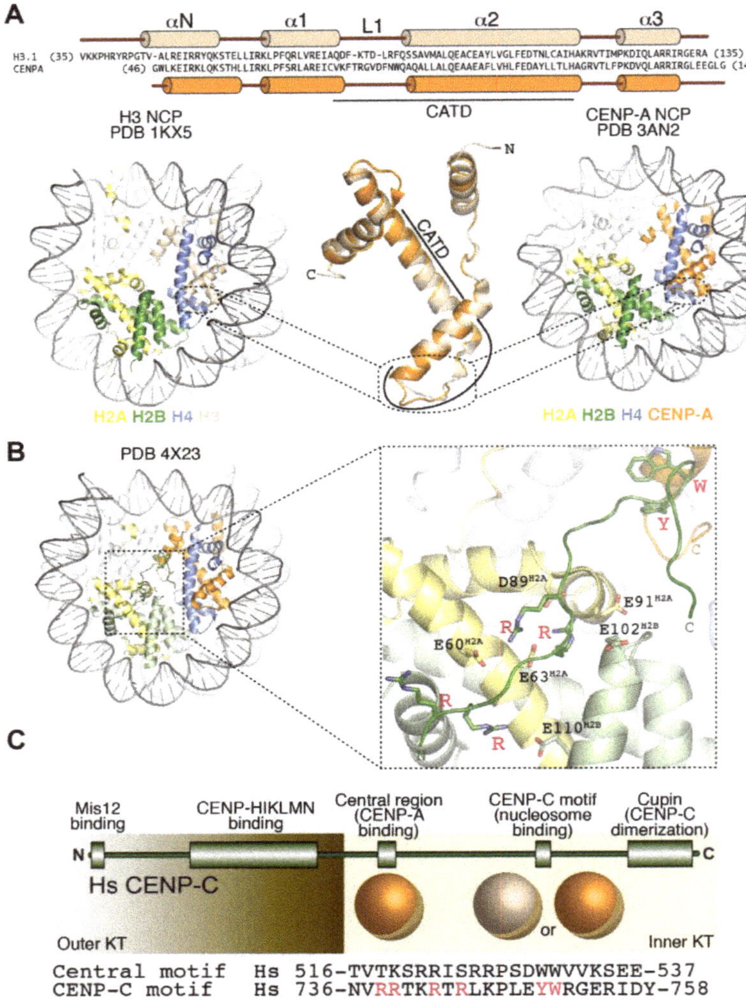

Figure 3. The CENP-A nucleosome and its specific recognition by CENP-C. (**A**) Comparison of H3 and CENP-A primary, secondary, tertiary, and quaternary structure. Sequence and structure changes concentrate in the N-terminal region, in the L1 segment of the CATD, and in the C-terminal region; (**B**) Structure of the complex of the CENP-C motif bound to a nucleosome containing a chimeric histone H3 with grafted hydrophobic C-terminal peptide of CENP-A [79]; (**C**) Scheme illustrating the organization of CENP-C as a "blueprint" for kinetochore assembly along the outer to inner kinetochore axis [80]. The H3 nucleosome structure is from *X. laevis*, the CENP-A nucleosome structure is human, and the CENP-C motif-bound structure has a *Drosophila* nucleosome core particle (in which the human CENP-A tail was grafted onto H3) bound to a rat CENP-C motif.

Why do centromeres in most species rely on an epigenetic identity and vary so significantly in sequence despite their essential role in cell division? And why do active centromeres accumulate repetitive DNA sequences? While definitive answers to these questions remain to be obtained, some current hypotheses are mentioned here in brief. With regards to the first question, the 'centromere drive' hypothesis posits that asymmetry of chromosome segregation in oocytes, where only a quarter of the genome is transmitted to the egg while the rest is discarded in polar bodies, leads to a genetic conflict that drives rapid centromere evolution [81]. This hypothesis has received support from evidence of adaptive evolution in centromere/kinetochore proteins [82,83] and from analysis of centromere activity in mouse strains with variation in centromeric repeats [84]. A different but not mutually exclusive idea is that the foundation for kinetochore assembly requires a chromatin state that is defined largely by architectural instead of sequence constraints, thereby reducing selection pressure to maintain specific sequences. Provided that this chromatin architecture can be inherited through division, such a model would explain centromere variation also in species that lack asymmetric segregation during meiosis. For the question regarding presence of repeats at centromeres, neocentromeres lacking repeats have been shown to be more sensitive to missegregation and to localize lower amounts of the error correction machinery [85], implicating repeats in segregation accuracy. One hypothesis, based on studies in *S. pombe* [86,87] is that repetitive sequences trigger heterochromatin formation, which in turn promotes cohesin complex enrichment that both mechanically strengthens the centromere and promotes localization of error correction machinery. Such a model would link optimal centromere functionality to repeat accumulation, potentially accounting for why repeats have independently accrued at centromeres of divergent species.

3. The Inner Kinetochore

3.1. The CCAN

Despite considerable diversity of centromere organization in different organisms, kinetochores share significant similarity in their biochemical composition in evolution [88,89]. Early inroads into the identification of proteins present at kinetochores were made when sera from patients diagnosed with the autoimmune syndrome CREST (Calcinosis, Reynaud's syndrome, Esophaegal dysmotility, Sclerodactyly, Telangiectasia) detected centromeres in cells [90]. Subsequent work with these anti-centromere antibodies (ACA) let to the identification of three antigens, which were named CENP-A, CENP-B, and CENP-C, where CENP stands for centromeric protein [57,58]. Subsequent work led to the identification of the coding sequence of the polypeptides to which these antigens belonged [61,91,92]. In the following years, additional human CENPs were identified, including CENP-H and CENP-I, the latter related to a previously identified fission yeast protein, Mis6 [93–95]. Subsequent analyses of the CENP-A, CENP-H, and CENP-I associated proteomes in vertebrate cells led to the identification of several new CENPs, including CENP-K, CENP-L, CENP-M, CENP-N, CENP-O, CENP-P, CENP-Q, CENP-R, CENP-S, CENP-T, CENP-U (also known as CENP-50), CENP-W, and CENP-X [96–101]. This set of CENPs is now collectively identified as constitutive centromere associated network (CCAN), a name emphasizing localization of at least a subset of CCAN subunits at centromeres throughout the cell cycle [100,102]. The CCAN proteins localize to the most chromatin-proximal region of the kinetochore [103–106] (Figure 2D).

Biochemical reconstitution and reciprocal dependency for kinetochore recruitment indicate that the majority of the CCAN assembly can be subdivided into 4 discrete entities (Figure 2D): the CENP-LN complex [107–110], the CENP-HIKM complex [80,110–115], the CENP-OPQRU complex [100,115,116] and the CENP-TWSX complex [100,101]. These CCAN sub-complexes are probably constitutive, i.e., the stability of their subunits depends critically on reciprocal interactions in their cognate complex. These building blocks further interact as discussed later in this section.

Most of the CCAN subunits have orthologs in *S. cerevisiae* [107,114,115,117–119], which are collectively identified as the Ctf19 complex (Figure 2A). A notable exception is the 4-subunit CBF3 complex (Figure 2B,C), a cognate binding partner of the CEN DNA of *S. cerevisiae* [34]. CEN DNA contains three major regions of sequence similarity, named CDEI, CDEII, and CDEIII [35]. CDEII, which has an AT content of ~90%, is the binding site for the CENP-A^{Cse4} nucleosome (discussed in more detail in Subsection 3.2) [120–122], whereas CDEI and CDEIII bind respectively to the general transcription factor Cbf1 and to CBF3 [34,123–125]. Furthermore, CBF3 and Cbf1 interact, establishing a bridge between CDEI and CDEIII that contains the CENP-A^{Cse4} nucleosome [126–129]. While the CENP-A^{Cse4} nucleosome may be intrinsically left-handed (see Section 3.2), it has been proposed that CBF3 may configure a right-handed DNA loop [130–132].

Surprisingly, CCAN subunits, with the notable exception of CENP-C, have not been found to date in certain lineages, e.g., in *D. melanogaster* or *C. elegans* and related species [133]. This apparent loss in species that rely on CENP-A-based kinetochores for chromosome segregation, together with variation in the phenotypic impact of removal of CCAN subunits in species where they are present, highlights that much still remains to be understood about the structural and functional contributions of these four CCAN complexes at the kinetochore.

3.2. Structural Organization of the CENP-A Nucleosome

CENP-A retains several properties of histone H3 (Figure 3A). It interacts tightly with histone H4, and is incorporated in vitro and in vivo into canonical octameric nucleosomes with histones H2A and H2B that share many structural features of the canonical H3-containing nucleosomes, including a left-handed DNA writhe [79,134–144]. CENP-A nucleosomes have looser terminal contacts in comparison to H3 nucleosomes and protect a shorter DNA core (~100–120 bps) in nuclease protection assays, a property enhanced by CENP-C binding [77,120,136,141,145–147].

Alternative models for the organization of CENP-A-containing nucleosomes have been proposed in recent years, and readers are referred to comprehensive recent discussions [148,149]. Given the importance that direct recognition of CENP-A plays in kinetochore assembly and stability, understanding the effective organization of the CENP-A nucleosome and its dynamic changes during the cell cycle is of great importance. To date, successful in vitro reconstitution of physical interaction of inner kinetochore proteins with CENP-A has been limited to octameric nucleosomes [79,110,141,150,151]. These high-affinity interactions occur at thermodynamic equilibrium and may account for the remarkable long-term stability of CENP-A in chromatin in vivo [141,152,153]. Thus, thermodynamic stability is a benchmark against which alternative models for the role of CENP-A in kinetochore assembly will have to be tested. This consideration does not detract from the possibility that structural changes in the organization of the CENP-A nucleosome occur during the cell cycle (e.g., during DNA replication) [148].

3.3. Recognition of CENP-A by CCAN Subunits

So far, two CCAN subunits, CENP-C and CENP-N (Figure 2D), have been found to interact directly with CENP-A and exhibits specificity for CENP-A versus H3 nucleosomes [76,79,110,140,150,151]. CENP-N binds directly to the CENP-A centromere-targeting domain (CATD, Figure 3A) of CENP-A [150]. The CATD comprises residues in the $\alpha1$–$\alpha2$ (L1) loop and the $\alpha2$ helix of CENP-A and harbors the highest concentration of sequence differences between CENP-A and H3, with a preponderance of these in the L1 loop, which is also the only solvent-exposed region of the CATD (Figure 3A). The CATD is required for incorporation in centromeric chromatin, and is also sufficient, when grafted onto the equivalent position of H3, for loading of the H3 chimera to centromeres [135,139]. The latter property likely reflects a second requirement of the CATD (besides CENP-N binding), the interaction with a specific CENP-A chaperone required for incorporation of CENP-A into chromatin (see Section 7).

CENP-C, on the other hand, interacts with the acidic patch of H2A and H2B as well as with the divergent C-terminal tail of CENP-A [76,79,140,154] (Figure 3B). Two sequence-related regions of CENP-C, the central region and the CENP-C motif, have been implicated, each on its own right, in the interaction with the CENP-A nucleosome (Figure 3C) [79]. The central region and the CENP-C motif each encompass ~25 residues, and contain several conserved positively charged residues near their N-terminus and two aromatic residues near their C-terminus. The N-terminal positively charged region interacts with the acidic patch of H2A and H2B on the CENP-A nucleosome, a region that has been implicated in the interaction of canonical H3 nucleosomes with different target proteins [155–157].

The aromatic residues, on the other hand, interact with the C-terminal tail of CENP-A, which is known to be necessary for CENP-C binding [69,70,76,151,154,158] (Figure 3C). Despite relatively modest evolutionary sequence conservation, a common trait of the CENP-A C-terminal tail is that its sequence is considerably more hydrophobic than that of H3 (Leu-Glu-Glu-Gly-Leu-Gly and Glu-Arg-Ala in human CENP-A versus H3, respectively, Figure 3A). Thus, rather than a specific amino acid sequence, the higher hydrophobicity of the C-terminal tail of CENP-A may be key for specific recognition by CENP-C [79].

Despite their being related in sequence, the central region and the CENP-C motif of CENP-C do not have the same potential for kinetochore recruitment. The central region is necessary and sufficient to promote CENP-A nucleosome binding in vitro and kinetochore targeting in vivo [79,151,159–166]. The CENP-C motif, on the other hand, is insufficient for kinetochore targeting, but can be recruited to kinetochores as part of a larger C-terminal fragment capable of homo-dimerization with endogenous CENP-C through a C-terminal 'Mif2-homology' cupin-like domain [162–164,167,168]. Furthermore, while not sufficient for centromere recruitment in the absence of endogenous CENP-C, the CENP-C motif and the dimerization domain contribute to the robustness of CENP-C recruitment to kinetochores [151,165].

In vitro, CENP-C and CENP-N show relatively modest selectivity for CENP-A over H3, with differences in dissociation constant of between 5- and 10-fold [110,151]. It is unlikely that these differences, in the absence of other factors, account for the exquisite selectivity of kinetochore targeting of these proteins to CENP-A nucleosomes, which are greatly outnumbered by H3 nucleosomes at centromeres and in the rest of the genome [169]. Dimerization of CENP-C through its C-terminal cupin-like domain (Figure 3C) suggests a role for multi-valency as a source of additional selectivity for the interaction of CENP-C with centromeric CENP-A nucleosomes [79]. A second source of selectivity may derive from the interaction, within the CCAN of CENP-C and CENP-N, which recognize distinct features of the CENP-A nucleosome [110,151]. Post-translational modifications of histones have also been implicated as a potential factor in the selective recognition of CENP-A nucleosomes [170,171].

In addition to CENP-C and CENP-N, the CENP-HIKM complex (Figure 2D) also contributes to CENP-A binding affinity, but this complex interacts equally well with CENP-A and H3 nucleosomes, and with linear DNA [110]. Importantly, however, CENP-C, CENP-HIKM, and CENP-LN interact in a tight 7-subunit complex, the CENP-CHIKMLN complex [110], whose stability builds on multiple interactions of its subunits, including direct interactions of CENP-HIKM or CENP-LN with CENP-C [2,108–110,166].

A comprehensive view of the structural organization of the CENP-CHIKMLN complex is currently missing. Crystal structures of the CENP-LN complex of *S. cerevisiae* and of human CENP-M have been obtained and negative-stain single particle EM reconstructions have been generated for CENP-HIKM [108,111]. CENP-M is structurally and evolutionary related to Ras family small GTPases. It has lost all signature motifs previously implicated in GTP binding and hydrolysis by small GTPases, and is therefore considered a pseudo-GTPase [111]. Biochemical reconstitution demonstrated that CENP-M is required to stabilize CENP-I, predicted to have a α-solenoid fold of β-karyopherins [111]. No CENP-M ortholog has been identified in *S. cerevisiae* whereas CENP-H, -I, and -K all have orthologs in this organism (Figure 2A,D) [88].

3.4. The CENP-TWSX Complex

The CENP-TW and CENP-SX subcomplexes (Figure 2D) associate to form the tetrameric CENP-TWSX complex. All four subunits in this tetrameric complex possess histone fold domains. CENP-T contains additional N-terminal sequences, whose function in kinetochore assembly is discussed in Section 4. While CENP-TW and CENP-SX form stable entities in isolation and can have distinct biological functions [172–174], the tetrameric CENP-TWSX assembly was proposed to form a nucleosome-like structure flanking the CENP-A nucleosome at centromeres [175]. However, rigorous structural and functional evidence for this hypothesis is lacking. In vitro, the CENP-TWSX complex induces positive DNA supercoiling, contrarily to H3 and CENP-A nucleosomes (and to the isolated CENP-TW and CENP-SX complexes), which induce negative supercoiling [176]. When incubated with in vitro reconstituted H3 or CENP-A di-nucleosomes and visualized by negative-stain EM, two tetramers of CENP-TWSX bound preferentially to the ~100 bp inter-nucleosome linker DNA rather than nucleosome-bound DNA, but the limited resolution did not allow discriminating whether the two CENP-TWSX tetramers formed a nucleosome-like structure on the linker DNA; in addition, nuclease cleavage did not identify the pattern normally observed with canonical nucleosomes [176].

DNA binding by CENP-TWSX requires the histone fold domains of CENP-T and CENP-W [100,175,176]. No evidence of DNA sequence selectivity for these domains has been reported. In contrast to CENP-A, the pool of CENP-TW turns over relatively rapidly at centromeres, and with a cell cycle-regulated pattern, with centromere incorporation in late S-phase and G2 [177]. Incorporation of CENP-TW at the kinetochore has not been shown to require histone chaperones. Rather, in both humans and yeast, kinetochore recruitment of CENP-T^{Cnn1} appears to depend, in addition to DNA binding, on a direct interaction with other CCAN subunits, and in particular with the CENP-HIKM complex [78,111,113,178]. CENP-T recruitment is also affected by the N-terminal tail of CENP-A, in both fission yeast and humans [70,179], although the biochemical basis for this effect is to date unclear. Furthermore, ablation of either CENP-TW or CENP-SX has a distinct effect on outer kinetochore stability [100,101,178]. These observations detract from the hypothesis of a nucleosome-like structure flanking the CENP-A nucleosome, and rather suggest that CENP-TW binds DNA weakly and requires concomitant binding to other CCAN subunits at the inner kinetochore for its recruitment.

3.5. The CENP-OPQRU Complex

CENP-O, -P, -Q, -R, and -U (Figure 2D) associate into a complex [109,180]. Recruitment of CENP-OPQRU to the kinetochore requires CENP-CHIKMLN [96,97]. Loss of this complex does not affect localization of other inner kinetochore components and the functional importance of this complex at vertebrate kinetochores appears to vary in different systems [109,181,182]. A role in chromosome congression, at least partly operating through microtubule-binding sites in the CENP-Q and CENP-U subunits, as well as through recruitment of the microtubule motor CENP-E to kinetochores, has been reported [182–184]. Furthermore, CENP-U has been implicated in kinetochore recruitment of Polo-like kinase 1 (Plk1), an important regulator of kinetochore-microtubule attachments [185,186]. Thus, this complex is relatively peripheral in the organization of the CCAN in vertebrates and may contribute to chromosome segregation via recruitment of motor and kinase activities.

In budding yeast, where the CENP-O/P/Q/U/R–related complex is known as the COMA complex (for Ctf19, Okp1, Ame1 and Mcm21), two of the subunits, Ctf19 and Mcm21 (homologous to CENP-P and CENP-O, respectively), are non-essential for viability but are required for accurate segregation [187]. These subunits harbor RWD domains related to those observed in the Spc24/Spc25 subunits of the Ndc80 complex and the Knl1 C-terminus [116]. Deletion of Ctf19 or Mcm21 disrupts proper replication timing and cohesin complex accumulation in the pericentromeric region [188–190]. Interestingly, the other two subunits, Ame1 and Okp1 (homologous to CENP-U and CENP-Q, respectively) are essential for viability and Ame1 has been shown to directly interact with the Mis12 complex of the KMN network via a motif whose selective mutation is lethal [191]. These results suggest a potentially more significant role for the COMA complex in inner-outer kinetochore linkage in budding yeast than has been observed for the CENP-O/P/Q/U/R complex in vertebrates. CENP-Q is essential in mouse embryonic stem cells but not in mouse fibroblasts [181]; whether this difference in phenotype arises from a differential role in kinetochore assembly is not known. It will be important to elucidate the functions of this peripheral CCAN complex and determine the reasons for distinct effects of its loss in different systems and in different contexts within the same species.

3.6. CENP-B

CENP-B, the only specific DNA binding protein at mammalian centromeres, binds to the conserved 17-bp CENP-B-box, many copies of which are disseminated in centromeric α-satellite DNA repeats [192] (Figure 2E). CENP-B shares sequence homology with transposases encoded by the *pogo* family of DNA transposons [193] and appears to have arisen from them. A role of CENP-B in centromere stability has been questioned because of its limited conservation (CENP-B like proteins are thought to have arisen from transposases independently in fungi, insects and mammals), because CENP-B boxes are absent from neocentromeres, and because there are chromosomes (such as the human Y chromosome) that lack CENP-B boxes altogether. Furthermore, deletion of CENP-B in mice does not affect viability [194–196]. Nonetheless, a role for CENP-B in centromere stability is suggested by its requirement for the de novo establishment of centromeres on human artificial chromosomes built using centromere-enriched satellite DNA [197,198], and from the fact that its deletion increases chromosome instability [199,200]. Several recent and older observations support a role of CENP-B in the stabilization of centromere structure. For instance, CENP-B appears to contribute to the phasing of CENP-A nucleosomes on centromeric DNA and to the typical unwrapping of its nucleosomal termini [146], and it increases the stability of reconstituted CENP-A nucleosomes [201], likely through a direct interaction with the N-terminal region of CENP-A [199]. Furthermore, CENP-B binds directly to CENP-C, supporting a second pathway of CENP-C recruitment in addition to that based on the interaction of CENP-C with the CENP-A C-terminal tail [159,199]. These interactions of CENP-B may be largely redundant with other stabilizing interactions at centromeres, but their importance is exposed following perturbation of normal CENP-A function [199].

3.7. Summary

This section illustrates the important concept that inner kinetochores are built by evolutionary conserved interactions of CCAN subunits with the CENP-A nucleosome. In line with the theory that centromere identity is determined epigenetically in most organisms (the notable exception being budding yeasts, where a specific sequence is recognized by the CBF3 complex to direct CENP-A loading), none of these interactions, appears to require specific DNA sequences, except for those made by CENP-B, which is not conserved even throughout vertebrates and is not essential but may contribute to kinetochore stability when present. The evolutionary presence/absence and phenotypic effect of CCAN subunit inhibitions, with the possible exception of CENP-C, are also surprisingly variable, with kinetochores of well-studied models such as *C. elegans* and *D. melanogaster* lacking the entire repertoire except for CENP-C [62,88,89,133]. These greatly simplified kinetochores appear to entirely rely on CENP-C being a linker between the CENP-A nucleosome and the outer kinetochore, as discussed in Section 5. This variation raises intriguing questions about the functional roles of this large group of inner kinetochore proteins that will be important to address in future studies.

4. The Outer Kinetochore

The outer kinetochore is the (main) platform for end-on microtubule binding by the kinetochore and responsible for transducing the force generated by depolymerizing microtubules to move chromosomes. The core of the outer kinetochore is a 10-subunit protein assembly known as KMN (for Knl1 complex, Mis12 complex, Ndc80 complex, described in Figure 2) [115,202–212]. The three sub-complexes (Knl1, Mis12, and Ndc80) exercise clearly distinct functions as summarized below.

4.1. The Ndc80 Complex

The 4-subunit Ndc80 complex is the primary microtubule receptor at the kinetochore [209,213]. Its four subunits contain large segments of coiled-coil, flanked by globular domains (Figure 4A). The complex is dumbbell-shaped and has a long axis of approximately 55 to 60 nm (Figure 4A,B). Microtubule-binding, mediated by the N-terminal regions of the Ndc80 and Nuf2 subunits, and kinetochore-targeting, mediated by the C-terminal regions of the Spc24 and Spc25 subunits, occupy opposite ends of the complex [214–218].

Crystal structures of isolated globular domains of Ndc80 complex subunits, and of engineered Ndc80 complexes lacking most of the coiled-coil (named Ndc80^{Bonsai} and Ndc80^{Dwarf}), revealed that the microtubule-binding region of the Ndc80 complex consists of a pair of tightly packed calponin-homology (CH) domains in the Nuf2 and Ndc80 (also called Hec1) subunits [216,218,219]. The latter are structural paralogs, whose overall domain organization, an N-terminal CH domain adjoined by a coiled-coil segment, is also found in three Intraflagellar Transport (IFT) complex B subunits, IFT81, IFT57, and CLUAP1 [220].

Visualization by cryo-electron microscopy of Ndc80^{Bonsai} bound to microtubules demonstrated a direct interaction of the CH domain of the Ndc80 subunit with the microtubule lattice with a spacing of 4 nm along each protofilament, indicative of interactions of the CH domain with both tubulin monomers [209,221,222] (Figure 4C). The interaction engages a region of the Ndc80 CH domain, designated toe, which gathers several positively charged residues previously shown to mediate high-affinity microtubule binding, and which has been proposed to act as a conformational sensor for straight protofilaments [221,223].

In addition to the CH domains, an ~80-residue, highly basic and structurally disordered N-terminal tail of the Ndc80 subunit has been functionally implicated in the Ndc80-microtubule interaction in vitro and in cells [213,216,218,224–226] (Figure 4A). The N-terminal tail may contain two distinct functional segments. One segment, running from residues 47–68 (of human Ndc80), has been implicated directly in microtubule binding through an interaction with E-hooks (the negatively charged C-terminal tails of α- and β-tubulin) of tubulin protomers in the adjacent protofilament [222]. Another segment, preceding the E-hook binding region, has been implicated in inter-Ndc80 complex interactions along the same protofilament [221,222]. Collectively, these interactions may be responsible for the ability of the Ndc80 complex to form clusters on the microtubule lattice [218,221,222], and suggest that binding of Ndc80 complexes to microtubules may be cooperative. As explained more thoroughly below, however, a microtubule-binding site in the kinetochore engages a relatively small number of Ndc80 complexes (probably 6 to 10) in an end-on configuration. Whether Ndc80 complexes can interact inter-molecularly in this setting, and whether their binding to microtubules is cooperative, remains controversial [227–229]. We additionally note that deletions of the Ndc80 tail in *S. cerevisiae* and in *C. elegans* do not exhibit the severe phenotypic consequences expected for a major defect in kinetochore-microtubule interactions [230–232]; in contrast, mutations in conserved CH domain residues have severe consequences in all systems where they have been analyzed. Thus, the precise role of the N-terminal tail of Ndc80 in kinetochore-microtubule interactions remains an important question for future investigation.

While the mechanism of the interaction of Ndc80 with microtubules and with itself requires further investigation, it is clear that the N-terminal tail of Ndc80 regulates microtubule binding. Aurora B kinase, a major regulator of kinetochore-microtubule attachment [14], phosphorylates up to nine sites in the human Ndc80 N-terminal tail [213,216,218,221,222,224–226,229,233]. Phosphorylation neutralizes the intrinsic positive charge of the Ndc80 N-terminal domain, greatly decreasing the binding affinity of the Ndc80 complex for microtubules in vitro [209,216,218,225,226]. The N-terminal tail of *C. elegans* Ndc80 is also the target of regulation by a protein complex that recruits and activates the dynein motor at kinetochores [232]; see Section 5.3 below.

4.2. The Mis12 and Knl1 Complexes

The Mis12 complex is an interaction hub that promotes KMN assembly through its binding sites for both the Ndc80 complex and the Knl1 complex, and that connects the KMN with the inner kinetochore through interactions with CENP-C and CENP-T. Previous low-resolution negative stain EM analyses depicted the Mis12 complex (known as MIND complex in *S. cerevisiae*) as a ~20 nm rod [234–237]. When Mis12 and Ndc80 are combined and their structure is examined by negative stain or rotary shadowing EM, they appear as ~90-nm particles, indicating that they interact 'in series' [238,239] (Figure 4D). Recent crystal structures of the human and yeast complexes demonstrate that the four subunits of the Mis12 complex are structural paralogs with high helical content (Figure 4E). They pair in Dsn1:Nsl1 and Mis12:Pmf1 sub-complexes, that meet in a central stalk domain. The N- and C-termini of all four subunits cluster at opposite ends of the rod [240,241]. Linear motifs near the C-termini of the Nsl1 and Dsn1 subunits of the Mis12 complex, invisible in the crystal structure, provide binding sites for the RWD domains in the C-terminal region of the Spc24 and Spc25 subunits of the Ndc80 complex [234,240–242].

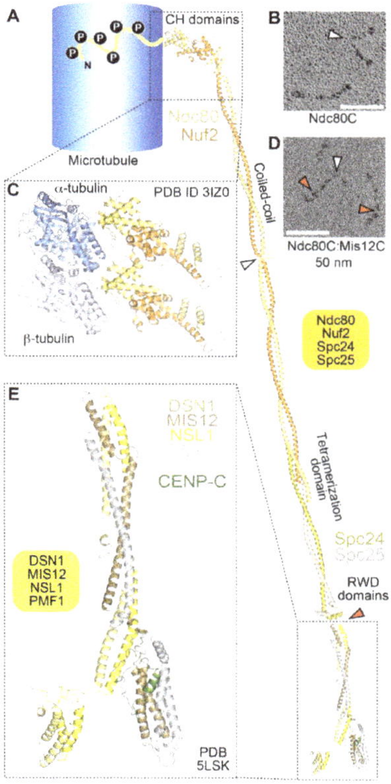

Figure 4. The NDC80 and MIS12 complexes of the KMN network. (**A**) The NDC80 complex is highly elongated and interacts with microtubules via calponin-homology (CH) domains in the N-terminal regions of NDC80 and NUF2. A basic N-terminal tail preceding the NDC80 CH domain (depicted as unstructured) is subject to Aurora kinase phosphorylation (*Ps in black circles*) and regulates microtubule binding. A long coiled-coil, interrupted by a loop (*white arrowhead*) terminates in a tetramerization domain with SPC24 and SPC25. The latter start with coiled-coils and terminate with RWD domains (*red arrowhead*), which interact with the MIS12 complex; (**B**) Rotary shadowing electron microscopy of the NDC80 complex, showing its characteristic dumbbell shape, and an overall length of ~65 nm. Images in (**B**,**D**) courtesy of Dr. Pim Huis in 't Veld, Max Planck Institute of Molecular Physiology, Dortmund (Germany) [239]; (**C**) Model from cryo-EM studies of the Ndc80^{Bonsai} complex bound to the microtubule lattice. Only a single α-tubulin:β-tubulin dimer is shown, with two Ndc80^{Bonsai} complexes bound via the toe region; (**D**) Complexes of the NDC80C and MIS12C are ~85 nm in length; (**E**) Structural organization of the MIS12 complex bound to the N-terminal region of CENP-C [241]. All structures shown are for the human complexes.

The stalk of the Mis12 complex, together with a ~20-residue C-terminal motif in Nsl1, also provides a binding site for Knl1, the largest outer kinetochore subunit (2316 residues in humans) (Figure 5). With the exception of the last ~500 residues, Knl1 is largely intrinsically disordered, and contains an array of protein docking motifs, including a canonical binding site for the PP1 phosphatase very near the N-terminus, and multiple Met-Glu-Leu-Thr (MELT) repeats, identified to act, after phosphorylation by the Mps1 kinase on the conserved Thr residue, as docking sites for the SAC protein complex Bub1:Bub3 [15,17] (Figure 5). The Knl1 C-terminal region, on the other hand, consists of a predicted coiled-coil followed by tandem RWD domains, and is therefore structurally related to Spc24, Spc25, CENP-O, and CENP-P, suggesting a common evolutionary origin of these proteins [116,235]. The RWD domain mediates a direct interaction of Knl1 with the Mis12 complex, whereas Zwint binds to a more extended domain additionally comprising the coiled-coil region [234,235,241].

With some variations, the description of the outer kinetochore in the previous paragraphs applies to yeast and human kinetochores alike. In addition, in the case of the outer kinetochore, exceptions have emerged in the course of evolution. In *D. melanogaster*, for instance, no ortholog of Dsn1 or Zwint was identified, whereas two closely related and functionally redundant paralogs of Pmf1^{Nnf1} (Nnf1a and Nnf1b) exist [243–246]. Furthermore, the unconventional SNARE family member Snap29 was recently shown to localize to kinetochores and to be required for KMN assembly in this organism [247]. In biochemical reconstitutions, the *Drosophila* Mis12 complex is highly stable in the absence of Dsn1 (unlike the human and yeast complexes) [245,246]. Significant adaptation of the Nsl1 sequence at residues implicated in Dsn1 binding by the structures of the human and yeast Mis12 complex explain this result (not shown). In *C. elegans*, where the KMN components are all present and were shown to self-associate in biochemical reconstitutions, there are some notable sequence variations, e.g., *C. elegans* Knl1 lacks the tandem RWD domains at the C-terminus and the RWD domains of Spc24 and Spc25 appear severely diminished. Knl1 family proteins also exhibit widespread and recurrent evolution of repeats in their N-terminal region [248]. The reasons for these variations are currently unclear. Nonetheless, relative to the CCAN, KMN is broadly conserved, likely reflecting its essential involvement in microtubule attachment and scaffolding of the SAC.

4.3. Complexity of the Kinetochore Microtubule Interfaces

Understanding the role of Ndc80 complex in the generation of dynamic load-bearing attachments during chromosome congression and segregation is a primary goal of current research (see review by Asbury and colleagues, reference [30]). Classic in vitro studies demonstrated that kinetochores can hold on to a depolymerizing microtubule end [249] and, more significantly, a depolymerizing microtubule generates significant force that is capable of moving chromosomes bound via their kinetochores [250,251]. These studies inspired efforts to analyze if the Ndc80 complex, the widely conserved outer kinetochore-localized microtubule-binding complex, acts as a coupler that is able to harness the force generated by a depolymerizing microtubule end. When immobilized on beads at sufficiently high concentrations, the Ndc80 complex is sufficient to create load-bearing attachments to depolymerizing microtubules [29,227,252]. Individual, soluble Ndc80 complexes, on the other hand, are unable to track depolymerizing microtubule ends [253]. These experiments suggest that clustering on beads enables the establishment of multiple microtubule attachments, and that the latter are required for microtubule plus-end tracking during depolymerization. Clustering of multiple Ndc80 complexes at kinetochores is likely to achieve a similar effect.

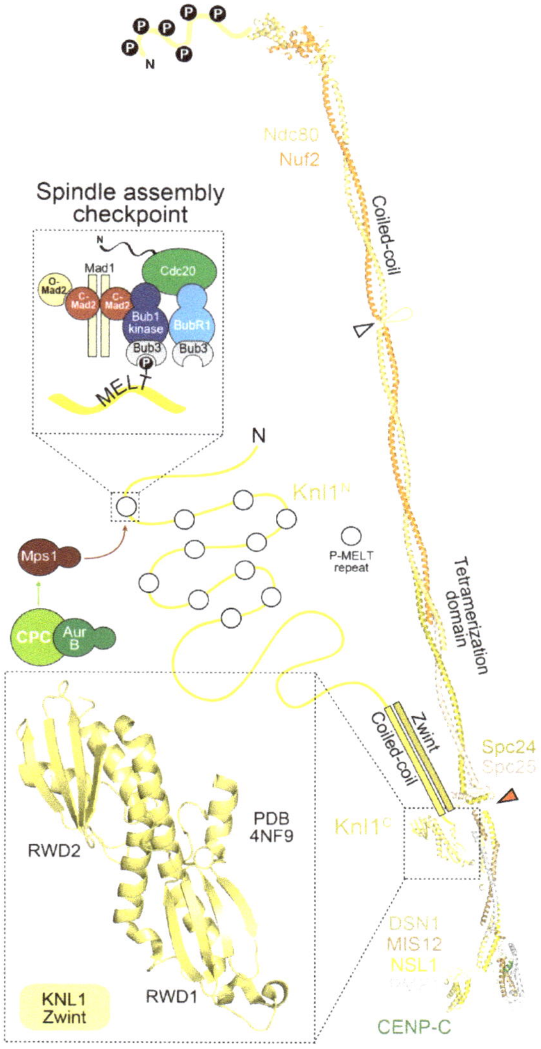

Figure 5. Orchestration of the spindle assembly checkpoint (SAC) by the KMN network SAC components are recruited via the Knl1 subunit that is 2316 residues in humans and is largely disordered. Exceptions are a predicted coiled-coil around residues 1850–2100, and the C-terminal tandem RWD domains, whose crystal structure is shown [235]. The RWD region of Knl1 binds directly to the MIS12 complex [234,235]. The N-terminal half of Knl1 contains multiple MELT repeats (Met-Glu-Leu-Thr) that are targeted by Mps1 kinase (which in turn requires Aurora B kinase to become activated). Each MELT repeat has the potential to assemble active SAC complexes that signal lack of microtubule attachment and arrest the cell cycle in mitosis.

It is important to note, however, that Ndc80 is not the sole player in linking kinetochores to microtubules. Additional microtubule-binding proteins and motors identified there include the SKA and Dam1 complexes, Kif18a, MCAK, SKAP:Astrin, XMAP215/CH-TOG, CENP-E, CENP-F, and Dynein [9,254]. The functions of these proteins at kinetochores are discussed in the essays from

Maiato, Lampson and Grishchuk, and Asbury and colleagues ([14,30,255]), and here we limit the discussion to a brief account of the SKA and the Dam1 complexes, two sequence and structurally unrelated microtubule binders with complementary phylogenetic distributions that have emerged as playing a fundamental role in microtubule coupling at kinetochores [256–259]. The human SKA complex and the budding yeast Dam1 complex can track dynamic microtubules, and interact with their cognate Ndc80 complex specifically when bound to microtubules [253,260–265]. The Aurora B kinase phosphorylates the SKA and Dam1 complexes to reduce their binding affinity for kinetochores [264–267], in line with a regulatory scheme that identifies the Aurora B kinase as a negative regulator of the strength of the attachment of kinetochores to microtubules, and as a crucial actor in the correction of improper kinetochore-microtubule attachments (see reviews by Lampson and Grishchuk and by Asbury and colleagues [14,30]). In a recent twist, the SKA complex was also shown to stimulate Aurora B activity [268].

Structural and biochemical work on the SKA complex has started to elucidate its organization and mechanism of action [253,269–271]. The SKA complex is a trimer of the Ska1, Ska2, and Ska3 subunits. It is 'W' shaped, and consists of dimers of triple helical bundles of the three subunits [269]. The C-terminal domain of the Ska1 subunit contains a winged-helix motif that interacts with surface-exposed regions of tubulin that are insensitive to microtubule curvature, while the unstructured C-terminal region of Ska3 facilitates the interaction of Ska1 with microtubules [271]. The Dam1 complex is a heterodecamer [272,273], and individual heterodecamers assemble into rings that encircle the microtubule surface [274–277]. Both the SKA and Dam1 complexes are dependent on the Ndc80 complex for their kinetochore localization in cells and enhance the microtubule coupling ability of the Ndc80 complex in vitro. These findings suggest that the concerted action of Ndc80 and SKA/Dam1 complexes underlies the load-bearing attachments made at kinetochores but the detailed mechanistic basis for their concerted action remains to be elucidated.

5. Linkages between the Inner and the Outer Kinetochore

5.1. Two Mechanisms Link Inner and Outer Kinetochores

The outer kinetochore is linked to the inner kinetochore via two different mechanisms (Figure 6). In the first mechanism, CENP-C directly binds to the Mis12 complex [165,238,240,241,245,278,279], which in turn binds to the Ndc80 complex and Knl1. In the second mechanism, the RWD domains in the Spc24 and Spc25 subunits of the Ndc80 complex directly interact with the intrinsically disordered N-terminal extension of CENP-T [75,113,239,242,279–282]. We summarize below the detailed understanding of these two mechanisms in yeast and vertebrates and their relative importance in outer kinetochore assembly in different systems.

CENP-C (and its yeast homolog Mif2) binds directly to the Mis12 complex through an ~45-residue N-terminal motif, an interaction captured in the recent co-crystal structures of the yeast and human Mis12^{MIND} complexes discussed in the previous section [240,241] (Figure 4E). In *S. cerevisiae*, the 4-subunit COMA complex (Ctf19:Okp1:Mcm21:Ame1), which is part of the yeast CCAN^{Ctf19} complex and whose subunits are related to those in the CENP-OPQRU complex, helps reinforce the interaction of CENP-C^{Mif2} with Mis12^{MIND} [191]. In both yeast and humans, Aurora B kinase regulates the Mis12^{MIND}:CENP-C^{Mif2} interaction by phosphorylation of two serine residues (Ser100 and Ser109 in humans) that reside in closely spaced, positively charged motifs in a disordered region of Dsn1 [280,283–287]. Aurora B phosphorylation of Dsn1 increases the binding affinity of CENP-C for the Mis12 complex by approximately two orders of magnitude, through relief of a competitive inhibitory mechanism in which the unphosphorylated Dsn1 region binds and masks in an intra-Mis12-complex manner the CENP-C binding region [240,241]. The significance of this regulation, which likely requires the presence of Aurora B kinase activity at centromeres, may be to stabilize the CENP-C:Mis12 complex interaction exclusively in the proximity of kinetochores.

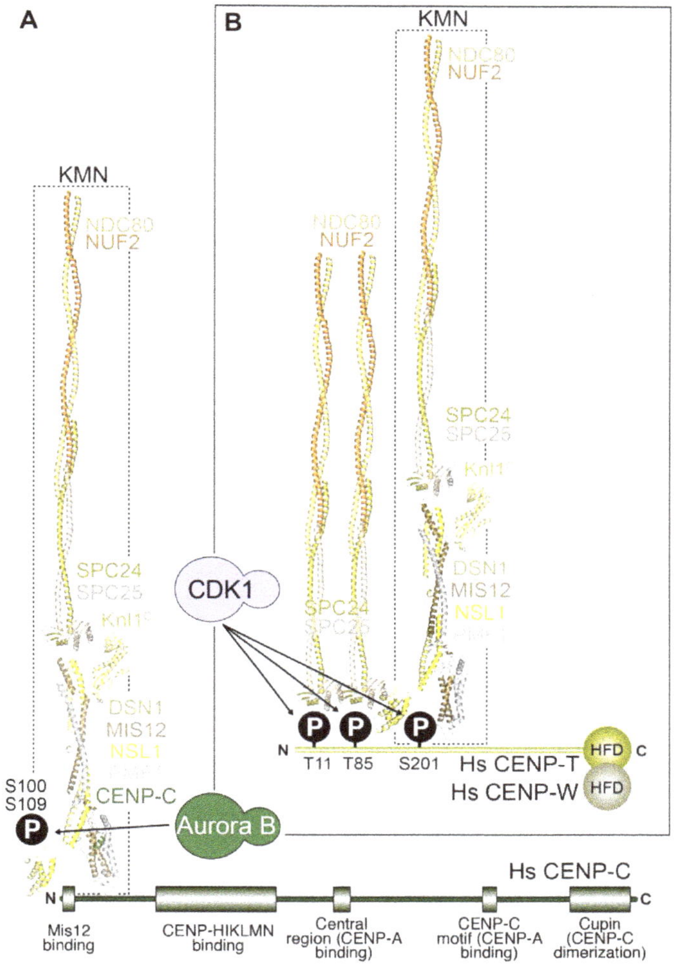

Figure 6. Linkages between the inner and outer kinetochore. Structures of a portion of the NDC80 complex (the N-terminal globular domains of NDC80 and NUF2 are not shown), the MIS12 complex and the C-terminal kinetochore-targeting domain of KNL1 are used to depict a KMN particle in humans. (**A**) The first linkage is formed by the interaction of a KMN particle with the N-terminal region of CENP-C. This interaction is enhanced by Aurora B phosphorylation of residues (S100 and S109) in the N-terminal region of the DSN1 subunit of the MIS12 complex; (**B**) The second linkage involves the interaction of up to two NDC80 complexes with two CDK1-phosphorylated residues (T11 & T85) in the N-terminal region of CENP-T, as well as of a second entire KMN recruited via a CDK1-dependent interaction of the MIS12 complex with S201 of CENP-T [239]. In vitro, CENP-C and CENP-T bind to the MIS12 complex within the KMN network competitively, implying that they cannot be bound to the same KMN [239]. All structures shown are for the human complexes.

The RWD domains in the Spc24 and Spc25 subunits of the Ndc80 complex interact directly with two related, short sequence motifs in the first 100 residues of the intrinsically disordered N-terminal extension of CENP-T (consisting of approximately 450 residues in humans) [75,113,239,242,279–282]. In humans, this interaction requires CDK phosphorylation of the CENP-T motifs, but the equivalent interaction of

Spc24:Spc25 with CENP-T^{Cnn1} in *S. cerevisiae* may not require phosphorylation [242,279,281]. At least in vitro, the two Ndc80 complex-binding motifs of CENP-T can be occupied concomitantly, suggesting that this mechanism can recruit up to two Ndc80 complexes per CENP-T molecule [113,239]. The motifs on CENP-T are closely related to the Spc24:Spc25-binding motif in Dsn1 (discussed in the previous section) [242]. Not surprisingly, therefore, recent crystal structures demonstrated that the Dsn1 and CENP-T motifs bind Spc24:Spc25 through a largely similar mechanism [240].

In an interesting recent twist, it was realized that CENP-T also contributes to kinetochore recruitment of the Mis12 complex [239,280,288]. This is promoted by a direct interaction of the Mis12 complex with a distinct, non-canonical CDK phosphorylation site on human CENP-T, Ser201 [239]. Whether this interaction is conserved in *S. cerevisiae* is currently unknown. Thus, at human kinetochores, a single N-terminal tail of CENP-T can, after appropriate phosphorylation, promote the localization of up to three Ndc80 complexes, two through a direct interaction, and one indirectly through the Mis12 complex (Figure 6). These biochemical data are consistent with analysis in human cells, where CENP-T depletion reduces Ndc80 complex localization at kinetochores to a third of that in controls, without affecting CENP-C localization (See Section 5.2 below).

The presence of two mechanisms for linking the outer and inner kinetochore raises the question why these two linkages are needed and whether they are widely employed. In all systems tested, CENP-C inhibition leads to severe defects and lethality (with the exception of *S. pombe*, where a suppressor mutation can improve growth of a CENP-C null mutant that is extremely sick and missegregates chromosomes at high frequency [166]. Depletion or deletion of CENP-T results in extensive outer kinetochore assembly and chromosome alignment defects in chicken and human cells [100,176,279,280,287,288]. In addition, chicken CENP-T can generate ectopic microtubule attachment sites that support chromosome segregation in the absence of CENP-C [75], and a chimeric construct in which the N-terminal region of CENP-T replaced the entire N-terminal domain of CENP-C appeared to support chromosome segregation [288]. Surprisingly, however, a deletion mutant of the CENP-T ortholog Cnn1 in *S. cerevisiae* is viable and the absence of CENP-T does not significantly reduce the amount of Ndc80 recruited to kinetochores in this system [242,282,289]. In addition, in *D. melanogaster* and *C. elegans*, which lack all CCAN subunits with the exception of CENP-C (see Section 3), the interaction of CENP-C with the Mis12 complex is likely the only linkage between the inner and outer kinetochore. Consistent with this notion, a tight CENP-C interaction with the Mis12 complex has been observed in biochemical reconstitutions of the *D. melanogaster* outer kinetochore [245,246]. Thus, from the analysis in different models to date, it appears that the CENP-C–outer kinetochore linkage is more commonly employed, although in the vertebrate species analyzed to date the CENP-T linkage makes the more dominant contribution to Ndc80 complex recruitment. Interestingly, the CENP-T ortholog in *S. pombe* (Cnp20) unlike Cnn1 in *S. cerevisiae*, is essential for viability [166]. Additional work on *S. pombe* CENP-T is needed to address whether its essential function relates to outer kinetochore assembly. More broadly, asking precisely why two types of linkages have evolved to link the inner and outer kinetochore and asking whether there is a functional specialization of these linkages are important questions for future studies.

In summary, the plan of kinetochore assembly from the chromatin layer to the outer kinetochore has been now delineated in significant detail. Crucial features of this assembly plan include: (1) Recruitment of all kinetochore proteins ultimately depends on specific interactions with CENP-A. CENP-C, CENP-N, and CENP-T, which have been implicated as the proteins at the base of the kinetochore, require CENP-A for their localization; (2) Both CENP-C and CENP-N bind directly to the CENP-A nucleosome; (3) CENP-T does not bind directly to the CENP-A nucleosome, but appears to recognize a combination of the CENP-HIKM complex (which interacts directly with CENP-C and CENP-NL) and naked DNA, possibly in a linker region neighboring the CENP-A nucleosome; (4) CENP-C creates a direct linkage between CENP-A (and its associated CCAN subunits) and the KMN network, binding concomitantly to both, and acting in analogy to a 'blueprint' to order kinetochore assembly; (5) CENP-T plays an analogous bridging function, and its N-terminal region can even replace the N-terminal region of CENP-C involved in Mis12 binding in an engineered context. The latter observation suggests that CENP-C and CENP-T might be distantly related in evolution.

These principles, which summarize a vast body of literature, were recently implemented in the biochemical reconstitution of a 21-subunit kinetochore particle containing the CENP-A nucleosome, the CENP-CHIKMLN complex, and the KMN network [110]. The reconstituted complex was shown to be sufficient to associate the CENP-A nucleosome with microtubules in vitro, demonstrating that its components can create a linkage between DNA associated with CENP-A and microtubules. Lacking from the reconstitution were the CENP-TW (and CENP-SX) complex and the CENP-OPQRU complex. The latter binds directly and with high affinity to the CENP-CHIKMLN complex, whereas incorporation of the former might require, as suggested above, a more complex chromatin template than a single CENP-A nucleosome.

A reconstitution approach also has the potential to fully define the CEN DNA-based kinetochore in *S. cerevisiae* that is built on a single well-positioned CENP-A nucleosome and binds to a single microtubule. Work on isolated kinetochore particles purified from *S. cerevisiae* has begun to illustrate the overall structural organization of this unit kinetochore and its microtubule binding modes [290,291] (Figure 7). When imaged by negative stain EM, the kinetochore particles had a central core of ~37 nm diameter, and were radially surrounded by 5 to 7 globular domains with ~21 nm diameter. When bound to microtubules, the particles appeared to contain a 50-nm ring structure surrounding the microtubule (likely the Dam1 complex), linked through a fibrous network (likely the Ndc80 complex) to the globular region. Two sites of microtubule attachment were visible, one coinciding with the ring structure and one at the junction of the fibrous structure with the globular region [291]. Continued analysis of yeast kinetochore particles and of human kinetochore reconstitutions of the type described recently should yield detailed insight into the structure and microtubule interaction properties of a unit kinetochore module in the foreseeable future.

5.2. Stoichiometry of Kinetochore Subunits

Estimates of the stoichiometry of human kinetochore composition were recently obtained through distinct experimental efforts, including biochemical reconstitution combined with analytical ultracentrifugation, and measurements of fluorescence intensity ratios in cells [110,288]. Biochemical reconstitution suggests that there are two CCAN complexes per CENP-A nucleosome [110]. As explained above, both CENP-C and CENP-T interact directly with the Mis12 complex, but their binding is mutually exclusive, implying that each of the two CCAN subunits has the potential to recruit the Mis12 complex independently [239]. Thus, if two copies of CENP-C and CENP-T associate with a CENP-A nucleosome, and each of them recruits Mis12, four Mis12 complexes will associated with the CENP-A nucleosome. Because each Mis12 complex also carries tightly bound Ndc80 and Knl1 complexes, as predicted by biochemical reconstitution experiments, at least four of each should be present. Furthermore, each CENP-T can also directly recruit up to two additional Ndc80 complexes, depending on the degree of saturation of phosphorylation and binding [113,239]. These numbers, summarized in Figure 8, are in excellent agreement with those obtained by quantification of fluorescence intensity at kinetochores [288]. In *S. cerevisiae*, early fluorescence measurements suggested ~8 KMN per centromeric CENP-A^{Cse4} nucleosome (or, more precisely, 8-fold higher fluorescence intensity for KMN subunits relative to CENP-A^{Cse4} in the cluster of 16 centromeres). As deletion of the CENP-T ortholog Cnn1 does not reduce kinetochore-localized Ndc80, which is in contrast to what is observed in human cells, how this stoichiometry is achieved remains at present unclear.

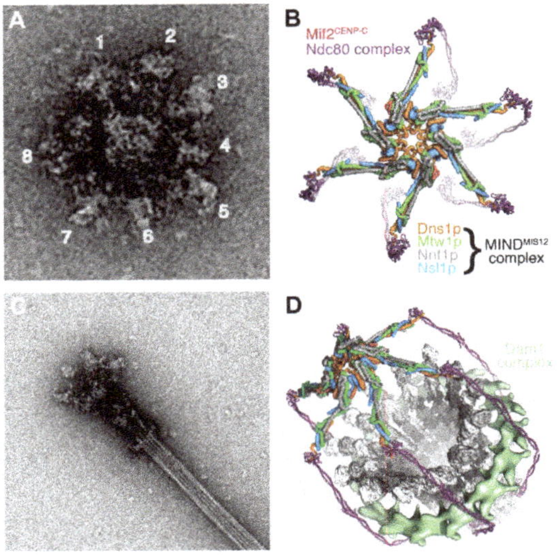

Figure 7. Images and structural model of budding yeast kinetochore particles. (**A**) Negative stain electron micrographs showing a kinetochore particle isolated from *S. cerevisiae* [291]. Images in this panel and in **C** courtesy of Sue Biggins and Tamir Gonen; (**B**) A rendered image with possible molecular interpretation of the negatively stained particles showing MIND^{MIS12}:CENP-C complexes departing from a central "hub" and connecting with Ndc80 complexes. Image reproduced with permission from reference [240]; (**C**) Negative stain electron micrographs showing a *S. cerevisiae* kinetochore particle bound to the end of a taxol-stabilized microtubule [291]. Image courtesy of Sue Biggins and Tamir Gonen; (**D**) The structure in (**B**) is shown to surround the microtubule in end-on configuration. The Dam1 complex stabilizes the arrangement by surrounding the microtubule. Image reproduced with permission from reference [240].

5.3. Temporal Framework of Kinetochore Assembly and Disassembly

The majority of CCAN subunits display continued centromere localization (defined as co-localization with CENP-A foci) during the cell cycle and, in most cases, negligible turnover rates [292–295]. Nonetheless, studies on the reciprocal dependencies of CCAN subunits during the cell cycle indicate clear differences between interphase and mitosis. For instance, CENP-C localization to centromeres appears to depend on CENP-HIKM subunits during interphase but not in mitosis (for instance see references [80,110,111,296]). New kinetochore incorporation of CENP-TW, CENP-N, and CENP-U may occur during DNA replication [177,297–299]. The kinetochore levels of CENP-OPQRU subunits, on the other hand, appear to decrease as cells enter mitosis [294,300]. The molecular basis for cell cycle-dependent regulation of CCAN subunit loading and stability is largely unknown. Phosphorylation likely plays a role in these processes [301].

Mitotic maturation of kinetochores focuses mainly on the creation of the outer kinetochore. In vertebrate cells, the KMN subunits are not localized with CENP-A foci in G1, but begin to be recruited in S-phase and G2, with the Ndc80 complex being the last to be recruited, due to its exclusion from the nuclear compartment and to its dependence on CDK activity for kinetochore localization [239,242,279,281,292,302]. In *D. melanogaster*, KMN assembly may only occur later, in prophase, but follows a similar assembly order, with the Mis12 complex and Knl1^{Spc105} assembly leading to Ndc80 complex recruitment after nuclear envelope breakdown [303]. As already clarified above, stabilization of the interaction of Mis12 with CENP-C might be an initiating trigger in KMN assembly on kinetochores. The components of the KMN network, on the other hand, disassemble from kinetochores at anaphase [292].

Probably the most dramatic physical transformation of regional kinetochores in metazoans is the formation of crescent-like shapes on their surface [7,8,24–26]. This phenomenon precedes end-on microtubule binding by the Ndc80 complex, and is believed to increase the likelihood of microtubule capture as well as to promote SAC signaling [8,27]. Proteins involved in this expansion had been previously localized to the kinetochore corona and include the microtubule motor CENP-E [26], the large (~400 kD) microtubule-binding protein CENP-F [8,304], and the dynein/dynactin motor complex along with its kinetochore targeting adaptors, the Rod-Zwilch-ZW10 (RZZ) complex [305]. RZZ's largest subunit, Rod, is structurally related to clathrin [306], pointing to its oligomerization as a possible driver of corona expansion.

As clarified in more detail in the chapter by Maiato and colleagues [255], the RZZ complex is required for kinetochore recruitment of the minus-end directed motor cytoplasmic Dynein. This function of RZZ requires an additional protein named Spindly, which additionally acts as an adaptor capable of stimulating Dynein motility [232,307–313]. Kinetochore localization of Spindly requires the RZZ complex and farnesylation on a Cys residue near the C-terminus of Spindly [314,315]. Interestingly, the motor protein CENP-E and the microtubule-binding component CENP-F that also localize to the corona region of the kinetochore are both farnesylated [316,317].

Upon conversion of kinetochore attachments from lateral to end-on, i.e., when Ndc80 gains the upper hand in the attachment mechanism, the shape of kinetochores converts from an extended crescent to a smaller, plate-like appearance [8,27]. This shape change is associated with the motor-dependent release of the Dynein:Dynactin:Spindly:RZZ complex from kinetochores towards spindle poles [318–325]. The central spindle assembly checkpoint component, the Mad1:Mad2 complex, is also removed from kinetochores through this mechanism, effectively terminating SAC signaling by the kinetochore [325–331].

This brief section highlights that kinetochore assembly is regulated in response to cell cycle cues and kinetochore composition changes in response to microtubule attachment. Defining precisely how regulation operates in these two contexts is an important challenge for the future.

6. Organization of the Chromatin Foundation of the Kinetochore in Regional Centromeres

The above sections have focused on discrete, high-affinity stoichiometric physical interactions of kinetochore subunits that build the kinetochore on its chromatin foundation. This type of approach is helping define the assembly unit of the human kinetochore, which we propose has a CENP-A and H3.3 dinucleosome as its foundation (Figure 8). The remarkable progress made on understanding high-affinity stoichiometric interactions masks a relative paucity of information on how the regional or holocentric centromeres of most metazoan species, where a small number of CENP-A nucleosomes are interspersed with a large excess of H3 nucleosomes, are organized to form a multi-microtubule binding kinetochore. While it is generally assumed that individual CENP-A nucleosomes will recruit machinery resembling the unit kinetochore module described above [332,333], how these modules are clustered and organized to build a 'surface' with a high density of kinetochore components on a CENP-A nucleosomal platform is not known. In *S. pombe*, ~10–15 CENP-A^{Cnp1} nucleosomes in the 10 kb centromeric central core region drive the assembly of kinetochores capable of binding three microtubules [334,335]. At human centromeres, which bind ~20 microtubules, current estimates indicate that ~100 CENP-A nucleosomes are present but are dispersed in a large centromeric DNA region in which CENP-A is largely substoichiometric relative to H3 (roughly 1 to 25) [169]. Within such large repetitive centromere regions, CENP-A may be enriched in arrays forming discrete sub-domains [336–338]. How centromeric chromatin folds to expose the CENP-A domains for kinetochore assembly is currently unclear, and several models have been discussed [6]. Of note, studies in *S. cerevisiae* have suggested that the high density of cohesin flanking centromeres may aid extrusion of loops on the termini of which CENP-A^{Cse4} nucleosomes are present. This line of thinking, stimulated by models of chromatin polymer properties, suggests that the 16 kinetochores of budding yeast may be organized in a manner that resembles multi-microtubule binding kinetochores of metazoans [339–341] (Figure 7). In the holocentric nematode *Caenorhabditis elegans*, CENP-A occupies non-repetitive regions of 10–12 kb dispersed across about half of the genome and is excluded from loci that are transcribed in the germline and early embryo [342]; for a different view that suggests the presence of focused spots of functional CENP-A in *C. elegans*, in addition to the pattern summarize above, see [343]). Notably, in this holocentric species, CENP-A removal, but not CENP-C removal, causes chromosome structure/condensation defects [344], suggesting that CENP-A chromatin, independently of its role in kinetochore assembly, has a propensity to coordinate structural organization of chromatin, potentially via its intrinsic properties or via effectors other than those required to build a kinetochore.

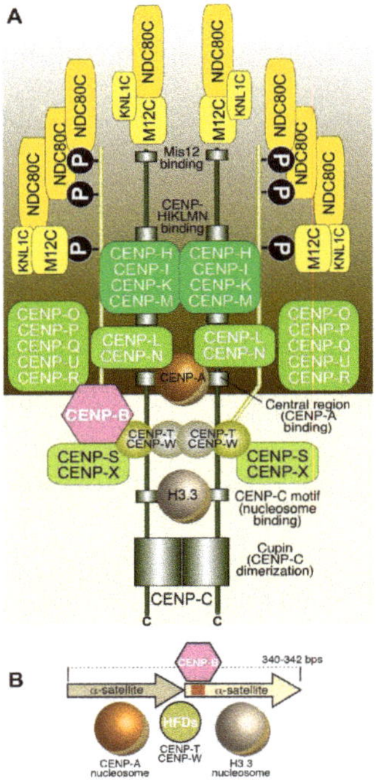

Figure 8. A model for the assembly unit of the human kinetochore. The kinetochore assembly unit is depicted as being organized on a CENP-A-H3.3 dinucleosome. (**A**) A primary determinant of the stoichiometry of kinetochore subunits is the valency of the CENP-A nucleosome, which confers the potential to interact with two CCAN complexes, as shown in work of in vitro reconstitution of kinetochore assembly [110]. Because CENP-C and CENP-T each carries a full KMN network, and CENP-T additionally carries two NDC80 complexes, there are four MIS12 and KNL1 complexes per CENP-A nucleosome, and up to 8 NDC80 complexes. CENP-C has the potential to interact with two nucleosomes (see Figure 3), with one of them (the one bound to the CENP-C central region) is a CENP-A nucleosome permanently marked by interactions with stably bound CCAN subunits. As clarified in Figure 9, we speculate that the identity of the second nucleosome, which binds to the CENP-C motif, varies during the cell cycle, alternating between CENP-A and H3.3. The C-terminal dimerization domain of CENP-C might "seal" this design. During mitosis, the second nucleosome is an H3.3 nucleosome; (**B**) This speculative design is compatible with the existence of the tandem α-satellite structures already discussed in Figure 2. CENP-TW is proposed to bind in the inter-nucleosomal linker region, near the CENP-B box.

As discussed in Section 5, the assembly unit of a kinetochore likely contains multiple Ndc80 complexes. Modulation by Aurora B phosphorylation of the binding affinity of each of these complexes for microtubules (see Section 4) likely gives rise to a considerable dynamic range of microtubule binding affinities. While kinetochores built on point centromeres bind a single microtubule, those built on regional centromeres bind multiple microtubules, but it is currently unclear whether a kinetochore module within regional kinetochores is associated with a single or multiple microtubules. In modeling studies, a random distribution of Ndc80 complexes on the kinetochore surface proves more versatile in comparison to the clustering into discrete regions binding a single microtubule [229,233]. Whether the scattering model is compatible with the kinetochore construction principles we have illustrated, however, remains to be clarified.

This section illustrates a major gap in our understanding of kinetochore structure in a majority of species—how the dispersed and rare CENP-A nucleosomes at repetitive centromeres are collected and organized to form a base for kinetochore assembly. The properties of this complex chromatin domain are also relatively poorly understood. This gap in turn leaves open the important question how microtubule-binding sites are organized to efficiently capture and couple dynamic microtubules and how multiple microtubule-binding sites are coordinated. Recent work is beginning to reveal the importance of transcriptional activity at centromeres that may contribute to centromere structural organization. Repeats at centromeres are transcribed and active RNA polymerase II is detected at centromeres [345–348]. In addition, complexes implicated in transcription, most notably the FACT complex that remodels chromatin during transcription, has been co-purified with CENP-A chromatin in multiple independent studies, has been implicated in CENP-A loading, and was recently reported to interact with CENP-T [97,99,112,349]. Numerous concepts are currently being explored on the role of transcriptional activity, including creation of paused transcription sites, generation of centromeric RNAs that somehow act in centromere assembly, and generation of a chromatin state that is conducive to CENP-A loading [350–352].

7. The Propagation of Centromeric Chromatin

A fundamental requirement for the epigenetic specification of centromeres is that the pool of CENP-A be maintained through cell division. In the absence of specific recognition of the underlying DNA sequence, this process is likely to be directed by the existing pool of CENP-A [72]. In vertebrates, incorporation of new CENP-A at centromeres occurs after mitotic exit, from mitotic telophase till early G1 phase [135,153,177,299,353]. The CENP-A pool is then distributed, without new incorporation, to the sister chromatids during DNA replication, with the resulting gaps being probably filled with histone H3.3 [354]. Thus, sister chromatids enter mitosis with half as much CENP-A as that present on the parental chromosome before replication. In the absence of sequence-specific interactions with DNA, a crucial unresolved question is if CENP-A positional information is retained during DNA replication, when the nucleosome structure of the centromere is likely to become temporarily perturbed, and if so, how. Once incorporated into chromatin, CENP-A is not further evicted, showing negligible dissociation kinetics [141,152,153,295,355].

Several factors involved in the deposition of new CENP-A have emerged [4,6] (Figure 9). The Mis18 complex, first identified based on a mutant in *S. pombe*, includes Mis18 (orthologous to human paralogs Mis18α and Mis18β) and Mis16 (orthologous to human proteins RbAp46/48, and acting as histone chaperones in several histone modification complexes) [339,356–359]. Mis18 forms oligomers in *S. pombe* and humans [360,361]. The Mis18 complex further interacts with M18BP1 (Mis18 binding protein 1, also known as Knl2) [356–358,362,363]. No Mis18BP1 has been identified in *S. pombe*, but two recently identified proteins, Mis19/Eic1 and Mis20/Eic2 may act as functional orthologs of M18BP1 in this organism [364,365].

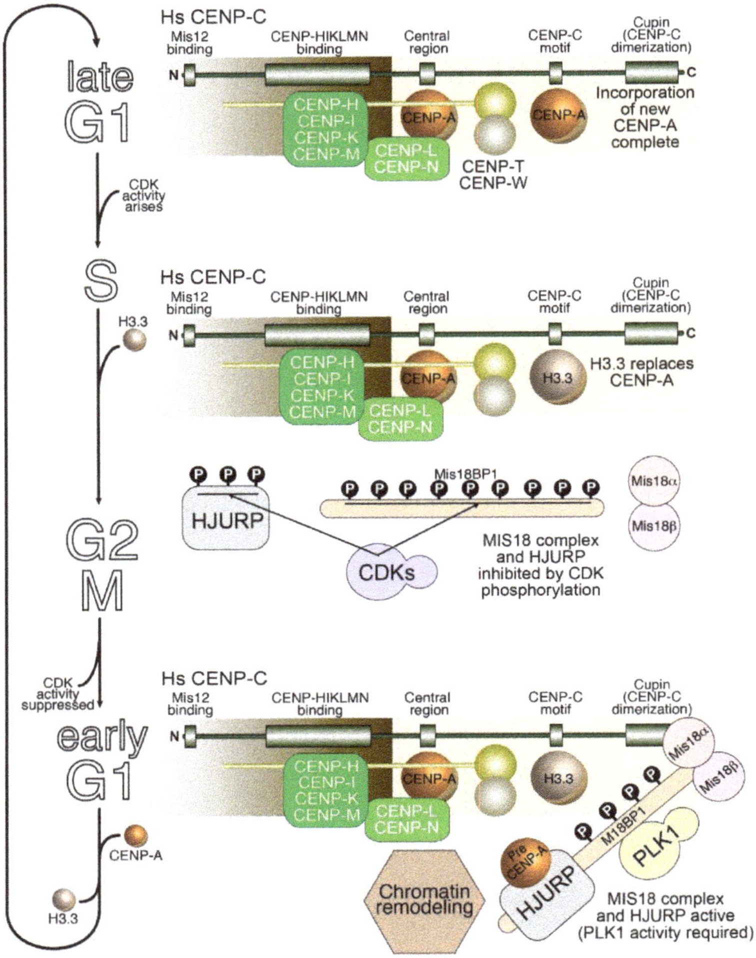

Figure 9. Cell cycle-regulated replenishment of CENP-A nucleosomes New. CENP-A incorporation takes place after mitotic exit, early in G1 phase (*bottom*). It is driven by an existing, active CENP-A nucleosome (e.g., because bound to CCAN subunits), and directed on a neighboring nucleosome. Here, we hypothesize that the neighboring nucleosome in humans is already bound to the CENP-C motif of CENP-C. An active MIS18 complex, including M18BP1, recruits HJURP, which binds to pre-nucleosomal CENP-A. The C-terminal region of CENP-C has been implicated in this reaction, which also requires PLK1 activity. A chromatin-remodeling factor and other chromatin-associated factors promote extraction of H3.3 and its replacement with CENP-A through an ATP-dependent reaction. The resulting configuration (*top*) persists until DNA replication (S-phase), when CENP-A "vacancies" caused by distribution of CENP-A to the sister chromatids during DNA replication, are filled with H3.3. This configuration then persists through the rest of the cell cycle until mitotic exit, because the CENP-A loading machinery is inhibited by CDK activity (*middle*).

The Mis18 complex, in conjunction with M18BP1, mediates a cell-cycle-regulated interaction with kinetochores that ultimately promotes the localization of a CENP-A selective histone chaperone known in vertebrates as HJURP (Holliday Junction-Recognizing Protein) [366–368]. Functional orthologs with limited sequence similarity to HJURP exist in *S. cerevisiae*, *S. pombe*, and *D. melanogaster* (named Scm3 and Cal1) [369–379]. HJURP, which may form functional dimers [380], contains an N-terminal Scm3-homology domain that binds pre-nucleosomal CENP-A [73,75,158,381–384]. Its central region, on the other hand, mediates binding to M18BP1 and is required for kinetochore recruitment [73,385,386]. The precise mechanism of kinetochore recruitment of the Mis18 complex and HJURP remains unclear, but interactions with CCAN subunits, including CENP-I^{Mis6} and CENP-C, have been identified [75,158,374,378,385,387–391]. In *S. cerevisiae*, on the other hand, Scm3 is recruited through an interaction with the CBF3 complex [371].

CENP-A deposition is tightly coordinated with cell cycle progression in a manner distinct from canonical histone H3, which is deposited concomitantly with DNA replication in S-phase. In vertebrates, incorporation of new CENP-A is limited to telophase and early G1, when CDK activity is suppressed [392–395]. The increase in CDK activity that precedes the initiation of DNA replication may be sufficient for inhibition of CENP-A deposition, implicating pre-mitotic Cdk2 kinase activity in complex with Cyclin E and Cyclin A as potential negative regulators of this process. A major target of this regulation is M18BP1, which CDKs phosphorylate at multiple sites, preventing its interaction with Mis18 subunits and its kinetochore recruitment [392,395,396]. Inhibition of CENP-A deposition through phosphorylation of HJURP and CENP-A has also been described [393,394,396,397].

The activity of another kinase, Polo-like Kinase 1 (Plk1), on the other hand, is required for the incorporation of new CENP-A [395]. Suppression of CDK activity in the G2 phase of the cell cycle results in ectopic and Plk1-dependent incorporation of CENP-A, suggesting that Plk1 is required for CENP-A deposition but does not contribute to cell cycle phase coordination. Like CDKs, Plk1 also targets the Mis18:M18BP1 complex, but the precise interactions controlled by this kinase remain to be discovered.

The molecular mechanism of new CENP-A deposition remains unclear [5]. Factors implicated in CENP-A deposition, or in its stable maintenance at centromeres, include the chromatin remodeling FACT complex, the histone chaperone NPM1/nucleophosmin, the GTPase-activating protein MgcRacGAP, And-1, and Condensin II [99,112,349,368,398–405]. The RSF and MgcRacGAP, the latter identified as a binding partner of [363] P1/Knl2, have been implicated specifically in the maintenance of newly incorporated CENP-A [400]. Several DNA and histone post-translational modifications have been implicated in CENP-A loading, including DNA methylation by DNMT3B, possibly through interactions with the C-terminal region of CENP-C and with Mis18α [406,407], acetylation [402,408,409], and histone H3K4 dimethylation [410,411].

As clarified above, during DNA replication the chromosome levels of CENP-A are reduced by 50%. It has been proposed that the CENP-A 'vacancies' generated at this stage are filled by placeholder histone H3.3 [354]. New CENP-A deposition in telophase and early G1 may therefore require the eviction of the H3.3 previously used to fill the CENP-A vacancy, and its replacement with CENP-A. Existing kinetochores provide a recruitment platform for the CENP-A deposition machinery, limiting its function to existing centromeres. The amount of CENP-A on each chromosome appears to be constant through generations, implying that each CENP-A may only trigger a single cycle of CENP-A incorporation [412,413]. Like other histone replacement reactions [414], rapid replacement of H3 with CENP-A is likely to require a source of energy. The ATPase carrying out the replacement has not yet been identified with certainty. Two chromatin-remodeling enzymes, Chd1 and RSF, have been implicated in CENP-A deposition in human cells [99,112,401], while Chd1 appears to be dispensable for CENP-A deposition in *D. melanogaster* [415].

If the CENP-A deposition reaction implies the substitution of H3.3 with CENP-A, a crucial question is how the CENP-A deposition machinery targets a specific H3.3 nucleosome neighboring CENP-A. Likely, there are specific molecular features that designate it as the nucleosome to be replaced. We suggest that the C-terminal region of CENP-C is responsible for this labeling function (Figure 9). As explained in Section 3, the N-terminal region of CENP-C, comprising a succession of binding sites for Mis12, the CENP-CHIKMLN complex, and CENP-A, is required for kinetochore assembly [110,165].

However, CENP-C contains a second nucleosome-binding motif, the CENP-C motif, and we surmise that it mediates the association with a second nucleosome. Our speculative working model is that the structure recognized by the CENP-A deposition machinery is a CENP-A:H3 dinucleosome in which CENP-C acts as a bridge between the two nucleosomes. After eviction of H3 and deposition of CENP-A, the deposition reaction is complete and the deposition machinery dissociates.

This model predicts that the core kinetochore module in regional kinetochores is a dinucleosome structure. One of the nucleosomes has permanent identity as CENP-A, being bound to the N-terminal region of CENP-C where all CCAN subunits stably assemble. The second one, bound to the C-terminal region of CENP-C (implicated in recruitment of the CENP-A deposition machinery, as observed above), toggles between two identities (H3 and CENP-A) during the cell cycle (Figure 9). While this hypothesis requires considerable further scrutiny, it is consistent with studies showing that human CENP-A nucleosomes are enriched on units of ~340 bps consisting of two α-satellite repeats separated by a CENP-B box, and on which CENP-B, CENP-C, and CENP-T also appear to co-exist with two CENP-A molecules in a single complex [77,78,416]. Dimeric repeat units are observed in other mammals [417]. In *S. cerevisiae*, where a specific DNA motif defines the centromere, only the organization of the "permanent" CENP-A nucleosome is preserved, while CBF3 functionally replaces the second nucleosome by providing direct DNA recognition and by recruiting the Scm3 chaperone. In this organism, deposition of new Cse4 occurs concomitantly with DNA replication [355,418].

8. Unconventional Kinetochores

Our discussion above has focused on kinetochore assembly that occurs on CENP-A nucleosomal chromatin. The picture we present, in which CENP-A nucleosomal chromatin recruits CCAN components that recruit KMN complexes that in turn recruit other microtubule-interacting proteins and SAC machinery, is true in budding and fission yeast, *D. melanogaster*, *C. elegans*, and vertebrates. However, notable exceptions to this picture have been found and have the potential to reveal core principles on how kinetochores are built to couple to microtubules and ensure chromosome segregation. The most striking exception found to date is in kinetoplastids, where proteins that are homologous to the majority of the kinetochore proteins we discussed here (CENP-A, CCAN, KMN) have not been found. Based on tagging and proteomic analysis, 20 kinetochore proteins have been identified in the kinetoplastid *Trypanosoma brucei* [419] and their detailed characterization should reveal how this divergent kinetochore is built and operates. Based on weak homology to the Ndc80 and Nuf2 coiled-coils, a very recent study identified a new kinetochore-localized protein in *T. brucei* that is important for chromosome segregation [420]. Although this protein lacks the calponin homology (CH) domains of Ndc80 and Nuf2 that mediate microtubule binding, the sequence homology in the coiled coil suggests divergence from an ancestral Ndc80/Nuf2-like protein. Kinetoplastids are classified as belonging to a diverse group of eukaryotes (known as the Excavata) that are distinct from the group that contains the commonly studied model organisms (known as the Opisthokonta). *Giardia intestinalis*, a diplomonad that is also classified as belonging to the Excavata group, does have a CENP-A-like histone and KMN proteins. Thus, kinetoplastids exhibit extreme divergence of ancestral components together with emergence of new kinetochore machinery, the reasons for which are currently mysterious. A second exception to the canonical kinetochore assembly pathway is observed in holocentric insect species, which have independently arisen multiple times. In at least four independent lineages of insects where holocentricity evolved, CENP-A and CENP-C appear to have been lost but KMN components are still present [62]. Thus, these holocentric insect lineages appear to build their kinetochores on a different foundation than CENP-A. This finding echoes earlier work in *C. elegans*, where the mitotic kinetochore requires CENP-A for assembly and follows the canonical pathway but the meiotic kinetochore in oocytes is built independently of CENP-A and CENP-C [421]. Defining how the holocentric insect kinetochore is built independently of CENP-A and assessing whether an epigenetic inheritance mechanism, similar to what is observed with a CENP-A based kinetochore foundation, is also operating in these species will be very revealing.

9. Conclusions

In this review, we have focused on the structural organization of the kinetochore. It is with this largely evolutionarily conserved scaffold that the kinetochore performs its complex functions, including error correction, spindle assembly checkpoint control, and chromosome alignment and segregation. While a detailed mechanistic understanding of kinetochore dynamics is still lacking, our grasp of kinetochore architecture inspires molecular hypotheses on the molecular changes to kinetochore architecture that distinguish microtubule-bound and -unbound states and their signaling properties.

Acknowledgments: Andrea Musacchio acknowledges funding by the European Research Council (ERC) AdG RECEPIANCE (No. 669686) and the DFG's Collaborative Research Centre (CRC) 1093. Arshad Desai acknowledges funding from the NIH (GM074215) and the Ludwig Institute for Cancer Research. Yoana Dimitrova, Stephen C. Harrison, Alexey Khodjakov, Sue Biggins, Tamir Gonen, Pim Huis in 't Veld, and J. Richard McIntosh contributed original figure panels.

Author Contributions: Andrea Musacchio and Arshad Desai discussed the content of the review and wrote it.

Conflicts of Interest: The authors declare no conflict of interest.

References

1. Cheeseman, I.M. The kinetochore. *Cold Spring Harb. Perspect. Biol.* **2014**, *6*, a015826. [CrossRef] [PubMed]
2. Nagpal, H.; Fukagawa, T. Kinetochore assembly and function through the cell cycle. *Chromosoma* **2016**, *125*, 645–659. [CrossRef] [PubMed]
3. Pesenti, M.E.; Weir, J.R.; Musacchio, A. Progress in the structural and functional characterization of kinetochores. *Curr. Opin. Struct. Biol.* **2016**, *37*, 152–163. [CrossRef] [PubMed]
4. McKinley, K.L.; Cheeseman, I.M. The molecular basis for centromere identity and function. *Nat. Rev. Mol. Cell Biol.* **2016**, *17*, 16–29. [CrossRef] [PubMed]
5. Muller, S.; Almouzni, G. A network of players in H3 histone variant deposition and maintenance at centromeres. *Biochim. Biophys. Acta* **2014**, *1839*, 241–250. [CrossRef] [PubMed]
6. Fukagawa, T.; Earnshaw, W.C. The centromere: Chromatin foundation for the kinetochore machinery. *Dev. Cell* **2014**, *30*, 496–508. [CrossRef] [PubMed]
7. Rieder, C.L. The formation, structure, and composition of the mammalian kinetochore and kinetochore fiber. *Int. Rev. Cytol.* **1982**, *79*, 1–58. [PubMed]
8. Magidson, V.; Paul, R.; Yang, N.; Ault, J.G.; O'Connell, C.B.; Tikhonenko, I.; McEwen, B.F.; Mogilner, A.; Khodjakov, A. Adaptive changes in the kinetochore architecture facilitate proper spindle assembly. *Nat. Cell Biol.* **2015**, *17*, 1134–1144. [CrossRef] [PubMed]
9. Foley, E.A.; Kapoor, T.M. Microtubule attachment and spindle assembly checkpoint signalling at the kinetochore. *Nat. Rev. Mol. Cell Biol.* **2013**, *14*, 25–37. [CrossRef] [PubMed]
10. Nicklas, R.B.; Koch, C.A. Chromosome micromanipulation. 3. Spindle fiber tension and the reorientation of mal-oriented chromosomes. *J. Cell Biol.* **1969**, *43*, 40–50. [CrossRef] [PubMed]
11. Uchida, K.S.; Takagaki, K.; Kumada, K.; Hirayama, Y.; Noda, T.; Hirota, T. Kinetochore stretching inactivates the spindle assembly checkpoint. *J. Cell Biol.* **2009**, *184*, 383–390. [CrossRef] [PubMed]
12. Maresca, T.J.; Salmon, E.D. Intrakinetochore stretch is associated with changes in kinetochore phosphorylation and spindle assembly checkpoint activity. *J. Cell Biol.* **2009**, *184*, 373–381. [CrossRef] [PubMed]
13. Maresca, T.J.; Salmon, E.D. Welcome to a new kind of tension: Translating kinetochore mechanics into a wait-anaphase signal. *J. Cell Sci.* **2010**, *123*, 825–835. [CrossRef] [PubMed]
14. Lampson, M.A.; Grishchuk, E.L. Mechanisms to Avoid and Correct Erroneous Kinetochore-Microtubule Attachments. *Biology (Basel)* **2017**, *6*. [CrossRef]
15. Musacchio, A. The Molecular Biology of Spindle Assembly Checkpoint Signaling Dynamics. *Curr. Biol.* **2015**, *25*, R1002–R1018. [CrossRef] [PubMed]
16. London, N.; Biggins, S. Signalling dynamics in the spindle checkpoint response. *Nat. Rev. Mol. Cell Biol.* **2014**, *15*, 736–747. [CrossRef] [PubMed]
17. Joglekar, A.P. A Cell Biological Perspective on Past, Present and Future Investigations of the Spindle Assembly Checkpoint. *Biology (Basel)* **2016**, *5*, E44. [CrossRef] [PubMed]

18. Brinkley, B.R.; Stubblefield, E. The fine structure of the kinetochore of a mammalian cell in vitro. *Chromosoma* **1966**, *19*, 28–43. [CrossRef] [PubMed]
19. Rieder, C.L.; Alexander, S.P. Kinetochores are transported poleward along a single astral microtubule during chromosome attachment to the spindle in newt lung cells. *J. Cell Biol.* **1990**, *110*, 81–95. [CrossRef] [PubMed]
20. McEwen, B.F.; Arena, J.T.; Frank, J.; Rieder, C.L. Structure of the colcemid-treated PtK1 kinetochore outer plate as determined by high voltage electron microscopic tomography. *J. Cell Biol.* **1993**, *120*, 301–312. [CrossRef] [PubMed]
21. Jokelainen, P.T. The ultrastructure and spatial organization of the metaphase kinetochore in mitotic rat cells. *J. Ultrastruct. Res.* **1967**, *19*, 19–44. [CrossRef]
22. Yao, X.; Anderson, K.L.; Cleveland, D.W. The microtubule-dependent motor centromere-associated protein E (CENP-E) is an integral component of kinetochore corona fibers that link centromeres to spindle microtubules. *J. Cell Biol.* **1997**, *139*, 435–447. [CrossRef] [PubMed]
23. Cooke, C.A.; Schaar, B.; Yen, T.J.; Earnshaw, W.C. Localization of CENP-E in the fibrous corona and outer plate of mammalian kinetochores from prometaphase through anaphase. *Chromosoma* **1997**, *106*, 446–455. [CrossRef] [PubMed]
24. Hoffman, D.B.; Pearson, C.G.; Yen, T.J.; Howell, B.J.; Salmon, E.D. Microtubule-dependent changes in assembly of microtubule motor proteins and mitotic spindle checkpoint proteins at PtK1 kinetochores. *Mol. Biol. Cell* **2001**, *12*, 1995–2009. [CrossRef] [PubMed]
25. Wynne, D.J.; Funabiki, H. Kinetochore function is controlled by a phospho-dependent coexpansion of inner and outer components. *J. Cell Biol.* **2015**, *210*, 899–916. [CrossRef] [PubMed]
26. Thrower, D.A.; Jordan, M.A.; Wilson, L. Modulation of CENP-E organization at kinetochores by spindle microtubule attachment. *Cell Motil. Cytoskelet.* **1996**, *35*, 121–133. [CrossRef]
27. Magidson, V.; He, J.; Ault, J.G.; O'Connell, C.B.; Yang, N.; Tikhonenko, I.; McEwen, B.F.; Sui, H.; Khodjakov, A. Unattached kinetochores rather than intrakinetochore tension arrest mitosis in taxol-treated cells. *J. Cell Biol.* **2016**, *212*, 307–319. [CrossRef] [PubMed]
28. Dong, Y.; Vanden Beldt, K.J.; Meng, X.; Khodjakov, A.; McEwen, B.F. The outer plate in vertebrate kinetochores is a flexible network with multiple microtubule interactions. *Nat. Cell Biol.* **2007**, *9*, 516–522. [CrossRef] [PubMed]
29. McIntosh, J.R.; Grishchuk, E.L.; Morphew, M.K.; Efremov, A.K.; Zhudenkov, K.; Volkov, V.A.; Cheeseman, I.M.; Desai, A.; Mastronarde, D.N.; Ataullakhanov, F.I. Fibrils connect microtubule tips with kinetochores: A mechanism to couple tubulin dynamics to chromosome motion. *Cell* **2008**, *135*, 322–333. [CrossRef] [PubMed]
30. Asbury, C.L. Anaphase A: Melting microtubules give mitosis its meaning. *Biology (Basel)* Submitted. **2017**.
31. Clarke, L.; Carbon, J. Genomic substitutions of centromeres in Saccharomyces cerevisiae. *Nature* **1983**, *305*, 23–28. [CrossRef] [PubMed]
32. Clarke, L.; Carbon, J. Isolation of a yeast centromere and construction of functional small circular chromosomes. *Nature* **1980**, *287*, 504–509. [CrossRef] [PubMed]
33. Fitzgerald-Hayes, M.; Clarke, L.; Carbon, J. Nucleotide sequence comparisons and functional analysis of yeast centromere DNAs. *Cell* **1982**, *29*, 235–244. [CrossRef]
34. Lechner, J.; Carbon, J. A 240 kd multisubunit protein complex, CBF3, is a major component of the budding yeast centromere. *Cell* **1991**, *64*, 717–725. [CrossRef]
35. Hieter, P.; Pridmore, D.; Hegemann, J.H.; Thomas, M.; Davis, R.W.; Philippsen, P. Functional selection and analysis of yeast centromeric DNA. *Cell* **1985**, *42*, 913–921. [CrossRef]
36. Pluta, A.F.; Mackay, A.M.; Ainsztein, A.M.; Goldberg, I.G.; Earnshaw, W.C. The centromere: Hub of chromosomal activities. *Science* **1995**, *270*, 1591–1594. [CrossRef] [PubMed]
37. Catania, S.; Allshire, R.C. Anarchic centromeres: Deciphering order from apparent chaos. *Curr. Opin. Cell Biol.* **2014**, *26*, 41–50. [CrossRef] [PubMed]
38. Aldrup-Macdonald, M.E.; Sullivan, B.A. The past, present, and future of human centromere genomics. *Genes (Basel)* **2014**, *5*, 33–50. [CrossRef] [PubMed]
39. Steiner, N.C.; Clarke, L. A novel epigenetic effect can alter centromere function in fission yeast. *Cell* **1994**, *79*, 865–874. [CrossRef]
40. Earnshaw, W.C.; Ratrie, H., 3rd; Stetten, G. Visualization of centromere proteins CENP-B and CENP-C on a stable dicentric chromosome in cytological spreads. *Chromosoma* **1989**, *98*, 1–12. [CrossRef] [PubMed]

41. Sullivan, B.A.; Schwartz, S. Identification of centromeric antigens in dicentric Robertsonian translocations: CENP-C and CENP-E are necessary components of functional centromeres. *Hum. Mol. Genet.* **1995**, *4*, 2189–2197. [CrossRef] [PubMed]
42. Voullaire, L.E.; Slater, H.R.; Petrovic, V.; Choo, K.H. A functional marker centromere with no detectable alpha-satellite, satellite III, or CENP-B protein: Activation of a latent centromere? *Am. J. Hum. Genet.* **1993**, *52*, 1153–1163. [PubMed]
43. Du Sart, D.; Cancilla, M.R.; Earle, E.; Mao, J.I.; Saffery, R.; Tainton, K.M.; Kalitsis, P.; Martyn, J.; Barry, A.E.; Choo, K.H. A functional neo-centromere formed through activation of a latent human centromere and consisting of non-alpha-satellite DNA. *Nat. Genet.* **1997**, *16*, 144–153. [CrossRef] [PubMed]
44. Amor, D.J.; Bentley, K.; Ryan, J.; Perry, J.; Wong, L.; Slater, H.; Choo, K.H. Human centromere repositioning "in progress". *Proc. Natl. Acad. Sci. USA* **2004**, *101*, 6542–6547. [CrossRef] [PubMed]
45. Murphy, T.D.; Karpen, G.H. Localization of centromere function in a Drosophila minichromosome. *Cell* **1995**, *82*, 599–609. [CrossRef]
46. Shang, W.H.; Hori, T.; Toyoda, A.; Kato, J.; Popendorf, K.; Sakakibara, Y.; Fujiyama, A.; Fukagawa, T. Chickens possess centromeres with both extended tandem repeats and short non-tandem-repetitive sequences. *Genome Res.* **2010**, *20*, 1219–1228. [CrossRef] [PubMed]
47. Locke, D.P.; Hillier, L.W.; Warren, W.C.; Worley, K.C.; Nazareth, L.V.; Muzny, D.M.; Yang, S.P.; Wang, Z.; Chinwalla, A.T.; Minx, P.; et al. Comparative and demographic analysis of orang-utan genomes. *Nature* **2011**, *469*, 529–533. [CrossRef] [PubMed]
48. Piras, F.M.; Nergadze, S.G.; Magnani, E.; Bertoni, L.; Attolini, C.; Khoriauli, L.; Raimondi, E.; Giulotto, E. Uncoupling of satellite DNA and centromeric function in the genus equus. *PLoS Genet.* **2010**, *6*, e1000845. [CrossRef] [PubMed]
49. Wade, C.M.; Giulotto, E.; Sigurdsson, S.; Zoli, M.; Gnerre, S.; Imsland, F.; Lear, T.L.; Adelson, D.L.; Bailey, E.; Bellone, R.R.; et al. Genome sequence, comparative analysis, and population genetics of the domestic horse. *Science* **2009**, *326*, 865–867. [CrossRef] [PubMed]
50. Sanyal, K.; Baum, M.; Carbon, J. Centromeric DNA sequences in the pathogenic yeast Candida albicans are all different and unique. *Proc. Natl. Acad. Sci. USA* **2004**, *101*, 11374–11379. [CrossRef] [PubMed]
51. Montefalcone, G.; Tempesta, S.; Rocchi, M.; Archidiacono, N. Centromere repositioning. *Genome Res.* **1999**, *9*, 1184–1188. [CrossRef] [PubMed]
52. Rocchi, M.; Archidiacono, N.; Schempp, W.; Capozzi, O.; Stanyon, R. Centromere repositioning in mammals. *Heredity (Edinb)* **2012**, *108*, 59–67. [CrossRef] [PubMed]
53. Ventura, M.; Antonacci, F.; Cardone, M.F.; Stanyon, R.; D'Addabbo, P.; Cellamare, A.; Sprague, L.J.; Eichler, E.E.; Archidiacono, N.; Rocchi, M. Evolutionary formation of new centromeres in macaque. *Science* **2007**, *316*, 243–246. [CrossRef] [PubMed]
54. Wiens, G.R.; Sorger, P.K. Centromeric chromatin and epigenetic effects in kinetochore assembly. *Cell* **1998**, *93*, 313–316. [CrossRef]
55. Karpen, G.H.; Allshire, R.C. The case for epigenetic effects on centromere identity and function. *Trends Genet.* **1997**, *13*, 489–496. [CrossRef]
56. Murphy, T.D.; Karpen, G.H. Centromeres take flight: Alpha satellite and the quest for the human centromere. *Cell* **1998**, *93*, 317–320. [CrossRef]
57. Earnshaw, W.C.; Rothfield, N. Identification of a family of human centromere proteins using autoimmune sera from patients with scleroderma. *Chromosoma* **1985**, *91*, 313–321. [CrossRef] [PubMed]
58. Earnshaw, W.C. Discovering centromere proteins: From cold white hands to the A, B, C of CENPs. *Nat. Rev. Mol. Cell Biol.* **2015**, *16*, 443–449. [CrossRef] [PubMed]
59. Palmer, D.K.; O'Day, K.; Wener, M.H.; Andrews, B.S.; Margolis, R.L. A 17-kD centromere protein (CENP-A) copurifies with nucleosome core particles and with histones. *J. Cell Biol.* **1987**, *104*, 805–815. [CrossRef] [PubMed]
60. Palmer, D.K.; O'Day, K.; Trong, H.L.; Charbonneau, H.; Margolis, R.L. Purification of the centromere-specific protein CENP-A and demonstration that it is a distinctive histone. *Proc. Natl. Acad. Sci. USA* **1991**, *88*, 3734–3738. [CrossRef] [PubMed]

61. Sullivan, K.F.; Hechenberger, M.; Masri, K. Human CENP-A contains a histone H3 related histone fold domain that is required for targeting to the centromere. *J. Cell Biol.* **1994**, *127*, 581–592. [CrossRef] [PubMed]
62. Drinnenberg, I.A.; deYoung, D.; Henikoff, S.; Malik, H.S. Recurrent loss of CenH3 is associated with independent transitions to holocentricity in insects. *eLife* **2014**, *3*. [CrossRef] [PubMed]
63. Warburton, P.E.; Cooke, C.A.; Bourassa, S.; Vafa, O.; Sullivan, B.A.; Stetten, G.; Gimelli, G.; Warburton, D.; Tyler-Smith, C.; Sullivan, K.F.; et al. Immunolocalization of cenp-a suggests a distinct nucleosome structure at the inner kinetochore plate of active centromeres. *Curr. Biol.* **1997**, *7*, 901–904. [CrossRef]
64. Stoler, S.; Keith, K.C.; Curnick, K.E.; Fitzgerald-Hayes, M. A mutation in CSE4, an essential gene encoding a novel chromatin-associated protein in yeast, causes chromosome nondisjunction and cell cycle arrest at mitosis. *Genes Dev.* **1995**, *9*, 573–586. [CrossRef] [PubMed]
65. Oegema, K.; Desai, A.; Rybina, S.; Kirkham, M.; Hyman, A.A. Functional analysis of kinetochore assembly in *Caenorhabditis elegans*. *J. Cell Biol.* **2001**, *153*, 1209–1226. [CrossRef] [PubMed]
66. Moore, L.L.; Roth, M.B. HCP-4, a CENP-C-like protein in *Caenorhabditis elegans*, is required for resolution of sister centromeres. *J. Cell Biol.* **2001**, *153*, 1199–1208. [CrossRef] [PubMed]
67. Howman, E.V.; Fowler, K.J.; Newson, A.J.; Redward, S.; MacDonald, A.C.; Kalitsis, P.; Choo, K.H. Early disruption of centromeric chromatin organization in centromere protein a (cenpa) null mice. *Proc. Natl. Acad. Sci. USA* **2000**, *97*, 1148–1153. [CrossRef] [PubMed]
68. Liu, S.T.; Rattner, J.B.; Jablonski, S.A.; Yen, T.J. Mapping the assembly pathways that specify formation of the trilaminar kinetochore plates in human cells. *J. Cell Biol.* **2006**, *175*, 41–53. [CrossRef] [PubMed]
69. Fachinetti, D.; Folco, H.D.; Nechemia-Arbely, Y.; Valente, L.P.; Nguyen, K.; Wong, A.J.; Zhu, Q.; Holland, A.J.; Desai, A.; Jansen, L.E.; et al. A two-step mechanism for epigenetic specification of centromere identity and function. *Nat. Cell Biol.* **2013**, *15*, 1056–1066. [CrossRef] [PubMed]
70. Logsdon, G.A.; Barrey, E.J.; Bassett, E.A.; DeNizio, J.E.; Guo, L.Y.; Panchenko, T.; Dawicki-McKenna, J.M.; Heun, P.; Black, B.E. Both tails and the centromere targeting domain of cenp-a are required for centromere establishment. *J. Cell Biol.* **2015**, *208*, 521–531. [CrossRef] [PubMed]
71. Heun, P.; Erhardt, S.; Blower, M.D.; Weiss, S.; Skora, A.D.; Karpen, G.H. Mislocalization of the Drosophila centromere-specific histone CID promotes formation of functional ectopic kinetochores. *Dev. Cell* **2006**, *10*, 303–315. [CrossRef] [PubMed]
72. Mendiburo, M.J.; Padeken, J.; Fulop, S.; Schepers, A.; Heun, P. Drosophila CENH3 is sufficient for centromere formation. *Science* **2011**, *334*, 686–690. [CrossRef] [PubMed]
73. Barnhart, M.C.; Kuich, P.H.; Stellfox, M.E.; Ward, J.A.; Bassett, E.A.; Black, B.E.; Foltz, D.R. Hjurp is a cenp-a chromatin assembly factor sufficient to form a functional de novo kinetochore. *J. Cell Biol.* **2011**, *194*, 229–243. [CrossRef] [PubMed]
74. Van Hooser, A.A.; Ouspenski, I.I.; Gregson, H.C.; Starr, D.A.; Yen, T.J.; Goldberg, M.L.; Yokomori, K.; Earnshaw, W.C.; Sullivan, K.F.; Brinkley, B.R. Specification of kinetochore-forming chromatin by the histone h3 variant cenp-a. *J. Cell Sci.* **2001**, *114*, 3529–3542. [PubMed]
75. Hori, T.; Shang, W.H.; Takeuchi, K.; Fukagawa, T. The CCAN recruits CENP-A to the centromere and forms the structural core for kinetochore assembly. *J. Cell Biol.* **2013**, *200*, 45–60. [CrossRef] [PubMed]
76. Guse, A.; Carroll, C.W.; Moree, B.; Fuller, C.J.; Straight, A.F. In vitro centromere and kinetochore assembly on defined chromatin templates. *Nature* **2011**, *477*, 354–358. [CrossRef] [PubMed]
77. Henikoff, J.G.; Thakur, J.; Kasinathan, S.; Henikoff, S. A unique chromatin complex occupies young alpha-satellite arrays of human centromeres. *Sci. Adv.* **2015**, *1*, e1400234. [CrossRef] [PubMed]
78. Thakur, J.; Henikoff, S. CENPT bridges adjacent CENPA nucleosomes on young human alpha-satellite dimers. *Genome Res.* **2016**, *26*, 1178–1187. [CrossRef] [PubMed]
79. Kato, H.; Jiang, J.; Zhou, B.R.; Rozendaal, M.; Feng, H.; Ghirlando, R.; Xiao, T.S.; Straight, A.F.; Bai, Y. A conserved mechanism for centromeric nucleosome recognition by centromere protein cenp-c. *Science* **2013**, *340*, 1110–1113. [CrossRef] [PubMed]
80. Klare, K.; Weir, J.R.; Basilico, F.; Zimniak, T.; Massimiliano, L.; Ludwigs, N.; Herzog, F.; Musacchio, A. Cenp-c is a blueprint for constitutive centromere-associated network assembly within human kinetochores. *J. Cell Biol.* **2015**, *210*, 11–22. [CrossRef] [PubMed]
81. Malik, H.S.; Henikoff, S. Conflict begets complexity: The evolution of centromeres. *Curr. Opin. Genet. Dev.* **2002**, *12*, 711–718. [CrossRef]

82. Talbert, P.B.; Bryson, T.D.; Henikoff, S. Adaptive evolution of centromere proteins in plants and animals. *J. Biol.* **2004**, *3*, 18. [CrossRef] [PubMed]
83. Schueler, M.G.; Swanson, W.; Thomas, P.J.; Program, N.C.S.; Green, E.D. Adaptive evolution of foundation kinetochore proteins in primates. *Mol. Biol. Evol.* **2010**, *27*, 1585–1597. [CrossRef] [PubMed]
84. Chmatal, L.; Gabriel, S.I.; Mitsainas, G.P.; Martinez-Vargas, J.; Ventura, J.; Searle, J.B.; Schultz, R.M.; Lampson, M.A. Centromere strength provides the cell biological basis for meiotic drive and karyotype evolution in mice. *Curr. Biol.* **2014**, *24*, 2295–2300. [CrossRef] [PubMed]
85. Bassett, E.A.; Wood, S.; Salimian, K.J.; Ajith, S.; Foltz, D.R.; Black, B.E. Epigenetic centromere specification directs aurora B accumulation but is insufficient to efficiently correct mitotic errors. *J. Cell Biol.* **2010**, *190*, 177–185. [CrossRef] [PubMed]
86. Nonaka, N.; Kitajima, T.; Yokobayashi, S.; Xiao, G.; Yamamoto, M.; Grewal, S.I.; Watanabe, Y. Recruitment of cohesin to heterochromatic regions by swi6/hp1 in fission yeast. *Nat. Cell Biol.* **2002**, *4*, 89–93. [CrossRef] [PubMed]
87. Bernard, P.; Maure, J.F.; Partridge, J.F.; Genier, S.; Javerzat, J.P.; Allshire, R.C. Requirement of heterochromatin for cohesion at centromeres. *Science* **2001**, *294*, 2539–2542. [CrossRef] [PubMed]
88. Westermann, S.; Schleiffer, A. Family matters: Structural and functional conservation of centromere-associated proteins from yeast to humans. *Trends Cell Biol.* **2013**, *23*, 260–269. [CrossRef] [PubMed]
89. Meraldi, P.; McAinsh, A.D.; Rheinbay, E.; Sorger, P.K. Phylogenetic and structural analysis of centromeric DNA and kinetochore proteins. *Genome Biol.* **2006**, *7*, R23. [CrossRef] [PubMed]
90. Moroi, Y.; Peebles, C.; Fritzler, M.J.; Steigerwald, J.; Tan, E.M. Autoantibody to centromere (kinetochore) in scleroderma sera. *Proc. Natl. Acad. Sci. USA* **1980**, *77*, 1627–1631. [CrossRef] [PubMed]
91. Earnshaw, W.C.; Sullivan, K.F.; Machlin, P.S.; Cooke, C.A.; Kaiser, D.A.; Pollard, T.D.; Rothfield, N.F.; Cleveland, D.W. Molecular cloning of cdna for cenp-b, the major human centromere autoantigen. *J. Cell Biol.* **1987**, *104*, 817–829. [CrossRef] [PubMed]
92. Saitoh, H.; Tomkiel, J.; Cooke, C.A.; Ratrie, H., 3rd; Maurer, M.; Rothfield, N.F.; Earnshaw, W.C. Cenp-c, an autoantigen in scleroderma, is a component of the human inner kinetochore plate. *Cell* **1992**, *70*, 115–125. [CrossRef]
93. Nishihashi, A.; Haraguchi, T.; Hiraoka, Y.; Ikemura, T.; Regnier, V.; Dodson, H.; Earnshaw, W.C.; Fukagawa, T. Cenp-i is essential for centromere function in vertebrate cells. *Dev. Cell* **2002**, *2*, 463–476. [CrossRef]
94. Sugata, N.; Munekata, E.; Todokoro, K. Characterization of a novel kinetochore protein, cenp-h. *J. Biol. Chem.* **1999**, *274*, 27343–27346. [CrossRef] [PubMed]
95. Saitoh, S.; Takahashi, K.; Yanagida, M. Mis6, a fission yeast inner centromere protein, acts during g1/s and forms specialized chromatin required for equal segregation. *Cell* **1997**, *90*, 131–143. [CrossRef]
96. Okada, M.; Cheeseman, I.M.; Hori, T.; Okawa, K.; McLeod, I.X.; Yates, J.R., 3rd; Desai, A.; Fukagawa, T. The cenp-h-i complex is required for the efficient incorporation of newly synthesized cenp-a into centromeres. *Nat. Cell Biol.* **2006**, *8*, 446–457. [CrossRef] [PubMed]
97. Foltz, D.R.; Jansen, L.E.; Black, B.E.; Bailey, A.O.; Yates, J.R., 3rd; Cleveland, D.W. The human CENP-A centromeric nucleosome-associated complex. *Nat. Cell Biol.* **2006**, *8*, 458–469. [CrossRef] [PubMed]
98. Obuse, C.; Yang, H.; Nozaki, N.; Goto, S.; Okazaki, T.; Yoda, K. Proteomics analysis of the centromere complex from HeLa interphase cells: UV-damaged DNA binding protein 1 (DDB-1) is a component of the CEN-complex, while BMI-1 is transiently co-localized with the centromeric region in interphase. *Genes Cells* **2004**, *9*, 105–120. [CrossRef]
99. Izuta, H.; Ikeno, M.; Suzuki, N.; Tomonaga, T.; Nozaki, N.; Obuse, C.; Kisu, Y.; Goshima, N.; Nomura, F.; Nomura, N.; et al. Comprehensive analysis of the icen (interphase centromere complex) components enriched in the cenp-a chromatin of human cells. *Genes Cells* **2006**, *11*, 673–684. [CrossRef] [PubMed]
100. Hori, T.; Amano, M.; Suzuki, A.; Backer, C.B.; Welburn, J.P.; Dong, Y.; McEwen, B.F.; Shang, W.H.; Suzuki, E.; Okawa, K.; et al. CCAN makes multiple contacts with centromeric DNA to provide distinct pathways to the outer kinetochore. *Cell* **2008**, *135*, 1039–1052. [CrossRef] [PubMed]
101. Amano, M.; Suzuki, A.; Hori, T.; Backer, C.; Okawa, K.; Cheeseman, I.M.; Fukagawa, T. The CENP-S complex is essential for the stable assembly of outer kinetochore structure. *J. Cell Biol.* **2009**, *186*, 173–182. [CrossRef] [PubMed]
102. Cheeseman, I.M.; Desai, A. Molecular architecture of the kinetochore-microtubule interface. *Nat. Rev. Mol. Cell Biol.* **2008**, *9*, 33–46. [CrossRef] [PubMed]

103. Joglekar, A.P.; Bloom, K.; Salmon, E.D. In vivo protein architecture of the eukaryotic kinetochore with nanometer scale accuracy. *Curr. Biol.* **2009**, *19*, 694–699. [CrossRef] [PubMed]
104. Wan, X.; O'Quinn, R.P.; Pierce, H.L.; Joglekar, A.P.; Gall, W.E.; DeLuca, J.G.; Carroll, C.W.; Liu, S.T.; Yen, T.J.; McEwen, B.F.; et al. Protein architecture of the human kinetochore microtubule attachment site. *Cell* **2009**, *137*, 672–684. [CrossRef] [PubMed]
105. Suzuki, A.; Badger, B.L.; Wan, X.; DeLuca, J.G.; Salmon, E.D. The architecture of ccan proteins creates a structural integrity to resist spindle forces and achieve proper intrakinetochore stretch. *Dev. Cell* **2014**, *30*, 717–730. [CrossRef] [PubMed]
106. Joglekar, A.P.; Bouck, D.C.; Molk, J.N.; Bloom, K.S.; Salmon, E.D. Molecular architecture of a kinetochore-microtubule attachment site. *Nat. Cell Biol.* **2006**, *8*, 581–585. [CrossRef] [PubMed]
107. Pot, I.; Measday, V.; Snydsman, B.; Cagney, G.; Fields, S.; Davis, T.N.; Muller, E.G.; Hieter, P. Chl4p and iml3p are two new members of the budding yeast outer kinetochore. *Mol. Biol. Cell* **2003**, *14*, 460–476. [CrossRef] [PubMed]
108. Hinshaw, S.M.; Harrison, S.C. An Iml3-Chl4 heterodimer links the core centromere to factors required for accurate chromosome segregation. *Cell Rep.* **2013**, *5*, 29–36. [CrossRef] [PubMed]
109. McKinley, K.L.; Sekulic, N.; Guo, L.Y.; Tsinman, T.; Black, B.E.; Cheeseman, I.M. The CENP-L-N Complex Forms a Critical Node in an Integrated Meshwork of Interactions at the Centromere-Kinetochore Interface. *Mol. Cell* **2015**, *60*, 886–898. [CrossRef] [PubMed]
110. Weir, J.R.; Faesen, A.C.; Klare, K.; Petrovic, A.; Basilico, F.; Fischbock, J.; Pentakota, S.; Keller, J.; Pesenti, M.E.; Pan, D.; et al. Insights from biochemical reconstitution into the architecture of human kinetochores. *Nature* **2016**, *537*, 249–253. [CrossRef] [PubMed]
111. Basilico, F.; Maffini, S.; Weir, J.R.; Prumbaum, D.; Rojas, A.M.; Zimniak, T.; De Antoni, A.; Jeganathan, S.; Voss, B.; van Gerwen, S.; et al. The pseudo gtpase cenp-m drives human kinetochore assembly. *eLife* **2014**, *3*, e02978. [CrossRef] [PubMed]
112. Okada, M.; Okawa, K.; Isobe, T.; Fukagawa, T. Cenp-h-containing complex facilitates centromere deposition of cenp-a in cooperation with fact and chd1. *Mol. Biol. Cell* **2009**, *20*, 3986–3995. [CrossRef] [PubMed]
113. Pekgoz Altunkaya, G.; Malvezzi, F.; Demianova, Z.; Zimniak, T.; Litos, G.; Weissmann, F.; Mechtler, K.; Herzog, F.; Westermann, S. Ccan assembly configures composite binding interfaces to promote cross-linking of ndc80 complexes at the kinetochore. *Curr. Biol.* **2016**, *26*, 2370–2378. [CrossRef] [PubMed]
114. Measday, V.; Hailey, D.W.; Pot, I.; Givan, S.A.; Hyland, K.M.; Cagney, G.; Fields, S.; Davis, T.N.; Hieter, P. Ctf3p, the mis6 budding yeast homolog, interacts with mcm22p and mcm16p at the yeast outer kinetochore. *Genes Dev.* **2002**, *16*, 101–113. [CrossRef] [PubMed]
115. De Wulf, P.; McAinsh, A.D.; Sorger, P.K. Hierarchical assembly of the budding yeast kinetochore from multiple subcomplexes. *Genes Dev.* **2003**, *17*, 2902–2921. [CrossRef] [PubMed]
116. Schmitzberger, F.; Harrison, S.C. RWD domain: A recurring module in kinetochore architecture shown by a Ctf19-Mcm21 complex structure. *EMBO Rep.* **2012**, *13*, 216–222. [CrossRef] [PubMed]
117. Ortiz, J.; Stemmann, O.; Rank, S.; Lechner, J. A putative protein complex consisting of Ctf19, Mcm21, and Okp1 represents a missing link in the budding yeast kinetochore. *Genes Dev.* **1999**, *13*, 1140–1155. [CrossRef] [PubMed]
118. Meluh, P.B.; Koshland, D. Evidence that the MIF2 gene of Saccharomyces cerevisiae encodes a centromere protein with homology to the mammalian centromere protein CENP-C. *Mol. Biol. Cell* **1995**, *6*, 793–807. [CrossRef] [PubMed]
119. Akiyoshi, B.; Nelson, C.R.; Ranish, J.A.; Biggins, S. Quantitative proteomic analysis of purified yeast kinetochores identifies a PP1 regulatory subunit. *Genes Dev.* **2009**, *23*, 2887–2899. [CrossRef] [PubMed]
120. Cole, H.A.; Howard, B.H.; Clark, D.J. The centromeric nucleosome of budding yeast is perfectly positioned and covers the entire centromere. *Proc. Natl. Acad. Sci. USA* **2011**, *108*, 12687–12692. [CrossRef] [PubMed]
121. Meluh, P.B.; Yang, P.; Glowczewski, L.; Koshland, D.; Smith, M.M. Cse4p is a component of the core centromere of Saccharomyces cerevisiae. *Cell* **1998**, *94*, 607–613. [CrossRef]
122. Furuyama, S.; Biggins, S. Centromere identity is specified by a single centromeric nucleosome in budding yeast. *Proc. Natl. Acad. Sci. USA* **2007**, *104*, 14706–14711. [CrossRef] [PubMed]
123. Bram, R.J.; Kornberg, R.D. Isolation of a Saccharomyces cerevisiae centromere DNA-binding protein, its human homolog, and its possible role as a transcription factor. *Mol. Cell. Biol.* **1987**, *7*, 403–409. [CrossRef] [PubMed]

124. Espelin, C.W.; Kaplan, K.B.; Sorger, P.K. Probing the architecture of a simple kinetochore using DNA-protein crosslinking. *J. Cell Biol.* **1997**, *139*, 1383–1396. [CrossRef] [PubMed]
125. Cheeseman, I.M.; Drubin, D.G.; Barnes, G. Simple centromere, complex kinetochore: Linking spindle microtubules and centromeric DNA in budding yeast. *J. Cell Biol.* **2002**, *157*, 199–203. [CrossRef] [PubMed]
126. Hemmerich, P.; Stoyan, T.; Wieland, G.; Koch, M.; Lechner, J.; Diekmann, S. Interaction of yeast kinetochore proteins with centromere-protein/transcription factor Cbf1. *Proc. Natl. Acad. Sci. USA* **2000**, *97*, 12583–12588. [CrossRef] [PubMed]
127. Krassovsky, K.; Henikoff, J.G.; Henikoff, S. Tripartite organization of centromeric chromatin in budding yeast. *Proc. Natl. Acad. Sci. USA* **2012**, *109*, 243–248. [CrossRef] [PubMed]
128. Cho, U.S.; Harrison, S.C. Ndc10 is a platform for inner kinetochore assembly in budding yeast. *Nat. Struct. Mol. Biol.* **2012**, *19*, 48–55. [CrossRef] [PubMed]
129. Xiao, H.; Mizuguchi, G.; Wisniewski, J.; Huang, Y.; Wei, D.; Wu, C. Nonhistone Scm3 binds to AT-rich DNA to organize atypical centromeric nucleosome of budding yeast. *Mol. Cell* **2011**, *43*, 369–380. [CrossRef] [PubMed]
130. Furuyama, T.; Henikoff, S. Centromeric nucleosomes induce positive DNA supercoils. *Cell* **2009**, *138*, 104–113. [CrossRef] [PubMed]
131. Huang, C.C.; Chang, K.M.; Cui, H.; Jayaram, M. Histone H3-variant Cse4-induced positive DNA supercoiling in the yeast plasmid has implications for a plasmid origin of a chromosome centromere. *Proc. Natl. Acad. Sci. USA* **2011**, *108*, 13671–13676. [CrossRef] [PubMed]
132. Diaz-Ingelmo, O.; Martinez-Garcia, B.; Segura, J.; Valdes, A.; Roca, J. DNA Topology and Global Architecture of Point Centromeres. *Cell Rep.* **2015**, *13*, 667–677. [CrossRef] [PubMed]
133. Drinnenberg, I.A.; Henikoff, S.; Malik, H.S. Evolutionary Turnover of Kinetochore Proteins: A Ship of Theseus? *Trends Cell Biol.* **2016**, *26*, 498–510. [CrossRef] [PubMed]
134. Palmer, D.K.; Margolis, R.L. Kinetochore components recognized by human autoantibodies are present on mononucleosomes. *Mol. Cell. Biol.* **1985**, *5*, 173–186. [CrossRef] [PubMed]
135. Shelby, R.D.; Vafa, O.; Sullivan, K.F. Assembly of CENP-A into centromeric chromatin requires a cooperative array of nucleosomal DNA contact sites. *J. Cell Biol.* **1997**, *136*, 501–513. [CrossRef] [PubMed]
136. Yoda, K.; Ando, S.; Morishita, S.; Houmura, K.; Hashimoto, K.; Takeyasu, K.; Okazaki, T. Human centromere protein a (cenp-a) can replace histone h3 in nucleosome reconstitution in vitro. *Proc. Natl. Acad. Sci. USA* **2000**, *97*, 7266–7271. [CrossRef] [PubMed]
137. Tachiwana, H.; Kagawa, W.; Shiga, T.; Osakabe, A.; Miya, Y.; Saito, K.; Hayashi-Takanaka, Y.; Oda, T.; Sato, M.; Park, S.Y.; et al. Crystal structure of the human centromeric nucleosome containing cenp-a. *Nature* **2011**, *476*, 232–235. [CrossRef] [PubMed]
138. Kingston, I.J.; Yung, J.S.; Singleton, M.R. Biophysical characterization of the centromere-specific nucleosome from budding yeast. *J. Biol. Chem.* **2011**, *286*, 4021–4026. [CrossRef] [PubMed]
139. Black, B.E.; Foltz, D.R.; Chakravarthy, S.; Luger, K.; Woods, V.L., Jr.; Cleveland, D.W. Structural determinants for generating centromeric chromatin. *Nature* **2004**, *430*, 578–582. [CrossRef] [PubMed]
140. Falk, S.J.; Lee, J.; Sekulic, N.; Sennett, M.A.; Lee, T.H.; Black, B.E. CENP-C directs a structural transition of CENP-A nucleosomes mainly through sliding of DNA gyres. *Nat. Struct. Mol. Biol.* **2016**, *23*, 204–208. [CrossRef] [PubMed]
141. Falk, S.J.; Guo, L.Y.; Sekulic, N.; Smoak, E.M.; Mani, T.; Logsdon, G.A.; Gupta, K.; Jansen, L.E.; Van Duyne, G.D.; Vinogradov, S.A.; et al. Chromosomes. Cenp-c reshapes and stabilizes cenp-a nucleosomes at the centromere. *Science* **2015**, *348*, 699–703. [CrossRef] [PubMed]
142. Sekulic, N.; Bassett, E.A.; Rogers, D.J.; Black, B.E. The structure of (CENP-A-H4)(2) reveals physical features that mark centromeres. *Nature* **2010**, *467*, 347–351. [CrossRef] [PubMed]
143. Conde e Silva, N.; Black, B.E.; Sivolob, A.; Filipski, J.; Cleveland, D.W.; Prunell, A. CENP-A-containing nucleosomes: Easier disassembly versus exclusive centromeric localization. *J. Mol. Biol.* **2007**, *370*, 555–573. [CrossRef] [PubMed]
144. Black, B.E.; Brock, M.A.; Bedard, S.; Woods, V.L., Jr.; Cleveland, D.W. An epigenetic mark generated by the incorporation of CENP-A into centromeric nucleosomes. *Proc. Natl. Acad. Sci. USA* **2007**, *104*, 5008–5013. [CrossRef] [PubMed]

145. Panchenko, T.; Sorensen, T.C.; Woodcock, C.L.; Kan, Z.Y.; Wood, S.; Resch, M.G.; Luger, K.; Englander, S.W.; Hansen, J.C.; Black, B.E. Replacement of histone h3 with cenp-a directs global nucleosome array condensation and loosening of nucleosome superhelical termini. *Proc. Natl. Acad. Sci. USA* **2011**, *108*, 16588–16593. [CrossRef] [PubMed]
146. Hasson, D.; Panchenko, T.; Salimian, K.J.; Salman, M.U.; Sekulic, N.; Alonso, A.; Warburton, P.E.; Black, B.E. The octamer is the major form of cenp-a nucleosomes at human centromeres. *Nat. Struct. Mol. Biol.* **2013**, *20*, 687–695. [CrossRef] [PubMed]
147. Miell, M.D.; Fuller, C.J.; Guse, A.; Barysz, H.M.; Downes, A.; Owen-Hughes, T.; Rappsilber, J.; Straight, A.F.; Allshire, R.C. Cenp-a confers a reduction in height on octameric nucleosomes. *Nat. Struct. Mol. Biol.* **2013**, *20*, 763–765. [CrossRef] [PubMed]
148. Black, B.E.; Cleveland, D.W. Epigenetic centromere propagation and the nature of CENP-a nucleosomes. *Cell* **2011**, *144*, 471–479. [CrossRef] [PubMed]
149. Padeganeh, A.; De Rop, V.; Maddox, P.S. Nucleosomal composition at the centromere: A numbers game. *Chromosome Res.* **2013**, *21*, 27–36. [CrossRef] [PubMed]
150. Carroll, C.W.; Silva, M.C.; Godek, K.M.; Jansen, L.E.; Straight, A.F. Centromere assembly requires the direct recognition of CENP-A nucleosomes by CENP-N. *Nat. Cell Biol.* **2009**, *11*, 896–902. [CrossRef] [PubMed]
151. Carroll, C.W.; Milks, K.J.; Straight, A.F. Dual recognition of CENP-A nucleosomes is required for centromere assembly. *J. Cell Biol.* **2010**, *189*, 1143–1155. [CrossRef] [PubMed]
152. Smoak, E.M.; Stein, P.; Schultz, R.M.; Lampson, M.A.; Black, B.E. Long-Term Retention of CENP-A Nucleosomes in Mammalian Oocytes Underpins Transgenerational Inheritance of Centromere Identity. *Curr. Biol.* **2016**, *26*, 1110–1116. [CrossRef] [PubMed]
153. Jansen, L.E.; Black, B.E.; Foltz, D.R.; Cleveland, D.W. Propagation of centromeric chromatin requires exit from mitosis. *J. Cell Biol.* **2007**, *176*, 795–805. [CrossRef] [PubMed]
154. Westhorpe, F.G.; Fuller, C.J.; Straight, A.F. A cell-free CENP-A assembly system defines the chromatin requirements for centromere maintenance. *J. Cell Biol.* **2015**, *209*, 789–801. [CrossRef] [PubMed]
155. Armache, K.J.; Garlick, J.D.; Canzio, D.; Narlikar, G.J.; Kingston, R.E. Structural basis of silencing: Sir3 BAH domain in complex with a nucleosome at 3.0 A resolution. *Science* **2011**, *334*, 977–982. [CrossRef] [PubMed]
156. Barbera, A.J.; Chodaparambil, J.V.; Kelley-Clarke, B.; Joukov, V.; Walter, J.C.; Luger, K.; Kaye, K.M. The nucleosomal surface as a docking station for kaposi's sarcoma herpesvirus lana. *Science* **2006**, *311*, 856–861. [CrossRef] [PubMed]
157. Makde, R.D.; England, J.R.; Yennawar, H.P.; Tan, S. Structure of RCC1 chromatin factor bound to the nucleosome core particle. *Nature* **2010**, *467*, 562–566. [CrossRef] [PubMed]
158. Tachiwana, H.; Muller, S.; Blumer, J.; Klare, K.; Musacchio, A.; Almouzni, G. HJURP involvement in de novo CenH3(CENP-A) and CENP-C recruitment. *Cell Rep.* **2015**, *11*, 22–32. [CrossRef] [PubMed]
159. Suzuki, N.; Nakano, M.; Nozaki, N.; Egashira, S.; Okazaki, T.; Masumoto, H. CENP-B interacts with CENP-C domains containing Mif2 regions responsible for centromere localization. *J. Biol. Chem.* **2004**, *279*, 5934–5946. [CrossRef] [PubMed]
160. Yang, C.H.; Tomkiel, J.; Saitoh, H.; Johnson, D.H.; Earnshaw, W.C. Identification of overlapping DNA-binding and centromere-targeting domains in the human kinetochore protein CENP-C. *Mol. Cell. Biol.* **1996**, *16*, 3576–3586. [CrossRef] [PubMed]
161. Politi, V.; Perini, G.; Trazzi, S.; Pliss, A.; Raska, I.; Earnshaw, W.C.; Della Valle, G. Cenp-c binds the alpha-satellite DNA in vivo at specific centromere domains. *J. Cell Sci.* **2002**, *115*, 2317–2327. [PubMed]
162. Trazzi, S.; Bernardoni, R.; Diolaiti, D.; Politi, V.; Earnshaw, W.C.; Perini, G.; Della Valle, G. In vivo functional dissection of human inner kinetochore protein cenp-c. *J. Struct. Biol.* **2002**, *140*, 39–48. [CrossRef]
163. Song, K.; Gronemeyer, B.; Lu, W.; Eugster, E.; Tomkiel, J.E. Mutational analysis of the central centromere targeting domain of human centromere protein C, (CENP-C). *Exp. Cell Res.* **2002**, *275*, 81–91. [CrossRef] [PubMed]
164. Lanini, L.; McKeon, F. Domains required for CENP-C assembly at the kinetochore. *Mol. Biol. Cell* **1995**, *6*, 1049–1059. [CrossRef] [PubMed]
165. Milks, K.J.; Moree, B.; Straight, A.F. Dissection of CENP-C-directed centromere and kinetochore assembly. *Mol. Biol. Cell* **2009**, *20*, 4246–4255. [CrossRef] [PubMed]
166. Tanaka, K.; Chang, H.L.; Kagami, A.; Watanabe, Y. CENP-C functions as a scaffold for effectors with essential kinetochore functions in mitosis and meiosis. *Dev. Cell* **2009**, *17*, 334–343. [CrossRef] [PubMed]

167. Cohen, R.L.; Espelin, C.W.; De Wulf, P.; Sorger, P.K.; Harrison, S.C.; Simons, K.T. Structural and functional dissection of Mif2p, a conserved DNA-binding kinetochore protein. *Mol. Biol. Cell* **2008**, *19*, 4480–4491. [CrossRef] [PubMed]
168. Sugimoto, K.; Kuriyama, K.; Shibata, A.; Himeno, M. Characterization of internal DNA-binding and C-terminal dimerization domains of human centromere/kinetochore autoantigen CENP-C in vitro: Role of DNA-binding and self-associating activities in kinetochore organization. *Chromosome Res.* **1997**, *5*, 132–141. [CrossRef] [PubMed]
169. Bodor, D.L.; Mata, J.F.; Sergeev, M.; David, A.F.; Salimian, K.J.; Panchenko, T.; Cleveland, D.W.; Black, B.E.; Shah, J.V.; Jansen, L.E. The quantitative architecture of centromeric chromatin. *eLife* **2014**, *3*, e02137. [CrossRef] [PubMed]
170. Hori, T.; Shang, W.H.; Toyoda, A.; Misu, S.; Monma, N.; Ikeo, K.; Molina, O.; Vargiu, G.; Fujiyama, A.; Kimura, H.; et al. Histone h4 lys 20 monomethylation of the cenp-a nucleosome is essential for kinetochore assembly. *Dev. Cell* **2014**, *29*, 740–749. [CrossRef] [PubMed]
171. Bailey, A.O.; Panchenko, T.; Sathyan, K.M.; Petkowski, J.J.; Pai, P.J.; Bai, D.L.; Russell, D.H.; Macara, I.G.; Shabanowitz, J.; Hunt, D.F.; et al. Posttranslational modification of cenp-a influences the conformation of centromeric chromatin. *Proc. Natl. Acad. Sci. USA* **2013**, *110*, 11827–11832. [CrossRef] [PubMed]
172. Yan, Z.; Delannoy, M.; Ling, C.; Daee, D.; Osman, F.; Muniandy, P.A.; Shen, X.; Oostra, A.B.; Du, H.; Steltenpool, J.; et al. A histone-fold complex and fancm form a conserved DNA-remodeling complex to maintain genome stability. *Mol. Cell* **2010**, *37*, 865–878. [CrossRef] [PubMed]
173. Singh, T.R.; Saro, D.; Ali, A.M.; Zheng, X.F.; Du, C.H.; Killen, M.W.; Sachpatzidis, A.; Wahengbam, K.; Pierce, A.J.; Xiong, Y.; et al. Mhf1-mhf2, a histone-fold-containing protein complex, participates in the fanconi anemia pathway via fancm. *Mol. Cell* **2010**, *37*, 879–886. [CrossRef] [PubMed]
174. Huang, M.; Kim, J.M.; Shiotani, B.; Yang, K.; Zou, L.; D'Andrea, A.D. The FANCM/FAAP24 complex is required for the DNA interstrand crosslink-induced checkpoint response. *Mol. Cell* **2010**, *39*, 259–268. [CrossRef] [PubMed]
175. Nishino, T.; Takeuchi, K.; Gascoigne, K.E.; Suzuki, A.; Hori, T.; Oyama, T.; Morikawa, K.; Cheeseman, I.M.; Fukagawa, T. Cenp-t-w-s-x forms a unique centromeric chromatin structure with a histone-like fold. *Cell* **2012**, *148*, 487–501. [CrossRef]
176. Takeuchi, K.; Nishino, T.; Mayanagi, K.; Horikoshi, N.; Osakabe, A.; Tachiwana, H.; Hori, T.; Kurumizaka, H.; Fukagawa, T. The centromeric nucleosome-like cenp-t-w-s-x complex induces positive supercoils into DNA. *Nucleic Acids Res.* **2014**, *42*, 1644–1655. [CrossRef] [PubMed]
177. Prendergast, L.; van Vuuren, C.; Kaczmarczyk, A.; Doering, V.; Hellwig, D.; Quinn, N.; Hoischen, C.; Diekmann, S.; Sullivan, K.F. Premitotic assembly of human cenps -t and -w switches centromeric chromatin to a mitotic state. *PLoS Biol.* **2011**, *9*, e1001082. [CrossRef] [PubMed]
178. Samejima, I.; Spanos, C.; Alves Fde, L.; Hori, T.; Perpelescu, M.; Zou, J.; Rappsilber, J.; Fukagawa, T.; Earnshaw, W.C. Whole-proteome genetic analysis of dependencies in assembly of a vertebrate kinetochore. *J. Cell Biol.* **2015**, *211*, 1141–1156. [CrossRef] [PubMed]
179. Folco, H.D.; Campbell, C.S.; May, K.M.; Espinoza, C.A.; Oegema, K.; Hardwick, K.G.; Grewal, S.I.; Desai, A. The cenp-a n-tail confers epigenetic stability to centromeres via the cenp-t branch of the ccan in fission yeast. *Curr. Biol.* **2015**, *25*, 348–356. [CrossRef] [PubMed]
180. Hori, T.; Okada, M.; Maenaka, K.; Fukagawa, T. CENP-O class proteins form a stable complex and are required for proper kinetochore function. *Mol. Biol. Cell* **2008**, *19*, 843–854. [CrossRef]
181. Kagawa, N.; Hori, T.; Hoki, Y.; Hosoya, O.; Tsutsui, K.; Saga, Y.; Sado, T.; Fukagawa, T. The cenp-o complex requirement varies among different cell types. *Chromosome Res.* **2014**, *22*, 293–303. [CrossRef] [PubMed]
182. Bancroft, J.; Auckland, P.; Samora, C.P.; McAinsh, A.D. Chromosome congression is promoted by CENP-Q- and CENP-E-dependent pathways. *J. Cell Sci.* **2015**, *128*, 171–184. [CrossRef] [PubMed]
183. Hua, S.; Wang, Z.; Jiang, K.; Huang, Y.; Ward, T.; Zhao, L.; Dou, Z.; Yao, X. Cenp-u cooperates with hec1 to orchestrate kinetochore-microtubule attachment. *J. Biol. Chem.* **2011**, *286*, 1627–1638. [CrossRef] [PubMed]
184. Amaro, A.C.; Samora, C.P.; Holtackers, R.; Wang, E.; Kingston, I.J.; Alonso, M.; Lampson, M.; McAinsh, A.D.; Meraldi, P. Molecular control of kinetochore-microtubule dynamics and chromosome oscillations. *Nat. Cell Biol.* **2010**, *12*, 319–329. [CrossRef] [PubMed]

185. Kang, Y.H.; Park, C.H.; Kim, T.S.; Soung, N.K.; Bang, J.K.; Kim, B.Y.; Park, J.E.; Lee, K.S. Mammalian polo-like kinase 1-dependent regulation of the pbip1-cenp-q complex at kinetochores. *J. Biol. Chem.* **2011**, *286*, 19744–19757. [CrossRef] [PubMed]
186. Kang, Y.H.; Park, J.E.; Yu, L.R.; Soung, N.K.; Yun, S.M.; Bang, J.K.; Seong, Y.S.; Yu, H.; Garfield, S.; Veenstra, T.D.; et al. Self-regulated plk1 recruitment to kinetochores by the plk1-pbip1 interaction is critical for proper chromosome segregation. *Mol. Cell* **2006**, *24*, 409–422. [CrossRef] [PubMed]
187. Hyland, K.M.; Kingsbury, J.; Koshland, D.; Hieter, P. Ctf19p: A novel kinetochore protein in Saccharomyces cerevisiae and a potential link between the kinetochore and mitotic spindle. *J. Cell Biol.* **1999**, *145*, 15–28. [CrossRef] [PubMed]
188. Fernius, J.; Marston, A.L. Establishment of cohesion at the pericentromere by the Ctf19 kinetochore subcomplex and the replication fork-associated factor, Csm3. *PLoS Genet.* **2009**, *5*, e1000629. [CrossRef] [PubMed]
189. Ng, T.M.; Waples, W.G.; Lavoie, B.D.; Biggins, S. Pericentromeric sister chromatid cohesion promotes kinetochore biorientation. *Mol. Biol. Cell* **2009**, *20*, 3818–3827. [CrossRef] [PubMed]
190. Natsume, T.; Muller, C.A.; Katou, Y.; Retkute, R.; Gierlinski, M.; Araki, H.; Blow, J.J.; Shirahige, K.; Nieduszynski, C.A.; Tanaka, T.U. Kinetochores coordinate pericentromeric cohesion and early DNA replication by cdc7-dbf4 kinase recruitment. *Mol. Cell* **2013**, *50*, 661–674. [CrossRef] [PubMed]
191. Hornung, P.; Troc, P.; Malvezzi, F.; Maier, M.; Demianova, Z.; Zimniak, T.; Litos, G.; Lampert, F.; Schleiffer, A.; Brunner, M.; et al. A cooperative mechanism drives budding yeast kinetochore assembly downstream of cenp-a. *J. Cell Biol.* **2014**, *206*, 509–524. [CrossRef] [PubMed]
192. Masumoto, H.; Masukata, H.; Muro, Y.; Nozaki, N.; Okazaki, T. A human centromere antigen (CENP-B) interacts with a short specific sequence in alphoid DNA, a human centromeric satellite. *J. Cell Biol.* **1989**, *109*, 1963–1973. [CrossRef] [PubMed]
193. Smit, A.F.; Riggs, A.D. Tiggers and DNA transposon fossils in the human genome. *Proc. Natl. Acad. Sci. USA* **1996**, *93*, 1443–1448. [CrossRef] [PubMed]
194. Hudson, D.F.; Fowler, K.J.; Earle, E.; Saffery, R.; Kalitsis, P.; Trowell, H.; Hill, J.; Wreford, N.G.; de Kretser, D.M.; Cancilla, M.R.; et al. Centromere protein b null mice are mitotically and meiotically normal but have lower body and testis weights. *J. Cell Biol.* **1998**, *141*, 309–319. [CrossRef]
195. Kapoor, M.; Montes de Oca Luna, R.; Liu, G.; Lozano, G.; Cummings, C.; Mancini, M.; Ouspenski, I.; Brinkley, B.R.; May, G.S. The cenpb gene is not essential in mice. *Chromosoma* **1998**, *107*, 570–576. [CrossRef] [PubMed]
196. Perez-Castro, A.V.; Shamanski, F.L.; Meneses, J.J.; Lovato, T.L.; Vogel, K.G.; Moyzis, R.K.; Pedersen, R. Centromeric protein b null mice are viable with no apparent abnormalities. *Dev. Biol.* **1998**, *201*, 135–143. [CrossRef] [PubMed]
197. Okada, T.; Ohzeki, J.; Nakano, M.; Yoda, K.; Brinkley, W.R.; Larionov, V.; Masumoto, H. Cenp-b controls centromere formation depending on the chromatin context. *Cell* **2007**, *131*, 1287–1300. [CrossRef] [PubMed]
198. Ohzeki, J.; Nakano, M.; Okada, T.; Masumoto, H. CENP-B box is required for de novo centromere chromatin assembly on human alphoid DNA. *J. Cell Biol.* **2002**, *159*, 765–775. [CrossRef] [PubMed]
199. Fachinetti, D.; Han, J.S.; McMahon, M.A.; Ly, P.; Abdullah, A.; Wong, A.J.; Cleveland, D.W. DNA sequence-specific binding of cenp-b enhances the fidelity of human centromere function. *Dev. Cell* **2015**, *33*, 314–327. [CrossRef] [PubMed]
200. Hoffmann, S.; Dumont, M.; Barra, V.; Ly, P.; Nechemia-Arbely, Y.; McMahon, M.A.; Herve, S.; Cleveland, D.W.; Fachinetti, D. Cenp-a is dispensable for mitotic centromere function after initial centromere/kinetochore assembly. *Cell Rep.* **2016**, *17*, 2394–2404. [CrossRef] [PubMed]
201. Fujita, R.; Otake, K.; Arimura, Y.; Horikoshi, N.; Miya, Y.; Shiga, T.; Osakabe, A.; Tachiwana, H.; Ohzeki, J.; Larionov, V.; et al. Stable complex formation of cenp-b with the cenp-a nucleosome. *Nucleic Acids Res.* **2015**, *43*, 4909–4922. [CrossRef] [PubMed]
202. Bharadwaj, R.; Qi, W.; Yu, H. Identification of two novel components of the human NDC80 kinetochore complex. *J. Biol. Chem.* **2004**, *279*, 13076–13085. [CrossRef] [PubMed]
203. McCleland, M.L.; Gardner, R.D.; Kallio, M.J.; Daum, J.R.; Gorbsky, G.J.; Burke, D.J.; Stukenberg, P.T. The highly conserved ndc80 complex is required for kinetochore assembly, chromosome congression, and spindle checkpoint activity. *Genes Dev.* **2003**, *17*, 101–114. [CrossRef] [PubMed]

204. Desai, A.; Rybina, S.; Muller-Reichert, T.; Shevchenko, A.; Shevchenko, A.; Hyman, A.; Oegema, K. Knl-1 directs assembly of the microtubule-binding interface of the kinetochore in c. Elegans. *Genes Dev.* **2003**, *17*, 2421–2435. [CrossRef] [PubMed]
205. Obuse, C.; Iwasaki, O.; Kiyomitsu, T.; Goshima, G.; Toyoda, Y.; Yanagida, M. A conserved Mis12 centromere complex is linked to heterochromatic HP1 and outer kinetochore protein Zwint-1. *Nat. Cell Biol.* **2004**, *6*, 1135–1141. [CrossRef] [PubMed]
206. Wigge, P.A.; Kilmartin, J.V. The Ndc80p complex from Saccharomyces cerevisiae contains conserved centromere components and has a function in chromosome segregation. *J. Cell Biol.* **2001**, *152*, 349–360. [CrossRef] [PubMed]
207. Westermann, S.; Cheeseman, I.M.; Anderson, S.; Yates, J.R., 3rd; Drubin, D.G.; Barnes, G. Architecture of the budding yeast kinetochore reveals a conserved molecular core. *J. Cell Biol.* **2003**, *163*, 215–222. [CrossRef] [PubMed]
208. Cheeseman, I.M.; Niessen, S.; Anderson, S.; Hyndman, F.; Yates, J.R., 3rd; Oegema, K.; Desai, A. A conserved protein network controls assembly of the outer kinetochore and its ability to sustain tension. *Genes Dev.* **2004**, *18*, 2255–2268. [CrossRef] [PubMed]
209. Cheeseman, I.M.; Chappie, J.S.; Wilson-Kubalek, E.M.; Desai, A. The conserved KMN network constitutes the core microtubule-binding site of the kinetochore. *Cell* **2006**, *127*, 983–997. [CrossRef]
210. Nekrasov, V.S.; Smith, M.A.; Peak-Chew, S.; Kilmartin, J.V. Interactions between centromere complexes in Saccharomyces cerevisiae. *Mol. Biol. Cell* **2003**, *14*, 4931–4946. [CrossRef] [PubMed]
211. Pinsky, B.A.; Tatsutani, S.Y.; Collins, K.A.; Biggins, S. An Mtw1 complex promotes kinetochore biorientation that is monitored by the Ipl1/Aurora protein kinase. *Dev. Cell* **2003**, *5*, 735–745. [CrossRef]
212. Kline, S.L.; Cheeseman, I.M.; Hori, T.; Fukagawa, T.; Desai, A. The human Mis12 complex is required for kinetochore assembly and proper chromosome segregation. *J. Cell Biol.* **2006**, *173*, 9–17. [CrossRef] [PubMed]
213. DeLuca, J.G.; Gall, W.E.; Ciferri, C.; Cimini, D.; Musacchio, A.; Salmon, E.D. Kinetochore microtubule dynamics and attachment stability are regulated by Hec1. *Cell* **2006**, *127*, 969–982. [CrossRef] [PubMed]
214. Wei, R.R.; Schnell, J.R.; Larsen, N.A.; Sorger, P.K.; Chou, J.J.; Harrison, S.C. Structure of a central component of the yeast kinetochore: The Spc24p/Spc25p globular domain. *Structure* **2006**, *14*, 1003–1009. [CrossRef] [PubMed]
215. Wei, R.R.; Sorger, P.K.; Harrison, S.C. Molecular organization of the Ndc80 complex, an essential kinetochore component. *Proc. Natl. Acad. Sci. USA* **2005**, *102*, 5363–5367. [CrossRef] [PubMed]
216. Wei, R.R.; Al-Bassam, J.; Harrison, S.C. The Ndc80/HEC1 complex is a contact point for kinetochore-microtubule attachment. *Nat. Struct. Mol. Biol.* **2007**, *14*, 54–59. [CrossRef] [PubMed]
217. Ciferri, C.; De Luca, J.; Monzani, S.; Ferrari, K.J.; Ristic, D.; Wyman, C.; Stark, H.; Kilmartin, J.; Salmon, E.D.; Musacchio, A. Architecture of the human ndc80-hec1 complex, a critical constituent of the outer kinetochore. *J. Biol. Chem.* **2005**, *280*, 29088–29095. [CrossRef] [PubMed]
218. Ciferri, C.; Pasqualato, S.; Screpanti, E.; Varetti, G.; Santaguida, S.; Dos Reis, G.; Maiolica, A.; Polka, J.; De Luca, J.G.; De Wulf, P.; et al. Implications for kinetochore-microtubule attachment from the structure of an engineered ndc80 complex. *Cell* **2008**, *133*, 427–439. [CrossRef] [PubMed]
219. Valverde, R.; Ingram, J.; Harrison, S.C. Conserved Tetramer Junction in the Kinetochore Ndc80 Complex. *Cell Rep.* **2016**, *17*, 1915–1922. [CrossRef] [PubMed]
220. Schou, K.B.; Andersen, J.S.; Pedersen, L.B. A divergent calponin homology (NN-CH) domain defines a novel family: Implications for evolution of ciliary IFT complex B proteins. *Bioinformatics* **2014**, *30*, 899–902. [CrossRef] [PubMed]
221. Alushin, G.M.; Ramey, V.H.; Pasqualato, S.; Ball, D.A.; Grigorieff, N.; Musacchio, A.; Nogales, E. The ndc80 kinetochore complex forms oligomeric arrays along microtubules. *Nature* **2010**, *467*, 805–810. [CrossRef] [PubMed]
222. Alushin, G.M.; Musinipally, V.; Matson, D.; Tooley, J.; Stukenberg, P.T.; Nogales, E. Multimodal microtubule binding by the Ndc80 kinetochore complex. *Nat. Struct. Mol. Biol.* **2012**, *19*, 1161–1167. [CrossRef] [PubMed]
223. DeLuca, J.G.; Musacchio, A. Structural organization of the kinetochore-microtubule interface. *Curr. Opin. Cell Biol.* **2012**, *24*, 48–56. [CrossRef] [PubMed]
224. DeLuca, K.F.; Lens, S.M.; DeLuca, J.G. Temporal changes in Hec1 phosphorylation control kinetochore-microtubule attachment stability during mitosis. *J. Cell Sci.* **2011**, *124*, 622–634. [CrossRef] [PubMed]

225. Miller, S.A.; Johnson, M.L.; Stukenberg, P.T. Kinetochore attachments require an interaction between unstructured tails on microtubules and Ndc80(Hec1). *Curr. Biol.* **2008**, *18*, 1785–1791. [CrossRef] [PubMed]
226. Guimaraes, G.J.; Dong, Y.; McEwen, B.F.; Deluca, J.G. Kinetochore-microtubule attachment relies on the disordered N-terminal tail domain of Hec1. *Curr. Biol.* **2008**, *18*, 1778–1784. [CrossRef] [PubMed]
227. Umbreit, N.T.; Gestaut, D.R.; Tien, J.F.; Vollmar, B.S.; Gonen, T.; Asbury, C.L.; Davis, T.N. The ndc80 kinetochore complex directly modulates microtubule dynamics. *Proc. Natl. Acad. Sci. USA* **2012**, *109*, 16113–16118. [CrossRef] [PubMed]
228. Aravamudhan, P.; Felzer-Kim, I.; Gurunathan, K.; Joglekar, A.P. Assembling the protein architecture of the budding yeast kinetochore-microtubule attachment using FRET. *Curr. Biol.* **2014**, *24*, 1437–1446. [CrossRef] [PubMed]
229. Zaytsev, A.V.; Mick, J.E.; Maslennikov, E.; Nikashin, B.; DeLuca, J.G.; Grishchuk, E.L. Multisite phosphorylation of the NDC80 complex gradually tunes its microtubule-binding affinity. *Mol. Biol. Cell* **2015**, *26*, 1829–1844. [CrossRef] [PubMed]
230. Kemmler, S.; Stach, M.; Knapp, M.; Ortiz, J.; Pfannstiel, J.; Ruppert, T.; Lechner, J. Mimicking ndc80 phosphorylation triggers spindle assembly checkpoint signalling. *EMBO J.* **2009**, *28*, 1099–1110. [CrossRef] [PubMed]
231. Lampert, F.; Mieck, C.; Alushin, G.M.; Nogales, E.; Westermann, S. Molecular requirements for the formation of a kinetochore-microtubule interface by Dam1 and Ndc80 complexes. *J. Cell Biol.* **2013**, *200*, 21–30. [CrossRef] [PubMed]
232. Cheerambathur, D.K.; Gassmann, R.; Cook, B.; Oegema, K.; Desai, A. Crosstalk between microtubule attachment complexes ensures accurate chromosome segregation. *Science* **2013**, *342*, 1239–1242. [CrossRef] [PubMed]
233. Zaytsev, A.V.; Sundin, L.J.; DeLuca, K.F.; Grishchuk, E.L.; DeLuca, J.G. Accurate phosphoregulation of kinetochore-microtubule affinity requires unconstrained molecular interactions. *J. Cell Biol.* **2014**, *206*, 45–59. [CrossRef] [PubMed]
234. Petrovic, A.; Pasqualato, S.; Dube, P.; Krenn, V.; Santaguida, S.; Cittaro, D.; Monzani, S.; Massimiliano, L.; Keller, J.; Tarricone, A.; et al. The mis12 complex is a protein interaction hub for outer kinetochore assembly. *J. Cell Biol.* **2010**, *190*, 835–852. [CrossRef] [PubMed]
235. Petrovic, A.; Mosalaganti, S.; Keller, J.; Mattiuzzo, M.; Overlack, K.; Krenn, V.; De Antoni, A.; Wohlgemuth, S.; Cecatiello, V.; Pasqualato, S.; et al. Modular assembly of rwd domains on the mis12 complex underlies outer kinetochore organization. *Mol. Cell* **2014**, *53*, 591–605. [CrossRef] [PubMed]
236. Maskell, D.P.; Hu, X.W.; Singleton, M.R. Molecular architecture and assembly of the yeast kinetochore mind complex. *J. Cell Biol.* **2010**, *190*, 823–834. [CrossRef] [PubMed]
237. Hornung, P.; Maier, M.; Alushin, G.M.; Lander, G.C.; Nogales, E.; Westermann, S. Molecular architecture and connectivity of the budding yeast mtw1 kinetochore complex. *J. Mol. Biol.* **2011**, *405*, 548–559. [CrossRef] [PubMed]
238. Screpanti, E.; De Antoni, A.; Alushin, G.M.; Petrovic, A.; Melis, T.; Nogales, E.; Musacchio, A. Direct binding of cenp-c to the mis12 complex joins the inner and outer kinetochore. *Curr. Biol.* **2011**, *21*, 391–398. [CrossRef] [PubMed]
239. Huis in 't Veld, P.J.; Jeganathan, S.; Petrovic, A.; John, J.; Singh, P.; Weissmann, F.; Bange, T.; Musacchio, A. Molecular basis of outer kinetochore assembly on cenp-t. *eLife* **2016**. [CrossRef] [PubMed]
240. Dimitrova, Y.N.; Jenni, S.; Valverde, R.; Khin, Y.; Harrison, S.C. Structure of the mind complex defines a regulatory focus for yeast kinetochore assembly. *Cell* **2016**, *167*, 1014–1027. [CrossRef] [PubMed]
241. Petrovic, A.; Keller, J.; Liu, Y.; Overlack, K.; John, J.; Dimitrova, Y.N.; Jenni, S.; van Gerwen, S.; Stege, P.; Wohlgemuth, S.; et al. Structure of the mis12 complex and molecular basis of its interaction with cenp-c at human kinetochores. *Cell* **2016**, *167*, 1028–1040. [CrossRef] [PubMed]
242. Malvezzi, F.; Litos, G.; Schleiffer, A.; Heuck, A.; Mechtler, K.; Clausen, T.; Westermann, S. A structural basis for kinetochore recruitment of the ndc80 complex via two distinct centromere receptors. *EMBO J.* **2013**, *32*, 409–423. [CrossRef] [PubMed]
243. Przewloka, M.R.; Zhang, W.; Costa, P.; Archambault, V.; D'Avino, P.P.; Lilley, K.S.; Laue, E.D.; McAinsh, A.D.; Glover, D.M. Molecular analysis of core kinetochore composition and assembly in *Drosophila melanogaster*. *PLoS ONE* **2007**, *2*, e478. [CrossRef] [PubMed]

244. Schittenhelm, R.B.; Heeger, S.; Althoff, F.; Walter, A.; Heidmann, S.; Mechtler, K.; Lehner, C.F. Spatial organization of a ubiquitous eukaryotic kinetochore protein network in drosophila chromosomes. *Chromosoma* **2007**, *116*, 385–402. [CrossRef] [PubMed]
245. Liu, Y.; Petrovic, A.; Rombaut, P.; Mosalaganti, S.; Keller, J.; Raunser, S.; Herzog, F.; Musacchio, A. Insights from the reconstitution of the divergent outer kinetochore of *Drosophila melanogaster*. *Open Biol.* **2016**, *6*. [CrossRef] [PubMed]
246. Richter, M.M.; Poznanski, J.; Zdziarska, A.; Czarnocki-Cieciura, M.; Lipinszki, Z.; Dadlez, M.; Glover, D.M.; Przewloka, M.R. Network of protein interactions within the drosophila inner kinetochore. *Open Biol.* **2016**, *6*. [CrossRef] [PubMed]
247. Morelli, E.; Mastrodonato, V.; Beznoussenko, G.V.; Mironov, A.A.; Tognon, E.; Vaccari, T. An essential step of kinetochore formation controlled by the snare protein snap29. *EMBO J.* **2016**, *35*, 2223–2237. [CrossRef] [PubMed]
248. Tromer, E.; Snel, B.; Kops, G.J. Widespread recurrent patterns of rapid repeat evolution in the kinetochore scaffold knl1. *Genome Biol. Evol.* **2015**, *7*, 2383–2393. [CrossRef] [PubMed]
249. Koshland, D.E.; Mitchison, T.J.; Kirschner, M.W. Polewards chromosome movement driven by microtubule depolymerization in vitro. *Nature* **1988**, *331*, 499–504. [CrossRef] [PubMed]
250. Coue, M.; Lombillo, V.A.; McIntosh, J.R. Microtubule depolymerization promotes particle and chromosome movement in vitro. *J. Cell Biol.* **1991**, *112*, 1165–1175. [CrossRef] [PubMed]
251. Grishchuk, E.L.; Molodtsov, M.I.; Ataullakhanov, F.I.; McIntosh, J.R. Force production by disassembling microtubules. *Nature* **2005**, *438*, 384–388. [CrossRef] [PubMed]
252. Powers, A.F.; Franck, A.D.; Gestaut, D.R.; Cooper, J.; Gracyzk, B.; Wei, R.R.; Wordeman, L.; Davis, T.N.; Asbury, C.L. The ndc80 kinetochore complex forms load-bearing attachments to dynamic microtubule tips via biased diffusion. *Cell* **2009**, *136*, 865–875. [CrossRef] [PubMed]
253. Schmidt, J.C.; Arthanari, H.; Boeszoermenyi, A.; Dashkevich, N.M.; Wilson-Kubalek, E.M.; Monnier, N.; Markus, M.; Oberer, M.; Milligan, R.A.; Bathe, M.; et al. The kinetochore-bound ska1 complex tracks depolymerizing microtubules and binds to curved protofilaments. *Dev. Cell* **2012**, *23*, 968–980. [CrossRef] [PubMed]
254. Maiato, H.; DeLuca, J.; Salmon, E.D.; Earnshaw, W.C. The dynamic kinetochore-microtubule interface. *J. Cell Sci.* **2004**, *117*, 5461–5477. [CrossRef] [PubMed]
255. Maiato, H.; Gomes, A.M.; Sousa, F.; Barisic, M. Mechanisms of chromosome congression during mitosis. *Biology (Basel)* **2017**.
256. Gaitanos, T.N.; Santamaria, A.; Jeyaprakash, A.A.; Wang, B.; Conti, E.; Nigg, E.A. Stable kinetochore-microtubule interactions depend on the ska complex and its new component ska3/c13orf3. *EMBO J.* **2009**, *28*, 1442–1452. [CrossRef] [PubMed]
257. Welburn, J.P.; Grishchuk, E.L.; Backer, C.B.; Wilson-Kubalek, E.M.; Yates, J.R., 3rd; Cheeseman, I.M. The human kinetochore ska1 complex facilitates microtubule depolymerization-coupled motility. *Dev. Cell* **2009**, *16*, 374–385. [CrossRef] [PubMed]
258. Theis, M.; Slabicki, M.; Junqueira, M.; Paszkowski-Rogacz, M.; Sontheimer, J.; Kittler, R.; Heninger, A.K.; Glatter, T.; Kruusmaa, K.; Poser, I.; et al. Comparative profiling identifies c13orf3 as a component of the ska complex required for mammalian cell division. *EMBO J.* **2009**, *28*, 1453–1465. [CrossRef] [PubMed]
259. Hofmann, C.; Cheeseman, I.M.; Goode, B.L.; McDonald, K.L.; Barnes, G.; Drubin, D.G. Saccharomyces cerevisiae duo1p and dam1p, novel proteins involved in mitotic spindle function. *J. Cell Biol.* **1998**, *143*, 1029–1040. [CrossRef] [PubMed]
260. Asbury, C.L.; Gestaut, D.R.; Powers, A.F.; Franck, A.D.; Davis, T.N. The dam1 kinetochore complex harnesses microtubule dynamics to produce force and movement. *Proc. Natl. Acad. Sci. USA* **2006**, *103*, 9873–9878. [CrossRef] [PubMed]
261. Grishchuk, E.L.; Efremov, A.K.; Volkov, V.A.; Spiridonov, I.S.; Gudimchuk, N.; Westermann, S.; Drubin, D.; Barnes, G.; McIntosh, J.R.; Ataullakhanov, F.I. The dam1 ring binds microtubules strongly enough to be a processive as well as energy-efficient coupler for chromosome motion. *Proc. Natl. Acad. Sci. USA* **2008**, *105*, 15423–15428. [CrossRef] [PubMed]
262. Grishchuk, E.L.; Spiridonov, I.S.; Volkov, V.A.; Efremov, A.; Westermann, S.; Drubin, D.; Barnes, G.; Ataullakhanov, F.I.; McIntosh, J.R. Different assemblies of the dam1 complex follow shortening microtubules by distinct mechanisms. *Proc. Natl. Acad. Sci. USA* **2008**, *105*, 6918–6923. [CrossRef] [PubMed]

263. Westermann, S.; Wang, H.W.; Avila-Sakar, A.; Drubin, D.G.; Nogales, E.; Barnes, G. The dam1 kinetochore ring complex moves processively on depolymerizing microtubule ends. *Nature* **2006**, *440*, 565–569. [CrossRef] [PubMed]
264. Tien, J.F.; Umbreit, N.T.; Gestaut, D.R.; Franck, A.D.; Cooper, J.; Wordeman, L.; Gonen, T.; Asbury, C.L.; Davis, T.N. Cooperation of the dam1 and ndc80 kinetochore complexes enhances microtubule coupling and is regulated by aurora b. *J. Cell Biol.* **2010**, *189*, 713–723. [CrossRef] [PubMed]
265. Lampert, F.; Hornung, P.; Westermann, S. The dam1 complex confers microtubule plus end-tracking activity to the ndc80 kinetochore complex. *J. Cell Biol.* **2010**, *189*, 641–649. [CrossRef] [PubMed]
266. Cheeseman, I.M.; Anderson, S.; Jwa, M.; Green, E.M.; Kang, J.; Yates, J.R., 3rd; Chan, C.S.; Drubin, D.G.; Barnes, G. Phospho-regulation of kinetochore-microtubule attachments by the aurora kinase ipl1p. *Cell* **2002**, *111*, 163–172. [CrossRef]
267. Chan, Y.W.; Jeyaprakash, A.A.; Nigg, E.A.; Santamaria, A. Aurora b controls kinetochore-microtubule attachments by inhibiting ska complex-kmn network interaction. *J. Cell Biol.* **2012**, *196*, 563–571. [CrossRef] [PubMed]
268. Redli, P.M.; Gasic, I.; Meraldi, P.; Nigg, E.A.; Santamaria, A. The ska complex promotes aurora b activity to ensure chromosome biorientation. *J. Cell Biol.* **2016**, *215*, 77–93. [CrossRef] [PubMed]
269. Jeyaprakash, A.A.; Santamaria, A.; Jayachandran, U.; Chan, Y.W.; Benda, C.; Nigg, E.A.; Conti, E. Structural and functional organization of the ska complex, a key component of the kinetochore-microtubule interface. *Mol. Cell* **2012**, *46*, 274–286. [CrossRef] [PubMed]
270. Abad, M.A.; Medina, B.; Santamaria, A.; Zou, J.; Plasberg-Hill, C.; Madhumalar, A.; Jayachandran, U.; Redli, P.M.; Rappsilber, J.; Nigg, E.A.; et al. Structural basis for microtubule recognition by the human kinetochore ska complex. *Nat. Commun.* **2014**, *5*. [CrossRef] [PubMed]
271. Abad, M.A.; Zou, J.; Medina-Pritchard, B.; Nigg, E.A.; Rappsilber, J.; Santamaria, A.; Jeyaprakash, A.A. Ska3 ensures timely mitotic progression by interacting directly with microtubules and ska1 microtubule binding domain. *Sci. Rep.* **2016**, *6*. [CrossRef] [PubMed]
272. Miranda, J.J.; De Wulf, P.; Sorger, P.K.; Harrison, S.C. The yeast dash complex forms closed rings on microtubules. *Nat. Struct. Mol. Biol.* **2005**, *12*, 138–143. [CrossRef] [PubMed]
273. Westermann, S.; Avila-Sakar, A.; Wang, H.W.; Niederstrasser, H.; Wong, J.; Drubin, D.G.; Nogales, E.; Barnes, G. Formation of a dynamic kinetochore- microtubule interface through assembly of the dam1 ring complex. *Mol. Cell* **2005**, *17*, 277–290. [CrossRef] [PubMed]
274. Ramey, V.H.; Wang, H.W.; Nakajima, Y.; Wong, A.; Liu, J.; Drubin, D.; Barnes, G.; Nogales, E. The dam1 ring binds to the e-hook of tubulin and diffuses along the microtubule. *Mol. Biol. Cell* **2011**, *22*, 457–466. [CrossRef] [PubMed]
275. Legal, T.; Zou, J.; Sochaj, A.; Rappsilber, J.; Welburn, J.P. Molecular architecture of the dam1 complex-microtubule interaction. *Open Biol.* **2016**, *6*. [CrossRef] [PubMed]
276. Wang, H.W.; Ramey, V.H.; Westermann, S.; Leschziner, A.E.; Welburn, J.P.; Nakajima, Y.; Drubin, D.G.; Barnes, G.; Nogales, E. Architecture of the dam1 kinetochore ring complex and implications for microtubule-driven assembly and force-coupling mechanisms. *Nat. Struct. Mol. Biol.* **2007**, *14*, 721–726. [CrossRef] [PubMed]
277. Ramey, V.H.; Wong, A.; Fang, J.; Howes, S.; Barnes, G.; Nogales, E. Subunit organization in the dam1 kinetochore complex and its ring around microtubules. *Mol. Biol. Cell* **2011**, *22*, 4335–4342. [CrossRef] [PubMed]
278. Przewloka, M.R.; Venkei, Z.; Bolanos-Garcia, V.M.; Debski, J.; Dadlez, M.; Glover, D.M. Cenp-c is a structural platform for kinetochore assembly. *Curr. Biol.* **2011**, *21*, 399–405. [CrossRef] [PubMed]
279. Gascoigne, K.E.; Takeuchi, K.; Suzuki, A.; Hori, T.; Fukagawa, T.; Cheeseman, I.M. Induced ectopic kinetochore assembly bypasses the requirement for cenp-a nucleosomes. *Cell* **2011**, *145*, 410–422. [CrossRef] [PubMed]
280. Rago, F.; Gascoigne, K.E.; Cheeseman, I.M. Distinct organization and regulation of the outer kinetochore kmn network downstream of cenp-c and cenp-t. *Curr. Biol.* **2015**, *25*, 671–677. [CrossRef] [PubMed]
281. Nishino, T.; Rago, F.; Hori, T.; Tomii, K.; Cheeseman, I.M.; Fukagawa, T. Cenp-t provides a structural platform for outer kinetochore assembly. *EMBO J.* **2013**, *32*, 424–436. [CrossRef] [PubMed]
282. Schleiffer, A.; Maier, M.; Litos, G.; Lampert, F.; Hornung, P.; Mechtler, K.; Westermann, S. Cenp-t proteins are conserved centromere receptors of the ndc80 complex. *Nat. Cell Biol.* **2012**, *14*, 604–613. [CrossRef] [PubMed]

283. Welburn, J.P.; Vleugel, M.; Liu, D.; Yates, J.R., 3rd; Lampson, M.A.; Fukagawa, T.; Cheeseman, I.M. Aurora b phosphorylates spatially distinct targets to differentially regulate the kinetochore-microtubule interface. *Mol. Cell* **2010**, *38*, 383–392. [CrossRef] [PubMed]
284. Akiyoshi, B.; Nelson, C.R.; Biggins, S. The aurora b kinase promotes inner and outer kinetochore interactions in budding yeast. *Genetics* **2013**, *194*, 785–789. [CrossRef] [PubMed]
285. Yang, Y.; Wu, F.; Ward, T.; Yan, F.; Wu, Q.; Wang, Z.; McGlothen, T.; Peng, W.; You, T.; Sun, M.; et al. Phosphorylation of hsmis13 by aurora b kinase is essential for assembly of functional kinetochore. *J. Biol. Chem.* **2008**, *283*, 26726–26736. [CrossRef] [PubMed]
286. Emanuele, M.J.; Lan, W.; Jwa, M.; Miller, S.A.; Chan, C.S.; Stukenberg, P.T. Aurora b kinase and protein phosphatase 1 have opposing roles in modulating kinetochore assembly. *J. Cell Biol.* **2008**, *181*, 241–254. [CrossRef] [PubMed]
287. Kim, S.; Yu, H. Multiple assembly mechanisms anchor the kmn spindle checkpoint platform at human mitotic kinetochores. *J. Cell Biol.* **2015**, *208*, 181–196. [CrossRef] [PubMed]
288. Suzuki, A.; Badger, B.L.; Salmon, E.D. A quantitative description of ndc80 complex linkage to human kinetochores. *Nat. Commun.* **2015**, *6*. [CrossRef] [PubMed]
289. Bock, L.J.; Pagliuca, C.; Kobayashi, N.; Grove, R.A.; Oku, Y.; Shrestha, K.; Alfieri, C.; Golfieri, C.; Oldani, A.; Dal Maschio, M.; et al. Cnn1 inhibits the interactions between the kmn complexes of the yeast kinetochore. *Nat. Cell Biol.* **2012**, *14*, 614–624. [CrossRef] [PubMed]
290. Akiyoshi, B.; Sarangapani, K.K.; Powers, A.F.; Nelson, C.R.; Reichow, S.L.; Arellano-Santoyo, H.; Gonen, T.; Ranish, J.A.; Asbury, C.L.; Biggins, S. Tension directly stabilizes reconstituted kinetochore-microtubule attachments. *Nature* **2010**, *468*, 576–579. [CrossRef] [PubMed]
291. Gonen, S.; Akiyoshi, B.; Iadanza, M.G.; Shi, D.; Duggan, N.; Biggins, S.; Gonen, T. The structure of purified kinetochores reveals multiple microtubule-attachment sites. *Nat. Struct. Mol. Biol.* **2012**, *19*, 925–929. [CrossRef] [PubMed]
292. Gascoigne, K.E.; Cheeseman, I.M. Cdk-dependent phosphorylation and nuclear exclusion coordinately control kinetochore assembly state. *J. Cell Biol.* **2013**, *201*, 23–32. [CrossRef] [PubMed]
293. Fukagawa, T.; Mikami, Y.; Nishihashi, A.; Regnier, V.; Haraguchi, T.; Hiraoka, Y.; Sugata, N.; Todokoro, K.; Brown, W.; Ikemura, T. Cenp-h, a constitutive centromere component, is required for centromere targeting of cenp-c in vertebrate cells. *EMBO J.* **2001**, *20*, 4603–4617. [CrossRef] [PubMed]
294. McClelland, S.E.; Borusu, S.; Amaro, A.C.; Winter, J.R.; Belwal, M.; McAinsh, A.D.; Meraldi, P. The cenp-a nac/cad kinetochore complex controls chromosome congression and spindle bipolarity. *EMBO J.* **2007**, *26*, 5033–5047. [CrossRef] [PubMed]
295. Hemmerich, P.; Weidtkamp-Peters, S.; Hoischen, C.; Schmiedeberg, L.; Erliandri, I.; Diekmann, S. Dynamics of inner kinetochore assembly and maintenance in living cells. *J. Cell Biol.* **2008**, *180*, 1101–1114. [CrossRef] [PubMed]
296. Kwon, M.S.; Hori, T.; Okada, M.; Fukagawa, T. Cenp-c is involved in chromosome segregation, mitotic checkpoint function, and kinetochore assembly. *Mol. Biol. Cell* **2007**, *18*, 2155–2168. [CrossRef] [PubMed]
297. Eskat, A.; Deng, W.; Hofmeister, A.; Rudolphi, S.; Emmerth, S.; Hellwig, D.; Ulbricht, T.; Doring, V.; Bancroft, J.M.; McAinsh, A.D.; et al. Step-wise assembly, maturation and dynamic behavior of the human cenp-p/o/r/q/u kinetochore sub-complex. *PLoS ONE* **2012**, *7*, e44717. [CrossRef] [PubMed]
298. Hellwig, D.; Emmerth, S.; Ulbricht, T.; Doring, V.; Hoischen, C.; Martin, R.; Samora, C.P.; McAinsh, A.D.; Carroll, C.W.; Straight, A.F.; et al. Dynamics of cenp-n kinetochore binding during the cell cycle. *J. Cell Sci.* **2011**, *124*, 3871–3883. [CrossRef] [PubMed]
299. Hellwig, D.; Munch, S.; Orthaus, S.; Hoischen, C.; Hemmerich, P.; Diekmann, S. Live-cell imaging reveals sustained centromere binding of cenp-t via cenp-a and cenp-b. *J. Biophotonics* **2008**, *1*, 245–254. [CrossRef] [PubMed]
300. McAinsh, A.D.; Meraldi, P.; Draviam, V.M.; Toso, A.; Sorger, P.K. The human kinetochore proteins nnf1r and mcm21r are required for accurate chromosome segregation. *EMBO J.* **2006**, *25*, 4033–4049. [CrossRef] [PubMed]
301. Ohta, S.; Kimura, M.; Takagi, S.; Toramoto, I.; Ishihama, Y. Identification of mitosis-specific phosphorylation in mitotic chromosome-associated proteins. *J Proteome Res* **2016**, *15*, 3331–3341. [CrossRef] [PubMed]
302. Cheeseman, I.M.; Hori, T.; Fukagawa, T.; Desai, A. Knl1 and the cenp-h/i/k complex coordinately direct kinetochore assembly in vertebrates. *Mol. Biol. Cell* **2008**, *19*, 587–594. [CrossRef] [PubMed]

303. Venkei, Z.; Przewloka, M.R.; Ladak, Y.; Albadri, S.; Sossick, A.; Juhasz, G.; Novak, B.; Glover, D.M. Spatiotemporal dynamics of spc105 regulates the assembly of the drosophila kinetochore. *Open Biol.* **2012**, *2*. [CrossRef] [PubMed]
304. Magidson, V.; O'Connell, C.B.; Loncarek, J.; Paul, R.; Mogilner, A.; Khodjakov, A. The spatial arrangement of chromosomes during prometaphase facilitates spindle assembly. *Cell* **2011**, *146*, 555–567. [CrossRef] [PubMed]
305. Karess, R. Rod-zw10-zwilch: A key player in the spindle checkpoint. *Trends Cell Biol.* **2005**, *15*, 386–392. [CrossRef] [PubMed]
306. Civril, F.; Wehenkel, A.; Giorgi, F.M.; Santaguida, S.; Di Fonzo, A.; Grigorean, G.; Ciccarelli, F.D.; Musacchio, A. Structural analysis of the rzz complex reveals common ancestry with multisubunit vesicle tethering machinery. *Structure* **2010**, *18*, 616–626. [CrossRef] [PubMed]
307. McKenney, R.J.; Huynh, W.; Tanenbaum, M.E.; Bhabha, G.; Vale, R.D. Activation of cytoplasmic dynein motility by dynactin-cargo adapter complexes. *Science* **2014**, *345*, 337–341. [CrossRef] [PubMed]
308. Barisic, M.; Sohm, B.; Mikolcevic, P.; Wandke, C.; Rauch, V.; Ringer, T.; Hess, M.; Bonn, G.; Geley, S. Spindly/ccdc99 is required for efficient chromosome congression and mitotic checkpoint regulation. *Mol. Biol. Cell* **2010**, *21*, 1968–1981. [CrossRef] [PubMed]
309. Chan, Y.W.; Fava, L.L.; Uldschmid, A.; Schmitz, M.H.; Gerlich, D.W.; Nigg, E.A.; Santamaria, A. Mitotic control of kinetochore-associated dynein and spindle orientation by human spindly. *J. Cell Biol.* **2009**, *185*, 859–874. [CrossRef] [PubMed]
310. Gassmann, R.; Essex, A.; Hu, J.S.; Maddox, P.S.; Motegi, F.; Sugimoto, A.; O'Rourke, S.M.; Bowerman, B.; McLeod, I.; Yates, J.R., 3rd; et al. A new mechanism controlling kinetochore-microtubule interactions revealed by comparison of two dynein-targeting components: Spdl-1 and the rod/zwilch/zw10 complex. *Genes Dev.* **2008**, *22*, 2385–2399. [CrossRef] [PubMed]
311. Starr, D.A.; Williams, B.C.; Hays, T.S.; Goldberg, M.L. Zw10 helps recruit dynactin and dynein to the kinetochore. *J. Cell Biol.* **1998**, *142*, 763–774. [CrossRef] [PubMed]
312. Griffis, E.R.; Stuurman, N.; Vale, R.D. Spindly, a novel protein essential for silencing the spindle assembly checkpoint, recruits dynein to the kinetochore. *J. Cell Biol.* **2007**, *177*, 1005–1015. [CrossRef] [PubMed]
313. Yamamoto, T.G.; Watanabe, S.; Essex, A.; Kitagawa, R. Spdl-1 functions as a kinetochore receptor for mdf-1 in caenorhabditis elegans. *J. Cell Biol.* **2008**, *183*, 187–194. [CrossRef] [PubMed]
314. Holland, A.J.; Reis, R.M.; Niessen, S.; Pereira, C.; Andres, D.A.; Spielmann, H.P.; Cleveland, D.W.; Desai, A.; Gassmann, R. Preventing farnesylation of the dynein adaptor spindly contributes to the mitotic defects caused by farnesyltransferase inhibitors. *Mol. Biol. Cell* **2015**, *26*, 1845–1856. [CrossRef] [PubMed]
315. Moudgil, D.K.; Westcott, N.; Famulski, J.K.; Patel, K.; Macdonald, D.; Hang, H.; Chan, G.K. A novel role of farnesylation in targeting a mitotic checkpoint protein, human spindly, to kinetochores. *J. Cell Biol.* **2015**, *208*, 881–896. [CrossRef] [PubMed]
316. Hussein, D.; Taylor, S.S. Farnesylation of cenp-f is required for g2/m progression and degradation after mitosis. *J. Cell Sci.* **2002**, *115*, 3403–3414. [PubMed]
317. Ashar, H.R.; James, L.; Gray, K.; Carr, D.; Black, S.; Armstrong, L.; Bishop, W.R.; Kirschmeier, P. Farnesyl transferase inhibitors block the farnesylation of cenp-e and cenp-f and alter the association of cenp-e with the microtubules. *J. Biol. Chem.* **2000**, *275*, 30451–30457. [CrossRef] [PubMed]
318. Basto, R.; Scaerou, F.; Mische, S.; Wojcik, E.; Lefebvre, C.; Gomes, R.; Hays, T.; Karess, R. In vivo dynamics of the rough deal checkpoint protein during drosophila mitosis. *Curr. Biol.* **2004**, *14*, 56–61. [CrossRef] [PubMed]
319. Howell, B.J.; McEwen, B.F.; Canman, J.C.; Hoffman, D.B.; Farrar, E.M.; Rieder, C.L.; Salmon, E.D. Cytoplasmic dynein/dynactin drives kinetochore protein transport to the spindle poles and has a role in mitotic spindle checkpoint inactivation. *J. Cell Biol.* **2001**, *155*, 1159–1172. [CrossRef] [PubMed]
320. Mische, S.; He, Y.; Ma, L.; Li, M.; Serr, M.; Hays, T.S. Dynein light intermediate chain: An essential subunit that contributes to spindle checkpoint inactivation. *Mol. Biol. Cell* **2008**, *19*, 4918–4929. [CrossRef] [PubMed]
321. Sivaram, M.V.; Wadzinski, T.L.; Redick, S.D.; Manna, T.; Doxsey, S.J. Dynein light intermediate chain 1 is required for progress through the spindle assembly checkpoint. *EMBO J.* **2009**, *28*, 902–914. [CrossRef] [PubMed]
322. Varma, D.; Monzo, P.; Stehman, S.A.; Vallee, R.B. Direct role of dynein motor in stable kinetochore-microtubule attachment, orientation, and alignment. *J. Cell Biol.* **2008**, *182*, 1045–1054. [CrossRef] [PubMed]

323. Williams, B.C.; Gatti, M.; Goldberg, M.L. Bipolar spindle attachments affect redistributions of zw10, a drosophila centromere/kinetochore component required for accurate chromosome segregation. *J. Cell Biol.* **1996**, *134*, 1127–1140. [CrossRef] [PubMed]
324. Wojcik, E.; Basto, R.; Serr, M.; Scaerou, F.; Karess, R.; Hays, T. Kinetochore dynein: Its dynamics and role in the transport of the rough deal checkpoint protein. *Nat. Cell Biol.* **2001**, *3*, 1001–1007. [CrossRef] [PubMed]
325. Gassmann, R.; Holland, A.J.; Varma, D.; Wan, X.; Civril, F.; Cleveland, D.W.; Oegema, K.; Salmon, E.D.; Desai, A. Removal of spindly from microtubule-attached kinetochores controls spindle checkpoint silencing in human cells. *Genes Dev.* **2010**, *24*, 957–971. [CrossRef] [PubMed]
326. Buffin, E.; Lefebvre, C.; Huang, J.; Gagou, M.E.; Karess, R.E. Recruitment of mad2 to the kinetochore requires the rod/zw10 complex. *Curr. Biol.* **2005**, *15*, 856–861. [CrossRef] [PubMed]
327. Caldas, G.V.; Lynch, T.R.; Anderson, R.; Afreen, S.; Varma, D.; DeLuca, J.G. The rzz complex requires the n-terminus of knl1 to mediate optimal mad1 kinetochore localization in human cells. *Open Biol.* **2015**, *5*. [CrossRef] [PubMed]
328. Kops, G.J.; Kim, Y.; Weaver, B.A.; Mao, Y.; McLeod, I.; Yates, J.R., 3rd; Tagaya, M.; Cleveland, D.W. Zw10 links mitotic checkpoint signaling to the structural kinetochore. *J. Cell Biol.* **2005**, *169*, 49–60. [CrossRef] [PubMed]
329. Matson, D.R.; Stukenberg, P.T. Cenp-i and aurora b act as a molecular switch that ties rzz/mad1 recruitment to kinetochore attachment status. *J. Cell Biol.* **2014**, *205*, 541–554. [CrossRef] [PubMed]
330. Silio, V.; McAinsh, A.D.; Millar, J.B. Knl1-bubs and rzz provide two separable pathways for checkpoint activation at human kinetochores. *Dev. Cell* **2015**, *35*, 600–613. [CrossRef] [PubMed]
331. Zhang, G.; Lischetti, T.; Hayward, D.G.; Nilsson, J. Distinct domains in bub1 localize rzz and bubr1 to kinetochores to regulate the checkpoint. *Nat. Commun.* **2015**, *6*. [CrossRef] [PubMed]
332. Zinkowski, R.P.; Meyne, J.; Brinkley, B.R. The centromere-kinetochore complex: A repeat subunit model. *J. Cell Biol.* **1991**, *113*, 1091–1110. [CrossRef] [PubMed]
333. Joglekar, A.P.; Bouck, D.; Finley, K.; Liu, X.; Wan, Y.; Berman, J.; He, X.; Salmon, E.D.; Bloom, K.S. Molecular architecture of the kinetochore-microtubule attachment site is conserved between point and regional centromeres. *J. Cell Biol.* **2008**, *181*, 587–594. [CrossRef] [PubMed]
334. Lawrimore, J.; Bloom, K.S.; Salmon, E.D. Point centromeres contain more than a single centromere-specific cse4 (cenp-a) nucleosome. *J. Cell Biol.* **2011**, *195*, 573–582. [CrossRef] [PubMed]
335. Lando, D.; Endesfelder, U.; Berger, H.; Subramanian, L.; Dunne, P.D.; McColl, J.; Klenerman, D.; Carr, A.M.; Sauer, M.; Allshire, R.C.; et al. Quantitative single-molecule microscopy reveals that cenp-a(cnp1) deposition occurs during g2 in fission yeast. *Open Biol.* **2012**, *2*. [CrossRef] [PubMed]
336. Blower, M.D.; Sullivan, B.A.; Karpen, G.H. Conserved organization of centromeric chromatin in flies and humans. *Dev. Cell* **2002**, *2*, 319–330. [CrossRef]
337. Sullivan, B.A.; Karpen, G.H. Centromeric chromatin exhibits a histone modification pattern that is distinct from both euchromatin and heterochromatin. *Nat. Struct. Mol. Biol.* **2004**, *11*, 1076–1083. [CrossRef] [PubMed]
338. Ribeiro, S.A.; Vagnarelli, P.; Dong, Y.; Hori, T.; McEwen, B.F.; Fukagawa, T.; Flors, C.; Earnshaw, W.C. A super-resolution map of the vertebrate kinetochore. *Proc. Natl. Acad. Sci. USA* **2010**, *107*, 10484–10489. [CrossRef] [PubMed]
339. Lawrimore, J.; Vasquez, P.A.; Falvo, M.R.; Taylor, R.M., 2nd; Vicci, L.; Yeh, E.; Forest, M.G.; Bloom, K. DNA loops generate intracentromere tension in mitosis. *J. Cell Biol.* **2015**, *210*, 553–564. [CrossRef] [PubMed]
340. Stephens, A.D.; Quammen, C.W.; Chang, B.; Haase, J.; Taylor, R.M., 2nd; Bloom, K. The spatial segregation of pericentric cohesin and condensin in the mitotic spindle. *Mol. Biol. Cell* **2013**, *24*, 3909–3919. [CrossRef] [PubMed]
341. Stephens, A.D.; Haase, J.; Vicci, L.; Taylor, R.M., 2nd; Bloom, K. Cohesin, condensin, and the intramolecular centromere loop together generate the mitotic chromatin spring. *J. Cell Biol.* **2011**, *193*, 1167–1180. [CrossRef] [PubMed]
342. Gassmann, R.; Rechtsteiner, A.; Yuen, K.W.; Muroyama, A.; Egelhofer, T.; Gaydos, L.; Barron, F.; Maddox, P.; Essex, A.; Monen, J.; et al. An inverse relationship to germline transcription defines centromeric chromatin in c. Elegans. *Nature* **2012**, *484*, 534–537. [CrossRef] [PubMed]
343. Steiner, F.A.; Henikoff, S. Holocentromeres are dispersed point centromeres localized at transcription factor hotspots. *eLife* **2014**, *3*, e02025. [CrossRef] [PubMed]

344. Maddox, P.S.; Portier, N.; Desai, A.; Oegema, K. Molecular analysis of mitotic chromosome condensation using a quantitative time-resolved fluorescence microscopy assay. *Proc. Natl. Acad. Sci. USA* **2006**, *103*, 15097–15102. [CrossRef] [PubMed]
345. Grenfell, A.W.; Heald, R.; Strzelecka, M. Mitotic noncoding rna processing promotes kinetochore and spindle assembly in xenopus. *J. Cell Biol.* **2016**, *214*, 133–141. [CrossRef] [PubMed]
346. Liu, H.; Qu, Q.; Warrington, R.; Rice, A.; Cheng, N.; Yu, H. Mitotic transcription installs sgo1 at centromeres to coordinate chromosome segregation. *Mol. Cell* **2015**, *59*, 426–436. [CrossRef] [PubMed]
347. Gent, J.I.; Dawe, R.K. Rna as a structural and regulatory component of the centromere. *Annu. Rev. Genet.* **2012**, *46*, 443–453. [CrossRef] [PubMed]
348. Chan, F.L.; Marshall, O.J.; Saffery, R.; Kim, B.W.; Earle, E.; Choo, K.H.; Wong, L.H. Active transcription and essential role of rna polymerase ii at the centromere during mitosis. *Proc. Natl. Acad. Sci. USA* **2012**, *109*, 1979–1984. [CrossRef] [PubMed]
349. Prendergast, L.; Muller, S.; Liu, Y.; Huang, H.; Dingli, F.; Loew, D.; Vassias, I.; Patel, D.J.; Sullivan, K.F.; Almouzni, G. The cenp-t/-w complex is a binding partner of the histone chaperone fact. *Genes Dev.* **2016**, *30*, 1313–1326. [CrossRef] [PubMed]
350. Quenet, D.; Dalal, Y. A long non-coding rna is required for targeting centromeric protein a to the human centromere. *eLife* **2014**, *3*, e03254. [CrossRef] [PubMed]
351. Rosic, S.; Kohler, F.; Erhardt, S. Repetitive centromeric satellite rna is essential for kinetochore formation and cell division. *J. Cell Biol.* **2014**, *207*, 335–349. [CrossRef] [PubMed]
352. Catania, S.; Pidoux, A.L.; Allshire, R.C. Sequence features and transcriptional stalling within centromere DNA promote establishment of cenp-a chromatin. *PLoS Genet.* **2015**, *11*, e1004986. [CrossRef] [PubMed]
353. Schuh, M.; Lehner, C.F.; Heidmann, S. Incorporation of drosophila cid/cenp-a and cenp-c into centromeres during early embryonic anaphase. *Curr. Biol.* **2007**, *17*, 237–243. [CrossRef] [PubMed]
354. Dunleavy, E.M.; Almouzni, G.; Karpen, G.H. H3.3 is deposited at centromeres in s phase as a placeholder for newly assembled cenp-a in g(1) phase. *Nucleus* **2011**, *2*, 146–157. [CrossRef] [PubMed]
355. Pearson, C.G.; Yeh, E.; Gardner, M.; Odde, D.; Salmon, E.D.; Bloom, K. Stable kinetochore-microtubule attachment constrains centromere positioning in metaphase. *Curr. Biol.* **2004**, *14*, 1962–1967. [CrossRef] [PubMed]
356. Hayashi, T.; Fujita, Y.; Iwasaki, O.; Adachi, Y.; Takahashi, K.; Yanagida, M. Mis16 and mis18 are required for cenp-a loading and histone deacetylation at centromeres. *Cell* **2004**, *118*, 715–729. [CrossRef] [PubMed]
357. Fujita, Y.; Hayashi, T.; Kiyomitsu, T.; Toyoda, Y.; Kokubu, A.; Obuse, C.; Yanagida, M. Priming of centromere for cenp-a recruitment by human hmis18alpha, hmis18beta, and m18bp1. *Dev. Cell* **2007**, *12*, 17–30. [CrossRef] [PubMed]
358. Shiroiwa, Y.; Hayashi, T.; Fujita, Y.; Villar-Briones, A.; Ikai, N.; Takeda, K.; Ebe, M.; Yanagida, M. Mis17 is a regulatory module of the mis6-mal2-sim4 centromere complex that is required for the recruitment of cenh3/cenp-a in fission yeast. *PLoS ONE* **2011**, *6*, e17761. [CrossRef] [PubMed]
359. Lee, B.C.; Lin, Z.; Yuen, K.W. Rbap46/48(lin-53) is required for holocentromere assembly in caenorhabditis elegans. *Cell Rep.* **2016**, *14*, 1819–1828. [CrossRef] [PubMed]
360. Subramanian, L.; Medina-Pritchard, B.; Barton, R.; Spiller, F.; Kulasegaran-Shylini, R.; Radaviciute, G.; Allshire, R.C.; Arockia Jeyaprakash, A. Centromere localization and function of mis18 requires yippee-like domain-mediated oligomerization. *EMBO Rep.* **2016**, *17*, 496–507. [CrossRef] [PubMed]
361. Nardi, I.K.; Zasadzinska, E.; Stellfox, M.E.; Knippler, C.M.; Foltz, D.R. Licensing of centromeric chromatin assembly through the mis18alpha-mis18beta heterotetramer. *Mol. Cell* **2016**, *61*, 774–787. [CrossRef] [PubMed]
362. Maddox, P.S.; Hyndman, F.; Monen, J.; Oegema, K.; Desai, A. Functional genomics identifies a myb domain-containing protein family required for assembly of cenp-a chromatin. *J. Cell Biol.* **2007**, *176*, 757–763. [CrossRef] [PubMed]
363. Pan, D.; Klare, K.; Petrovic, A.; Take, A.; Walstein, K.; Singh, P.; Rondelet, A.; Bird, A.W.; Musacchio, A. Cdk-regulated dimerization of m18bp1 on a mis18 hexamer is necessary for cenp-a loading. *eLife* **2017**, *6*. [CrossRef] [PubMed]
364. Hayashi, T.; Ebe, M.; Nagao, K.; Kokubu, A.; Sajiki, K.; Yanagida, M. Schizosaccharomyces pombe centromere protein mis19 links mis16 and mis18 to recruit cenp-a through interacting with nmd factors and the swi/snf complex. *Genes Cells* **2014**, *19*, 541–554. [CrossRef] [PubMed]

365. Subramanian, L.; Toda, N.R.; Rappsilber, J.; Allshire, R.C. Eic1 links mis18 with the ccan/mis6/ctf19 complex to promote cenp-a assembly. *Open Biol.* **2014**, *4*. [CrossRef] [PubMed]
366. Dunleavy, E.M.; Roche, D.; Tagami, H.; Lacoste, N.; Ray-Gallet, D.; Nakamura, Y.; Daigo, Y.; Nakatani, Y.; Almouzni-Pettinotti, G. Hjurp is a cell-cycle-dependent maintenance and deposition factor of cenp-a at centromeres. *Cell* **2009**, *137*, 485–497. [CrossRef] [PubMed]
367. Foltz, D.R.; Jansen, L.E.; Bailey, A.O.; Yates, J.R., 3rd; Bassett, E.A.; Wood, S.; Black, B.E.; Cleveland, D.W. Centromere-specific assembly of cenp-a nucleosomes is mediated by hjurp. *Cell* **2009**, *137*, 472–484. [CrossRef] [PubMed]
368. Bernad, R.; Sanchez, P.; Rivera, T.; Rodriguez-Corsino, M.; Boyarchuk, E.; Vassias, I.; Ray-Gallet, D.; Arnaoutov, A.; Dasso, M.; Almouzni, G.; et al. Xenopus hjurp and condensin ii are required for cenp-a assembly. *J. Cell Biol.* **2011**, *192*, 569–582. [CrossRef] [PubMed]
369. Aravind, L.; Iyer, L.M.; Wu, C. Domain architectures of the scm3p protein provide insights into centromere function and evolution. *Cell Cycle* **2007**, *6*, 2511–2515. [CrossRef] [PubMed]
370. Stoler, S.; Rogers, K.; Weitze, S.; Morey, L.; Fitzgerald-Hayes, M.; Baker, R.E. Scm3, an essential saccharomyces cerevisiae centromere protein required for g2/m progression and cse4 localization. *Proc. Natl. Acad. Sci. USA* **2007**, *104*, 10571–10576. [CrossRef] [PubMed]
371. Camahort, R.; Li, B.; Florens, L.; Swanson, S.K.; Washburn, M.P.; Gerton, J.L. Scm3 is essential to recruit the histone h3 variant cse4 to centromeres and to maintain a functional kinetochore. *Mol. Cell* **2007**, *26*, 853–865. [CrossRef] [PubMed]
372. Pidoux, A.L.; Choi, E.S.; Abbott, J.K.; Liu, X.; Kagansky, A.; Castillo, A.G.; Hamilton, G.L.; Richardson, W.; Rappsilber, J.; He, X.; et al. Fission yeast scm3: A cenp-a receptor required for integrity of subkinetochore chromatin. *Mol. Cell* **2009**, *33*, 299–311. [CrossRef] [PubMed]
373. Mizuguchi, G.; Xiao, H.; Wisniewski, J.; Smith, M.M.; Wu, C. Nonhistone scm3 and histones cenh3-h4 assemble the core of centromere-specific nucleosomes. *Cell* **2007**, *129*, 1153–1164. [CrossRef] [PubMed]
374. Chen, C.C.; Dechassa, M.L.; Bettini, E.; Ledoux, M.B.; Belisario, C.; Heun, P.; Luger, K.; Mellone, B.G. Cal1 is the drosophila cenp-a assembly factor. *J. Cell Biol.* **2014**, *204*, 313–329. [CrossRef] [PubMed]
375. Williams, J.S.; Hayashi, T.; Yanagida, M.; Russell, P. Fission yeast scm3 mediates stable assembly of cnp1/cenp-a into centromeric chromatin. *Mol. Cell* **2009**, *33*, 287–298. [CrossRef] [PubMed]
376. Sanchez-Pulido, L.; Pidoux, A.L.; Ponting, C.P.; Allshire, R.C. Common ancestry of the cenp-a chaperones scm3 and hjurp. *Cell* **2009**, *137*, 1173–1174. [CrossRef] [PubMed]
377. Erhardt, S.; Mellone, B.G.; Betts, C.M.; Zhang, W.; Karpen, G.H.; Straight, A.F. Genome-wide analysis reveals a cell cycle-dependent mechanism controlling centromere propagation. *J. Cell Biol.* **2008**, *183*, 805–818. [CrossRef] [PubMed]
378. Schittenhelm, R.B.; Althoff, F.; Heidmann, S.; Lehner, C.F. Detrimental incorporation of excess cenp-a/cid and cenp-c into drosophila centromeres is prevented by limiting amounts of the bridging factor cal1. *J. Cell Sci.* **2010**, *123*, 3768–3779. [CrossRef] [PubMed]
379. Rosin, L.; Mellone, B.G. Co-evolving cenp-a and cal1 domains mediate centromeric cenp-a deposition across drosophila species. *Dev. Cell* **2016**, *37*, 136–147. [CrossRef] [PubMed]
380. Zasadzinska, E.; Barnhart-Dailey, M.C.; Kuich, P.H.; Foltz, D.R. Dimerization of the cenp-a assembly factor hjurp is required for centromeric nucleosome deposition. *EMBO J.* **2013**, *32*, 2113–2124. [CrossRef] [PubMed]
381. Shuaib, M.; Ouararhni, K.; Dimitrov, S.; Hamiche, A. Hjurp binds cenp-a via a highly conserved n-terminal domain and mediates its deposition at centromeres. *Proc. Natl. Acad. Sci. USA* **2010**, *107*, 1349–1354. [CrossRef] [PubMed]
382. Cho, U.S.; Harrison, S.C. Recognition of the centromere-specific histone cse4 by the chaperone scm3. *Proc. Natl. Acad. Sci. USA* **2011**, *108*, 9367–9371. [CrossRef] [PubMed]
383. Hu, H.; Liu, Y.; Wang, M.; Fang, J.; Huang, H.; Yang, N.; Li, Y.; Wang, J.; Yao, X.; Shi, Y.; et al. Structure of a cenp-a-histone h4 heterodimer in complex with chaperone hjurp. *Genes Dev.* **2011**, *25*, 901–906. [CrossRef] [PubMed]
384. Zhou, Z.; Feng, H.; Zhou, B.R.; Ghirlando, R.; Hu, K.; Zwolak, A.; Miller Jenkins, L.M.; Xiao, H.; Tjandra, N.; Wu, C.; et al. Structural basis for recognition of centromere histone variant cenh3 by the chaperone scm3. *Nature* **2011**, *472*, 234–237. [CrossRef] [PubMed]

385. Perpelescu, M.; Hori, T.; Toyoda, A.; Misu, S.; Monma, N.; Ikeo, K.; Obuse, C.; Fujiyama, A.; Fukagawa, T. Hjurp is involved in the expansion of centromeric chromatin. *Mol. Biol. Cell* **2015**, *26*, 2742–2754. [CrossRef] [PubMed]
386. Wang, J.; Liu, X.; Dou, Z.; Chen, L.; Jiang, H.; Fu, C.; Fu, G.; Liu, D.; Zhang, J.; Zhu, T.; et al. Mitotic regulator mis18beta interacts with and specifies the centromeric assembly of molecular chaperone holliday junction recognition protein (hjurp). *J. Biol. Chem.* **2014**, *289*, 8326–8336. [CrossRef] [PubMed]
387. Moree, B.; Meyer, C.B.; Fuller, C.J.; Straight, A.F. Cenp-c recruits m18bp1 to centromeres to promote cenp-a chromatin assembly. *J. Cell Biol.* **2011**, *194*, 855–871. [CrossRef] [PubMed]
388. Stellfox, M.E.; Nardi, I.K.; Knippler, C.M.; Foltz, D.R. Differential binding partners of the mis18alpha/beta yippee domains regulate mis18 complex recruitment to centromeres. *Cell Rep.* **2016**, *15*, 2127–2135. [CrossRef] [PubMed]
389. Dambacher, S.; Deng, W.; Hahn, M.; Sadic, D.; Frohlich, J.; Nuber, A.; Hoischen, C.; Diekmann, S.; Leonhardt, H.; Schotta, G. Cenp-c facilitates the recruitment of m18bp1 to centromeric chromatin. *Nucleus* **2012**, *3*, 101–110. [CrossRef] [PubMed]
390. Takahashi, K.; Chen, E.S.; Yanagida, M. Requirement of mis6 centromere connector for localizing a cenp-a-like protein in fission yeast. *Science* **2000**, *288*, 2215–2219. [CrossRef] [PubMed]
391. Shono, N.; Ohzeki, J.; Otake, K.; Martins, N.M.; Nagase, T.; Kimura, H.; Larionov, V.; Earnshaw, W.C.; Masumoto, H. Cenp-c and cenp-i are key connecting factors for kinetochore and cenp-a assembly. *J. Cell Sci.* **2015**, *128*, 4572–4587. [CrossRef] [PubMed]
392. Silva, M.C.; Bodor, D.L.; Stellfox, M.E.; Martins, N.M.; Hochegger, H.; Foltz, D.R.; Jansen, L.E. Cdk activity couples epigenetic centromere inheritance to cell cycle progression. *Dev. Cell* **2012**, *22*, 52–63. [CrossRef] [PubMed]
393. Yu, Z.; Zhou, X.; Wang, W.; Deng, W.; Fang, J.; Hu, H.; Wang, Z.; Li, S.; Cui, L.; Shen, J.; et al. Dynamic phosphorylation of cenp-a at ser68 orchestrates its cell-cycle-dependent deposition at centromeres. *Dev. Cell* **2015**, *32*, 68–81. [CrossRef] [PubMed]
394. Muller, S.; Montes de Oca, R.; Lacoste, N.; Dingli, F.; Loew, D.; Almouzni, G. Phosphorylation and DNA binding of hjurp determine its centromeric recruitment and function in cenh3(cenp-a) loading. *Cell Rep.* **2014**, *8*, 190–203. [CrossRef] [PubMed]
395. McKinley, K.L.; Cheeseman, I.M. Polo-like kinase 1 licenses cenp-a deposition at centromeres. *Cell* **2014**, *158*, 397–411. [CrossRef] [PubMed]
396. Stankovic, A.; Guo, L.Y.; Mata, J.F.; Bodor, D.L.; Cao, X.J.; Bailey, A.O.; Shabanowitz, J.; Hunt, D.F.; Garcia, B.A.; Black, B.E.; et al. A dual inhibitory mechanism sufficient to maintain cell-cycle-restricted cenp-a assembly. *Mol. Cell* **2016**. [CrossRef] [PubMed]
397. Miell, M.D.; Straight, A.F. Regulating the timing of cenp-a nucleosome assembly by phosphorylation. *Dev. Cell* **2015**, *32*, 1–2. [CrossRef] [PubMed]
398. Deyter, G.M.; Biggins, S. The fact complex interacts with the e3 ubiquitin ligase psh1 to prevent ectopic localization of cenp-a. *Genes Dev.* **2014**, *28*, 1815–1826. [CrossRef] [PubMed]
399. Chen, C.C.; Bowers, S.; Lipinszki, Z.; Palladino, J.; Trusiak, S.; Bettini, E.; Rosin, L.; Przewloka, M.R.; Glover, D.M.; O'Neill, R.J.; et al. Establishment of centromeric chromatin by the cenp-a assembly factor cal1 requires fact-mediated transcription. *Dev. Cell* **2015**, *34*, 73–84. [CrossRef] [PubMed]
400. Lagana, A.; Dorn, J.F.; De Rop, V.; Ladouceur, A.M.; Maddox, A.S.; Maddox, P.S. A small gtpase molecular switch regulates epigenetic centromere maintenance by stabilizing newly incorporated cenp-a. *Nat. Cell Biol.* **2010**, *12*, 1186–1193. [CrossRef] [PubMed]
401. Perpelescu, M.; Nozaki, N.; Obuse, C.; Yang, H.; Yoda, K. Active establishment of centromeric cenp-a chromatin by rsf complex. *J. Cell Biol.* **2009**, *185*, 397–407. [CrossRef] [PubMed]
402. Boltengagen, M.; Huang, A.; Boltengagen, A.; Trixl, L.; Lindner, H.; Kremser, L.; Offterdinger, M.; Lusser, A. A novel role for the histone acetyltransferase hat1 in the cenp-a/cid assembly pathway in *Drosophila melanogaster*. *Nucleic Acids Res.* **2016**, *44*, 2145–2159. [CrossRef] [PubMed]
403. Jaramillo-Lambert, A.; Hao, J.; Xiao, H.; Li, Y.; Han, Z.; Zhu, W. Acidic nucleoplasmic DNA-binding protein (and-1) controls chromosome congression by regulating the assembly of centromere protein a (cenp-a) at centromeres. *J. Biol. Chem.* **2013**, *288*, 1480–1488. [CrossRef] [PubMed]

404. Choi, E.S.; Stralfors, A.; Catania, S.; Castillo, A.G.; Svensson, J.P.; Pidoux, A.L.; Ekwall, K.; Allshire, R.C. Factors that promote h3 chromatin integrity during transcription prevent promiscuous deposition of cenp-a(cnp1) in fission yeast. *PLoS Genet.* **2012**, *8*, e1002985. [CrossRef] [PubMed]
405. Mamnun, Y.M.; Katayama, S.; Toda, T. Fission yeast mcl1 interacts with scf(pof3) and is required for centromere formation. *Biochem. Biophys. Res. Commun.* **2006**, *350*, 125–130. [CrossRef] [PubMed]
406. Gopalakrishnan, S.; Sullivan, B.A.; Trazzi, S.; Della Valle, G.; Robertson, K.D. Dnmt3b interacts with constitutive centromere protein cenp-c to modulate DNA methylation and the histone code at centromeric regions. *Hum. Mol. Genet.* **2009**, *18*, 3178–3193. [CrossRef] [PubMed]
407. Kim, I.S.; Lee, M.; Park, K.C.; Jeon, Y.; Park, J.H.; Hwang, E.J.; Jeon, T.I.; Ko, S.; Lee, H.; Baek, S.H.; et al. Roles of mis18alpha in epigenetic regulation of centromeric chromatin and cenp-a loading. *Mol. Cell* **2012**, *46*, 260–273. [CrossRef] [PubMed]
408. Vernarecci, S.; Ornaghi, P.; Bagu, A.; Cundari, E.; Ballario, P.; Filetici, P. Gcn5p plays an important role in centromere kinetochore function in budding yeast. *Mol. Cell. Biol.* **2008**, *28*, 988–996. [CrossRef] [PubMed]
409. Ohzeki, J.; Shono, N.; Otake, K.; Martins, N.M.; Kugou, K.; Kimura, H.; Nagase, T.; Larionov, V.; Earnshaw, W.C.; Masumoto, H. Kat7/hbo1/myst2 regulates cenp-a chromatin assembly by antagonizing suv39h1-mediated centromere inactivation. *Dev. Cell* **2016**, *37*, 413–427. [CrossRef] [PubMed]
410. Bergmann, J.H.; Rodriguez, M.G.; Martins, N.M.; Kimura, H.; Kelly, D.A.; Masumoto, H.; Larionov, V.; Jansen, L.E.; Earnshaw, W.C. Epigenetic engineering shows h3k4me2 is required for hjurp targeting and cenp-a assembly on a synthetic human kinetochore. *EMBO J.* **2011**, *30*, 328–340. [CrossRef] [PubMed]
411. Ohzeki, J.; Bergmann, J.H.; Kouprina, N.; Noskov, V.N.; Nakano, M.; Kimura, H.; Earnshaw, W.C.; Larionov, V.; Masumoto, H. Breaking the hac barrier: Histone h3k9 acetyl/methyl balance regulates cenp-a assembly. *EMBO J.* **2012**, *31*, 2391–2402. [CrossRef] [PubMed]
412. Raychaudhuri, N.; Dubruille, R.; Orsi, G.A.; Bagheri, H.C.; Loppin, B.; Lehner, C.F. Transgenerational propagation and quantitative maintenance of paternal centromeres depends on cid/cenp-a presence in drosophila sperm. *PLoS Biol.* **2012**, *10*, e1001434. [CrossRef] [PubMed]
413. Ross, J.E.; Woodlief, K.S.; Sullivan, B.A. Inheritance of the cenp-a chromatin domain is spatially and temporally constrained at human centromeres. *Epigenet. Chromatin* **2016**, *9*. [CrossRef] [PubMed]
414. Mizuguchi, G.; Shen, X.; Landry, J.; Wu, W.H.; Sen, S.; Wu, C. Atp-driven exchange of histone h2az variant catalyzed by swr1 chromatin remodeling complex. *Science* **2004**, *303*, 343–348. [CrossRef] [PubMed]
415. Podhraski, V.; Campo-Fernandez, B.; Worle, H.; Piatti, P.; Niederegger, H.; Bock, G.; Fyodorov, D.V.; Lusser, A. Cenh3/cid incorporation is not dependent on the chromatin assembly factor chd1 in drosophila. *PLoS ONE* **2010**, *5*, e10120. [CrossRef] [PubMed]
416. Thompson, J.D.; Sylvester, J.E.; Gonzalez, I.L.; Costanzi, C.C.; Gillespie, D. Definition of a second dimeric subfamily of human alpha satellite DNA. *Nucleic Acids Res.* **1989**, *17*, 2769–2782. [CrossRef] [PubMed]
417. Cellamare, A.; Catacchio, C.R.; Alkan, C.; Giannuzzi, G.; Antonacci, F.; Cardone, M.F.; Della Valle, G.; Malig, M.; Rocchi, M.; Eichler, E.E.; et al. New insights into centromere organization and evolution from the white-cheeked gibbon and marmoset. *Mol. Biol. Evol.* **2009**, *26*, 1889–1900. [CrossRef] [PubMed]
418. Camahort, R.; Shivaraju, M.; Mattingly, M.; Li, B.; Nakanishi, S.; Zhu, D.; Shilatifard, A.; Workman, J.L.; Gerton, J.L. Cse4 is part of an octameric nucleosome in budding yeast. *Mol. Cell* **2009**, *35*, 794–805. [CrossRef] [PubMed]
419. Akiyoshi, B.; Gull, K. Discovery of unconventional kinetochores in kinetoplastids. *Cell* **2014**, *156*, 1247–1258. [CrossRef] [PubMed]
420. D'Archivio, S.; Wickstead, B. Trypanosome outer kinetochore proteins suggest conservation of chromosome segregation machinery across eukaryotes. *J. Cell Biol.* **2016**. [CrossRef] [PubMed]
421. Monen, J.; Maddox, P.S.; Hyndman, F.; Oegema, K.; Desai, A. Differential role of cenp-a in the segregation of holocentric c. Elegans chromosomes during meiosis and mitosis. *Nat. Cell Biol.* **2005**, *7*, 1248–1255. [CrossRef] [PubMed]

biology

MDPI

Review

Mechanisms of Chromosome Congression during Mitosis

Helder Maiato [1,2,3,*], Ana Margarida Gomes [1,2], Filipe Sousa [1,2,3] and Marin Barisic [4]

1. Chromosome Instability & Dynamics Laboratory, Instituto de Biologia Molecular e Celular, Universidade do Porto, Rua Alfredo Allen 208, 4200-135 Porto, Portugal; margarida.gomes@ibmc.up.pt (A.M.G.); filipe.sousa@ibmc.up.pt (F.S.)
2. Instituto de Investigação e Inovação em Saúde—i3S, Universidade do Porto, Rua Alfredo Allen 208, 4200-135 Porto, Portugal
3. Cell Division Group, Experimental Biology Unit, Department of Biomedicine, Faculdade de Medicina, Universidade do Porto, Alameda Prof. Hernâni Monteiro, 4200-319 Porto, Portugal
4. Danish Cancer Society Research Center, Cell Division Laboratory, Strandboulevarden 49, 2100 Copenhagen, Denmark; barisic@cancer.dk

* Correspondence: maiato@i3s.up.pt; Tel.: +351-22-408-800

Academic Editor: J. Richard McIntosh
Received: 1 October 2016; Accepted: 28 January 2017; Published: 17 February 2017

Abstract: Chromosome congression during prometaphase culminates with the establishment of a metaphase plate, a hallmark of mitosis in metazoans. Classical views resulting from more than 100 years of research on this topic have attempted to explain chromosome congression based on the balance between opposing pulling and/or pushing forces that reach an equilibrium near the spindle equator. However, in mammalian cells, chromosome bi-orientation and force balance at kinetochores are not required for chromosome congression, whereas the mechanisms of chromosome congression are not necessarily involved in the maintenance of chromosome alignment after congression. Thus, chromosome congression and maintenance of alignment are determined by different principles. Moreover, it is now clear that not all chromosomes use the same mechanism for congressing to the spindle equator. Those chromosomes that are favorably positioned between both poles when the nuclear envelope breaks down use the so-called "direct congression" pathway in which chromosomes align after bi-orientation and the establishment of end-on kinetochore-microtubule attachments. This favors the balanced action of kinetochore pulling forces and polar ejection forces along chromosome arms that drive chromosome oscillatory movements during and after congression. The other pathway, which we call "peripheral congression", is independent of end-on kinetochore microtubule-attachments and relies on the dominant and coordinated action of the kinetochore motors Dynein and Centromere Protein E (CENP-E) that mediate the lateral transport of peripheral chromosomes along microtubules, first towards the poles and subsequently towards the equator. How the opposite polarities of kinetochore motors are regulated in space and time to drive congression of peripheral chromosomes only now starts to be understood. This appears to be regulated by position-dependent phosphorylation of both Dynein and CENP-E and by spindle microtubule diversity by means of tubulin post-translational modifications. This so-called "tubulin code" might work as a navigation system that selectively guides kinetochore motors with opposite polarities along specific spindle microtubule populations, ultimately leading to the congression of peripheral chromosomes. We propose an integrated model of chromosome congression in mammalian cells that depends essentially on the following parameters: (1) chromosome position relative to the spindle poles after nuclear envelope breakdown; (2) establishment of stable end-on kinetochore-microtubule attachments and bi-orientation; (3) coordination between kinetochore- and arm-associated motors; and (4) spatial signatures associated with post-translational modifications of specific spindle microtubule populations. The physiological consequences of abnormal chromosome congression, as well as the therapeutic potential of inhibiting chromosome congression are also discussed.

Keywords: mitosis; microtubule; kinetochore; mitotic spindle; polar ejection forces; Kinesin; Dynein; CENP-E; Chromokinesin; chromosome; tubulin code

---

1. Introduction

1.1. What is Chromosome Congression?

In preparation for cell division, two poles and an equator start to be defined by the mitotic spindle axis. Precisely at the onset of mitosis, when chromosomes start condensing and the nuclear envelope breaks down, dispersed chromosomes initiate directed movements that culminate with their position at the spindle equator before migrating to the poles after sister chromatid separation. This stochastic motion towards the equator coincides with the beginning of prometaphase and is known as "chromosome congression" (from the English "to come together"; terminology first introduced by Darlington [1]). Chromosome congression truly represents the first challenge of mitosis and culminates with the formation of a metaphase plate, a hallmark of mitosis in metazoans, and occurs in tight spatiotemporal coordination with the assembly of the mitotic spindle that mediates the microtubule-chromosome interactions required for chromosome movement.

1.2. Why do Chromosomes Congress?

At first glance, it may seem counterintuitive that before chromosomes segregate to the poles (during anaphase), they first meet at the equator. This likely reflects millions of years of evolution aiming to improve chromosome segregation fidelity. For instance, if one imagines a mitotic cell in which chromosomes do not congress, the risk of chromosome missegregation after sister chromatid separation at anaphase would be too high, unless all chromatids are extensively moved apart, like in the budding yeast S. cerevisiae, in which the anaphase spindle elongates about 5-fold relative to the metaphase spindle length [2]. In contrast, metazoan spindles only elongate less than 2-fold the metaphase spindle length [3] and thus must rely on different strategies to ensure faithful chromosome segregation during anaphase. One of these strategies is precisely the formation of a metaphase plate, forcing all chromosomes to start subsequent poleward motion from the same position relative to the spindle axis, i.e., from the equator. The other is to trigger an abrupt cleavage of cohesin by separase-mediated degradation of securin, leading to the synchronous separation and movement of sister chromatids towards the pole. This anaphase synchrony has been shown to depend on the uniform distribution of spindle forces acting on all chromosomes prior to anaphase [4]. Aligning chromosomes at the equator also maximizes the chances of kinetochore capture by microtubules emanating from both spindle poles leading to chromosome bi-orientation, which is required to satisfy the spindle assembly checkpoint (SAC; see [5]). Finally, chromosome congression is important to prevent unstable/erroneous kinetochore-microtubule attachments because the proximity to the poles promotes microtubule destabilization at kinetochores due to high Aurora A kinase activity that leads to phosphorylation of Ndc80 (among others), thereby reducing its affinity for microtubules [6–8]. In addition, tension generated by opposing pulling forces on aligned bi-oriented chromosomes is required and sufficient to stabilize correct attachments [9].

2. Mechanisms of Chromosome Congression

2.1. Historical Perspective

In contrast to many other fundamental concepts behind cell division, if one looks for references to the problem of chromosome congression in the early compilations about "The Cell" by E. B. Wilson at the turn of the 20th century, one finds a huge gap in knowledge between the so-called "prophases", which dealt essentially with the condensation and resolution of visible threads/chromatids, and metaphase, by which chromosomes already lie at the equator. The very few references to what happens between these two stages can be resumed in a single sentence: "After definite formation of the chromosomes the nuclear membrane usually disappears and the chromosomes (...) are set free in the protoplasm (and) take up their position in the equatorial plane of the spindle" [10]. Most of the attention at that time was focused on the mechanisms of anaphase and, due to the lack of live-cell studies, the longest stage of mitosis in vertebrates that comprises the entire prometaphase (a term that was only later introduced by Lawrence [11]) was completely left out of the equation.

The first ideas that attempted to explain the process of chromosome congression date back to 1895 from the works of Drüner [12], and later further developed in the works of Belar, Darlington, Rashevsky, Wada and Östergren [1,13–17] (reviewed in [18]). These models conceived that chromosomes are either repelled from the pole by a pushing force that decreases with distance, or attracted to the pole by a pulling force that increases with distance, until all chromosomes eventually reach an equilibrium condition at the equator (Figure 1). One key conceptual difference between these models was the assumption (by some authors) of the existence of kinetochore-to-pole connections from the very beginning of prometaphase. For instance, Belar conceived unaligned chromosomes attached to a "traction fiber" sliding along continuous fibers (most likely interpolar microtubules, as we know them today) until chromosomes eventually reach the equator. However, it was unclear whether bi-orientation and the formation of effective kinetochore-microtubule attachments that connect unaligned chromosomes with the poles was required for initial chromosome congression towards the equator. Moreover, it had been naively assumed that the mechanisms required for initial chromosome congression also play a role in maintaining the equatorial position of chromosomes (see Section 2.10). This is particularly evident in the model proposed by Östergren, who explained chromosome congression by a model in which pulling forces on a given kinetochore act as a linear function of kinetochore-fiber (k-fiber) length. Östergren based his arguments on work with naturally occurring trivalents during meiosis I that were often found positioned off the equator, with their two-kinetochore side closer to the pole, based on the assumption that the pulling force on two kinetochores is higher than on single kinetochores [17,19].

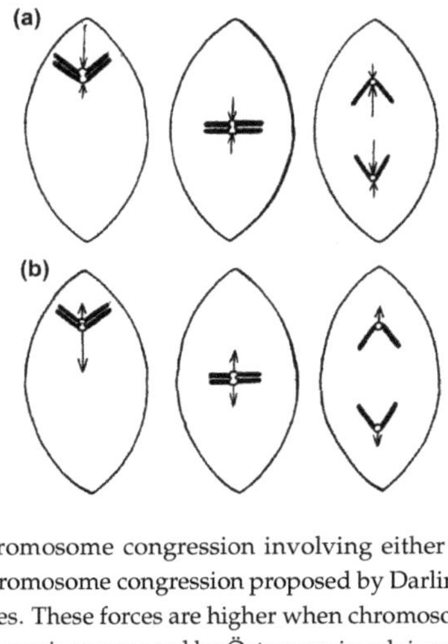

Figure 1. First models of chromosome congression involving either pushing or pulling forces on chromosomes. (**a**) Model of chromosome congression proposed by Darlington [1] involving a balance of pushing forces on chromosomes. These forces are higher when chromosomes are closer to spindle poles; (**b**) Model of chromosome congression proposed by Östergren involving pulling forces on chromosomes that are proportional to k-fiber length. Adapted from Östergren, 1950 [13] and displayed under a Creative Commons Attribution-Noncommercial-Share Alike 4.0 International license, as described at https://creativecommons.org/licenses/by-nc-sa/4.0/legalcode.

Direct evidence that the equatorial position of (already aligned) chromosomes is determined by antagonistic pulling forces on opposing kinetochores was provided by the works of Izutsu and colleagues. They irradiated one kinetochore region of a grasshopper bivalent chromosome in metaphase I using a focused UV microbeam, resulting in the gradual motion of the irradiated bivalent towards the spindle pole facing the non-irradiated kinetochore [20–22] (Figure 2a). Similar findings were subsequently reported by McNeal and Berns for mitotic chromosomes in cultured PtK2 cells [23] (see Figure 2b for a representative example using *Drosophila* S2 cells). Hays and colleagues also estimated the force-length relationship on experimentally generated trivalents in living grasshopper spermatocytes and found it to be consistent with Östergren's hypothesis [24]. However, ideas that the pulling force on kinetochores is not a function of k-fiber length, but rather of their diameter (as function of the number of microtubules attached) started to emerge [25], but even this view has been controversial. For instance, a balance of microtubule numbers on opposite kinetochores has been suggested by elegant experiments using laser microsurgery combined with correlative light and electron microscopy of meiosis I spermatocytes [26], but recent work that measured birefringence retardation of k-fibers of maloriented bivalents challenged this model [27]. In addition, no positive correlation between the number of kinetochore microtubules and the direction of chromosome movement could be observed in vertebrate cells [28]. Overall, these pioneering studies provided definitive demonstration that chromosome position at the equator is maintained (but not necessarily achieved) through a balance of pulling forces acting on opposite kinetochores from the same chromosome that do not strictly depend on k-fiber length or kinetochore microtubule number.

2.2. Polar Ejection Forces

Several subsequent works have challenged aspects of Östergren's hypothesis based on the prediction that kinetochore-pulling forces depend on k-fiber length. If that were the case, one would expect that severing a k-fiber on a metaphase chromosome should lead to a significant displacement of the aligned chromosome towards the pole facing the undamaged k-fiber. However, several experiments that aimed to cut through k-fibers in different systems (from plant to human cells in culture) have revealed that chromosomes either do not shift at all or shift only slightly towards the pole of the unperturbed k-fiber [21,22,29–38].

Important observations that shed light on the mechanism of chromosome congression came from studies of chromosome behavior during transient monopolar spindle formation in newt cells by Bajer and Mole-Bajer. They astutely noticed that " ... the chromosomes approached the pole only up to a certain distance and it was evident that they could not come closer to the pole." [39]. These observations further challenged Östergren's hypothesis based exclusively on pulling forces acting on kinetochores from the same chromosome, as it would have been predicted that a mono-oriented chromosome would travel all the way to the pole, which was not the case. Overall, these data indicate that although kinetochore pulling forces are important to position chromosomes at the equator, as proposed by Östergren, their magnitude is independent of k-fiber length, implying the existence of additional mechanisms.

Based on their observations on transient monopolar spindles, Bajer and Mole-Bajer proposed that "The only logical explanation for the behavior of chromosomes in monopolar division is that the chromosomes approach the center of the aster only to the point at which there is equilibrium between the aster elimination property and the pulling of kinetochore fibers." [39]. Although this "aster elimination property" or "polar ejection force (PEF)" has been noted more than a century ago by Drüner, who refers to a pressure by "growing beams" [i.e., microtubules] from the poles when they encounter an obstacle such as chromosomes [12] (Figure 3a), and was quite evident in the invaginations of the nuclear envelope as the aster develops in prophase (see [10]; Figure 3b) and found to exclude large organelles (e.g., mitochondria) from the centriolar region (reviewed in [40]; Figure 3c), it was Darlington that firmly proposed its involvement in chromosome congression (although he assumed this was essentially due to electrostatic repulsions). This view was based on the analysis of pollen-grain mitosis, in which the distance of peripheral centromeres relative to the spindle pole was highly variable [1] (Figure 3d). This irregular pattern likely reflected the dynamic behavior of chromosomes on monopolar spindles, which was subsequently extensively characterized by Bajer and colleagues [39,41,42] (Figure 3e,f). Together, these studies supported a new view of chromosome congression involving a balance of PEFs and kinetochore-pulling forces.

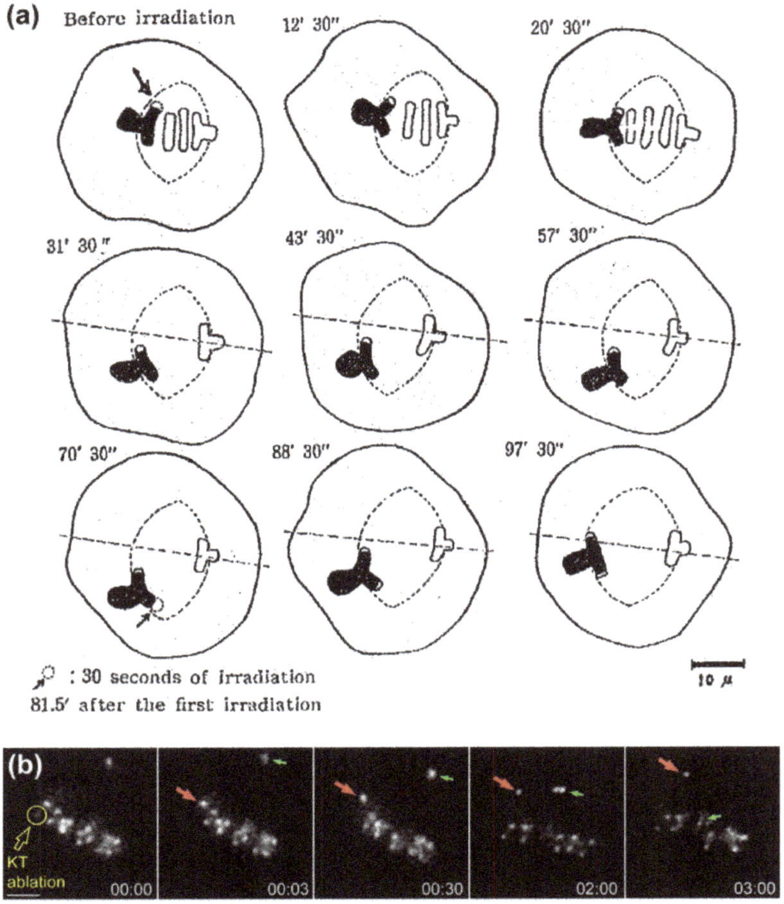

Figure 2. Evidence that forces on kinetochores are required to position chromosomes at the equator. (**a**) Original drawings from Izutzu depicting the loss of equatorial position when one of the kinetochore regions from a bivalent chromosome was irradiated with an UV microbeam. Note the displacement of the bivalent from the metaphase plate towards the pole facing the non-irradiated kinetochore after irradiation. Scale bar is 10 μm. Reprinted from Izutsu et al., 1959 [20]; (**b**) Laser microsurgery of one of the kinetochores from an equatorially-aligned chromosome in a *Drosophila* S2 cell. Kinetochores were directly labelled with the Centromere Protein A (CENP-A) homologue Cid fused with Green Fluorescent Protein (GFP). Likewise, the chromosome was displaced from the equator after surgery and underwent poleward migration towards the pole facing the undisturbed kinetochore from the pair. Red arrows track the undisturbed kinetochore from the irradiated pair. Green arrows track the congression of an undisturbed chromosome. Laser microsurgery was performed as described in [29]. Scale bar is 2 μm.

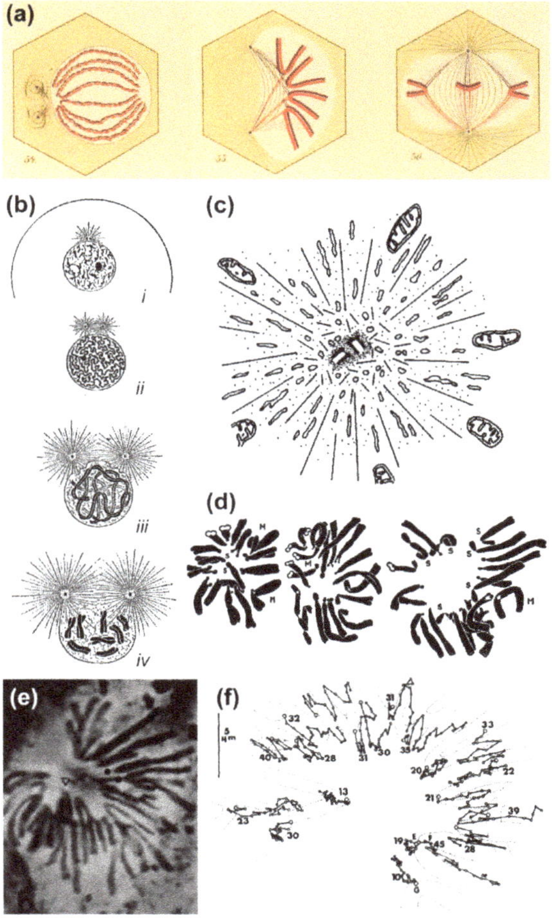

Figure 3. Evidence that centrosome-derived microtubules can exert pushing forces. (**a**) Original Drawings by Drüner depicting the invasion of the chromosomal region by microtubules, which exert a pushing force that assists chromosome alignment at the spindle equator. Reprinted from Drüner, 1895 [12]. Image courtesy of Biodiversity Heritage Library. http://www.biodiversitylibrary.org; (**b**) Schematic drawing by E. B. Wilson illustrating the pushing action of centrosomal microtubules on the nuclear envelope and subsequent rupture. Reprinted from Wilson, 1925 [10]. Image displayed under a Creative Commons Attribution-Noncommercial-Share Alike 4.0 International license, as described at https://creativecommons.org/licenses/by-nc-sa/4.0/legalcode. Image courtesy of the Wellcome Library. http://wellcomelibrary.org; (**c**) Schematic drawing by Luykx illustrating the repulsive action of centrosomal microtubules over large organelles (mitochondria). Reprinted from Luykx, 1970 [40]. Courtesy of Elsevier; (**d**) Original drawings by Darlington illustrating the variability in chromosome positioning in pollen grain cells. Reprinted from Darlington, 1937 [1]. Image courtesy of Biodiversity Heritage Library. http://www.biodiversitylibrary.org; (**e**,**f**) Phase contrast image of a newt lung cell undergoing transient monopolar configuration. Kinetochore position was tracked over time, clearly demonstrating the oscillatory behavior of mono-oriented chromosomes in this system. Note that chromosomes do not travel all the way towards the pole. Reprinted from Bajer et al., 1982 [41] and displayed under a Creative Commons Attribution-Noncommercial-Share Alike 4.0 International license, as described at https://creativecommons.org/licenses/by-nc-sa/4.0/legalcode.

The exact nature and mode of action of PEFs was only elucidated by Rieder and colleagues using an elegant combination of laser microsurgery and correlative light-electron microscopy experiments [43]. First, they demonstrated that the distal kinetochore from an oscillating mono-oriented chromosome was indeed devoid of microtubules and consequently was not under opposing kinetochore pulling forces. Second, by cutting near the kinetochore regions of mono-oriented chromosomes to generate acentric fragments (i.e., without kinetochore), they found that kinetochore-free chromosome arms were immediately ejected away from the spindle pole with velocities similar to the outward movement of an oscillating chromosome [44], whereas the remaining kinetochore-containing fragment moved closer to the pole [43] (see also [44,45]; Figure 4a,b). Subsequent studies by Salmon, Rieder and colleagues have further demonstrated that when astral microtubules were reversibly depolymerized/polymerized, mono-oriented chromosomes moved closer to or were pushed away from the pole, respectively [44,46,47]. These studies revealed no difference in the mechanism of chromosome positioning between monopolar and bipolar spindles, including average distances from the pole. Finally, it was shown that kinetochores moving away from their associated pole do not exert a significant pushing force on the chromosome [48,49] and PEFs determine the amplitude of chromosome oscillations near the pole [50]. Thus, PEFs derived from astral microtubules acting along chromosome arms oppose kinetochore-pulling forces. This "push-pull" mechanism was proposed to account for chromosome oscillations, while determining chromosome position relative to the spindle pole. In the context of a bipolar spindle, chromosome congression could now be explained in light of the balance of four forces on a chromosome: two antagonistic poleward forces acting at the kinetochores and two opposing PEFs acting along chromosome arms. As so, formation of a metaphase plate equidistant to the spindle poles would result from the net forces applied to the chromosomes being zero [44,47]. An integrated view of these studies can be found in a landmark essay that firmly established the contribution of PEFs and kinetochore directional instability (i.e., kinetochores can switch from poleward to anti-poleward motion) for chromosome congression in vertebrates [51].

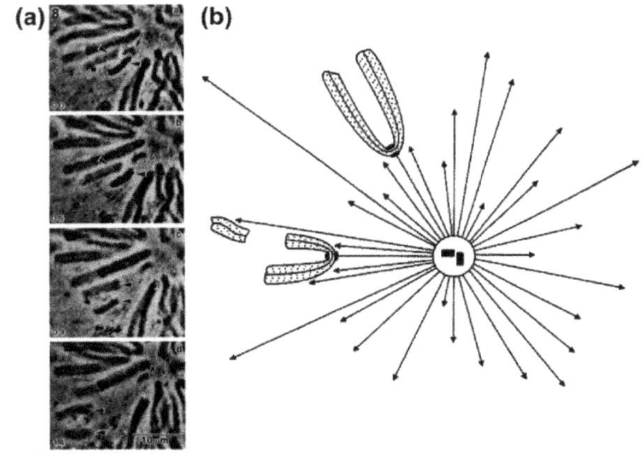

Figure 4. Demonstration that polar ejection forces act along the entire chromosome. (**a**) Phase contrast image of a newt lung cell in which the chromosome arms on one chromosome (arrowheads) were physically separated from the kinetochore region using laser microsurgery. Note the ejection of the acentric chromosome arms away from the polar region. In contrast, the kinetochore-containing region (arrow) moves closer to the polar region. Reprinted from Rieder et al., 1986 [43] and displayed under a Creative Commons Attribution-Noncommercial-Share Alike 4.0 International license, as described at https://creativecommons.org/licenses/by-nc-sa/4.0/legalcode; (**b**) Schematic representation of the experiment illustrated in (a). Reprinted from Salmon, 1989 [44]. Courtesy of Elsevier.

PEFs are likely associated with the pushing action of elongating astral microtubules undergoing dynamic instability along the length of the chromosome. Consistent with this idea, taxol-induced polymerization of polar microtubules can push chromosome arms away from the pole [45,52], whereas nocodazole or colcemid treatment completely abolished PEFs on chromosomes [45]. Importantly, dynamic microtubules were shown to be required for continuous ejection of chromosome arms away from the poles [45]. Theoretical predictions and calculation of PEFs distribution further indicate that PEFs are stronger closer to the center of the aster, where microtubule density is higher, and depend on chromosome size [47,50]. In vitro measurements of the force produced by a polymerizing microtubule against a rigid surface or inside lipid vesicles have determined maximal forces between 2–4 pN [53,54]. Interestingly, it was found that forces on short buckling microtubules tend to be higher than those on long buckling microtubules, likely reflecting the length-dependent stiffness of microtubules [53]. Attempts to measure the scale of PEFs by individual microtubules on chromosomes using either in vitro reconstitution or in vivo systems have estimated a force between 0.5–1 pN per microtubule [55,56] and ~100 pN near the pole where microtubule density is higher [56]. While the PEF produced by individual microtubules is compatible with that generated by polymerizing microtubules in vitro [53,54], it was also consistent with the force generated by single Kinesin motors [57–59], suggesting their involvement in PEFs [44,60].

2.3. The Role of Chromosome Arm-Associated Motors in the Generation of Polar Ejection Forces

Chromokinesins are Kinesin-like motor proteins that have DNA-binding properties and associate with chromosomes during mitosis [61,62]. The best characterized mammalian Chromokinesins are Kif4A and Kid, which belong to two distinct families: Kinesin-4 and Kinesin-10, respectively (reviewed in [63]). Functional analysis revealed a combined role for Kinesin-4 and Kinesin-10 in chromosome congression, arm-orientation and normal chromosome oscillations, consistent with an active role of Kinesin-4 and Kinesin-10 in the generation of PEFs [6,62,64–72]. Both Kinesin-4 and Kinesin-10 were shown to have microtubule plus-end directed motility [73–75], but they appeared to be non- or weakly-processive motors under load [74,75]. Nevertheless, antibody-blocking experiments in vitro suggested that Kinesin-10 is a major contributor for PEFs [56]. In vitro reconstitution experiments have indicated that, despite of its slower motility compared to Kinesin-4 [74,75], Kinesin-10 binds more strongly to microtubules and dominates over Kinesin-4 during cooperative microtubule motility associated with chromatin [76]. Similar findings have been reported upon functional perturbation of these two Chromokinesin families in *Drosophila* and human cells, which suggested a combined role during chromosome congression, with Kinesin-10 providing the major PEF required for arm orientation and Kinesin-4 mainly regulating microtubule dynamics [68,71]. Altogether, these data can be reconciled in light of the "soft" nature of the chromosomes. If strong and highly processive motors worked as PEF generators, this would likely lead to chromatin deformations/damage and loss of chromosome structure. Indeed, overexpression of Kinesin-10 in *Drosophila* S2 cells was shown to stretch and deform chromatin when microtubules impact or pass by the chromosomes [77]. As discussed by Brouhard and Hunt for Kinesin-10 [56], the combined action of Kinesin-10 and Kinesin-4 on chromosome arms is ideal for exerting PEFs against microtubules through slow, weak, and discontinuous action, which would be sufficient to bias chromosome ejection away from the poles without inducing damage. Finally, direct demonstration that Kinesin-4 and Kinesin-10 collectively mediate PEFs on chromosome arms in human cells was only recently obtained. By combining RNAi-mediated depletion of Kid and Kif4A with laser microsurgery to generate acentric chromosome fragments in human culture cells, it was shown that arm ejection forces operating in the absence of kinetochore-pulling forces relied on the cooperative action between Kinesin-4 and Kinesin-10, with only a minor fraction that could be attributed to the pushing force of polymerizing microtubules impacting on chromosome arms [6]. Most importantly, this work revealed that PEFs operating on acentric fragments caused the ejection of chromosome arms in random directions, including towards the cortex. This indicated that although PEFs mediated by Kinesin-4 and Kinesin-10 are sufficient to exert a pushing force on chromosome arms that leads to chromosome ejection away from the pole, they are not the critical players that conduct chromosome movement exclusively towards the equator.

A critical aspect of the model proposed by Rieder and Salmon was that congressing mono-oriented chromosomes experience tension at kinetochores as result from the push-pull between PEFs along chromosome arms and kinetochore-pulling forces [51,78]. This was a reasonable assumption based on findings that mono-oriented chromosomes during transient monopolar formation in newt cells showed robust k-fibers on the attached kinetochore [43,46]. However, kinetochore-microtubule attachments on mono-oriented chromosomes are highly unstable, unless constant tension away from the pole is applied [9,79,80] (see also [47]). This apparent paradox could only be solved if PEFs produce sufficient kinetochore tension independently of opposing kinetochore-pulling forces that result from chromosome bi-orientation. This hypothesis has been recently tested in *Drosophila* culture cells. Elegant experiments involving overexpression of Kinesin-10 have first indicated that elevated PEFs could indeed stabilize kinetochore-microtubule attachments [77]. These proof-of-concept experiments were followed by studies of *Drosophila* cultured cells undergoing mitosis with unreplicated genomes (MUGs), where the function of individual kinetochores could be investigated in the context of single chromatids that are unable to bi-orient [81]. In this work it was shown that PEFs mediated by Kinesin-4 and Kinesin-10 stabilize kinetochore-microtubule attachments on mono-oriented chromosomes. Over time, mono-oriented chromosomes were also shown to experience significant intra-kinetochore stretch or structural deformation (see discussion in [82–84]) comparable with those typically experienced by bi-oriented chromosomes [81]. Taken together, these data indicate that Chromokinesin-mediated PEFs oppose kinetochore-pulling forces and contribute to tension-dependent stabilization of microtubule attachments on mono-oriented chromosomes.

2.4. Coordination between PEFs and Kinetochore-Pulling Forces Drives Chromosome Congression after Bi-Orientation

A related problem that derives from the existence of kinetochore-pulling forces on attached chromosomes concerns their nature. One model is based on the action of pulling forces resulting from depolymerization of attached kinetochore microtubules. This model stems from original work by Shinya Inoue on the effect of colchicine on spindle microtubules and chromosome movement using oocytes from the marine annelid worm *Chaetopterus pergamentaceous*. In this system, the metaphase arrested spindle is anchored by one of its poles to the cell cortex and, upon addition of colchicine or cold treatment (now well established treatments that induce spindle microtubule depolymerization), the aligned chromosomes at the spindle equator were observed to move towards the anchored pole [31,85]. Based on these observations, Inoue concluded that the spindle affected by colchicine or cold is able to perform mechanical work and exert a pulling force on chromosomes (reviewed in [86]). In vitro reconstitution works have provided additional evidence that microtubule depolymerization at their plus-ends can exert a pulling force on the kinetochore that is independent of ATP hydrolysis and is sufficient to move chromosomes [87–89]. In agreement, nocodazole-induced microtubule depolymerization has been shown to occur near the kinetochore during poleward chromosome movement in prometaphase [90]. Moreover, oscillating mono-oriented chromosomes have been proposed to switch from microtubule depolymerization and polymerization states, as inferred by accumulation of EB proteins at growing microtubule plus-ends at kinetochores [91]. However, based on the analysis of the profile of individual microtubule plus-ends within a k-fiber, it has been proposed that two-thirds adopt a conformation compatible with a microtubule depolymerizing state, regardless of the directional instability associated with poleward and anti-poleward chromosome oscillations [92]. These apparently contradicting findings have recently been reconciled by the observation that EB protein bursts near kinetochores are rather infrequent and only represent a small bias for microtubule polymerization within an incoherent k-fiber that contains a mixture of polymerizing and depolymerizing microtubules [93]. Overall, these data support a model in which regulation of microtubule dynamics favoring depolymerization can generate pulling forces on attached kinetochores.

Any model of chromosome congression involving kinetochore-pulling forces implies that any perturbation of end-on kinetochore-microtubule attachments or defects in spindle assembly/organization would lead to chromosome alignment problems. Indeed, an extensive survey of the literature revealed more than 100 proteins that have been implicated in chromosome alignment (Table 1), yet it is only for a select handful that we know the mechanism and thus will represent our focus in this review. Probably the best studied case is the one involving the KMN network, which forms the core microtubule interface at kinetochores and all respective regulatory proteins, such as Aurora B and Plk1 kinases (reviewed in [94]). Additionally, proteins that modulate kinetochore-microtubule attachments and their dynamic state are also likely to play an important role. Among these, microtubule plus-end-tracking proteins (+TIPs) are of special interest due to their specific accumulation at the plus-ends of microtubules [95–97] where they promote microtubule growth by catalyzing the addition of tubulin subunits to microtubule plus-ends [98], by inducing rescue [99], or by stabilizing microtubules [100,101]. CLIP-170 was the first +TIP reported [102] and was initially associated with microtubule rescue [99]. Functional inhibition of CLIP-170 during mitosis results in chromosome alignment defects, possibly associated with defective kinetochore-microtubule attachments [103,104]. However, CLIP-170 inhibition does not seem to affect kinetochore microtubule dynamics or stability, possibly because it is stripped from the kinetochore by Dynein upon the establishment of end-on kinetochore-microtubule attachments [103,104]. Moreover, phosphorylation of CLIP-170 at S312 by Plk1 regulates its binding to microtubules and is crucial for chromosome alignment [105]. CLIP-170 appears to promote kinetochore-microtubule attachments and chromosome congression by counteracting Dynein/Dynactin [106]. The XMAP215/Ch-TOG and CLASP families of +TIPs have also been implicated in chromosome congression. The XMAP215/Ch-TOG proteins act as microtubule polymerases at microtubule plus-ends and promote microtubule assembly [98,107,108], whereas CLASPs promote microtubule rescue and suppress catastrophe [109,110]. Depletion of proteins from the XMAP215/Ch-TOG family results in the presence of unattached kinetochores and chromosome alignment defects [111–114]. Moreover, XMAP215/Ch-TOG contributes to chromosome oscillations [115]. Recruitment of CLASPs to microtubule plus-ends requires interactions with CLIP-170 and EB1 [100,101]. Importantly, CLASPs also localize to kinetochores in a microtubule-independent manner and remain at kinetochores upon microtubule attachment [116,117]. This localization at the kinetochore-microtubule interface favors a role of CLASPs in the regulation of microtubule dynamics at the kinetochore [118,119], thereby contributing for chromosome congression [116]. Surprisingly, perturbation of either CLASPs or XMAP215/Ch-TOG increases the stability of kinetochore-microtubule attachments [115,119]. One possibility might be that during mitosis the activity of these proteins is regulated by phosphorylation and/or binding to other proteins that promote microtubule depolymerization [120,121].

The members of the Kinesin-13 family Kif2a, Kif2b and Kif2c/MCAK are also important regulators of microtubule dynamics, including at kinetochores [122]. Kinesin-13 proteins are non-motile but use the energy from ATP hydrolysis to promote microtubule depolymerization by binding both the plus- and the minus-ends of microtubules and inducing a conformational change that leads to a catastrophe event [123–125]. In the context of the mitotic spindle, Kinesin-13 proteins associate with both spindle poles and kinetochores where they play distinct roles [124,126]. Kif2b and MCAK regulate microtubule plus-end dynamics at the kinetochore where they play an important role in the correction of erroneous microtubule attachments [124,127–130], while Kif2a appears to have a preference for microtubule minus-ends where it plays an important role in the regulation of spindle microtubule flux [131,132]. Interestingly, Kif2a and MCAK are dispensable for chromosome congression [132], whereas Kif2b appears to be required for proper chromosome oscillation on a monopolar spindle configuration [124]. However, because Kif2b only transiently associates with kinetochores before microtubule attachments [124] it is unlikely to play an important role assisting chromosome congression after bi-orientation, suggesting the involvement of other players.

The widely conserved Kinesin-8 family has been proposed to function both as plus-end-directed motors and as microtubule depolymerases [133–136]. However, the depolymerase activity of human Kif18A remains controversial. Although Kif18A was initially proposed as a microtubule depolymerase [134], further studies suggested that Kif18A suppresses microtubule growth by capping the microtubule plus-ends [137,138]. This would be consistent with the emerging role of Kinesin-8 motors as negative regulators of microtubule length, since loss of Kinesin-8 activity generally leads to longer cellular microtubules [134,139–142]. Importantly, genetic and siRNA-based studies demonstrate that Kinesin-8 motors are necessary for proper chromosome alignment by suppressing chromosome oscillations on bi-oriented chromosomes [68,70,134,139,143,144]. Accordingly, in the absence of functional Kif18A, kinetochores exhibit an increase in the oscillation amplitude leading to a deregulation of metaphase plate organization [144]. Furthermore, loss of Kif18A leads to a modest increase in spindle size and longer microtubules [134,144]. In agreement, overexpression of Kif18A decreases chromosome oscillations, favoring chromosome alignment at the metaphase plate [144,145]. Overall, these data are consistent with a model of chromosome congression after bi-orientation, in which Kif18A forms a gradient along attached kinetochore microtubules, directly regulating their length and dynamics to facilitate chromosome alignment at the spindle equator [144].

The co-existence of PEFs acting along the entire chromosome arms and kinetochore-pulling forces driven by microtubule depolymerization suggests that they might work in parallel to regulate chromosome oscillations during congression after bi-orientation. Disruption of PEFs by inhibition of Chromokinesin function in cultured cells altered chromosome oscillations on both monopolar and bipolar spindles [65–67,71]. Although perturbation of Chromokinesin functions did not fully compromise chromosome congression, few monooriented chromosomes remained close to the poles, suggesting that PEFs might increase the efficiency of chromosome congression by facilitating the stabilization of end-on kinetochore microtubule attachments and biorientation [77,81]. Furthermore, despite having opposite effects on chromosome movement, PEFs and Kif18A synergistically promote the position of bi-oriented chromosomes near the spindle equator [146]. Overall, these findings suggest that the coordinated activities of Kif18A and PEFs regulate chromosome oscillations and are important for chromosome congression after bi-orientation.

Table 1. Proteins that have been implicated in chromosome alignment.

Protein Name	Subcellular Localization	Misaligned Chromosomes/ Chromatids	Chromosome Congression Defects (by Live Cell Imaging)	References
Astrin	Spindle pole; kinetochores	Yes	Yes	[147–150]
HICE1/HAUS8	Centrosome; mitotic spindle; spindle midzone; midbody	Yes	ND	[151]
Aurora A	Centrosome; central spindle	Yes	Yes	[152–154]
CENP-E	Kinetochore	Yes	Yes	[6,155–157]
CEP57	Centrosome	Yes	ND	[158]
Cep72	Centrosome	Yes	ND	[159]
Cep90	Centrosome; Pericentriolar satellites	Yes	ND	[160]
ChTOG	Centrosome; spindle pole	Yes	Yes	[112,150,161]
CLASPs	Centrosome; kinetochore; microtubule plus ends; central spindle	Yes	Yes	[150,162]
Aurora-B	Centromere; spindle; spindle midzone	Yes	Yes	[163,164]
Haspin	Chromosome; centrosome	Yes	Yes	[165–167]
ILK	Plasma membrane; focal adhesion; cytosol	Yes	ND	[168]
Kinastrin/SKAP	Spindle pole; kinetochore; microtubule plus ends	Yes	yes	[148,149,169]
HEC1	Kinetochore	Yes	Yes	[170–173]

Table 1. *Cont.*

Protein Name	Subcellular Localization	Misaligned Chromosomes/ Chromatids	Chromosome Congression Defects (by Live Cell Imaging)	References
Spc24	Kinetochore	Yes	ND	[174]
Spc25	Kinetochore	Yes	ND	[174]
Nuf2	Kinetochore	Yes	Yes	[174,175]
NuMA	Nucleus; spindle pole	Yes	ND	[176]
Sgo1/Shugoshin	Centromere; kinetochore; centrosome; spindle pole	Yes	Yes	[177]
Spindly	Kinetochore; spindle pole	Yes	Yes	[178,179]
TACC3	Centrosome	Yes	Yes	[161,180–182]
CHC (Clathrin heavy chain)	Mitotic spindle	Yes	Yes	[181,183]
4.1r	Mature centriole	Yes	ND	[184]
Ska1	Kinetochore; mitotic spindle	Yes	Yes	[185–188]
Ska2	Kinetochore; mitotic spindle	Yes	Yes	[185–188]
Ska3/RAMA1	Kinetochore; mitotic spindle	Yes	Yes	[186–190]
Kid	Chromosome arms; spindle poles	Yes	Yes	[70,71,191]
Kif4A	Chromosome arms; spindle midzone	Yes	Yes	[69–71]
Kif18A	Plus-ends of kMTs	Yes	Yes	[134,144,146,192,193]
Kif18B	Astral microtubule plus ends	Yes	Yes	[194–196]
MCAK	Spindle poles; spindle midzone; kinetochore	Yes	Yes	[70,124,197]
HURP	Kinetochore	Yes	Yes	[198–200]
CENP-L	Kinetochore	Yes	Yes	[201]
NuSAP1	Central spindle	Yes	Yes	[202,203]
SAF-A/hnRNP-U	Spindle microtubules; spindle midzone	Yes	Yes	[204]
Bub1	Kinetochore	Yes	Yes	[164,205]
BubR1	Kinetochore	Yes	Yes	[164,206–208]
NUP188	Centrosomes	Yes	Yes	[209]
CENP-F/mitosin	Kinetochore	Yes	Yes	[210–212]
Plk1	Centrosome	Yes	Yes	[213–215]
NudC	Kinetochore	Yes	Yes	[216,217]
RRS1	Chromosome periphery	Yes	Yes	[218]
Nucleolin	Nucleoli; chromosome periphery	Yes	Yes	[219]
KIBRA	ND	Yes	ND	[220]
DDA3	Spindle microtubules; kinetochores; midbody	Yes	Yes	[221,222]
HIP1r	Mitotic spindle	Yes	Yes	[223]
Nucleophosmin	Perichromosomal region	Yes	Yes	[224]
Kif2a	Spindle poles	Yes	Yes	[124,221]
Beclin-1	Kinetochore	Yes	Yes	[225]
CLIP-170	Kinetochore; mitotic spindle	Yes	Yes	[104,106]
ATRX	Pericentromeric heterochromatin	Yes	Yes	[226]
CHICA	Mitotic spindle	Yes	Yes	[227,228]
p38γ	Kinetochore; spindle poles	Yes	Yes	[229]
SPICE	Mitotic spindle; centrioles	Yes	Yes	[230]
Zw10	Kinetochore	Yes	Yes	[231,232]

Table 1. *Cont.*

Protein Name	Subcellular Localization	Misaligned Chromosomes/ Chromatids	Chromosome Congression Defects (by Live Cell Imaging)	References
DHC/DYNC1H1	Kinetochore; mitotic spindle	Yes	Yes	[6,178]
DIC2/DYNC1I2	Kinetochore; mitotic spindle	Yes	Yes	[178]
Roadblock-1/DYNLRB1	Kinetochore; mitotic spindle	Yes	Yes	[178]
Lis1/PAFAH1B1	Kinetochore; mitotic spindle	Yes	Yes	[178]
Nde1	Kinetochore; mitotic spindle	Yes	Yes	[178]
Ndel1	Kinetochore; mitotic spindle	Yes	Yes	[178]
ARP1	Kinetochore; mitotic spindle	Yes	Yes	[178]
TAO1/MARKK	Microtubules	Yes	Yes	[233]
Kif14	Spindle poles; mitotic spindle; midbody	Yes	Yes	[70,234]
CENP-W	Kinetochore	yes	yes	[235–237]
CENP-T	Kinetochore	Yes	ND	[235,238]
CENP-H	Kinetochore	Yes	Yes	[239]
Chl4r	Kinetochore	Yes	Yes	[239]
Nnf1R	Kinetochore	Yes	Yes	[239,240]
CENP-Q	Kinetochore	Yes	Yes	[241]
CENP-U	Kinetochore	Yes	Yes	[238,242]
CENP-N	Kinetochore	Yes	ND	[238]
CENP-M	Kinetochore	Yes	ND	[238,243]
Septin 7	Spindle poles; mitotic spindle; midbody	Yes	ND	[244]
TRAMM	Perinuclear region	Yes	Yes	[245]
Shp2	Kinetochore; centrosome; spindle midzone; midbody	Yes	Yes	[246,247]
Bod1	Centrosomes; kinetochores	Yes	Yes	[248,249]
PTEN	Centrosome; mitotic spindle; midbody	Yes	Yes	[250]
RSK2/RPS6KA3	Centrosomes; mitotic spindle; midbody; kinetochore	Yes	Yes	[251–253]
Nup62	Nuclear envelope; cytoplasm; centrosomes	Yes	ND	[254,255]
Mdp3	Mitotic spindle	Yes	Yes	[256]
ANKRD53	Spindle poles	Yes	Yes	[257]
NF-1 (neurofibromatosis type 1)	Astral microtubules; mitotic spindle; centrosomes; midbody	Yes	ND	[258]
Hsp72	Mitotic spindle; midbody	Yes	Yes	[259]
RGS2	Centrosome; mitotic spindle; astral microtubules	Yes	ND	[260]
B56	Centromere	Yes	Yes	[174,207,261,262]
And-1 (acidic nucleoplasmic DNA-binding protein 1)	Cytoplasm	Yes	ND	[263]
ASURA (PHB2)	Cytoplasm	Yes	ND	[264]
Rab5	Early endosomes	Yes	Yes	[211]
MST1	ND	Yes	Yes	[265]
GAK	Trans-Golgi network	Yes	ND	[266]
Usp16	Cytoplasmic in interphase; kinetochore	Yes	Yes	[267]
TTL	Mitotic spindle	Yes	Yes	[345]
TCP	ND	Yes	Yes	[345]

ND (not determined).

2.5. The Role of Kinetochore Motors in Chromosome Congression

A concurrent model for the explanation of kinetochore-pulling forces is based on the presence of ATP-dependent motor proteins at kinetochores. The best candidate for such force generator is the cytoplasmic form of the microtubule minus-end-directed motor Dynein, which has been shown to localize to kinetochores [268,269] and was proposed to counteract the action of PEFs on chromosome arms by generating kinetochore poleward motion [51]. However, despite some evidence (mostly from studies in anaphase) supporting a requirement for kinetochore Dynein in chromosome poleward motion, this remains a highly controversial issue (reviewed in [270]). The strongest arguments against such a role are based on the fact that chromosome-to-pole velocities in anaphase are about one order of magnitude slower than those typically observed by Dynein-dependent transport and Dynein accumulation at kinetochores is negatively regulated by microtubule attachments [271,272]. Moreover, inhibition of Dynein motor activity did not affect minus-end-directed chromosome motion driven by microtubule depolymerization in vitro [88,273,274]. Although it remains possible that few molecules of Dynein are able to generate kinetochore-pulling forces after the establishment of end-on microtubule attachments during chromosome congression, the rate of motion is likely governed by other processes, such as microtubule depolymerization.

Although a major role played by kinetochore Dynein in the generation of kinetochore-pulling forces after the establishment of end-on microtubule attachments is disputable, its role in the stages that precede chromosome congression is well supported. It has long been noticed by Schneider that some chromosomes tend to move toward the poles before congressing to the spindle equator [275]. Bajer and Mole-Bajer, in their classic cinematographic studies of mitosis also clearly demonstrate and recognize that some chromosomes undergo poleward motion before migrating to the equator [276,277]. Similar findings have been reported in cultured newt cells by Zirkle and colleagues, who first recognized the frequent appearance of "centrophilic" chromosomes (i.e., that lie near the centrosomes) that do not migrate straightaway to the equator [278–280], as well as in insect spermatocytes [281] and PtK1 cells [282]. These sharp observations have indicated that the process of chromosome congression is complex and that not all chromosomes follow the same path, suggesting the existence of concurrent mechanisms.

The implication of Dynein in the poleward movement of chromosomes that precede congression of some chromosomes was proposed even a few months before the report of its localization to kinetochores [268,269], based on the characterization of initial kinetochore-microtubule interactions during early prometaphase [283]. This study showed that a single astral microtubule extending well beyond the kinetochore region was sufficient to mediate the initial attachment and subsequent poleward movement of some chromosomes. Importantly, this association involved the tangential interaction between the microtubule and the kinetochore fibrous corona (the outermost domain of the kinetochore that expands into crescents in the absence of attached microtubules) and was independent of microtubule depolymerization. Based on the recorded velocities of chromosomes during this fast poleward movement after initial lateral interaction between kinetochores and microtubules (typically ranging between 25–55 μm/min in newt lung cells in culture), Rieder and Alexander proposed that Dynein at kinetochores could account for this behavior. This proposal was seconded by Merdes and De Mey (after the discovery of Dynein at kinetochores) who reported similar findings [284]. Shortly thereafter, it was shown that kinetochore Dynein is indeed a component of the fibrous corona [285], but direct demonstration of this hypothesis came only several years later. By studying the specific role of kinetochore Dynein by RNAi-mediated depletion of its kinetochore-targeting factor ZW10, as well as injection of function-blocking antibodies against Dynein Intermediate Chain, or injection of Dynamitin protein that disrupts the Dynein/Dynactin complex, several laboratories reported a role for Dynein in the fast poleward movement of chromosomes during the initial encounters between microtubules and kinetochores, but not in k-fiber formation [231,232,286]. Consequently, in some of these perturbations, particularly evident after ZW10 RNAi, some chromosomes failed to complete congression and remained outside the spindle pole with mono-oriented or unattached kinetochores [231,232]. Similar findings were also reported after RNAi of Spindly, a protein that is required to recruit Dynein to kinetochores without affecting the SAC [179]. Overall, these data indicated a role for kinetochore Dynein in the poleward movement of chromosomes during early prometaphase, with possible implications for the mechanism of congression in a subset of chromosomes.

In addition to a microtubule minus-end-directed motor activity, in vitro studies have also revealed the existence of a microtubule plus-end-directed activity at kinetochores from purified chromosomes [287,288]. Independent work by Yen and colleagues led to the discovery of CENP-E, which is enriched at prometaphase kinetochores [289] and was subsequently shown to be a Kinesin-like (Kinesin-7) motor protein [290] associated with the kinetochore fibrous corona [291,292]. Direct demonstration of microtubule plus-end-directed activity was obtained after characterization of CENP-E in *Xenopus*, where immunodepletion/immunoblocking experiments in oocyte extracts revealed a role in chromosome alignment [293]. Similar findings were reported after microinjection of function-blocking antibodies, expression of a dominant-negative motor-less CENP-E construct and antisense oligonucleotide blocking in human cells in culture [294,295] or analysis of CENP-E mutants in *Drosophila* [296]. However, these experiments were unable to make a clear distinction whether CENP-E motor activity was required for chromosome congression or maintenance of chromosome alignment after reaching the equator. This was only firmly established by live-cell recordings from nuclear envelope breakdown (NEB) after perturbation of CENP-E function by antibody microinjection in human cells in culture, where some chromosomes that were found to undergo initial poleward movement were unable to complete congression within the next 2h after NEB [297]. Overall, these studies demonstrated the existence of a Kinesin-like motor protein with microtubule plus-end-directed activity that is associated with the kinetochore fibrous corona and plays a role in chromosome congression. Importantly, because most chromosomes are able to align at the equator after perturbation of CENP-E function, it was concluded that the dependence on CENP-E for chromosome congression must be critically linked to chromosome position within the spindle (see Section 2.6), further demonstrating the existence of concurrent mechanisms.

For years, it was believed that CENP-E function at kinetochores required for chromosome congression and bi-orientation was related to the regulation of end-on kinetochore microtubule attachments [297–299], in part through a contribution of CENP-E in maintaining attachment of kinetochores to the end of a depolymerizing microtubule [273]. However, this capacity to couple kinetochores to depolymerizing microtubule plus-ends does not require ATP, suggesting that the role of CENP-E in chromosome congression relies on a different mechanism. The paradigm shift occurred after the demonstration that chromosomes can congress to the spindle equator before bi-orientation [300]. In this work, Khodjakov and colleagues demonstrated that mono-oriented chromosomes located near the poles could glide towards the equator along pre-existing spindle microtubules, including k-fibers, in a CENP-E-dependent manner. These observations provided an explanation for the involvement of CENP-E microtubule plus-end-directed motility at the kinetochore fibrous corona for chromosome congression (Figure 5).

One controversial issue has been related with CENP-E processivity. In vitro microtubule gliding assays with recombinant CENP-E motor domain revealed a velocity around 5 μm/min [293,301]. Similar microtubule gliding assays with the full-length protein reported velocities around 1 μm/min [301,302]. More recently, single CENP-E molecule measurements (either the full length or motor domain only) have indicated a much faster velocity in the order of 20 μm/min [303,304], suggesting that CENP-E binding to the coverslip in traditional gliding assays is partially inhibitory of its function. Interestingly, the measured chromosome velocity during CENP-E-dependent congression of polar chromosomes in human cells was around 1.5 μm/min [6,305] indicating that, in vivo, cumulative CENP-E processivity is significantly attenuated by a yet unknown mechanism. One possibility could be related with the presence of non-motile microtubule-associated proteins (MAPs) or residual Dynein activity on microtubules that could slow down CENP-E-dependent transport of chromosomes during congression in vivo.

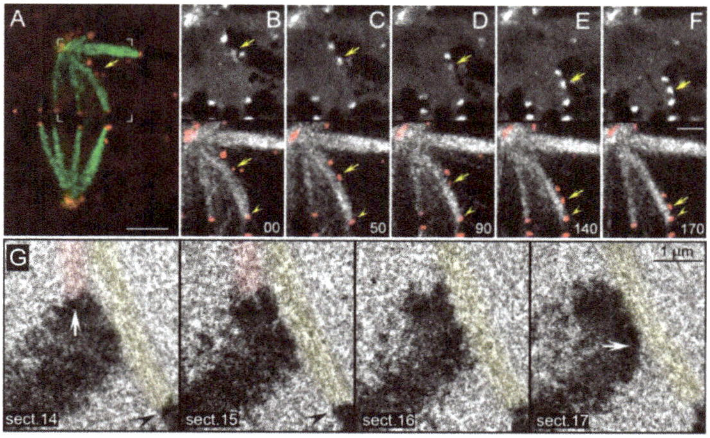

Figure 5. Demonstration that chromosome congression is independent of bi-orientation. From A-F, the movement of a polar chromosome along a pre-existing k-fiber is illustrated in a PtK1 cell. The leading kinetochore is indicated (yellow arrows). The kinetochore of a neighbor k-fiber on a bi-oriented chromosome is also indicated (yellow arrowheads). Time is in sec. In G, serial sections of a sliding mono-oriented chromosome with the leading kinetochore laterally attached to a neighbor k-fiber. Kinetochores of the congressing chromosome are indicated (white arrows), as well as the kinetochore of a neighbor k-fiber (black arrowheads). Images adapted from Kapoor et al., 2006 [300]. Reprinted with permission from The American Association for the Advancement of Science (AAAS).

2.6. Chromosome Positioning Relative to Spindle Poles at NEB Defines the Mechanism of Congression

Another critical question has been what determines that some chromosomes use (or not) the motor-dependent pathway for congression. Classical correlative light and electron microscopy studies in PtK1 cells at the onset of prometaphase have suggested that chromosomes that were equidistant from the two spindle poles immediately bi-orient (the so-called "direct congression"), whereas chromosomes that were closer to only one of the spindle poles become mono-oriented before congressing to the equator [282,306]. Interestingly, inhibition of CENP-E function in human cultured cells only prevents congression of about 20% of the chromosomes [6,241], suggesting that most chromosomes utilize a motor-independent pathway to align at the equator. By back-tracking those chromosomes that were found locked at the spindle poles after CENP-E inhibition, it was found that they were mostly located outside the interpolar region at NEB [6], suggesting that chromosomes that are favorably positioned between the two spindle poles at NEB undergo direct motor-independent congression involving PEFs and kinetochore-pulling forces after bi-orientation. This might be facilitated by the organization of chromosomes in a ring-like configuration and by the expansion of the outer kinetochore, thereby facilitating microtubule capture and immediate bi-orientation during early prometaphase [72,307]. Interestingly, early embryonic divisions in the nematode *C. elegans*, which lacks a CENP-E orthologue but has holocentric centromeres extending along the entire chromosome length, occur in a stereotypical manner, always with two fully separated centrosomes at NEB [308]. The combination of large kinetochores with fully separated centrosomes at NEB might favor the direct congression of chromosomes in this system, where PEFs mediated by Chromokinesins also appear to play a critical role [309]. Thus, the action of Dynein and CENP-E motors at kinetochores appears to be only critical to align peripheral chromosomes that lie much closer to one of the spindle poles, where bi-orientation at NEB is unlikely to occur. A corollary of this hypothesis is that the action of kinetochore Dynein in bringing peripheral chromosomes to the vicinity of the spindle poles after initial lateral attachments, followed by CENP-E-mediated congression, increases the chances of bi-orientation as chromosomes approach the equator.

2.7. Coordination between Kinetochore- and Arm-Associated Motors

As all great solutions to a problem, they usually open up more questions. The existence of two distinct motor activities operated by Dynein and CENP-E, both localized at the kinetochore fibrous corona, but with opposite directional preferences along microtubules, posed obvious questions regarding their coordination to mediate chromosome congression (see Sections 2.8 and 2.9). In addition, the identification of microtubule plus-end-directed activities at kinetochores and chromosome arms demanded clarification of their relative contribution in moving chromosomes away from the pole. The critical role of kinetochores for chromosome movement towards the equator is known since the works of Zirkle and colleagues using focused UV or proton microbeams on parts of chromosomes in cultured newt cells [278–280]. They found that "centrophilic" chromosomes in which the kinetochore region was irradiated lost their ability to move in a directed fashion, drifted about until anaphase and never joined the metaphase plate. Similar findings were later reported in PtK1 and PtK2 cells [23,310,311]. These observations indicate that despite the action of PEFs on chromosome arms [43], they are not sufficient to drive the congression of "centrophilic" chromosomes. Moreover, these observations demonstrate that kinetochores are essential for this process, suggesting a dominant role over PEFs. Work by Brinkley and colleagues using CHO cells undergoing MUGs, in which kinetochores completely detach from chromatin, has further demonstrated that kinetochores are not only required, but they are also sufficient to ensure chromosome migration to the equator [312,313] (see also [314] for similar findings in HeLa cells undergoing MUGs). However, it should be noted that, under these circumstances, chromatin-detached kinetochores frequently establish unorthodox attachments with spindle microtubules, mostly resulting in merotelic attachments in which the same kinetochore binds microtubules from opposite poles [313,314]. In agreement, merotelic attachments on chromosome fragments with only one kinetochore have been shown to support chromosome congression [315].

A systematic dissection of the respective roles of kinetochore- and arm-associated motors for chromosome congression in human cells has been recently performed [6]. Accordingly, by combining molecular perturbations of the different motor functions with laser microsurgery of chromosome arms, it was shown that "centrophilic" chromosomes rely on CENP-E motor activity at kinetochores to counteract Dynein-mediated poleward force and move towards the equator. When chromosome arms were released from the kinetochore region by laser microsurgery, about 20% of them did not move towards the equator. Instead, they moved away towards the cortex in a Chromokinesin-dependent manner. Thus, although Chromokinesin-mediated PEFs can mediate chromosome ejection away from the poles, CENP-E-mediated forces at kinetochores are dominant and required to bias chromosome motion exclusively towards the equator. This work further demonstrated that kinetochore Dynein activity is dominant over PEFs along chromosome arms and this is required for poleward motion after initial lateral kinetochore-microtubule attachments. This role of Dynein prevents random chromosome ejection and stabilization of end-on kinetochore-microtubule attachments on chromosomes positioned near the poles due to the action of PEFs along chromosome arms, while bringing chromosomes close to the highest Aurora A activity near the poles [6,7,77,316]. This explains why "centrophilic" chromosomes after perturbation of CENP-E function move abnormally close to the pole and are mostly devoid of end-on attached microtubules [297,298] and lack any detectable oscillatory motion [295,297]. Overall, Dynein activity was proposed to prevent the formation of premature/erroneous kinetochore-microtubule attachments, thereby allowing CENP-E to undergo processive motion necessary to transport polar chromosomes along pre-existing spindle microtubules towards the equator [6,316].

Interestingly, CENP-E activity at kinetochores was shown to be required for chromosome ejection from the poles, including in monopolar spindles in which chromosome bi-orientation does not take place [6], probably by mediating the motion of leading kinetochores [300], since trailing kinetochores do not seem to exert a significant pushing force [48]. Intriguingly, CENP-E activity required for chromosome congression is independent of the establishment of stable end-on kinetochore-microtubule attachments and the formation of k-fibers, but appears to require spindle microtubule stabilization [305,317]. In contrast, Dynein was found to counteract PEFs also in monopolar spindles [6,179,231]. Thus, both CENP-E and Dynein are dominant over PEFs and play antagonistic roles at the kinetochore, independently of the establishment of stable end-on kinetochore-microtubule attachments and chromosome bi-orientation. Finally, simultaneous inhibition of all kinetochore and arm-associated motors did not prevent congression of all chromosomes [6], further demonstrating the existence of motor-dependent and -independent pathways that ultimately mediate the alignment of all chromosomes at the spindle equator.

2.8. Motor Regulators

The mechanism of chromosome congression independent of chromosome bi-orientation requires the spatial and temporal coordination of different motor activities. For instance, the direction of motor movement at kinetochores in vitro has long been known to be regulated by phosphorylation, namely by the activation of the plus-end-directed and/or inactivation of the minus-end-directed motor activities at kinetochores [288]. The kinetochore motor CENP-E is extensively phosphorylated during mitosis [318], although the functional significance of many of these phosphorylation events is not completely understood. CENP-E phosphorylation at its C-terminal tail by Cdk1 and MAPK regulates CENP-E interaction with microtubules [319,320]. This C-terminal tail is able to completely block CENP-E motility in vitro due to a direct interaction with the motor domain [301]. This auto-inhibition of CENP-E can be reversed by Mps1- or Cdk1-mediated phosphorylation of its C-terminal tail, thereby restoring normal CENP-E motility in vitro [301]. Additionally, CENP-E is phosphorylated in a conserved residue (T422) close to the motor domain by Aurora A and B [321]. This phosphorylation reduces the affinity of CENP-E for microtubules and is required for congression of polar chromosomes. However, it remains unclear how a reduction in microtubule affinity would promote CENP-E processivity necessary to overcome Dynein-mediated poleward motion. Importantly, dephosphorylation of CENP-E at T422 by PP1 phosphatase is required for stable chromosome bi-orientation after congression [321]. The recent demonstration of the existence of an Aurora A activity gradient from the spindle poles [7] has provided the necessary positional cues to control the extent of CENP-E phosphorylation at T422 as polar chromosomes approach the equator. Interestingly, Dynein intermediate chain is phosphorylated by Plk1 on T89 also in a chromosome position-dependent manner and this appears to be counteracted by PP1 phosphatase [322,323]. This phosphorylation is required for normal Dynein recruitment to kinetochores and inhibits its association with Dynactin, as well as Dynein poleward streaming along attached microtubules. Since Dynactin is required for cytoplasmic Dynein processivity [324,325], these results suggest that Dynein phosphorylation at T89 is inhibitory of its motor-mediated transport functions, as originally predicted by in vitro studies [288].

The role of CENP-E in polar chromosome congression is also regulated by sumoylation and farnesylation. When sumoylation is inhibited by overexpressing the SUMO isopeptidase SENP2, CENP-E no longer localizes to kinetochores and chromosome congression is impaired [326]. Interestingly, cells treated with farnesyltransferase inhibitors (FTIs) exhibit a prometaphase delay, suggesting the involvement of farnesylated proteins in chromosome alignment [327–330] (see also Section 4.2). These mitotic defects observed after treatment with FTIs were initially attributed to the inhibition of CENP-E and CENP-F farnesylation [327,328,331]. While inhibition of farnesylation appears to interfere with CENP-E association with microtubules [327], the role of farnesylation in regulating CENP-E localization and function at kinetochores remains controversial. Treatment of cells with FTIs was reported to deplete CENP-E and CENP-F from metaphase, but not from prometaphase kinetochores [328]. CENP-E is also degraded shortly after mitotic exit [332], and its degradation requires farnesylation [333]. Interestingly, it was suggested that farnesylation of Spindly is also involved in the regulation of kinetochore Dynein, since mutation of a potential farnesylation site in Spindly prevented its localization at the kinetochore [179]. More recently, two independent studies confirmed Spindly as a farnesylation substrate [334,335]. In one study, FTI treatment resulted in loss of Spindly at kinetochores without affecting the RZZ complex or CENP-E and CENP-F kinetochore localization [335]. In contrast, in another study, CENP-E and CENP-F kinetochore levels were also affected by FTI treatment, but to a less extent compared to Spindly [334]. Both studies have shown that preventing farnesylation of Spindly delays chromosome congression, producing a similar phenotype observed in cells treated with FTIs. Taking these findings together, it seems that the role of farnesylation in regulating CENP-E function during chromosome congression rather represents a minor effect, while loss of Spindly kinetochore localization (and consequently Dynein) after farnesylation inhibition appears to be the major contributing factor to the congression defects observed in cells treated with FTIs.

Different studies have implicated Mps1 in chromosome alignment, but the underlying molecular mechanism remains unclear [336–339]. Initially it was proposed that regulation of chromosome alignment by Mps1 acts through modulation of Aurora B kinase activity [337]. However, recent studies have provided evidence that regulation of chromosome alignment by Mps1 is independent of Aurora B [336,340,341]. The regulation of chromosome alignment by Mps1 may be through CENP-E phosphorylation [301], as this is necessary to recruit CENP-E to kinetochores [336,342]. These results suggest that the role of CENP-E in polar chromosome congression might be regulated by Mps1.

Finally, motor proteins involved in chromosome congression are also regulated by proteolysis. For instance, the Kinesin-10 Kid and the kinetochore motor CENP-E are degraded at the end of mitosis, consistent with down-regulation of PEFs at the metaphase-anaphase transition to allow chromosome poleward movement [44,65,332].

2.9. The Role of Tubulin PTMs as a Navigation System for Kinetochore-Based Motility of Chromosomes

In addition to the regulation of kinetochore motor activities, the possibility that tubulin post-translational modifications (PTMs), as part of the so-called "tubulin code" [343,344], additionally contribute with spatial cues required for chromosome congression has recently been proposed [345,346]. Tubulin, the building unit of microtubules, can be enzymatically processed to undergo different PTMs, including detyrosination, (poly)glutamylation, glycylation, phosphorylation, acetylation and the recently-discovered methylation [344,347]. Some of these modifications have been already shown to regulate the motor activity of Kinesin-1, affecting its binding and transport in neurons [348–351]. In vitro reconstitution assays have further dissected the impact of tubulin PTMs on the performance of motor proteins such as Kinesin-1, Kinesin-2, Kinesin-13, and Dynein [352,353]. Therefore, it is plausible that the activities of the motor proteins involved in the directed transport of chromosomes along distinct microtubule populations, before and during chromosome congression, are also regulated by PTMs that differentiate the microtubule tracks on which they move [346]. Indeed, it has been known for decades that different PTMs label distinct microtubule populations within the mitotic spindle [354–357]. For instance, the dynamic, short-lived astral microtubules that extend from the spindle poles towards the cell cortex are highly tyrosinated (i.e. they contain a tyrosine as the last amino acid on the α-tubulin C-terminal tail), while more stable spindle microtubules, such as k-fibers and possibly interpolar microtubules, are detyrosinated, acetylated and polyglutamylated [354–357]. Therefore, this patterned distribution of different tubulin PTMs within the mitotic spindle could work as a navigation system for kinetochore-based motor proteins involved in the critical steps that anticipate and mediate chromosome congression [345,346].

Such a navigation system would have particular implications for the congression of peripheral chromosomes that are unable to bi-orient soon after NEB. According to this model, the Dynein-mediated poleward movement of peripheral chromosomes upon the initial interaction with astral microtubules would be regulated by their high tyrosinated state [355,356]. In support of this concept, recent in vitro reconstitution studies of Dynein/Dynactin activity have indicated that tubulin C-terminal tail tyrosination is of great importance for Dynactin-mediated initiation of Dynein motion on microtubules [358]. Similar findings have been reported in vivo, where the Dynactin subunit p150 and tubulin tyrosination were shown to mediate the initiation of retrograde vesicle transport in neurons [359]. Finally, these data are in line with previous studies reporting that p150/Dynactin has higher affinity for tyrosinated microtubules [325,360] and that the motility of both cytoplasmic and axonemal Dyneins highly depends on tubulin C-terminal tails [325,361–363].

After the initial Dynein dominance during the poleward transport of peripheral chromosomes along tyrosinated astral microtubules, Dynein is overtaken by CENP-E to drive the congression of polar chromosomes to the equator [6]. In concert with Aurora A kinase-mediated activation of CENP-E by phosphorylation near the poles [321], and in agreement with the slow association of CENP-E with microtubules observed in vitro [364], recent work revealed that CENP-E has a preference for the more stable detyrosinated spindle microtubules, and this is important to guide polar chromosomes towards the equator [345]. Accordingly, this study showed that, similar to CENP-E depletion/inhibition, attenuation of tubulin detyrosination either by inhibition of the tubulin carboxypeptidase (TCP) (the enzyme that removes the last tyrosine from the α-tubulin C-terminal tail on polymerized microtubules), or by overexpression of the tubulin tyrosine ligase (TTL) (the enzyme that adds back tyrosine to soluble α-tubulin), prevented polar chromosomes from congressing. In vitro reconstitution experiments confirmed that CENP-E motility is enhanced on detyrosinated microtubules [345]. Moreover, RNAi-mediated depletion of TTL, which increases overall detyrosination of the mitotic spindle, including astral microtubules, prevented peripheral chromosomes from reaching the spindle pole [345]. Since this could only be partially rescued by co-depletion of CENP-E [345], it suggests that increased detyrosination of astral microtubules further prevents kinetochore Dynein-mediated poleward transport. Altogether, these data support that the state of α-tubulin detyrosination provides important spatial cues for the regulation of chromosome movements during mitosis [346]. As so, the difference in detyrosination levels between highly dynamic astral and more stable spindle microtubules mediates an activity switch that enables the fine spatiotemporal regulation of the opposite motility of Dynein and CENP-E at kinetochores. This ensures that peripheral chromosomes are first transported poleward by Dynein along tyrosinated astral microtubules, followed by CENP-E-mediated congression along more detyrosinated microtubules pointing to the equator.

This activity switch seems to be very finely regulated, since in vitro studies showed that tubulin (de)tyrosination induced less than 2- and up to 4-fold changes in the processivity of CENP-E and Dynein motors, respectively [345,358]. Importantly, a recent in vitro reconstitution study demonstrated that single Kinesin and Dynein motors produce approximately similar forces [365], which helps to explain how slight differences in tubulin (de)tyrosination can influence motor kinetics and determine the directionality of chromosome movements. This is further supported by recent theoretical work, which demonstrated that tubulin PTMs are sufficient to generate a 2-fold difference on motor kinetics and target cargoes to specific locations along microtubules [366].

A critical emerging question is how a single amino acid change at the α-tubulin C-terminal tail selectively affects motor recognition and function at the structural level. It is well established that tubulin C-terminal tails regulate the binding and processivity of Kinesin-1 and Dynein in vitro [362,367]. CryoEM, backed-up by crystallographic studies, have allowed the visualization of the CENP-E motor domain in complex with microtubules [368,369]. Although the exact interaction between CENP-E and tubulin C-terminal tails has not been determined due to their flexible nature, these works indicate that the CENP-E motor domain might interact with helix 12 from α-tubulin, close to the C-terminal tail. Because the association of the CENP-E C-terminal kinetochore-binding domain with microtubules depends little (20% reduction) on tubulin C-terminal tails [370], these results suggest that microtubule detyrosination directly regulates recognition by the CENP-E motor domain. In contrast, the recognition of tyrosinated microtubules by Dynein has been shown to involve p150/Dynactin [358,360] and structural reconstructions have indicated that this interaction is mediated by the GKNDG motif on the CAP-Gly domain of p150/Dynactin [371,372].

2.10. Chromosome Congression vs. Maintenance of Alignment

One poorly understood aspect of mitosis is whether the mechanisms that mediate chromosome congression consist of the same principles that ensure the maintenance of a bi-oriented chromosome at the equator after completing congression. Clearly, motor-dependent chromosome congression does not rely on a force balance on a given kinetochore pair, as chromosome bi-orientation is not required to complete congression [300]. Moreover, end-on kinetochore-microtubule attachments are not even required for motor-driven congression to the equator, but are essential to maintain aligned chromosomes at the metaphase plate [305]. This is corroborated by microsurgery experiments in which the kinetochore region of a once aligned chromosome is irradiated with a focused UV or laser microbeam, causing the chromosome to immediately move towards the direction of the undisturbed kinetochore [20–23]. In contrast, when k-fibers are cut on a bi-oriented chromosome positioned at the equator, chromosomes either do not shift at all or shift only slightly towards the pole of the unperturbed k-fiber [21,22,29–38]. Interestingly, inter-kinetochore tension in vertebrate and insect cells is proportional to k-fiber length [37,38] (Figure 6). Overall, these data indicate that while force at kinetochores is proportional to k-fiber length, maintenance of chromosome position near the equator is not.

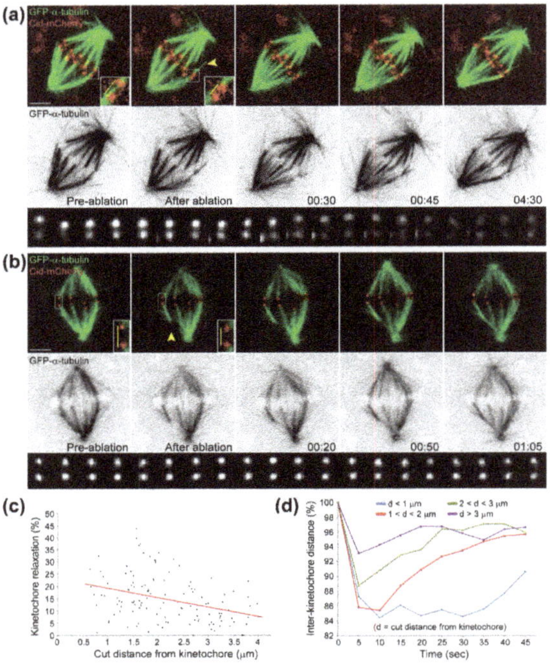

Figure 6. Forces at kinetochores are proportional with k-fiber length, but chromosome position at the equator is independent of k-fiber length. (**a**,**b**) Laser microsurgery of k-fibers in *Drosophila* S2 cells stably expressing GFP-α-tubulin to label microtubules (green) and Cid-mCherry to label kinetochores (red). K-fibers were cut (yellow arrowhead) and grew back as described previously [29]. Inverted contrast of GFP-α-tubulin is also shown, as well as the variation of inter-kinetochore distance over time (kymograph; first frame corresponds to pre-surgery distance; second frame onwards are after surgery). Measurement of the inter-kinetochore distance before and after laser surgery ablation of k-fibers (yellow bars) indicates that kinetochores relax after surgery, and this relaxation is more evident the closer the cut is to the kinetochore. Time is in min:sec. White scale bars are 2 μm; (**c**) Quantification of the percentage of kinetochore relaxation after surgery (determined by the difference between initial inter-kinetochore distance and the minimum observed distance after surgery) indicates a negative correlation ($R^2 = -0.361$; $p < 0.001$) with the cut distance from the kinetochore (n = 125 cells); (**d**) Corresponding quantification of the inter-kinetochore distance over time as a function of the cut distance from the kinetochore. Each group was normalized against its initial distance such that one hundred percent corresponds to the average initial distance. The closer the cut is to the kinetochore, the longer the recovery of inter-kinetochore distance and the higher is the relaxation. The inclusion of a kinetochore marker in this study and the observed variability of inter-kinetochore distance after k-fiber cut explains previous observations in which no detectable kinetochore relaxation was observed without the use of a kinetochore marker [29]. Laser microsurgery was performed essentially as described in [373].

Several theoretical and experimental studies have predicted or provided evidence for mechanical coupling between kinetochore and non-kinetochore (interpolar) microtubules [4,37,38,374–381], which might account for the maintenance of chromosome positioning at the equator independently of k-fiber length. While the molecular nature of this spindle microtubule coupling system remains unknown, it is likely to involve multiple players that possess the necessary molecular properties to serve this purpose. These include several MAPs and motors with microtubule cross-linking properties, such as PRC1, Kinesin-5, Kinesin-15, CLASPs, Clathrin/Ch-TOG/TACC3, Asp, NuMa, Kinesin-14 and Dynein [382–384]. In addition, Chromokinesins, Kif4A in particular, might also work as a coupling element between k-fibers and interpolar microtubules interacting with chromosome arms [71].

Interestingly, many loss-of-function studies of Chromokinesins revealed only a very minor role during chromosome congression, while being critical to maintain chromosomes aligned at the equator [6,71]. These results suggest that Chromokinesins might additionally contribute to the stabilization of kinetochore-microtubule attachments of aligned chromosomes, possibly in coordination with the activity of Kinesin-8 [146]. Indeed, recent works in *Drosophila* S2 cells have shown that Chromokinesins promote kinetochore-microtubule stabilization and the conversion from lateral to end-on attachments, independently of chromosome bi-orientation [77,81], which might be important to maintain chromosomes aligned at the equator after congression. This implies that CENP-E is no longer dominant over Chromokinesins once chromosome bi-orientation and equatorial alignment is achieved. This would be consistent with the finding that CENP-E levels at the kinetochore decrease significantly due to Dynein-mediated stripping upon microtubule attachment and chromosome bi-orientation [385]. However, whether CENP-E plays a role in maintaining chromosome positioning at the equator after alignment has been controversial. For instance, CENP-E has been proposed to play a role in stabilizing end-on kinetochore-microtubule attachments [297–299]. This model is supported by electron microscopy studies after inactivation of CENP-E function, which showed a reduced microtubule number at kinetochores of aligned bi-oriented chromosomes, supporting a role for CENP-E after chromosome congression [297,298]. Importantly, the observed differences relative to controls appear to be attenuated during a prolonged mitosis where the range of microtubule binding was similar to controls, indicating that CENP-E is not essential for binding of a full complement of microtubules at kinetochores of bi-oriented chromosomes [297]. Interestingly, original antibody micro-injection experiments in metaphase cells have indicated that CENP-E is not required for maintenance of chromosome alignment [289]. In contrast, treatment of metaphase cells with a CENP-E inhibitor that forces CENP-E to bind tightly to microtubules (a "*rigor*" state) caused the displacement of chromosomes from the equator, supporting a role of CENP-E in maintaining chromosome alignment after bi-orientation, in addition to mediating chromosome congression [303]. The availability of a second generation of CENP-E inhibitors that compromise ATPase activity without interfering with microtubule binding [386] will be important to clarify the role of CENP-E after chromosome alignment.

Finally, many studies have reported chromosome misalignment problems after functional perturbation of several proteins (see Table 1). However, since live-cell imaging was not used in many of these studies, it remains unclear whether it truly reflects a direct role of these proteins in chromosome congression or in the maintenance of chromosome alignment. The recent discovery that apparently unrelated experimental perturbations associated with a metaphase delay often lead to "cohesion fatigue" (i.e., the uncoordinated loss of sister chromatid cohesion after chromosome congression but prior to anaphase onset, due to the action of mitotic spindle forces) [155,387,388] incites for a systematic re-evalution of proteins formerly associated with chromosome alignment using state-of-the-art live-cell imaging techniques.

2.11. An Integrated Model of Chromosome Congression

Based on the arguments expressed in the previous sections, we propose that chromosome congression in humans can essentially be explained by two main mechanisms that operate in parallel (Figure 7), meaning that not all chromosomes rely on the same mechanism to complete congression. A key aspect that determines which mechanism is used depends essentially on whether chromosomes establish lateral or end-on attachments at their kinetochores on their way towards the equator. This is influenced by the position of chromosomes relative to the spindle poles at NEB. Those chromosomes that are able to bi-orient soon after NEB would use a "direct congression" mechanism in which opposite kinetochore-pulling forces, resulting from the tight regulation of microtubule dynamics and length at the kinetochores, in coordination with PEFs along chromosome arms, drive chromosome oscillations until net force is zero near the equator. A corollary from this model is that the establishment of stable end-on attachments inhibits the other congression mechanism relying on lateral interactions between microtubules and kinetochores. This second mechanism would take advantage of the high processivity of the Dynein/Dynactin motor localized on unattached kinetochores to capture peripheral chromosomes, which are unable to bi-orient at NEB and establish stable end-on kinetochore microtubule attachments. The minus-end directed motion of Dynein/Dynactin along tyrosinated astral microtubules transports peripheral chromosomes close to one of the spindle poles, where Aurora A activity is highest and prevents the stabilization of end-on kinetochore-microtubule attachments. This configuration also imposes a dominance of kinetochore Dynein/Dynactin over the action of Chromokinesin-mediated PEFs along chromosome arms that would otherwise promote the premature stabilization of end-on kinetochore-microtubule attachments and lead to errors resulting in chromosome missegregation. In addition, while travelling along tyrosinated astral microtubules, Dynein/Dynactin will be dominant over the other kinetochore motor, CENP-E, with plus-end-directed motility and a preference for more stable detyrosinated microtubules. Once at the poles, phosphorylation by Aurora A will activate CENP-E, (while other centrosome kinases, such as Plk1, inactivate Dynein/Dynactin), favoring the lateral transport of chromosomes by CENP-E along detyrosinated microtubules (either k-fibers or interpolar microtubule bundles) towards the equator, where the chances for bi-orientation are maximal. At the equator, Chromokinesins promote the conversion from lateral to end-on attachments, which further downregulates CENP-E and Dynein, thereby ensuring the maintenance of chromosome position at the metaphase plate. Once aligned and bi-oriented at the metaphase plate, the coordination between kinetochore-pulling forces and PEFs continue to determine the amplitude of chromosome oscillations, but maintenance of chromosome position near the equator will depend on additional factors that mediate the cross-linking between kinetochore and non-kinetochore microtubules.

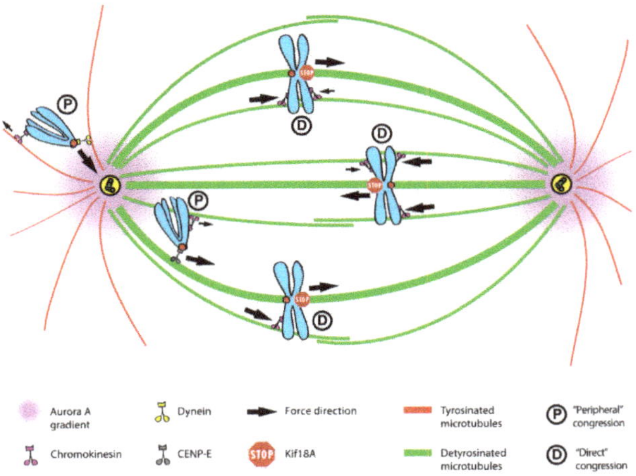

Figure 7. Integrated model of chromosome congression in human cells. In this representation, Kif18A is shown to restrict k-fiber length, thereby contributing to a directional switch and regulating chromosome oscillations after bi-orientation. See text for a detailed description.

2.12. A Note about Chromosome Congression in Acentrosomal Systems

The problem of chromosome congression in acentrosomal systems such as animal oocytes and land plants is not less complex than in mammalian somatic cells. While the lack of centrosomes could in principle simplify the process and decrease microtubule heterogeneity within the context of the spindle, these systems have developed alternative microtubule organizing structures or mechanisms that, in a way, functionally resemble the centrosomes. For instance, land plants assemble a "prophase spindle" on opposite sides of the nucleus before NEB. These prophase spindle microtubules undergo "search-and-capture" and eventually interact with chromosomes and assist their motion (reviewed in [39,389,390]). There is, however, good evidence that canonical PEFs are rather weak or absent in plants [39,391]. Mammalian oocytes form acentriolar microtubule-organizing centers (aMTOCs) that assemble transient "multipolar" spindles that ultimately cluster into a bipolar structure (and show astral-like microtubules) and mediate interactions with chromosomes towards bi-orientation [392,393]. Therefore, "direct congression" of at least some chromosomes after NEB is likely to take place in mammalian oocytes and land plants. In contrast, in *Xenopus* oocyte extracts, microtubules organize "inside-out" in the vicinity of chromatin and in a Ran-GTP-dependent manner (reviewed in [394]). As so, chromosomes already start "congressed" during spindle assembly and do not need to be transported from the poles. Nevertheless, CENP-E and Kid/Chromokinesin motors appear to be necessary to maintain chromosomes equidistant from the poles in this system, either by promoting chromosome bi-orientation or simply by mediating persistent microtubule plus-end-directed chromosome motion, such as in PEFs [65,66,293].

Recent insight from live-cell imaging of mammalian oocytes has revealed unprecedented details about the process of chromosome congression in this system [395]. It was found that chromosome congression is completed before bi-orientation due to the establishment of an intermediate configuration, the "prometaphase belt", in which chromosomes are organized around the spindle. During congression, chromosomes that were located far from the equator moved towards it by sliding along spindle microtubules, whereas chromosomes that were already located near the equator remained stationary. Subsequently, chromosomes invaded the spindle area establishing the final metaphase plate organization and bi-orientation. Interestingly, very similar findings have been reported for human somatic cells in culture [72], suggesting conservation of the mechanisms of chromosome congression between mammalian centrosomal and acentrosomal systems. In support of this idea, chromosome congression in mammalian oocytes also does not seem to depend on the Chromokinesin Kid [395,396], but CENP-E activity appears to be required, possibly by facilitating bi-orientation [397]. Similar findings were also recently reported in *Drosophila* and *C. elegans* oocytes, in which prometaphase chromosome motion and bi-orientation was shown to depend essentially on lateral attachments [398,399]. However, while in *Drosophila* oocytes chromosome bi-orientation and lateral attachments were shown to rely on CENP-E [398], in the case of *C. elegans* oocytes the process might involve the Chromokinesin KLP-19 [399]. It should be noted that chromosomes in *Drosophila* oocytes are compacted into a karyosome and, similar to *Xenopus* oocyte extracts, congression is unnecessary, whereas in *C. elegans* KLP-19 is only required for chromosome alignment in metaphase I-arrested, but not normally progressing oocytes [399,400]. Therefore, CENP-E and Chromokinesin activities in these systems might only be required to maintain chromosomes at the equator. In the case of land plants, they appear to lack cytoplasmic Dynein motors [390,401], but CENP-E-like Kinesin-7 motors and Chromokinesins are conserved [390,402] and the former has been implicated in chromosome congression in moss, even though it does not seem to localize at kinetochores [403]. Finally, it is worth remarking that even in animal somatic cells in which centrosome function was genetically perturbed, chromosome congression was delayed but not prevented, further supporting a marginal role for centrosome-mediated PEFs in chromosome alignment in metazoans [404].

3. Consequences of Abnormal Congression

3.1. Aneuploidy, Tumor Suppression and Oncogenic Potential

Aneuploidy is defined as a karyotype state with a chromosome number that deviates from a multiple of the haploid, and is a hallmark of human cancers. Aneuploidy is often accompanied by high rates of chromosome missegregation, a phenomenon called chromosomal instability (CIN), in which chromosomes are permanently gained and lost during multiple divisions [405]. Therefore, CIN might contribute to tumorigenesis by changing the dosage of oncogenes and tumor suppressors required for tissue homeostasis. CIN has also been associated with both poor patient prognosis and resistance to some chemotherapeutic agents [406–410]. Paradoxically, there is also evidence that excessive CIN is a disadvantage for tumor progression and is associated with better prognosis [411]. Whatever the case may be, and despite all controversy, direct targeting of CIN as a potential anti-cancer therapy is now the subject of active research [412,413].

Chromosome congression defects are amongst the multiple pathways that could lead to CIN [405,414,415]. Different studies reported that cell and animal models with reduced levels of CENP-E generate high levels of aneuploidy. CENP-E deletion in mouse embryonic fibroblasts (MEFs) and in liver tissues resulted in cells with several mitotic defects, including chromosome misalignment and increased levels of lagging chromosomes, an indication of chromosome missegregation [298,416]. Homozygous disruption of the CENP-E gene causes early embryonic lethality [298], while heterozygous loss of CENP-E causes aneuploidy and CIN that can both promote or suppress tumor formation, depending on the context [417,418]. Mice heterozygous for CENP-E show a mild increase in the rate of spontaneous lung and spleen tumors, but exhibit a decreased incidence of liver tumors [418]. CENP-E heterozygosity did not accelerate tumor initiation or progression after treatment with the chemical carcinogen DMBA [417,418]. Moreover, when CENP-E heterozygosity was combined with the loss of the tumor suppressor p19ARF (CENP-E$^{+/-}$ p19ARF$^{-/-}$), most of the animals showed a strong delay in tumorigenesis [417,418]. Furthermore, exacerbating the level of CIN in CENP-E$^{+/-}$ mice by crossing them with Mad2$^{+/-}$ or APC$^{Min/+}$ resulted in increased cell death and reduced tumor progression [417,419]. These findings suggest that low levels of CIN caused by minor chromosome congression and segregation defects could potentially lead to transformation, whereas an elevated rate of CIN inhibits tumor formation.

Drosophila models have also been generated to investigate whether induction of aneuploidy by knocking down CENP-E is tumorigenic . In one study, CENP-E depletion alone was not sufficient to drive tumorigenesis [420]. However, another study found that knockdown of CENP-E and Nsl1 (which targets Bub3 to the kinetochore, compromising the SAC) induced a tumorigenic response [421]. These results suggest that, per se, minor chromosome congression defects are insufficient to drive tumor formation in flies and that a significant level of aneuploidy is required.

Altered expression or mutations in CENP-E have been reported in some human diseases. CENP-E is upregulated in individuals with rheumatoid arthritis [422] and with breast cancer [423]. Moreover, CENP-E expression negatively correlated with disease-specific survival in patients with breast cancer [423]. In contrast, human hepatocellular carcinoma exhibits abnormally low levels of CENP-E [424]. Several non-synonymous single nucleotide polymorphisms were also reported in CENP-E and the Y63H point mutation, which disrupts the native conformation of the ATP-binding region in the CENP-E motor domain, was found to be associated with cancer [425]. Finally, mutations in CENP-E leading to chromosome congression problems were also associated with microcephalic primordial dwarfism (MPD) [426].

Kif18A is overexpressed in human colorectal [427] and human breast cancers [428]. Kif18A expression in breast cancers correlates with tumor grade, metastasis and survival, whilst suppression of Kif18A expression in breast cancer cells inhibits tumor growth in vivo [428]. In addition, proteomic analysis identified Kif18A as a potential biomarker of cholangiocarcinoma and lung cancer [429,430]. Genetic studies in mice demonstrated that disrupting Kif18A function affects male, but not female, fertility [193]. Kif18A$^{-/-}$ male mice develop relatively normally and exhibit defects in the testis, but not in other organs. Testis atrophy in these mice is caused by impaired microtubule dynamics and loss of spindle pole integrity associated with chromosome congression defects during mitosis and meiosis. Another study showed that depletion of Kif18A protects animals from colitis-associated colorectal (CAC) cancers [431]. Although suggestive, the involvement of Kif18A in cancer requires further investigation.

Besides its function during chromosome congression, the Chromokinesin Kif4A plays several other roles throughout mitosis, and loss of this protein leads to various mitotic defects including chromosome hypercondensation, aberrant spindle formation, anaphase bridges, defective cytokinesis and aneuploidy [69,432]. Kif4A is absent or expressed at low levels in 35% of human cancers [433]. Kif4A is also downregulated in gastric carcinoma tissues and Kif4A expression levels correlate with tumor differentiation [434]. Interestingly, overexpression of Kif4A in gastric cancer cells inhibits proliferation in vitro, as well as the ability to form tumors in vivo [434]. Kif4A is also overexpressed in cervical cancer [435] and non-small cell lung cancer associated with poor patient outcome [436]. Furthermore, loss of Kif4A in murine embryonic stem cell results in several mitotic defects, including chromosome misalignment, spindle defects and aberrant cytokinesis [433]. Additionally, a high percentage of cells lacking Kif4A are aneuploid and injection of these cells into nude mice has the ability to form tumors. Based on these findings, the aneuploidy associated with aberrant mitosis after Kif4A depletion can promote tumor formation, but it remains unclear whether this is a direct consequence of its role in chromosome congression. Altogether, these findings demonstrate that loss of different Kinesin-like proteins involved in different aspects of chromosome congression might lead to aneuploidy.

4. Targeting Chromosome Congression for Cancer Therapy

4.1. *CENP-E Inhibitors*

Microtubule poisons that disrupt spindle assembly and function have demonstrated to be powerful tools in the treatment of many human cancers [437], but their efficacy is limited by side effects such as neurotoxicity, neutropenia and acquisition of resistance [438–440]. Taxanes and vinca alkaloids are amongst the most successful microtubule drugs and are known to compromise chromosome congression by preventing the formation of proper kinetochore-microtubule attachments that nevertheless satisfy the SAC, leading to an abnormal mitotic exit and apoptosis [441–445]. The discovery of new mitotic targets for cancer therapy has raised interest in developing antimitotic agents that do not target microtubules [446,447]. The most notable targets are the Aurora kinases A and B, as well as Plk1 [448]. Although there are obvious drawbacks (and the main reason for failure in clinical trials) related with cytotoxicity of normal fast dividing cells, such as those in the bone marrow, gut, and hair follicles, protein targets that are only expressed in dividing cells are attractive for cancer therapy, since non-dividing differentiated cells should not be affected. CENP-E is expressed predominantly in mitosis (and G2) [290] and plays an important role in peripheral chromosome congression [293,295,300], thereby representing an attractive target for cancer therapeutics. GSK923295 is an allosteric inhibitor of CENP-E that blocks its microtubule stimulated ATPase activity and stabilizes the interaction between the motor domain and microtubules [449,450]. GSK923295 has demonstrated both in vitro and in vivo antitumor activity against various malignancies [449,451–455]. Cells treated with GSK923925 assemble bipolar spindles and the majority of chromosomes align at the spindle equator. However, some chromosomes remain clustered near the spindle poles, leading to mitotic arrest and apoptosis [449,456]. The antitumor activity of GSK923925 has been evaluated in combination with standard chemotherapies, as well as with other emerging targeted drugs [454]. Inhibition of ERK1 revealed a significant synergistic proliferation inhibition activity when combined with GSK923225 in neuroblastoma, lung, pancreatic and colon carcinoma cell lines [454]. Combination of GSK923225 with Pgp-pump modulators also appeared to improve the antitumor effects against cells with Pgp overexpression, thereby overcoming the resistance to Pgp inhibitors [457].

Another CENP-E inhibitor, PF-2771, selectively inhibits proliferation of basal breast cancer cell lines compared with normal and premalignant cells. Moreover, the sensitivity to this inhibitor correlates with the degree of CIN, suggesting that cancers with elevated CIN may benefit from CENP-E-targeted therapy [423]. Finally, inhibition of CENP-E motor function by PF-2771 resulted in tumor regression in a patient-derived basal-like breast cancer xenograft tumor model [423]. More recently, a new inhibitor of CENP-E directly targeting its ATPase activity, known as compound A, was found to have anti-proliferative activity in multiple cancer cell lines and in a xenograft nude mouse model [386,458]. CENP-E inhibition using compound A resulted in p53-dependent post-mitotic apoptosis triggered by elevated chromosome missegregation [458]. Interestingly, both CENP-E inhibitors PF-2771 and GSK923295 were found to increase CIN levels in a recent large-scale screen [459]. Taken together, these data suggest that CENP-E may be an effective therapeutic target for cancer cells with high levels of CIN.

Other compounds have been claimed to specifically inhibit CENP-E, but turned out to target other proteins. For instance, the compound UA62784 was initially described to be a specific inhibitor of the ATPase activity of CENP-E and highly cytotoxic against human pancreatic cancer cell lines with a deletion of the DPC4 gene [460]. However, a subsequent study demonstrated that this compound does not exert its cellular activity by inhibiting CENP-E and rather binds microtubules tightly [461,462]. Another study that tested the antitumor activity of UA62784 and 80 analogs against pancreatic cancer cell lines revealed that these compounds potently inhibit several protein kinases that are overexpressed in these cancer cells, but not mitotic Kinesins (Kinesin-5, CENP-E, MKLP-1, and MCAK) [463]. Another compound, Syntelin, was also reported to be a highly selective CENP-E inhibitor [464]. Inhibition of CENP-E by Syntelin caused misaligned chromosomes with syntelic attachments, in which sister kinetochores stably attached to microtubules near the same spindle pole [464]. This was surprising, since perturbation of CENP-E produces polar chromosomes that are mostly devoid of microtubules at kinetochores [6,297,298], suggesting that Syntelin also targets other proteins (e.g., Aurora B).

To date, only one of the CENP-E inhibitors, GSK923295, has been evaluated in a Phase I clinical trial [465]. In this trial, peripheral neuropathy, a well-known taxane adverse effect, was not evident. As such, the use of CENP-E inhibitors as anticancer drugs could be better tolerated than taxanes and possibly easier to use in combination with other cancer therapies. Thus, better understanding of the molecular mechanisms behind CENP-E inhibition might help to find optimal clinical strategies for certain human cancers.

4.2. *Farnesyltransferase Inhibitors (FTIs)*

FTIs are promising agents for therapeutic intervention in several diseases, including cancer, malaria and progeria [466–473]. Due to the clinical relevance of these drugs it became important the identification of the cellular substrates of the farnesyltransferase. There are several proteins that are prone to be farnesylated [474] and several studies have shown that FTIs prevent the farnesylation of Ras family and some mitotic proteins involved in chromosome congression (such as CENP-E, CENP-F and Spindly) [327,334,335,475,476]. Since farnesylation is required for the recruitment of Ras proteins to the plasma membrane and many tumors exhibit mutations in Ras, FTIs were initially developed as therapeutic agents that target Ras activity in cancer cells [477]. Indeed, FTIs exhibited a potent inhibitory effect on the proliferation and invasive capabilities of breast cancer cells with active H-Ras in culture [478]. However, it became evident that the target of FTIs might not be only Ras proteins [479], and there was some evidence that FTIs demonstrated activity in cancer cells irrespective of Ras mutations [330,480–482]. Moreover, some studies have shown that treatment of different cancer cells with FTIs enhanced the anti-proliferative and apoptotic effects of cisplatin [483], 5-fluorouracil [484], MEK inhibitors [485], Cdk inhibitors [486], mTOR inhibitor (rapamycin) [487] and taxol [488]. Finally, and most relevant for our purposes, FTIs were shown to affect bipolar spindle assembly and chromosome congression [328,329,335].

Some FTIs, such as Tipifarnib (or R115777), Lonafarnib (or SCH66336), BMS-214662, L-778123 and SCH44342 are currently in clinical trials for the treatment of various solid tumors and hematological malignancies [471,489–493]. Although FTIs have been extensively tested in the clinics, their mechanism of cytotoxicity is not fully understood. In some clinical trials, treatment with FTIs alone or in combination with chemotherapeutic agents failed to improve the overall outcome of patients with solid tumors and leukemia [494–501]. However, other clinical trials demonstrated that the combination of FTIs with conventional chemotherapeutic agents might be useful in hematologic and some solid tumors [502–508]. Moreover, patients with poor-risk acute myeloid leukemia may benefit from FTIs maintenance therapy following cytotoxic induction and consolidation therapies [509]. Understanding the mechanisms by which these drugs inhibit cell proliferation and induce cell death might facilitate the development of new therapeutic strategies.

4.3. Inhibitors of Tubulin PTMs

The levels of various tubulin PTMs, including acetylation, detyrosination, Δ2 deglutamylation, polyglutamylation and glycylation, are altered in different cancer cell lines and tissues, contributing to tumor growth and enhancing their metastatic potential [510–521]. α-tubulin acetylation and detyrosination are increased in breast cancer cells and correlate with tumor aggressiveness and poor prognosis in patients [511,517]. A balance of tubulin acetylation and deacetylation by α-TAT1 and HDAC6 enzymes with opposite activities was proposed to regulate the migratory and invasive capacities of breast tumor cells [510]. Low expression of TTL, the enzyme responsible for tubulin retyrosination, leads to increased microtubule detyrosination and is correlated with inhibition of neuronal differentiation and increased cell growth in neuroblastoma with poor prognosis [518]. TTL expression was found to be suppressed during tumor growth in mice [516], as well as during epithelial-to-mesenchymal transition in human mammary epithelial cells in vitro [521], implicating the tubulin tyrosination cycle in both tumor propagation and metastasis. Such highly acetylated and detyrosinated microtubules can indeed form microtentacle protrusions that enhance cellular invasive migration and re-attachment [511,517]. Experimental microtubule deacetylation, achieved by mutating the α-tubulin acetylation site at Lysine 40, decreased the incidence of microtentacles and inhibited cellular migration and invasiveness, confirming the interdependence between cancer progression and tubulin PTMs [511].

Because of their correlation with cancer, tubulin PTMs present a very promising target for novel therapeutic approaches in human cancers. One of the most obvious strategies would rely on the pharmacological inhibition of the enzymes responsible for tubulin PTMs. A promising group of potential anti-cancer drugs that target tubulin detyrosination are sesquiterpene lactones, a series of bioactive compounds isolated from the Asteraceae family of plants [522]. The most studied compound is parthenolide, which has already been used in cancer clinical trials [523,524] and suppresses several different steps within the nuclear factor kappa B (NF-κB) signaling pathway [525–528]. In addition, parthenolide prevents microtubule detyrosination by inhibiting TCP, independently from its effect on NF-κB [529]. Therefore, parthenolide-mediated targeting of TCP and microtubule detyrosination might have a preventive effect on tumor growth, aneuploidy and metastasis, independently from its interference with the NF-κB pathway. Indeed, parthenolide-mediated suppression of cell invasiveness and re-attachment of breast cancer metastatic cells was shown to be independent of NF-κB [530]. Interestingly, several studies reported that various sesquiterpene lactones induced a G2 or M arrest [531–533], which might account for their anti-cancer activity. More recently, the effect of parthenolide over TCP inhibition was found to cause chromosome congression defects during mitosis [345,346], reinforcing the potential of targeting chromosome congression for cancer therapy.

The great advantage of parthenolide as an anti-cancer drug is that it appears to selectively target cancer cells, as documented by several different in vitro studies [524]. Moreover, parthenolide was the first small molecule shown to selectively kill cancer stem cells, while leaving normal stem cells intact [524,534]. This is of enormous therapeutic importance, since the presence of cancer stem cells is considered as one of the main reasons underlying chemotherapy resistance and tumor relapse due to their capacity of self-renewal and differentiation into multiple cell types [535,536]. The mechanism behind parthenolide selectivity towards cancer stem cells is not completely understood, but it is believed that the reason lies in its ability to target multiple major pathways required for cancer stem cell survival and self-renewal, such as MAPK, JAK/STAT, PI3K and NF-κB signaling [524,537]. Whether TCP inhibition by parthenolide contributes to cancer stem cell eradication remains to be elucidated.

The biggest disadvantage of parthenolide as a therapeutic drug is its high hydrophobicity, which limits its bioavailability for oral usage and solubility in plasma [523]. This is partially circumvented by the synthesis of a more water-soluble analog dimethylamino-parthenolide (DMAPT), which possesses an increased oral bioavailability [524]. DMAPT has already proved effective in selective eradication of human acute myeloid leukemia primary cultured stem cells [538] and breast cancer stem-like cultured cells [539], and has been shown to inhibit tumor growth and metastasis of prostate, lung and bladder cancer xenografts in mice [531,540]. However, although parthenolide and DMAPT demonstrated high potential in prevention of metastasis and treatment of cancer stem cells, they were not able to reduce tumor volumes. In contrast, radiotherapy and more conventional chemotherapeutic drugs, including the microtubule poisons taxanes, are able to reduce tumor volume, but usually fail to target cancer stem cells. Therefore, a therapy that includes radiotherapy or conventional chemotherapeutics, in combination with parthenolide/DMAPT could simultaneously target all types of cancer cells. Indeed, a synergistic effect of parthenolide in combination with either taxanes [541,542] or vinca alkaloids [543] was observed in breast cancer xenograft models in mice, affecting both tumor cells and cancer stem cells, while preventing metastasis. The development of new drugs that more specifically target enzymes that account for tubulin PTMs might reveal useful in evaluating potential clinical applications in the future.

5. Conclusions and Future Perspectives

Overall, we conclude that chromosome congression in mammalian cells relies on the concerted action of motor-dependent and -independent mechanisms, which are determined by the establishment of end-on or lateral kinetochore-microtubule interactions. Therefore, any perturbation that introduces alterations of microtubule dynamics or kinetochore function will likely compromise the congression of at least some chromosomes during mitosis. In addition, the recent discovery that tubulin PTMs have an impact on kinetochore motors and might work as a navigation system during chromosome congression brings together two old research fields, while opening up new and exciting avenues for investigation in the future. To date, more than 100 proteins have been implicated in chromosome alignment (Table 1), but their exact role in the activities necessary for either congression or maintenance of alignment remains unknown for >90% of them. A systematic analysis of the respective role of these proteins in chromosome congression will be an important challenge for future studies of mitosis. Moreover, the functional relationship between forces involved in chromosome congression and mitotic spindle architecture remains poorly understood and deserves further attention [415]. Finally, it will be important to firmly establish whether problems in chromosome congression are directly responsible for human diseases, such as cancer, and whether targeting chromosome congression represents a valid therapeutic approach.

Acknowledgments: We thank the precious and patient help of Nina Schweizer with translation of classical German works about chromosome congression. We also thank Liam Cheeseman for the critical reading and proofreading of the manuscript. Helder Maiato acknowledges funding support (including open access costs) provided by the CODECHECK grant from the European Research Council (ERC) under the European Union's Horizon 2020 research and innovation programme (grant agreement No. 681443), FLAD Life Science 2020, and the Louis-Jeantet Young Investigator Career Award. Marin Barisic is supported by a Junior Group Leader start-up research grant from the Danish Cancer Society and a Lundbeck Foundation Fellowship.

Author Contributions: H. Maiato conceived the structure of the review, wrote Sections 1, 2 and 5, prepared Figures 1–5 and performed the experiment in Figure 2b. A. M. Gomes wrote Sections 2.4, 2.8, 3 and 4 and prepared Table 1. M. Barisic wrote Sections 2.9 and 4.3 and drew Figure 7. F. Sousa and M. Barisic performed the experiments in Figure 6. F. Sousa prepared Figure 6. H. Maiato contributed to all the sections, edited and revised the work.

Conflicts of Interest: The authors declare no conflict of interest.

Abbreviations

The following abbreviations are used in this manuscript:

SAC	Spindle Assembly Checkpoint
UV	Ultra-violet
PEF	Polar Ejection Force
GFPMUGs	Green Fluorescent ProteinMitosis with Unreplicated Genomes
+TIPs	Microtubule Plus-End-Tracking Proteins
ATP	Adenosine Triphosphate
RNAi	RNA interference
MAPs	Microtubule-Associated Proteins
NEB	Nuclear Envelope Breakdown
PTMs	Post-Translational Modifications
TCP	Tubulin Carboxypeptidase
TTL	Tubulin Tyrosine Ligase
aMTOCs	acentriolar microtubule-organizing centers
CIN	Chromosomal Instability
MPD	Microcephalic primordial dwarfism
CAC	Colitis-Associated Cancer
DMAPT	Dimethylamino-parthenolide

References

1. Darlington, C.D. *Recent Advances in Cytology*, 2nd ed.; The Blakiston Company: Philadelphia, PA, USA, 1937.
2. Straight, A.F.; Marshall, W.F.; Sedat, J.W.; Murray, A.W. Mitosis in living budding yeast: Anaphase a but no metaphase plate. *Science* **1997**, *277*, 574–578. [CrossRef] [PubMed]
3. Goshima, G.; Scholey, J.M. Control of mitotic spindle length. *Annu. Rev. Cell Dev. Biol.* **2010**, *26*, 21–57. [CrossRef] [PubMed]
4. Matos, I.; Pereira, A.J.; Lince-Faria, M.; Cameron, L.A.; Salmon, E.D.; Maiato, H. Synchronizing chromosome segregation by flux-dependent force equalization at kinetochores. *J. Cell Biol.* **2009**, *186*, 11–26. [CrossRef] [PubMed]
5. Joglekar, A.P. A cell biological perspective on past, present and future investigations of the spindle assembly checkpoint. *Biology* **2016**, *5*, 44. [CrossRef] [PubMed]
6. Barisic, M.; Aguiar, P.; Geley, S.; Maiato, H. Kinetochore motors drive congression of peripheral polar chromosomes by overcoming random arm-ejection forces. *Nat. Cell Biol.* **2014**, *16*, 1249–1256. [CrossRef] [PubMed]
7. Ye, A.A.; Deretic, J.; Hoel, C.M.; Hinman, A.W.; Cimini, D.; Welburn, J.P.; Maresca, T.J. Aurora A Kinase Contributes to a Pole-Based Error Correction Pathway. *Curr. Biol.* **2015**, *25*, 1842–1851. [CrossRef] [PubMed]
8. Chmatal, L.; Yang, K.; Schultz, R.M.; Lampson, M.A. Spatial Regulation of Kinetochore Microtubule Attachments by Destabilization at Spindle Poles in Meiosis I. *Curr. Biol.* **2015**, *25*, 1835–1841. [CrossRef] [PubMed]
9. King, J.M.; Nicklas, R.B. Tension on chromosomes increases the number of kinetochore microtubules but only within limits. *J. Cell Sci.* **2000**, *113*, 3815–3823. [PubMed]
10. Wilson, E.B. *The Cell in Development and Heredity*, 3rd ed.; Macmillan: New York, NY, USA, 1925.
11. Lawrence, W.J.C. The genetics and cytology of Dahlia variabilis. *J. Genet.* **1931**, *24*, 257–306. [CrossRef]
12. Drüner, L. Studien über den mechanismus der zellteilung. *Jenaische Ztschr. Naturw.* **1895**, *29*, 271–344.
13. Östergren, G. Considerations on some elementary features of mitosis. *Hereditas* **1950**, *36*, 1–18. [CrossRef]
14. Belar, K. Beiträge zur kausalanalyse der mitose II. *Arch. Entwickl.* **1929**, *118*, 359–480. [CrossRef]

15. Rashevsky, N. Some remarks on the movement of chromosomes during cell division. *Bull. Math. Biophys.* **1941**, *3*, 1–3. [CrossRef]
16. Wada, B. The mechanism of mitosis based on studies of the submicroscopic structure and of the living state of the Tradescantia cell. *Cytologia* **1950**, *16*, 1–26. [CrossRef]
17. Östergren, G. Equilibrium of trivalents and the mechanism of chromosome movements. *Hereditas* **1945**, *31*, 498.
18. Schrader, F. *Mitosis—The Movements of Chromosomes in Cell Division*, 2nd ed.; Columbia University Press: New York, NY, USA, 1953.
19. Böök, J.A. Equilibrium of trivalents at metaphase. *Hereditas* **1945**, *31*, 499. [PubMed]
20. Izutsu, K. Irradiation of parts of single mitotic apparatus in grasshopper spermatocytes with an ultraviolet-microbeam. *Mie Med. J.* **1959**, *9*, 15–29.
21. Takeda, S.; Izutsu, K. Partial irradiation of individual mitotic cells with ultraviolet microbeam. *Symposia Cell Chem.* **1960**, *10*, 245–259.
22. Izutsu, K. Effects of ultraviolet microbeam irradiation upon division in grasshoper spermatocytes. II. Results of irradiation during metaphase and anaphase I. *Mie Med. J.* **1961**, *11*, 213–232.
23. McNeill, P.A.; Berns, M.W. Chromosome behavior after laser microirradiation of a single kinetochore in mitotic PtK2 cells. *J. Cell Biol.* **1981**, *88*, 543–553. [CrossRef] [PubMed]
24. Hays, T.S.; Wise, D.; Salmon, E.D. Traction force on a kinetochore at metaphase acts as a linear function of kinetochore fiber length. *J. Cell Biol.* **1982**, *93*, 374–389. [CrossRef] [PubMed]
25. Dietz, R. Anaphase behaviour of inversions in living crane-fly spermatocytes. *Chromosom. Today* **1972**, *3*, 70–85.
26. Hays, T.S.; Salmon, E.D. Poleward force at the kinetochore in metaphase depends on the number of kinetochore microtubules. *J. Cell Biol.* **1990**, *110*, 391–404. [CrossRef] [PubMed]
27. LaFountain, J.R., Jr.; Oldenbourg, R. Maloriented bivalents have metaphase positions at the spindle equator with more kinetochore microtubules to one pole than to the other. *Mol. Biol. Cell* **2004**, *15*, 5346–5355. [CrossRef] [PubMed]
28. McEwen, B.F.; Heagle, A.B.; Cassels, G.O.; Buttle, K.F.; Rieder, C.L. Kinetochore fiber maturation in PtK1 cells and its implications for the mechanisms of chromosome congression and anaphase onset. *J. Cell Biol.* **1997**, *137*, 1567–1580. [CrossRef] [PubMed]
29. Maiato, H.; Rieder, C.L.; Khodjakov, A. Kinetochore-driven formation of kinetochore fibers contributes to spindle assembly during animal mitosis. *J. Cell Biol.* **2004**, *167*, 831–840. [CrossRef] [PubMed]
30. Forer, A. Local Reduction of Spindle Fiber Birefringence in Living Nephrotoma Suturalis (Loew) Spermatocytes Induced by Ultraviolet Microbeam Irradiation. *J. Cell Biol.* **1965**, *25*, 95–117. [CrossRef]
31. Inoue, S. Organization and function of the mitotic spindle. In *Primitive Motile Systems in Cell Biology*; Academic Press: New York, NY, USA, 1964.
32. Spurck, T.P.; Stonington, O.G.; Snyder, J.A.; Pickett-Heaps, J.D.; Bajer, A.; Mole-Bajer, J. UV microbeam irradiations of the mitotic spindle. II. Spindle fiber dynamics and force production. *J. Cell Biol.* **1990**, *111*, 1505–1518. [CrossRef] [PubMed]
33. Nicklas, R.B. The motor for poleward chromosome movement in anaphase is in or near the kinetochore. *J. Cell Biol.* **1989**, *109*, 2245–2255. [CrossRef] [PubMed]
34. Czaban, B.B.; Forer, A.; Bajer, A.S. Ultraviolet microbeam irradiation of chromosomal spindle fibres in Haemanthus katherinae endosperm. I. Behaviour of the irradiated region. *J. Cell Sci.* **1993**, *105*, 571–578. [PubMed]
35. Sikirzhytski, V.; Magidson, V.; Steinman, J.B.; He, J.; Le Berre, M.; Tikhonenko, I.; Ault, J.G.; McEwen, B.F.; Chen, J.K.; Sui, H.; et al. Direct kinetochore-spindle pole connections are not required for chromosome segregation. *J. Cell Biol.* **2014**, *206*, 231–243. [CrossRef] [PubMed]
36. Elting, M.W.; Hueschen, C.L.; Udy, D.B.; Dumont, S. Force on spindle microtubule minus ends moves chromosomes. *J. Cell Biol.* **2014**, *206*, 245–256. [CrossRef] [PubMed]
37. Kajtez, J.; Solomatina, A.; Novak, M.; Polak, B.; Vukusic, K.; Rudiger, J.; Cojoc, G.; Milas, A.; Sumanovac Sestak, I.; Risteski, P.; et al. Overlap microtubules link sister k-fibres and balance the forces on bi-oriented kinetochores. *Nat. Commun.* **2016**, *7*, 10298. [CrossRef] [PubMed]
38. Milas, A.; Tolic, I.M.; Zür, M. Relaxation of interkinetochore tension after severing of a k-fiber depends on the length of the k-fiber stub. *Matters* **2016**. [CrossRef]

39. Bajer, A.S.; Molè-Bajer, J. Spindle dynamics and chromosome movements. *Int. Rev. Cytol.* **1972**, Supplement 3, 1–271.
40. Luykx, P. Cellular mechanisms of chromosome distribution. *Int. Rev. Cytol.* **1970**, Supplement 2, 1–173.
41. Bajer, A.S. Functional autonomy of monopolar spindle and evidence for oscillatory movement in mitosis. *J. Cell Biol.* **1982**, *93*, 33–48. [CrossRef] [PubMed]
42. Molè-Bajer, J.; Bajer, A.; Owczarzak, A. Chromosome movements in prometaphase and aster transport in the newt. *Cytobios* **1975**, *13*, 45–65.
43. Rieder, C.L.; Davison, E.A.; Jensen, L.C.; Cassimeris, L.; Salmon, E.D. Oscillatory movements of monooriented chromosomes and their position relative to the spindle pole result from the ejection properties of the aster and half-spindle. *J. Cell Biol.* **1986**, *103*, 581–591. [CrossRef] [PubMed]
44. Salmon, E.D. Microtubule dynamics and chromosome movement. In *Mitosis: Molecules and Mechanisms*; Hyams, J.S., Brinkley, B.R., Eds.; Academic Press Limited: London, UK, 1989.
45. Ault, J.G.; DeMarco, A.J.; Salmon, E.D.; Rieder, C.L. Studies on the ejection properties of asters: Astral microtubule turnover influences the oscillatory behavior and positioning of mono-oriented chromosomes. *J. Cell Sci.* **1991**, *99*, 701–710. [PubMed]
46. Cassimeris, L.; Rieder, C.L.; Salmon, E.D. Microtubule assembly and kinetochore directional instability in vertebrate monopolar spindles: Implications for the mechanism of chromosome congression. *J. Cell Sci.* **1994**, *107*, 285–297. [PubMed]
47. Salmon, E.D. Metaphase chromosome congression and anaphase poleward movement. In *Cell Movement: Kinesin, Dynein and Microtubule Dynamics*; Warner, F.D., McIntosh, J.R., Eds.; Alan R. Liss, Inc.: New York, NY, USA, 1989; pp. 431–440.
48. Khodjakov, A.; Rieder, C.L. Kinetochores moving away from their associated pole do not exert a significant pushing force on the chromosome. *J. Cell Biol.* **1996**, *135*, 315–327. [CrossRef] [PubMed]
49. Waters, J.C.; Skibbens, R.V.; Salmon, E.D. Oscillating mitotic newt lung cell kinetochores are, on average, under tension and rarely push. *J. Cell Sci.* **1996**, *109*, 2823–2831. [PubMed]
50. Ke, K.; Cheng, J.; Hunt, A.J. The distribution of polar ejection forces determines the amplitude of chromosome directional instability. *Curr. Biol.* **2009**, *19*, 807–815. [CrossRef] [PubMed]
51. Rieder, C.L.; Salmon, E.D. Motile kinetochores and polar ejection forces dictate chromosome position on the vertebrate mitotic spindle. *J. Cell Biol.* **1994**, *124*, 223–233. [CrossRef] [PubMed]
52. Bajer, A.S.; Cypher, C.; Mole-Bajer, J.; Howard, H.M. Taxol-induced anaphase reversal: Evidence that elongating microtubules can exert a pushing force in living cells. *Proc. Natl. Acad. Sci. USA* **1982**, *79*, 6569–6573. [CrossRef] [PubMed]
53. Dogterom, M.; Yurke, B. Measurement of the force-velocity relation for growing microtubules. *Science* **1997**, *278*, 856–860. [CrossRef] [PubMed]
54. Fygenson, D.K.; Marko, J.F.; Libchaber, A. Mechanics of microtubule-based membrane extension. *Phys. Rev. Lett.* **1997**, *79*, 4497–4500. [CrossRef]
55. Marshall, W.F.; Marko, J.F.; Agard, D.A.; Sedat, J.W. Chromosome elasticity and mitotic polar ejection force measured in living Drosophila embryos by four-dimensional microscopy-based motion analysis. *Curr. Biol.* **2001**, *11*, 569–578. [CrossRef]
56. Brouhard, G.J.; Hunt, A.J. Microtubule movements on the arms of mitotic chromosomes: Polar ejection forces quantified in vitro. *Proc. Natl. Acad. Sci. USA* **2005**, *102*, 13903–13908. [CrossRef] [PubMed]
57. Kuo, S.C.; Sheetz, M.P. Force of single kinesin molecules measured with optical tweezers. *Science* **1993**, *260*, 232–234. [CrossRef] [PubMed]
58. Svoboda, K.; Block, S.M. Force and velocity measured for single kinesin molecules. *Cell* **1994**, *77*, 773–784. [CrossRef]
59. Hall, K.; Cole, D.; Yeh, Y.; Baskin, R.J. Kinesin force generation measured using a centrifuge microscope sperm-gliding motility assay. *Biophys. J.* **1996**, *71*, 3467–3476. [CrossRef]
60. Carpenter, A.T. Distributive segregation: Motors in the polar wind? *Cell* **1991**, *64*, 885–890. [CrossRef]
61. Wang, S.Z.; Adler, R. Chromokinesin: A DNA-binding, kinesin-like nuclear protein. *J. Cell Biol.* **1995**, *128*, 761–768. [CrossRef] [PubMed]
62. Vernos, I.; Raats, J.; Hirano, T.; Heasman, J.; Karsenti, E.; Wylie, C. Xklp1, a chromosomal Xenopus kinesin-like protein essential for spindle organization and chromosome positioning. *Cell* **1995**, *81*, 117–127. [CrossRef]

63. Vanneste, D.; Ferreira, V.; Vernos, I. Chromokinesins: Localization-dependent functions and regulation during cell division. *Biochem. Soc. Trans.* **2011**, *39*, 1154–1160. [CrossRef] [PubMed]
64. Theurkauf, W.E.; Hawley, R.S. Meiotic spindle assembly in Drosophila females: Behavior of nonexchange chromosomes and the effects of mutations in the nod kinesin-like protein. *J. Cell Biol.* **1992**, *116*, 1167–1180. [CrossRef] [PubMed]
65. Funabiki, H.; Murray, A.W. The Xenopus chromokinesin Xkid is essential for metaphase chromosome alignment and must be degraded to allow anaphase chromosome movement. *Cell* **2000**, *102*, 411–424. [CrossRef]
66. Antonio, C.; Ferby, I.; Wilhelm, H.; Jones, M.; Karsenti, E.; Nebreda, A.R.; Vernos, I. Xkid, a chromokinesin required for chromosome alignment on the metaphase plate. *Cell* **2000**, *102*, 425–435. [CrossRef]
67. Levesque, A.A.; Compton, D.A. The chromokinesin Kid is necessary for chromosome arm orientation and oscillation, but not congression, on mitotic spindles. *J. Cell Biol.* **2001**, *154*, 1135–1146. [CrossRef] [PubMed]
68. Goshima, G.; Vale, R.D. The roles of microtubule-based motor proteins in mitosis: Comprehensive RNAi analysis in the Drosophila S2 cell line. *J. Cell Biol.* **2003**, *162*, 1003–1016. [CrossRef] [PubMed]
69. Mazumdar, M.; Sundareshan, S.; Misteli, T. Human chromokinesin KIF4A functions in chromosome condensation and segregation. *J. Cell Biol.* **2004**, *166*, 613–620. [CrossRef] [PubMed]
70. Zhu, C.; Zhao, J.; Bibikova, M.; Leverson, J.D.; Bossy-Wetzel, E.; Fan, J.B.; Abraham, R.T.; Jiang, W. Functional analysis of human microtubule-based motor proteins, the kinesins and dyneins, in mitosis/cytokinesis using RNA interference. *Mol. Biol. Cell* **2005**, *16*, 3187–3199. [CrossRef] [PubMed]
71. Wandke, C.; Barisic, M.; Sigl, R.; Rauch, V.; Wolf, F.; Amaro, A.C.; Tan, C.H.; Pereira, A.J.; Kutay, U.; Maiato, H.; et al. Human chromokinesins promote chromosome congression and spindle microtubule dynamics during mitosis. *J. Cell Biol.* **2012**, *198*, 847–863. [CrossRef] [PubMed]
72. Magidson, V.; O'Connell, C.B.; Loncarek, J.; Paul, R.; Mogilner, A.; Khodjakov, A. The spatial arrangement of chromosomes during prometaphase facilitates spindle assembly. *Cell* **2011**, *146*, 555–567. [CrossRef] [PubMed]
73. Sekine, Y.; Okada, Y.; Noda, Y.; Kondo, S.; Aizawa, H.; Takemura, R.; Hirokawa, N. A novel microtubule-based motor protein (KIF4) for organelle transports, whose expression is regulated developmentally. *J. Cell Biol.* **1994**, *127*, 187–201. [CrossRef] [PubMed]
74. Bringmann, H.; Skiniotis, G.; Spilker, A.; Kandels-Lewis, S.; Vernos, I.; Surrey, T. A kinesin-like motor inhibits microtubule dynamic instability. *Science* **2004**, *303*, 1519–1522. [CrossRef] [PubMed]
75. Yajima, J.; Edamatsu, M.; Watai-Nishii, J.; Tokai-Nishizumi, N.; Yamamoto, T.; Toyoshima, Y.Y. The human chromokinesin Kid is a plus end-directed microtubule-based motor. *EMBO J.* **2003**, *22*, 1067–1074. [CrossRef] [PubMed]
76. Bieling, P.; Kronja, I.; Surrey, T. Microtubule motility on reconstituted meiotic chromatin. *Curr. Biol.* **2010**, *20*, 763–769. [CrossRef] [PubMed]
77. Cane, S.; Ye, A.A.; Luks-Morgan, S.J.; Maresca, T.J. Elevated polar ejection forces stabilize kinetochore-microtubule attachments. *J. Cell Biol.* **2013**, *200*, 203–218. [CrossRef] [PubMed]
78. Skibbens, R.V.; Skeen, V.P.; Salmon, E.D. Directional instability of kinetochore motility during chromosome congression and segregation in mitotic newt lung cells: A push-pull mechanism. *J. Cell Biol.* **1993**, *122*, 859–875. [CrossRef] [PubMed]
79. Nicklas, R.B.; Koch, C.A. Chromosome micromanipulation. 3. Spindle fiber tension and the reorientation of mal-oriented chromosomes. *J. Cell Biol.* **1969**, *43*, 40–50. [CrossRef] [PubMed]
80. Nicklas, R.B.; Ward, S.C. Elements of error correction in mitosis: Microtubule capture, release, and tension. *J. Cell Biol.* **1994**, *126*, 1241–1253. [CrossRef] [PubMed]
81. Drpic, D.; Pereira, A.J.; Barisic, M.; Maresca, T.J.; Maiato, H. Polar Ejection Forces Promote the Conversion from Lateral to End-on Kinetochore-Microtubule Attachments on Mono-oriented Chromosomes. *Cell Rep.* **2015**, *13*, 460–469. [CrossRef] [PubMed]
82. Maresca, T.J.; Salmon, E.D. Intrakinetochore stretch is associated with changes in kinetochore phosphorylation and spindle assembly checkpoint activity. *J. Cell Biol.* **2009**, *184*, 373–381. [CrossRef] [PubMed]
83. Uchida, K.S.; Takagaki, K.; Kumada, K.; Hirayama, Y.; Noda, T.; Hirota, T. Kinetochore stretching inactivates the spindle assembly checkpoint. *J. Cell Biol.* **2009**, *184*, 383–390. [CrossRef] [PubMed]

84. Magidson, V.; He, J.; Ault, J.G.; O'Connell, C.B.; Yang, N.; Tikhonenko, I.; McEwen, B.F.; Sui, H.; Khodjakov, A. Unattached kinetochores rather than intrakinetochore tension arrest mitosis in taxol-treated cells. *J. Cell Biol.* **2016**, *212*, 307–319. [CrossRef] [PubMed]
85. Inoue, S. The effect of colchicine on the microscopic and submicroscopic structure of the mitotic spindle. *Exp. Cell Res.* **1952**, *2*, 305–318.
86. Inoue, S.; Salmon, E.D. Force generation by microtubule assembly/disassembly in mitosis and related movements. *Mol. Biol. Cell* **1995**, *6*, 1619–1640. [CrossRef] [PubMed]
87. Koshland, D.E.; Mitchison, T.J.; Kirschner, M.W. Polewards chromosome movement driven by microtubule depolymerization in vitro. *Nature* **1988**, *331*, 499–504. [CrossRef] [PubMed]
88. Coue, M.; Lombillo, V.A.; McIntosh, J.R. Microtubule depolymerization promotes particle and chromosome movement in vitro. *J. Cell Biol.* **1991**, *112*, 1165–1175. [CrossRef] [PubMed]
89. Grishchuk, E.L.; Molodtsov, M.I.; Ataullakhanov, F.I.; McIntosh, J.R. Force production by disassembling microtubules. *Nature* **2005**, *438*, 384–388. [CrossRef] [PubMed]
90. Cassimeris, L.; Salmon, E.D. Kinetochore microtubules shorten by loss of subunits at the kinetochores of prometaphase chromosomes. *J. Cell Sci.* **1991**, *98*, 151–158. [PubMed]
91. Tirnauer, J.S.; Canman, J.C.; Salmon, E.D.; Mitchison, T.J. EB1 targets to kinetochores with attached, polymerizing microtubules. *Mol. Biol. Cell* **2002**, *13*, 4308–4316. [CrossRef] [PubMed]
92. VandenBeldt, K.J.; Barnard, R.M.; Hergert, P.J.; Meng, X.; Maiato, H.; McEwen, B.F. Kinetochores use a novel mechanism for coordinating the dynamics of individual microtubules. *Curr. Biol.* **2006**, *16*, 1217–1223. [CrossRef] [PubMed]
93. Armond, J.W.; Vladimirou, E.; Erent, M.; McAinsh, A.D.; Burroughs, N.J. Probing microtubule polymerisation state at single kinetochores during metaphase chromosome motion. *J. Cell Sci.* **2015**, *128*, 1991–2001. [CrossRef] [PubMed]
94. Cheeseman, I.M.; Desai, A. Molecular architecture of the kinetochore-microtubule interface. *Nat. Rev. Mol. Cell Biol.* **2008**, *9*, 33–46. [CrossRef] [PubMed]
95. Schuyler, S.C.; Pellman, D. Microtubule "plus-end-tracking proteins": The end is just the beginning. *Cell* **2001**, *105*, 421–424. [CrossRef]
96. Akhmanova, A.; Steinmetz, M.O. Microtubule +TIPs at a glance. *J. Cell Sci.* **2010**, *123*, 3415–3419. [CrossRef] [PubMed]
97. Akhmanova, A.; Steinmetz, M.O. Tracking the ends: A dynamic protein network controls the fate of microtubule tips. *Nat. Rev. Mol. Cell Biol.* **2008**, *9*, 309–322. [CrossRef] [PubMed]
98. Brouhard, G.J.; Stear, J.H.; Noetzel, T.L.; Al-Bassam, J.; Kinoshita, K.; Harrison, S.C.; Howard, J.; Hyman, A.A. XMAP215 is a processive microtubule polymerase. *Cell* **2008**, *132*, 79–88. [CrossRef] [PubMed]
99. Komarova, Y.A.; Akhmanova, A.S.; Kojima, S.; Galjart, N.; Borisy, G.G. Cytoplasmic linker proteins promote microtubule rescue in vivo. *J. Cell Biol.* **2002**, *159*, 589–599. [CrossRef] [PubMed]
100. Mimori-Kiyosue, Y.; Grigoriev, I.; Lansbergen, G.; Sasaki, H.; Matsui, C.; Severin, F.; Galjart, N.; Grosveld, F.; Vorobjev, I.; Tsukita, S.; et al. CLASP1 and CLASP2 bind to EB1 and regulate microtubule plus-end dynamics at the cell cortex. *J. Cell Biol.* **2005**, *168*, 141–153. [CrossRef] [PubMed]
101. Akhmanova, A.; Hoogenraad, C.C.; Drabek, K.; Stepanova, T.; Dortland, B.; Verkerk, T.; Vermeulen, W.; Burgering, B.M.; De Zeeuw, C.I.; Grosveld, F.; et al. Clasps are CLIP-115 and -170 associating proteins involved in the regional regulation of microtubule dynamics in motile fibroblasts. *Cell* **2001**, *104*, 923–935. [CrossRef]
102. Perez, F.; Diamantopoulos, G.S.; Stalder, R.; Kreis, T.E. CLIP-170 highlights growing microtubule ends in vivo. *Cell* **1999**, *96*, 517–527. [CrossRef]
103. Dujardin, D.; Wacker, U.I.; Moreau, A.; Schroer, T.A.; Rickard, J.E.; De Mey, J.R. Evidence for a role of CLIP-170 in the establishment of metaphase chromosome alignment. *J. Cell Biol.* **1998**, *141*, 849–862. [CrossRef] [PubMed]
104. Tanenbaum, M.E.; Galjart, N.; van Vugt, M.A.T.M.; Medema, R.H. CLIP-170 facilitates the formation of kinetochore-microtubule attachments. *EMBO J.* **2006**, *25*, 45–57. [CrossRef] [PubMed]
105. Kakeno, M.; Matsuzawa, K.; Matsui, T.; Akita, H.; Sugiyama, I.; Ishidate, F.; Nakano, A.; Takashima, S.; Goto, H.; Inagaki, M.; et al. Plk1 phosphorylates CLIP-170 and regulates its binding to microtubules for chromosome alignment. *Cell Struct. Funct.* **2014**, *39*, 45–59. [CrossRef] [PubMed]

106. Amin, M.A.; Kobayashi, K.; Tanaka, K. CLIP-170 tethers kinetochores to microtubule plus ends against poleward force by dynein for stable kinetochore-microtubule attachment. *FEBS Lett.* **2015**, *589*, 2739–2746. [CrossRef] [PubMed]
107. Bonfils, C.; Bec, N.; Lacroix, B.; Harricane, M.C.; Larroque, C. Kinetic analysis of tubulin assembly in the presence of the microtubule-associated protein TOGp. *J. Biol. Chem.* **2007**, *282*, 5570–5581. [CrossRef] [PubMed]
108. Gard, D.L.; Kirschner, M.W. A microtubule-associated protein from Xenopus eggs that specifically promotes assembly at the plus-end. *J. Cell Biol.* **1987**, *105*, 2203–2215. [CrossRef] [PubMed]
109. Al-Bassam, J.; Chang, F. Regulation of microtubule dynamics by TOG-domain proteins XMAP215/Dis1 and CLASP. *Trends Cell Biol.* **2011**, *21*, 604–614. [CrossRef] [PubMed]
110. Al-Bassam, J.; Kim, H.; Brouhard, G.; van Oijen, A.; Harrison, S.C.; Chang, F. CLASP promotes microtubule rescue by recruiting tubulin dimers to the microtubule. *Dev. Cell* **2010**, *19*, 245–258. [CrossRef] [PubMed]
111. Gandhi, S.R.; Gierlinski, M.; Mino, A.; Tanaka, K.; Kitamura, E.; Clayton, L.; Tanaka, T.U. Kinetochore-dependent microtubule rescue ensures their efficient and sustained interactions in early mitosis. *Dev. Cell* **2011**, *21*, 920–933. [CrossRef] [PubMed]
112. Gergely, F.; Draviam, V.M.; Raff, J.W. The ch-TOG/XMAP215 protein is essential for spindle pole organization in human somatic cells. *Genes Dev.* **2003**, *17*, 336–341. [CrossRef] [PubMed]
113. Kitamura, E.; Tanaka, K.; Komoto, S.; Kitamura, Y.; Antony, C.; Tanaka, T.U. Kinetochores generate microtubules with distal plus ends: Their roles and limited lifetime in mitosis. *Dev. Cell* **2010**, *18*, 248–259. [CrossRef] [PubMed]
114. Miller, M.P.; Asbury, C.L.; Biggins, S. A TOG Protein Confers Tension Sensitivity to Kinetochore-Microtubule Attachments. *Cell* **2016**, *165*, 1428–1439. [CrossRef] [PubMed]
115. Cassimeris, L.; Becker, B.; Carney, B. TOGp regulates microtubule assembly and density during mitosis and contributes to chromosome directional instability. *Cell Motil. Cytoskeleton* **2009**, *66*, 535–545. [CrossRef] [PubMed]
116. Maiato, H.; Fairley, E.A.; Rieder, C.L.; Swedlow, J.R.; Sunkel, C.E.; Earnshaw, W.C. Human CLASP1 is an outer kinetochore component that regulates spindle microtubule dynamics. *Cell* **2003**, *113*, 891–904. [CrossRef]
117. Pereira, A.L.; Pereira, A.J.; Maia, A.R.; Drabek, K.; Sayas, C.L.; Hergert, P.J.; Lince-Faria, M.; Matos, I.; Duque, C.; Stepanova, T.; et al. Mammalian CLASP1 and CLASP2 cooperate to ensure mitotic fidelity by regulating spindle and kinetochore function. *Mol. Biol. Cell* **2006**, *17*, 4526–4542. [CrossRef] [PubMed]
118. Maiato, H.; Khodjakov, A.; Rieder, C.L. Drosophila CLASP is required for the incorporation of microtubule subunits into fluxing kinetochore fibres. *Nat. Cell Biol.* **2005**, *7*, 42–47. [CrossRef] [PubMed]
119. Maffini, S.; Maia, A.R.; Manning, A.L.; Maliga, Z.; Pereira, A.L.; Junqueira, M.; Shevchenko, A.; Hyman, A.; Yates, J.R., 3rd; Galjart, N.; et al. Motor-independent targeting of CLASPs to kinetochores by CENP-E promotes microtubule turnover and poleward flux. *Curr. Biol.* **2009**, *19*, 1566–1572. [CrossRef] [PubMed]
120. Manning, A.L.; Bakhoum, S.F.; Maffini, S.; Correia-Melo, C.; Maiato, H.; Compton, D.A. CLASP1, astrin and Kif2b form a molecular switch that regulates kinetochore-microtubule dynamics to promote mitotic progression and fidelity. *EMBO J.* **2010**, *29*, 3531–3543. [CrossRef] [PubMed]
121. Maia, A.R.; Garcia, Z.; Kabeche, L.; Barisic, M.; Maffini, S.; Macedo-Ribeiro, S.; Cheeseman, I.M.; Compton, D.A.; Kaverina, I.; Maiato, H. Cdk1 and Plk1 mediate a CLASP2 phospho-switch that stabilizes kinetochore-microtubule attachments. *J. Cell Biol.* **2012**, *199*, 285–301. [CrossRef] [PubMed]
122. Walczak, C.E.; Gayek, S.; Ohi, R. Microtubule-depolymerizing kinesins. *Annu. Rev. Cell Dev. Biol.* **2013**, *29*, 417–441. [CrossRef] [PubMed]
123. Desai, A.; Verma, S.; Mitchison, T.J.; Walczak, C.E. Kin I kinesins are microtubule-destabilizing enzymes. *Cell* **1999**, *96*, 69–78. [CrossRef]
124. Manning, A.L.; Ganem, N.J.; Bakhoum, S.F.; Wagenbach, M.; Wordeman, L.; Compton, D.A. The kinesin-13 proteins Kif2a, Kif2b, and Kif2c/MCAK have distinct roles during mitosis in human cells. *Mol. Biol. Cell* **2007**, *18*, 2970–2979. [CrossRef] [PubMed]
125. Walczak, C.E. The Kin I kinesins are microtubule end-stimulated ATPases. *Mol. Cell* **2003**, *11*, 286–288. [CrossRef]
126. Ganem, N.J.; Compton, D.A. The KinI kinesin Kif2a is required for bipolar spindle assembly through a functional relationship with MCAK. *J. Cell Biol.* **2004**, *166*, 473–478. [CrossRef] [PubMed]

127. Walczak, C.E.; Mitchison, T.J.; Desai, A. XKCM1: A Xenopus kinesin-related protein that regulates microtubule dynamics during mitotic spindle assembly. *Cell* **1996**, *84*, 37–47. [CrossRef]
128. Kline-Smith, S.L.; Walczak, C.E. The microtubule-destabilizing kinesin XKCM1 regulates microtubule dynamic instability in cells. *Mol. Biol. Cell* **2002**, *13*, 2718–2731. [CrossRef] [PubMed]
129. Wordeman, L.; Wagenbach, M.; von Dassow, G. MCAK facilitates chromosome movement by promoting kinetochore microtubule turnover. *J. Cell Biol.* **2007**, *179*, 869–879. [CrossRef] [PubMed]
130. Bakhoum, S.F.; Thompson, S.L.; Manning, A.L.; Compton, D.A. Genome stability is ensured by temporal control of kinetochore-microtubule dynamics. *Nat. Cell Biol.* **2009**, *11*, 27–35. [CrossRef] [PubMed]
131. Gaetz, J.; Kapoor, T.M. Dynein/dynactin regulate metaphase spindle length by targeting depolymerizing activities to spindle poles. *J. Cell Biol.* **2004**, *166*, 465–471. [CrossRef] [PubMed]
132. Ganem, N.J.; Upton, K.; Compton, D.A. Efficient mitosis in human cells lacking poleward microtubule flux. *Curr. Biol.* **2005**, *15*, 1827–1832. [CrossRef] [PubMed]
133. Gupta, M.L.; Carvalho, P.; Roof, D.M.; Pellman, D. Plus end-specific depolymerase activity of Kip3, a kinesin-8 protein, explains its role in positioning the yeast mitotic spindle. *Nat. Cell Biol.* **2006**, *8*, 913–923. [CrossRef] [PubMed]
134. Mayr, M.I.; Hümmer, S.; Bormann, J.; Grüner, T.; Adio, S.; Woehlke, G.; Mayer, T.U. The human kinesin Kif18A is a motile microtubule depolymerase essential for chromosome congression. *Curr. Biol.* **2007**, *17*, 488–498. [CrossRef] [PubMed]
135. Varga, V.; Helenius, J.; Tanaka, K.; Hyman, A.A.; Tanaka, T.U.; Howard, J. Yeast kinesin-8 depolymerizes microtubules in a length-dependent manner. *Nat. Cell Biol.* **2006**, *8*, 957–962. [CrossRef] [PubMed]
136. Varga, V.; Leduc, C.; Bormuth, V.; Diez, S.; Howard, J. Kinesin-8 motors act cooperatively to mediate length-dependent microtubule depolymerization. *Cell* **2009**, *138*, 1174–1183. [CrossRef] [PubMed]
137. Du, Y.; English, C.A.; Ohi, R. The kinesin-8 Kif18A dampens microtubule plus-end dynamics. *Curr. Biol.* **2010**, *20*, 374–380. [CrossRef] [PubMed]
138. Stumpff, J.; Du, Y.; English, C.A.; Maliga, Z.; Wagenbach, M.; Asbury, C.L.; Wordeman, L.; Ohi, R. A tethering mechanism controls the processivity and kinetochore-microtubule plus-end enrichment of the kinesin-8 Kif18A. *Mol. Cell* **2011**, *43*, 764–775. [CrossRef] [PubMed]
139. Gandhi, R.; Bonaccorsi, S.; Wentworth, D.; Doxsey, S.; Gatti, M.; Pereira, A. The Drosophila kinesin-like protein KLP67A is essential for mitotic and male meiotic spindle assembly. *Mol. Biol. Cell* **2004**, *15*, 121–131. [CrossRef] [PubMed]
140. Goshima, G.; Wollman, R.; Stuurman, N.; Scholey, J.M.; Vale, R.D. Length control of the metaphase spindle. *Curr. Biol.* **2005**, *15*, 1979–1988. [CrossRef] [PubMed]
141. Rischitor, P.E.; Konzack, S.; Fischer, R. The Kip3-like kinesin KipB moves along microtubules and determines spindle position during synchronized mitoses in Aspergillus nidulans hyphae. *Eukaryotic Cell* **2004**, *3*, 632–645. [CrossRef] [PubMed]
142. Straight, A.F.; Sedat, J.W.; Murray, A.W. Time-lapse microscopy reveals unique roles for kinesins during anaphase in budding yeast. *J. Cell Biol.* **1998**, *143*, 687–694. [CrossRef] [PubMed]
143. West, R.R.; Malmstrom, T.; McIntosh, J.R. Kinesins klp5(+) and klp6(+) are required for normal chromosome movement in mitosis. *J. Cell Sci.* **2002**, *115*, 931–940. [PubMed]
144. Stumpff, J.; von Dassow, G.; Wagenbach, M.; Asbury, C.; Wordeman, L. The kinesin-8 motor Kif18A suppresses kinetochore movements to control mitotic chromosome alignment. *Dev. Cell* **2008**, *14*, 252–262. [CrossRef] [PubMed]
145. Jaqaman, K.; King, E.M.; Amaro, A.C.; Winter, J.R.; Dorn, J.F.; Elliott, H.L.; McHedlishvili, N.; McClelland, S.E.; Porter, I.M.; Posch, M.; et al. Kinetochore alignment within the metaphase plate is regulated by centromere stiffness and microtubule depolymerases. *J. Cell Biol.* **2010**, *188*, 665–679. [CrossRef] [PubMed]
146. Stumpff, J.; Wagenbach, M.; Franck, A.; Asbury, C.L.; Wordeman, L. Kif18A and chromokinesins confine centromere movements via microtubule growth suppression and spatial control of kinetochore tension. *Dev. Cell* **2012**, *22*, 1017–1029. [CrossRef] [PubMed]
147. Thein, K.H.; Kleylein-Sohn, J.; Nigg, E.A.; Gruneberg, U. Astrin is required for the maintenance of sister chromatid cohesion and centrosome integrity. *J. Cell Biol.* **2007**, *178*, 345–354. [CrossRef] [PubMed]

148. Schmidt, J.C.; Kiyomitsu, T.; Hori, T.; Backer, C.B.; Fukagawa, T.; Cheeseman, I.M. Aurora B kinase controls the targeting of the Astrin-SKAP complex to bioriented kinetochores. *J. Cell Biol.* **2010**, *191*, 269–280. [CrossRef] [PubMed]
149. Dunsch, A.K.; Linnane, E.; Barr, F.A.; Gruneberg, U. The astrin-kinastrin/SKAP complex localizes to microtubule plus ends and facilitates chromosome alignment. *J. Cell Biol.* **2011**, *192*, 959–968. [CrossRef] [PubMed]
150. Logarinho, E.; Maffini, S.; Barisic, M.; Marques, A.; Toso, A.; Meraldi, P.; Maiato, H. CLASPs prevent irreversible multipolarity by ensuring spindle-pole resistance to traction forces during chromosome alignment. *Nat. Cell Biol.* **2012**, *14*, 295–303. [CrossRef] [PubMed]
151. Wu, G.; Lin, Y.-T.; Wei, R.; Chen, Y.; Shan, Z.; Lee, W.-H. Hice1, a novel microtubule-associated protein required for maintenance of spindle integrity and chromosomal stability in human cells. *Mol. Cell. Biol.* **2008**, *28*, 3652–3662. [CrossRef] [PubMed]
152. Hoar, K.; Chakravarty, A.; Rabino, C.; Wysong, D.; Bowman, D.; Roy, N.; Ecsedy, J.A. MLN8054, a Small-Molecule Inhibitor of Aurora A, Causes Spindle Pole and Chromosome Congression Defects Leading to Aneuploidy. *Mol. Cell. Biol.* **2007**, *27*, 4513–4525. [CrossRef] [PubMed]
153. Sasai, K.; Parant, J.M.; Brandt, M.E.; Carter, J.; Adams, H.P.; Stass, S.A.; Killary, A.M.; Katayama, H.; Sen, S. Targeted disruption of Aurora A causes abnormal mitotic spindle assembly, chromosome misalignment and embryonic lethality. *Oncogene* **2008**, *27*, 4122–4127. [CrossRef] [PubMed]
154. Kesisova, I.A.; Nakos, K.C.; Tsolou, A.; Angelis, D.; Lewis, J.; Chatzaki, A.; Agianian, B.; Giannis, A.; Koffa, M.D. Tripolin A, a novel small-molecule inhibitor of aurora A kinase, reveals new regulation of HURP's distribution on microtubules. *PLoS ONE* **2013**, *8*, e58485. [CrossRef] [PubMed]
155. Stevens, D.; Gassmann, R.; Oegema, K.; Desai, A. Uncoordinated loss of chromatid cohesion is a common outcome of extended metaphase arrest. *PLoS ONE* **2011**, *6*, e22969. [CrossRef] [PubMed]
156. Tanudji, M.; Shoemaker, J.; L'Italien, L.; Russell, L.; Chin, G.; Schebye, X.M. Gene silencing of CENP-E by small interfering RNA in HeLa cells leads to missegregation of chromosomes after a mitotic delay. *Mol. Biol. Cell* **2004**, *15*, 3771–3781. [CrossRef] [PubMed]
157. Maia, A.F.; Feijão, T.; Vromans, M.J.M.; Sunkel, C.E.; Lens, S.M.A. Aurora B kinase cooperates with CENP-E to promote timely anaphase onset. *Chromosoma* **2010**, *119*, 405–413. [CrossRef] [PubMed]
158. Wu, Q.; He, R.; Zhou, H.; Yu, A.C.H.; Zhang, B.; Teng, J.; Chen, J. Cep57, a NEDD1-binding pericentriolar material component, is essential for spindle pole integrity. *Cell Res.* **2012**, *22*, 1390–1401. [CrossRef] [PubMed]
159. Oshimori, N.; Li, X.; Ohsugi, M.; Yamamoto, T. Cep72 regulates the localization of key centrosomal proteins and proper bipolar spindle formation. *EMBO J.* **2009**, *28*, 2066–2076. [CrossRef] [PubMed]
160. Kim, K.; Rhee, K. The pericentriolar satellite protein CEP90 is crucial for integrity of the mitotic spindle pole. *J. Cell Sci.* **2011**, *124*, 338–347. [CrossRef] [PubMed]
161. Kimura, M.; Yoshioka, T.; Saio, M.; Banno, Y.; Nagaoka, H.; Okano, Y. Mitotic catastrophe and cell death induced by depletion of centrosomal proteins. *Cell Death Dis.* **2013**, *4*, e603. [CrossRef] [PubMed]
162. Mimori-Kiyosue, Y.; Grigoriev, I.; Sasaki, H.; Matsui, C.; Akhmanova, A.; Tsukita, S.; Vorobjev, I. Mammalian CLASPs are required for mitotic spindle organization and kinetochore alignment. *Genes Cells* **2006**, *11*, 845–857. [CrossRef] [PubMed]
163. Hauf, S.; Cole, R.W.; LaTerra, S.; Zimmer, C.; Schnapp, G.; Walter, R.; Heckel, A.; van Meel, J.; Rieder, C.L.; Peters, J.-M. The small molecule Hesperadin reveals a role for Aurora B in correcting kinetochore-microtubule attachment and in maintaining the spindle assembly checkpoint. *J. Cell Biol.* **2003**, *161*, 281–294. [CrossRef] [PubMed]
164. Johnson, V.L.; Scott, M.I.F.; Holt, S.V.; Hussein, D.; Taylor, S.S. Bub1 is required for kinetochore localization of BubR1, Cenp-E, Cenp-F and Mad2, and chromosome congression. *J. Cell Sci.* **2004**, *117*, 1577–1589. [CrossRef] [PubMed]
165. Dai, J.; Sultan, S.; Taylor, S.S.; Higgins, J.M.G. The kinase haspin is required for mitotic histone H3 Thr 3 phosphorylation and normal metaphase chromosome alignment. *Genes Dev.* **2005**, *19*, 472–488. [CrossRef] [PubMed]
166. Dai, J.; Sullivan, B.A.; Higgins, J.M.G. Regulation of mitotic chromosome cohesion by Haspin and Aurora B. *Dev. Cell* **2006**, *11*, 741–750. [CrossRef] [PubMed]
167. Dai, J.; Kateneva, A.V.; Higgins, J.M.G. Studies of haspin-depleted cells reveal that spindle-pole integrity in mitosis requires chromosome cohesion. *J. Cell Sci.* **2009**, *122*, 4168–4176. [CrossRef]

168. Fielding, A.B.; Dobreva, I.; McDonald, P.C.; Foster, L.J.; Dedhar, S. Integrin-linked kinase localizes to the centrosome and regulates mitotic spindle organization. *J. Cell Biol.* **2008**, *180*, 681–689. [CrossRef] [PubMed]
169. Fang, L.; Seki, A.; Fang, G. SKAP associates with kinetochores and promotes the metaphase-to-anaphase transition. *Cell Cycle* **2009**, *8*, 2819–2827. [CrossRef] [PubMed]
170. Martin-Lluesma, S.; Stucke, V.M.; Nigg, E.A. Role of Hec1 in spindle checkpoint signaling and kinetochore recruitment of Mad1/Mad2. *Science* **2002**, *297*, 2267–2270. [CrossRef] [PubMed]
171. Joseph, J.; Liu, S.-T.; Jablonski, S.A.; Yen, T.J.; Dasso, M. The RanGAP1-RanBP2 complex is essential for microtubule-kinetochore interactions in vivo. *Curr. Biol.* **2004**, *14*, 611–617. [CrossRef] [PubMed]
172. Li, L.; Yang, L.; Scudiero, D.A.; Miller, S.A.; Yu, Z.X.; Stukenberg, P.T.; Shoemaker, R.H.; Kotin, R.M. Development of recombinant adeno-associated virus vectors carrying small interfering RNA (shHec1)-mediated depletion of kinetochore Hec1 protein in tumor cells. *Gene Ther.* **2007**, *14*, 814–827. [CrossRef] [PubMed]
173. Sundin, L.J.; Guimaraes, G.J.; Deluca, J.G. The NDC80 complex proteins Nuf2 and Hec1 make distinct contributions to kinetochore-microtubule attachment in mitosis. *Mol. Biol. Cell* **2011**, *22*, 759–768. [CrossRef] [PubMed]
174. Xu, P.; Virshup, D.M.; Lee, S.H. B56-PP2A regulates motor dynamics for mitotic chromosome alignment. *J. Cell Sci.* **2014**, *127*, 4567–4573. [CrossRef] [PubMed]
175. DeLuca, J.G.; Moree, B.; Hickey, J.M.; Kilmartin, J.V.; Salmon, E.D. hNuf2 inhibition blocks stable kinetochore-microtubule attachment and induces mitotic cell death in HeLa cells. *J. Cell Biol.* **2002**, *159*, 549–555. [CrossRef] [PubMed]
176. Haren, L.; Gnadt, N.; Wright, M.; Merdes, A. NuMA is required for proper spindle assembly and chromosome alignment in prometaphase. *BMC Res.* **2009**, *2*, 64. [CrossRef] [PubMed]
177. McGuinness, B.E.; Hirota, T.; Kudo, N.R.; Peters, J.-M.; Nasmyth, K. Shugoshin prevents dissociation of cohesin from centromeres during mitosis in vertebrate cells. *PLoS Biol.* **2005**, *3*, e86. [CrossRef] [PubMed]
178. Raaijmakers, J.A.; Tanenbaum, M.E.; Medema, R.H. Systematic dissection of dynein regulators in mitosis. *J. Cell Biol.* **2013**, *201*, 201–215. [CrossRef] [PubMed]
179. Barisic, M.; Sohm, B.; Mikolcevic, P.; Wandke, C.; Rauch, V.; Ringer, T.; Hess, M.; Bonn, G.; Geley, S. Spindly/CCDC99 is required for efficient chromosome congression and mitotic checkpoint regulation. *Mol. Biol. Cell* **2010**, *21*, 1968–1981. [CrossRef] [PubMed]
180. Schneider, L.; Essmann, F.; Kletke, A.; Rio, P.; Hanenberg, H.; Wetzel, W.; Schulze-Osthoff, K.; Nurnberg, B.; Piekorz, R.P. The transforming acidic coiled coil 3 protein is essential for spindle-dependent chromosome alignment and mitotic survival. *J. Biol. Chem.* **2007**, *282*, 29273–29283. [CrossRef] [PubMed]
181. Lin, C.H.; Hu, C.K.; Shih, H.M. Clathrin heavy chain mediates TACC3 targeting to mitotic spindles to ensure spindle stability. *J. Cell Biol.* **2010**, *189*, 1097–1105. [CrossRef] [PubMed]
182. Cheeseman, L.P.; Harry, E.F.; McAinsh, A.D.; Prior, I.A.; Royle, S.J. Specific removal of TACC3-ch-TOG-clathrin at metaphase deregulates kinetochore fiber tension. *J. Cell Sci.* **2013**, *126*, 2102–2113. [CrossRef] [PubMed]
183. Royle, S.J.; Bright, N.A.; Lagnado, L. Clathrin is required for the function of the mitotic spindle. *Nature* **2005**, *434*, 1152–1157. [CrossRef] [PubMed]
184. Krauss, S.W.; Spence, J.R.; Bahmanyar, S.; Barth, A.I.M.; Go, M.M.; Czerwinski, D.; Meyer, A.J. Downregulation of protein 4.1R, a mature centriole protein, disrupts centrosomes, alters cell cycle progression, and perturbs mitotic spindles and anaphase. *Mol. Cell. Biol.* **2008**, *28*, 2283–2294. [CrossRef] [PubMed]
185. Hanisch, A.; Silljé, H.H.W.; Nigg, E.A. Timely anaphase onset requires a novel spindle and kinetochore complex comprising Ska1 and Ska2. *EMBO J.* **2006**, *25*, 5504–5515. [CrossRef] [PubMed]
186. Sivakumar, S.; Daum, J.R.; Tipton, A.R.; Rankin, S.; Gorbsky, G.J. The spindle and kinetochore-associated (Ska) complex enhances binding of the anaphase-promoting complex/cyclosome (APC/C) to chromosomes and promotes mitotic exit. *Mol. Biol. Cell* **2014**, *25*, 594–605. [CrossRef] [PubMed]
187. Gaitanos, T.N.; Santamaria, A.; Jeyaprakash, A.A.; Wang, B.; Conti, E.; Nigg, E.A. Stable kinetochore-microtubule interactions depend on the Ska complex and its new component Ska3/C13Orf3. *EMBO J.* **2009**, *28*, 1442–1452. [CrossRef] [PubMed]

188. Welburn, J.P.I.; Grishchuk, E.L.; Backer, C.B.; Wilson-Kubalek, E.M.; Yates, J.R.; Cheeseman, I.M. The human kinetochore Ska1 complex facilitates microtubule depolymerization-coupled motility. *Dev. Cell* **2009**, *16*, 374–385. [CrossRef] [PubMed]
189. Daum, J.R.; Wren, J.D.; Daniel, J.J.; Sivakumar, S.; McAvoy, J.N.; Potapova, T.A.; Gorbsky, G.J. Ska3 is required for spindle checkpoint silencing and the maintenance of chromosome cohesion in mitosis. *Curr. Biol.* **2009**, *19*, 1467–1472. [CrossRef] [PubMed]
190. Raaijmakers, J.A.; Tanenbaum, M.E.; Maia, A.F.; Medema, R.H. RAMA1 is a novel kinetochore protein involved in kinetochore-microtubule attachment. *J. Cell Sci.* **2009**, *122*, 2436–2445. [CrossRef] [PubMed]
191. Tokai-Nishizumi, N.; Ohsugi, M.; Suzuki, E.; Yamamoto, T. The chromokinesin Kid is required for maintenance of proper metaphase spindle size. *Mol. Biol. Cell* **2005**, *16*, 5455–5463. [CrossRef] [PubMed]
192. Huang, Y.; Yao, Y.; Xu, H.-Z.; Wang, Z.-G.; Lu, L.; Dai, W. Defects in chromosome congression and mitotic progression in KIF18A-deficient cells are partly mediated through impaired functions of CENP-E. *Cell Cycle* **2009**, *8*, 2643–2649. [CrossRef] [PubMed]
193. Liu, X.-S.; Zhao, X.-D.; Wang, X.; Yao, Y.-X.; Zhang, L.-L.; Shu, R.-Z.; Ren, W.-H.; Huang, Y.; Huang, L.; Gu, M.-M.; et al. Germinal Cell Aplasia in Kif18a Mutant Male Mice Due to Impaired Chromosome Congression and Dysregulated BubR1 and CENP-E. *Genes Cancer* **2010**, *1*, 26–39. [CrossRef] [PubMed]
194. Tanenbaum, M.E.; Macurek, L.; van der Vaart, B.; Galli, M.; Akhmanova, A.; Medema, R.H. A complex of Kif18b and MCAK promotes microtubule depolymerization and is negatively regulated by Aurora kinases. *Curr. Biol.* **2011**, *21*, 1356–1365. [CrossRef] [PubMed]
195. Stout, J.R.; Yount, A.L.; Powers, J.A.; Leblanc, C.; Ems-McClung, S.C.; Walczak, C.E. Kif18B interacts with EB1 and controls astral microtubule length during mitosis. *Mol. Biol. Cell* **2011**, *22*, 3070–3080. [CrossRef] [PubMed]
196. Walczak, C.E.; Zong, H.; Jain, S.; Stout, J.R. Spatial regulation of astral microtubule dynamics by Kif18B in PtK cells. *Mol. Biol. Cell* **2016**, *27*, 3021–3030. [CrossRef] [PubMed]
197. Kline-Smith, S.L.; Khodjakov, A.; Hergert, P.; Walczak, C.E. Depletion of centromeric MCAK leads to chromosome congression and segregation defects due to improper kinetochore attachments. *Mol. Biol. Cell* **2004**, *15*, 1146–1159. [CrossRef] [PubMed]
198. Sillјé, H.H.W.; Nagel, S.; Körner, R.; Nigg, E.A. HURP is a Ran-importin beta-regulated protein that stabilizes kinetochore microtubules in the vicinity of chromosomes. *Curr. Biol.* **2006**, *16*, 731–742. [CrossRef] [PubMed]
199. Wong, J.; Fang, G. HURP controls spindle dynamics to promote proper interkinetochore tension and efficient kinetochore capture. *J. Cell Biol.* **2006**, *173*, 879–891. [CrossRef] [PubMed]
200. Ye, F.; Tan, L.; Yang, Q.; Xia, Y.; Deng, L.-W.; Murata-Hori, M.; Liou, Y.-C. HURP regulates chromosome congression by modulating kinesin Kif18A function. *Curr. Biol.* **2011**, *21*, 1584–1591. [CrossRef] [PubMed]
201. McHedlishvili, N.; Wieser, S.; Holtackers, R.; Mouysset, J.; Belwal, M.; Amaro, A.C.; Meraldi, P. Kinetochores accelerate centrosome separation to ensure faithful chromosome segregation. *J. Cell Sci.* **2012**, *125*, 906–918. [CrossRef] [PubMed]
202. Raemaekers, T.; Ribbeck, K.; Beaudouin, J.; Annaert, W.; Van Camp, M.; Stockmans, I.; Smets, N.; Bouillon, R.; Ellenberg, J.; Carmeliet, G. NuSAP, a novel microtubule-associated protein involved in mitotic spindle organization. *J. Cell Biol.* **2003**, *162*, 1017–1029. [CrossRef] [PubMed]
203. Li, C.; Xue, C.; Yang, Q.; Low, B.C.; Liou, Y.C. NuSAP governs chromosome oscillation by facilitating the Kid-generated polar ejection force. *Nat. Commun.* **2016**, *7*, 10597. [CrossRef] [PubMed]
204. Ma, N.; Matsunaga, S.; Morimoto, A.; Sakashita, G.; Urano, T.; Uchiyama, S.; Fukui, K. The nuclear scaffold protein SAF-A is required for kinetochore-microtubule attachment and contributes to the targeting of Aurora-A to mitotic spindles. *J. Cell Sci.* **2011**, *124*, 394–404. [CrossRef] [PubMed]
205. Meraldi, P.; Sorger, P.K. A dual role for Bub1 in the spindle checkpoint and chromosome congression. *EMBO J.* **2005**, *24*, 1621–1633. [CrossRef] [PubMed]
206. Ditchfield, C.; Johnson, V.L.; Tighe, A.; Ellston, R.; Haworth, C.; Johnson, T.; Mortlock, A.; Keen, N.; Taylor, S.S. Aurora B couples chromosome alignment with anaphase by targeting BubR1, Mad2, and Cenp-E to kinetochores. *J. Cell Biol.* **2003**, *161*, 267–280. [CrossRef] [PubMed]
207. Xu, P.; Raetz, E.A.; Kitagawa, M.; Virshup, D.M.; Lee, S.H. BUBR1 recruits PP2A via the B56 family of targeting subunits to promote chromosome congression. *Biol. Open* **2013**, *2*, 479–486. [CrossRef] [PubMed]
208. Elowe, S.; Dulla, K.; Uldschmid, A.; Li, X.; Dou, Z.; Nigg, E.A. Uncoupling of the spindle-checkpoint and chromosome-congression functions of BubR1. *J. Cell Sci.* **2010**, *123*, 84–94. [CrossRef]

209. Itoh, G.; Sugino, S.; Ikeda, M.; Mizuguchi, M.; Kanno, S.-i.; Amin, M.A.; Iemura, K.; Yasui, A.; Hirota, T.; Tanaka, K. Nucleoporin Nup188 is required for chromosome alignment in mitosis. *Cancer Sci.* **2013**, *104*, 871–879. [CrossRef] [PubMed]
210. Holt, S.V.; Vergnolle, M.A.S.; Hussein, D.; Wozniak, M.J.; Allan, V.J.; Taylor, S.S. Silencing Cenp-F weakens centromeric cohesion, prevents chromosome alignment and activates the spindle checkpoint. *J. Cell Sci.* **2005**, *118*, 4889–4900. [CrossRef] [PubMed]
211. Serio, G.; Margaria, V.; Jensen, S.; Oldani, A.; Bartek, J.; Bussolino, F.; Lanzetti, L. Small GTPase Rab5 participates in chromosome congression and regulates localization of the centromere-associated protein CENP-F to kinetochores. *Proc. Natl. Acad. Sci. USA* **2011**, *108*, 17337–17342. [CrossRef] [PubMed]
212. Yang, Z.; Guo, J.; Chen, Q.; Ding, C.; Du, J.; Zhu, X. Silencing mitosin induces misaligned chromosomes, premature chromosome decondensation before anaphase onset, and mitotic cell death. *Mol. Cell. Biol.* **2005**, *25*, 4062–4074. [CrossRef] [PubMed]
213. De Luca, M.; Lavia, P.; Guarguaglini, G. A functional interplay between Aurora-A, Plk1 and TPX2 at spindle poles: Plk1 controls centrosomal localization of Aurora-A and TPX2 spindle association. *Cell Cycle* **2006**, *5*, 296–303. [CrossRef] [PubMed]
214. Sumara, I.; Giménez-Abián, J.F.; Gerlich, D.; Hirota, T.; Kraft, C.; de la Torre, C.; Ellenberg, J.; Peters, J.-M. Roles of polo-like kinase 1 in the assembly of functional mitotic spindles. *Curr. Biol.* **2004**, *14*, 1712–1722. [CrossRef] [PubMed]
215. Neumann, B.; Held, M.; Liebel, U.; Erfle, H.; Rogers, P.; Pepperkok, R.; Ellenberg, J. High-throughput RNAi screening by time-lapse imaging of live human cells. *Nat. Methods* **2006**, *3*, 385–390. [CrossRef] [PubMed]
216. Nishino, M.; Kurasawa, Y.; Evans, R.; Lin, S.-H.; Brinkley, B.R.; Yu-Lee, L.-Y. NudC is required for Plk1 targeting to the kinetochore and chromosome congression. *Curr. Biol.* **2006**, *16*, 1414–1421. [CrossRef] [PubMed]
217. Chuang, C.; Pan, J.; Hawke, D.H.; Lin, S.H.; Yu-Lee, L.Y. NudC deacetylation regulates mitotic progression. *PLoS ONE* **2013**, *8*, e73841. [CrossRef] [PubMed]
218. Gambe, A.E.; Matsunaga, S.; Takata, H.; Ono-Maniwa, R.; Baba, A.; Uchiyama, S.; Fukui, K. A nucleolar protein RRS1 contributes to chromosome congression. *FEBS Lett.* **2009**, *583*, 1951–1956. [CrossRef] [PubMed]
219. Ma, N.; Matsunaga, S.; Takata, H.; Ono-Maniwa, R.; Uchiyama, S.; Fukui, K. Nucleolin functions in nucleolus formation and chromosome congression. *J. Cell Sci.* **2007**, *120*, 2091–2105. [CrossRef] [PubMed]
220. Zhang, L.; Iyer, J.; Chowdhury, A.; Ji, M.; Xiao, L.; Yang, S.; Chen, Y.; Tsai, M.-Y.; Dong, J. KIBRA regulates aurora kinase activity and is required for precise chromosome alignment during mitosis. *J. Biol. Chem.* **2012**, *287*, 34069–34077. [CrossRef] [PubMed]
221. Jang, C.-Y.; Wong, J.; Coppinger, J.A.; Seki, A.; Yates, J.R.; Fang, G. DDA3 recruits microtubule depolymerase Kif2a to spindle poles and controls spindle dynamics and mitotic chromosome movement. *J. Cell Biol.* **2008**, *181*, 255–267. [CrossRef]
222. Jang, C.-Y.; Fang, G. DDA3 associates with MCAK and controls chromosome congression. *Biochem. Biophys. Res. Commun.* **2011**, *407*, 610–614. [CrossRef] [PubMed]
223. Park, S.J. Huntingtin-interacting protein 1-related is required for accurate congression and segregation of chromosomes. *BMB Rep.* **2010**, *43*, 795–800. [CrossRef] [PubMed]
224. Amin, M.A.; Matsunaga, S.; Uchiyama, S.; Fukui, K. Depletion of nucleophosmin leads to distortion of nucleolar and nuclear structures in HeLa cells. *Biochem. J.* **2008**, *415*, 345–351. [CrossRef] [PubMed]
225. Frémont, S.; Gérard, A.; Galloux, M.; Janvier, K.; Karess, R.E.; Berlioz-Torrent, C. Beclin-1 is required for chromosome congression and proper outer kinetochore assembly. *EMBO Rep.* **2013**, *14*, 364–372.
226. Ritchie, K.; Seah, C.; Moulin, J.; Isaac, C.; Dick, F.; Bérubé, N.G. Loss of ATRX leads to chromosome cohesion and congression defects. *J. Cell Biol.* **2008**, *180*, 315–324. [CrossRef] [PubMed]
227. Santamaria, A.; Nagel, S.; Sillje, H.H.W.; Nigg, E.A. The spindle protein CHICA mediates localization of the chromokinesin Kid to the mitotic spindle. *Curr. Biol.* **2008**, *18*, 723–729. [CrossRef] [PubMed]
228. Dunsch, A.K.; Hammond, D.; Lloyd, J.; Schermelleh, L.; Gruneberg, U.; Barr, F.A. Dynein light chain 1 and a spindle-associated adaptor promote dynein asymmetry and spindle orientation. *J. Cell Biol.* **2012**, *198*, 1039–1054. [CrossRef] [PubMed]
229. Kukkonen-Macchi, A.; Sicora, O.; Kaczynska, K.; Oetken-Lindholm, C.; Pouwels, J.; Laine, L.; Kallio, M.J. Loss of p38gamma MAPK induces pleiotropic mitotic defects and massive cell death. *J. Cell Sci.* **2011**, *124*, 216–227. [CrossRef] [PubMed]

230. Archinti, M.; Lacasa, C.; Teixidó-Travesa, N.; Lüders, J. SPICE—A previously uncharacterized protein required for centriole duplication and mitotic chromosome congression. *J. Cell Sci.* **2010**, *123*, 3039–3046. [CrossRef] [PubMed]
231. Li, Y.; Yu, W.; Liang, Y.; Zhu, X. Kinetochore dynein generates a poleward pulling force to facilitate congression and full chromosome alignment. *Cell Res.* **2007**, *17*, 701–712. [CrossRef] [PubMed]
232. Yang, Z.; Tulu, U.S.; Wadsworth, P.; Rieder, C.L. Kinetochore dynein is required for chromosome motion and congression independent of the spindle checkpoint. *Curr. Biol.* **2007**, *17*, 973–980. [CrossRef] [PubMed]
233. Shrestha, R.L.; Tamura, N.; Fries, A.; Levin, N.; Clark, J.; Draviam, V.M. TAO1 kinase maintains chromosomal stability by facilitating proper congression of chromosomes. *Open Biol.* **2014**, *4*, 130108. [CrossRef] [PubMed]
234. Carleton, M.; Mao, M.; Biery, M.; Warrener, P.; Kim, S.; Buser, C.; Marshall, C.G.; Fernandes, C.; Annis, J.; Linsley, P.S. RNA Interference-Mediated Silencing of Mitotic Kinesin KIF14 Disrupts Cell Cycle Progression and Induces Cytokinesis Failure. *Mol. Cell. Biol.* **2006**, *26*, 3853–3863. [CrossRef] [PubMed]
235. Prendergast, L.; van Vuuren, C.; Kaczmarczyk, A.; Doering, V.; Hellwig, D.; Quinn, N.; Hoischen, C.; Diekmann, S.; Sullivan, K.F. Premitotic Assembly of Human CENPs -T and -W Switches Centromeric Chromatin to a Mitotic State. *PLoS Biol.* **2011**, *9*, e1001082. [CrossRef] [PubMed]
236. Kaczmarczyk, A.; Sullivan, K.F. CENP-W Plays a Role in Maintaining Bipolar Spindle Structure. *PLoS ONE* **2014**, *9*, e106464. [CrossRef] [PubMed]
237. Chun, Y.; Kim, R.; Lee, S. Centromere Protein (CENP)-W Interacts with Heterogeneous Nuclear Ribonucleoprotein (hnRNP) U and May Contribute to Kinetochore-Microtubule Attachment in Mitotic Cells. *PLoS ONE* **2016**, *11*, e0149127. [CrossRef] [PubMed]
238. Foltz, D.R.; Jansen, L.E.; Black, B.E.; Bailey, A.O.; Yates, J.R., 3rd; Cleveland, D.W. The human CENP-A centromeric nucleosome-associated complex. *Nat. Cell Biol.* **2006**, *8*, 458–469. [CrossRef] [PubMed]
239. McClelland, S.E.; Borusu, S.; Amaro, A.C.; Winter, J.R.; Belwal, M.; McAinsh, A.D.; Meraldi, P. The CENP-A NAC/CAD kinetochore complex controls chromosome congression and spindle bipolarity. *EMBO J.* **2007**, *26*, 5033–5047. [CrossRef] [PubMed]
240. McAinsh, A.D.; Meraldi, P.; Draviam, V.M.; Toso, A.; Sorger, P.K. The human kinetochore proteins Nnf1R and Mcm21R are required for accurate chromosome segregation. *EMBO J.* **2006**, *25*, 4033–4049. [CrossRef] [PubMed]
241. Bancroft, J.; Auckland, P.; Samora, C.P.; McAinsh, A.D. Chromosome congression is promoted by CENP-Q- and CENP-E-dependent pathways. *J. Cell Sci.* **2015**, *128*, 171–184. [CrossRef] [PubMed]
242. Hua, S.; Wang, Z.; Jiang, K.; Huang, Y.; Ward, T.; Zhao, L.; Dou, Z.; Yao, X. CENP-U Cooperates with Hec1 to Orchestrate Kinetochore-Microtubule Attachment. *J. Biol. Chem.* **2011**, *286*, 1627–1638. [CrossRef] [PubMed]
243. Basilico, F.; Maffini, S.; Weir, J.R.; Prumbaum, D.; Rojas, A.M.; Zimniak, T.; De Antoni, A.; Jeganathan, S.; Voss, B.; van Gerwen, S.; et al. The pseudo GTPase CENP-M drives human kinetochore assembly. *Elife* **2014**, *3*, e02978. [CrossRef] [PubMed]
244. Zhu, M.; Wang, F.; Yan, F.; Yao, P.Y.; Du, J.; Gao, X.; Wang, X.; Wu, Q.; Ward, T.; Li, J.; et al. Septin 7 Interacts with Centromere-associated Protein E and Is Required for Its Kinetochore Localization. *J. Biol. Chem.* **2008**, *283*, 18916–18925. [CrossRef]
245. Milev, M.P.; Hasaj, B.; Saint-Dic, D.; Snounou, S.; Zhao, Q.; Sacher, M. TRAMM/TrappC12 plays a role in chromosome congression, kinetochore stability, and CENP-E recruitment. *J. Cell Biol.* **2015**, *209*, 221–234. [CrossRef] [PubMed]
246. Liu, X.; Zheng, H.; Qu, C.-K. Protein tyrosine phosphatase Shp2 (Ptpn11) plays an important role in maintenance of chromosome stability. *Cancer Res.* **2012**, *72*, 5296–5306. [CrossRef] [PubMed]
247. Liu, X.; Zheng, H.; Li, X.; Wang, S.; Meyerson, H.J.; Yang, W.; Neel, B.G.; Qu, C.-K. Gain-of-function mutations of Ptpn11 (Shp2) cause aberrant mitosis and increase susceptibility to DNA damage-induced malignancies. *Proc. Natl. Acad. Sci. USA* **2016**, *113*, 984–989. [CrossRef] [PubMed]
248. Porter, I.M.; McClelland, S.E.; Khoudoli, G.A.; Hunter, C.J.; Andersen, J.S.; McAinsh, A.D.; Blow, J.J.; Swedlow, J.R. Bod1, a novel kinetochore protein required for chromosome biorientation. *J. Cell Biol.* **2007**, *179*, 187–197. [CrossRef] [PubMed]
249. Porter, I.M.; Schleicher, K.; Porter, M.; Swedlow, J.R. Bod1 regulates protein phosphatase 2A at mitotic kinetochores. *Nat. Commun.* **2013**, *4*, 2677. [CrossRef] [PubMed]

250. He, J.; Zhang, Z.; Ouyang, M.; Yang, F.; Hao, H.; Lamb, K.L.; Yang, J.; Yin, Y.; Shen, W.H. PTEN regulates EG5 to control spindle architecture and chromosome congression during mitosis. *Nat. Commun.* **2016**, *7*, 12335. [CrossRef] [PubMed]
251. Park, Y.Y.; Nam, H.-J.; Do, M.; Lee, J.-H. The p90 ribosomal S6 kinase 2 specifically affects mitotic progression by regulating the basal level, distribution and stability of mitotic spindles. *Exp. Mol. Med.* **2016**, *48*, e250. [CrossRef] [PubMed]
252. Vigneron, S.; Brioudes, E.; Burgess, A.; Labbé, J.C.; Lorca, T.; Castro, A. RSK2 is a kinetochore-associated protein that participates in the spindle assembly checkpoint. *Oncogene* **2010**, *29*, 3566–3574. [CrossRef] [PubMed]
253. Willard, F.S.; Crouch, M.F. MEK, ERK, and p90RSK are present on mitotic tubulin in Swiss 3T3 cells: A role for the MAP kinase pathway in regulating mitotic exit. *Cell. Signal.* **2001**, *13*, 653–664. [CrossRef]
254. Hashizume, C.; Moyori, A.; Kobayashi, A.; Yamakoshi, N.; Endo, A.; Wong, R.W. Nucleoporin Nup62 maintains centrosome homeostasis. *Cell Cycle* **2013**, *12*, 3804–3816. [CrossRef] [PubMed]
255. Wu, Z.; Jin, Z.; Zhang, X.; Shen, N.; Wang, J.; Zhao, Y.; Mei, L. Nup62, associated with spindle microtubule rather than spindle matrix, is involved in chromosome alignment and spindle assembly during mitosis. *Cell Biol. Int.* **2016**, *40*, 968–975. [CrossRef] [PubMed]
256. Kwon, H.J.; Park, J.E.; Song, H.; Jang, C.-Y. DDA3 and Mdp3 modulate Kif2a recruitment onto the mitotic spindle to control minus-end spindle dynamics. *J. Cell Sci.* **2016**, *129*, 2719–2725. [CrossRef] [PubMed]
257. Kim, S.; Jang, C.-Y. ANKRD53 interacts with DDA3 and regulates chromosome integrity during mitosis. *Biochem. Biophys. Res. Commun.* **2016**, *470*, 484–491. [CrossRef] [PubMed]
258. Koliou, X.; Fedonidis, C.; Kalpachidou, T.; Mangoura, D. Nuclear import mechanism of neurofibromin for localization on the spindle and function in chromosome congression. *J. Neurochem.* **2016**, *136*, 78–91. [CrossRef] [PubMed]
259. O'Regan, L.; Sampson, J.; Richards, M.W.; Knebel, A.; Roth, D.; Hood, F.E.; Straube, A.; Royle, S.J.; Bayliss, R.; Fry, A.M. Hsp72 is targeted to the mitotic spindle by Nek6 to promote K-fiber assembly and mitotic progression. *J. Cell Biol.* **2015**, *209*, 349–358. [CrossRef] [PubMed]
260. de Souza, E.E.; Hehnly, H.; Perez, A.M.; Meirelles, G.V.; Smetana, J.H.C.; Doxsey, S.; Kobarg, J. Human Nek7-interactor RGS2 is required for mitotic spindle organization. *Cell Cycle* **2015**, *14*, 656–667. [CrossRef] [PubMed]
261. Foley, E.A.; Maldonado, M.; Kapoor, T.M. Formation of stable attachments between kinetochores and microtubules depends on the B56-PP2A phosphatase. *Nat. Cell Biol.* **2011**, *13*, 1265–1271. [CrossRef] [PubMed]
262. Kitajima, T.S.; Sakuno, T.; Ishiguro, K.-i.; Iemura, S.-i.; Natsume, T.; Kawashima, S.A.; Watanabe, Y. Shugoshin collaborates with protein phosphatase 2A to protect cohesin. *Nature* **2006**, *441*, 46–52. [CrossRef] [PubMed]
263. Jaramillo-Lambert, A.; Hao, J.; Xiao, H.; Li, Y.; Han, Z.; Zhu, W. Acidic nucleoplasmic DNA-binding protein (And-1) controls chromosome congression by regulating the assembly of centromere protein A (CENP-A) at centromeres. *J. Biol. Chem.* **2013**, *288*, 1480–1488. [CrossRef] [PubMed]
264. Lee, M.H.; Lin, L.; Equilibrina, I.; Uchiyama, S.; Matsunaga, S.; Fukui, K. ASURA (PHB2) Is Required for Kinetochore Assembly and Subsequent Chromosome Congression. *Acta Histochem. Cytochem.* **2011**, *44*, 247–258. [CrossRef] [PubMed]
265. Oh, H.J.; Kim, M.J.; Song, S.J.; Kim, T.; Lee, D.; Kwon, S.-H.; Choi, E.-J.; Lim, D.-S. MST1 limits the kinase activity of aurora B to promote stable kinetochore-microtubule attachment. *Curr. Biol.* **2010**, *20*, 416–422. [CrossRef] [PubMed]
266. Shimizu, H.; Nagamori, I.; Yabuta, N.; Nojima, H. GAK, a regulator of clathrin-mediated membrane traffic, also controls centrosome integrity and chromosome congression. *J. Cell Sci.* **2009**, *122*, 3145–3152. [CrossRef] [PubMed]
267. Zhuo, X.; Guo, X.; Zhang, X.; Jing, G.; Wang, Y.; Chen, Q.; Jiang, Q.; Liu, J.; Zhang, C. Usp16 regulates kinetochore localization of Plk1 to promote proper chromosome alignment in mitosis. *J. Cell Biol.* **2015**, *210*, 727–735. [CrossRef] [PubMed]
268. Pfarr, C.M.; Coue, M.; Grissom, P.M.; Hays, T.S.; Porter, M.E.; McIntosh, J.R. Cytoplasmic dynein is localized to kinetochores during mitosis. *Nature* **1990**, *345*, 263–265. [CrossRef] [PubMed]
269. Steuer, E.R.; Wordeman, L.; Schroer, T.A.; Sheetz, M.P. Localization of cytoplasmic dynein to mitotic spindles and kinetochores. *Nature* **1990**, *345*, 266–268. [CrossRef] [PubMed]
270. Maiato, H.; Lince-Faria, M. The perpetual movements of anaphase. *Cell Mol. Life Sci.* **2010**, *67*, 2251–2269. [CrossRef] [PubMed]

271. King, J.M.; Hays, T.S.; Nicklas, R.B. Dynein is a transient kinetochore component whose binding is regulated by microtubule attachment, not tension. *J. Cell Biol.* **2000**, *151*, 739–748. [CrossRef] [PubMed]
272. Wojcik, E.; Basto, R.; Serr, M.; Scaerou, F.; Karess, R.; Hays, T. Kinetochore dynein: Its dynamics and role in the transport of the Rough deal checkpoint protein. *Nat. Cell Biol.* **2001**, *3*, 1001–1007. [CrossRef] [PubMed]
273. Lombillo, V.A.; Nislow, C.; Yen, T.J.; Gelfand, V.I.; McIntosh, J.R. Antibodies to the kinesin motor domain and CENP-E inhibit microtubule depolymerization-dependent motion of chromosomes in vitro. *J. Cell Biol.* **1995**, *128*, 107–115. [CrossRef] [PubMed]
274. Lombillo, V.A.; Stewart, R.J.; McIntosh, J.R. Minus-end-directed motion of kinesin-coated microspheres driven by microtubule depolymerization. *Nature* **1995**, *373*, 161–164. [CrossRef] [PubMed]
275. Schneider, B. Über die umordunug der chromosomen bei der mitose. *Z. Zellf Mikr Anat.* **1933**, *17*, 255–312. [CrossRef]
276. Bajer, A. Cine-micrographic studies on mitosis in endosperm I. *Acta Soc. Bot. Poloniae* **1954**, *23*, 383–412. [CrossRef]
277. Bajer, A.; Molè-Bajer, J. Cine-micrographic studies on mitosis in endosperm. II. *Chromosoma* **1956**, *7*, 558–607. [CrossRef]
278. Uretz, R.B.; Bloom, W.; Zirkle, R.E. Irradiation of parts of individual cells. II. Effects of an ultraviolet microbeam focused on parts of chromosomes. *Science* **1954**, *120*, 197–199. [CrossRef] [PubMed]
279. Bloom, W.; Zirkle, R.E.; Uretz, R.B. Irradiation of parts of individual cells. III. Effects of chromosomal and extrachromosomal irradiation on chromosome movements. *Ann. N. Y. Acad. Sci.* **1955**, *59*, 503–513. [CrossRef] [PubMed]
280. Zirkle, R.E. Partial-cell irradiation. *Adv. Biol. Med. Phys.* **1957**, *5*, 103–146. [PubMed]
281. Rickards, G.K. Prophase chromosome movements in living house cricket spermatocytes and their relationship to prometaphase, anaphase and granule movements. *Chromosoma* **1975**, *49*, 407–455. [CrossRef] [PubMed]
282. Roos, U.P. Light and electron microscopy of rat kangaroo cells in mitosis. III. Patterns of chromosome behavior during prometaphase. *Chromosoma* **1976**, *54*, 363–385. [CrossRef] [PubMed]
283. Rieder, C.L.; Alexander, S.P. Kinetochores are transported poleward along a single astral microtubule during chromosome attachment to the spindle in newt lung cells. *J. Cell Biol.* **1990**, *110*, 81–95. [CrossRef] [PubMed]
284. Merdes, A.; De Mey, J. The mechanism of kinetochore-spindle attachment and polewards movement analyzed in PtK2 cells at the prophase-prometaphase transition. *Eur. J. Cell Biol.* **1990**, *53*, 313–325. [PubMed]
285. Wordeman, L.; Steuer, E.R.; Sheetz, M.P.; Mitchison, T. Chemical subdomains within the kinetochore domain of isolated CHO mitotic chromosomes. *J. Cell Biol.* **1991**, *114*, 285–294. [CrossRef] [PubMed]
286. Vorozhko, V.V.; Emanuele, M.J.; Kallio, M.J.; Stukenberg, P.T.; Gorbsky, G.J. Multiple mechanisms of chromosome movement in vertebrate cells mediated through the Ndc80 complex and dynein/dynactin. *Chromosoma* **2008**, *117*, 169–179. [CrossRef] [PubMed]
287. Mitchison, T.J.; Kirschner, M.W. Properties of the kinetochore in vitro. II. Microtubule capture and ATP-dependent translocation. *J. Cell Biol.* **1985**, *101*, 766–777. [CrossRef] [PubMed]
288. Hyman, A.A.; Mitchison, T.J. Two different microtubule-based motor activities with opposite polarities in kinetochores. *Nature* **1991**, *351*, 206–211. [CrossRef] [PubMed]
289. Yen, T.J.; Compton, D.A.; Wise, D.; Zinkowski, R.P.; Brinkley, B.R.; Earnshaw, W.C.; Cleveland, D.W. CENP-E, a novel human centromere-associated protein required for progression from metaphase to anaphase. *EMBO J.* **1991**, *10*, 1245–1254. [PubMed]
290. Yen, T.J.; Li, G.; Schaar, B.T.; Szilak, I.; Cleveland, D.W. CENP-E is a putative kinetochore motor that accumulates just before mitosis. *Nature* **1992**, *359*, 536–539. [CrossRef] [PubMed]
291. Yao, X.; Anderson, K.L.; Cleveland, D.W. The microtubule-dependent motor centromere-associated protein E (CENP-E) is an integral component of kinetochore corona fibers that link centromeres to spindle microtubules. *J. Cell Biol.* **1997**, *139*, 435–447. [CrossRef] [PubMed]
292. Cooke, C.A.; Schaar, B.; Yen, T.J.; Earnshaw, W.C. Localization of CENP-E in the fibrous corona and outer plate of mammalian kinetochores from prometaphase through anaphase. *Chromosoma* **1997**, *106*, 446–455. [CrossRef]
293. Wood, K.W.; Sakowicz, R.; Goldstein, L.S.; Cleveland, D.W. CENP-E is a plus end-directed kinetochore motor required for metaphase chromosome alignment. *Cell* **1997**, *91*, 357–366. [CrossRef]
294. Yao, X.; Abrieu, A.; Zheng, Y.; Sullivan, K.F.; Cleveland, D.W. CENP-E forms a link between attachment of spindle microtubules to kinetochores and the mitotic checkpoint. *Nat. Cell Biol.* **2000**, *2*, 484–491. [PubMed]

295. Schaar, B.T.; Chan, G.K.; Maddox, P.; Salmon, E.D.; Yen, T.J. CENP-E function at kinetochores is essential for chromosome alignment. *J. Cell Biol.* **1997**, *139*, 1373–1382. [CrossRef] [PubMed]
296. Yucel, J.K.; Marszalek, J.D.; McIntosh, J.R.; Goldstein, L.S.; Cleveland, D.W.; Philp, A.V. CENP-meta, an essential kinetochore kinesin required for the maintenance of metaphase chromosome alignment in Drosophila. *J. Cell Biol.* **2000**, *150*, 1–11. [CrossRef] [PubMed]
297. McEwen, B.F.; Chan, G.K.; Zubrowski, B.; Savoian, M.S.; Sauer, M.T.; Yen, T.J. CENP-E is essential for reliable bioriented spindle attachment, but chromosome alignment can be achieved via redundant mechanisms in mammalian cells. *Mol. Biol. Cell* **2001**, *12*, 2776–2789. [CrossRef] [PubMed]
298. Putkey, F.R.; Cramer, T.; Morphew, M.K.; Silk, A.D.; Johnson, R.S.; McIntosh, J.R.; Cleveland, D.W. Unstable kinetochore-microtubule capture and chromosomal instability following deletion of CENP-E. *Dev. Cell* **2002**, *3*, 351–365. [CrossRef]
299. Cleveland, D.W.; Mao, Y.; Sullivan, K.F. Centromeres and kinetochores: from epigenetics to mitotic checkpoint signaling. *Cell* **2003**, *112*, 407–421. [CrossRef]
300. Kapoor, T.M.; Lampson, M.A.; Hergert, P.; Cameron, L.; Cimini, D.; Salmon, E.D.; McEwen, B.F.; Khodjakov, A. Chromosomes can congress to the metaphase plate before biorientation. *Science* **2006**, *311*, 388–391. [CrossRef] [PubMed]
301. Espeut, J.; Gaussen, A.; Bieling, P.; Morin, V.; Prieto, S.; Fesquet, D.; Surrey, T.; Abrieu, A. Phosphorylation relieves autoinhibition of the kinetochore motor Cenp-E. *Mol. Cell* **2008**, *29*, 637–643. [CrossRef] [PubMed]
302. Kim, Y.; Heuser, J.E.; Waterman, C.M.; Cleveland, D.W. CENP-E combines a slow, processive motor and a flexible coiled coil to produce an essential motile kinetochore tether. *J. Cell Biol.* **2008**, *181*, 411–419. [CrossRef] [PubMed]
303. Gudimchuk, N.; Vitre, B.; Kim, Y.; Kiyatkin, A.; Cleveland, D.W.; Ataullakhanov, F.I.; Grishchuk, E.L. Kinetochore kinesin CENP-E is a processive bi-directional tracker of dynamic microtubule tips. *Nat. Cell Biol.* **2013**, *15*, 1079–1088. [CrossRef] [PubMed]
304. Vitre, B.; Gudimchuk, N.; Borda, R.; Kim, Y.; Heuser, J.E.; Cleveland, D.W.; Grishchuk, E.L. Kinetochore-microtubule attachment throughout mitosis potentiated by the elongated stalk of the kinetochore kinesin CENP-E. *Mol. Biol. Cell* **2014**, *25*, 2272–2281. [CrossRef] [PubMed]
305. Cai, S.; O'Connell, C.B.; Khodjakov, A.; Walczak, C.E. Chromosome congression in the absence of kinetochore fibres. *Nat. Cell Biol.* **2009**, *11*, 832–838. [CrossRef] [PubMed]
306. Roos, U.P. Light and electron microscopy of rat kangaroo cells in mitosis. II. Kinetochore structure and function. *Chromosoma* **1973**, *41*, 195–220. [CrossRef] [PubMed]
307. Magidson, V.; Paul, R.; Yang, N.; Ault, J.G.; O'Connell, C.B.; Tikhonenko, I.; McEwen, B.F.; Mogilner, A.; Khodjakov, A. Adaptive changes in the kinetochore architecture facilitate proper spindle assembly. *Nat. Cell Biol.* **2015**, *17*, 1134–1144. [CrossRef] [PubMed]
308. Maddox, P.S.; Oegema, K.; Desai, A.; Cheeseman, I.M. Holoer than thou: Chromosome segregation and kinetochore function in *C. elegans*. *Chromosome Res.* **2004**, *12*, 641–653. [CrossRef] [PubMed]
309. Powers, J.; Rose, D.J.; Saunders, A.; Dunkelbarger, S.; Strome, S.; Saxton, W.M. Loss of KLP-19 polar ejection force causes misorientation and missegregation of holocentric chromosomes. *J. Cell Biol.* **2004**, *166*, 991–1001. [CrossRef] [PubMed]
310. Rieder, C.L.; Cole, R.W.; Khodjakov, A.; Sluder, G. The checkpoint delaying anaphase in response to chromosome monoorientation is mediated by an inhibitory signal produced by unattached kinetochores. *J. Cell Biol.* **1995**, *130*, 941–948. [CrossRef]
311. Brenner, S.L.; Liaw, L.H.; Berns, M.W. Laser microirradiation of kinetochores in mitotic PtK2 cells: Chromatid separation and micronucleus formation. *Cell Biophys.* **1980**, *2*, 139–152. [CrossRef] [PubMed]
312. Brinkley, B.R.; Zinkowski, R.P.; Mollon, W.L.; Davis, F.M.; Pisegna, M.A.; Pershouse, M.; Rao, P.N. Movement and segregation of kinetochores experimentally detached from mammalian chromosomes. *Nature* **1988**, *336*, 251–254. [CrossRef] [PubMed]
313. Wise, D.A.; Brinkley, B.R. Mitosis in cells with unreplicated genomes (MUGs): Spindle assembly and behavior of centromere fragments. *Cell Motil Cytoskeleton* **1997**, *36*, 291–302. [CrossRef]
314. O'Connell, C.B.; Loncarek, J.; Hergert, P.; Kourtidis, A.; Conklin, D.S.; Khodjakov, A. The spindle assembly checkpoint is satisfied in the absence of interkinetochore tension during mitosis with unreplicated genomes. *J. Cell Biol.* **2008**, *183*, 29–36. [CrossRef] [PubMed]

315. Khodjakov, A.; Cole, R.W.; McEwen, B.F.; Buttle, K.F.; Rieder, C.L. Chromosome fragments possessing only one kinetochore can congress to the spindle equator. *J. Cell Biol.* **1997**, *136*, 229–240. [CrossRef] [PubMed]
316. Barisic, M.; Maiato, H. Dynein prevents erroneous kinetochore-microtubule attachments in mitosis. *Cell Cycle* **2015**, *14*, 3356–3361. [CrossRef] [PubMed]
317. Iemura, K.; Tanaka, K. Chromokinesin Kid and kinetochore kinesin CENP-E differentially support chromosome congression without end-on attachment to microtubules. *Nat. Commun.* **2015**, *6*, 6447. [CrossRef] [PubMed]
318. Nousiainen, M.; Sillјé, H.H.W.; Sauer, G.; Nigg, E.A.; Körner, R. Phosphoproteome analysis of the human mitotic spindle. *Proc. Natl. Acad. Sci. USA* **2006**, *103*, 5391–5396. [CrossRef] [PubMed]
319. Liao, H.; Li, G.; Yen, T.J. Mitotic regulation of microtubule cross-linking activity of CENP-E kinetochore protein. *Science* **1994**, *265*, 394–398. [CrossRef] [PubMed]
320. Zecevic, M.; Catling, A.D.; Eblen, S.T.; Renzi, L.; Hittle, J.C.; Yen, T.J.; Gorbsky, G.J.; Weber, M.J. Active MAP kinase in mitosis: localization at kinetochores and association with the motor protein CENP-E. *J. Cell Biol.* **1998**, *142*, 1547–1558. [CrossRef] [PubMed]
321. Kim, Y.; Holland, A.J.; Lan, W.; Cleveland, D.W. Aurora kinases and protein phosphatase 1 mediate chromosome congression through regulation of CENP-E. *Cell* **2010**, *142*, 444–455. [CrossRef] [PubMed]
322. Whyte, J.; Bader, J.R.; Tauhata, S.B.; Raycroft, M.; Hornick, J.; Pfister, K.K.; Lane, W.S.; Chan, G.K.; Hinchcliffe, E.H.; Vaughan, P.S.; et al. Phosphorylation regulates targeting of cytoplasmic dynein to kinetochores during mitosis. *J. Cell Biol.* **2008**, *183*, 819–834. [CrossRef] [PubMed]
323. Bader, J.R.; Kasuboski, J.M.; Winding, M.; Vaughan, P.S.; Hinchcliffe, E.H.; Vaughan, K.T. Polo-like kinase1 is required for recruitment of dynein to kinetochores during mitosis. *J Biol. Chem.* **2011**, *286*, 20769–20777. [CrossRef] [PubMed]
324. Kardon, J.R.; Reck-Peterson, S.L.; Vale, R.D. Regulation of the processivity and intracellular localization of Saccharomyces cerevisiae dynein by dynactin. *Proc. Natl. Acad. Sci. USA* **2009**, *106*, 5669–5674. [CrossRef] [PubMed]
325. McKenney, R.J.; Huynh, W.; Tanenbaum, M.E.; Bhabha, G.; Vale, R.D. Activation of cytoplasmic dynein motility by dynactin-cargo adapter complexes. *Science* **2014**, *345*, 337–341. [CrossRef] [PubMed]
326. Zhang, X.-D.; Goeres, J.; Zhang, H.; Yen, T.J.; Porter, A.C.G.; Matunis, M.J. SUMO-2/3 modification and binding regulate the association of CENP-E with kinetochores and progression through mitosis. *Mol. Cell* **2008**, *29*, 729–741. [CrossRef] [PubMed]
327. Ashar, H.R.; James, L.; Gray, K.; Carr, D.; Black, S.; Armstrong, L.; Bishop, W.R.; Kirschmeier, P. Farnesyl transferase inhibitors block the farnesylation of CENP-E and CENP-F and alter the association of CENP-E with the microtubules. *J. Biol. Chem.* **2000**, *275*, 30451–30457. [CrossRef] [PubMed]
328. Schafer-Hales, K.; Iaconelli, J.; Snyder, J.P.; Prussia, A.; Nettles, J.H.; El-Naggar, A.; Khuri, F.R.; Giannakakou, P.; Marcus, A.I. Farnesyl transferase inhibitors impair chromosomal maintenance in cell lines and human tumors by compromising CENP-E and CENP-F function. *Mol. Cancer Ther.* **2007**, *6*, 1317–1328. [CrossRef] [PubMed]
329. Crespo, N.C.; Ohkanda, J.; Yen, T.J.; Hamilton, A.D.; Sebti, S.M. The farnesyltransferase inhibitor, FTI-2153, blocks bipolar spindle formation and chromosome alignment and causes prometaphase accumulation during mitosis of human lung cancer cells. *J. Biol. Chem.* **2001**, *276*, 16161–16167. [CrossRef] [PubMed]
330. Crespo, N.C.; Delarue, F.; Ohkanda, J.; Carrico, D.; Hamilton, A.D.; Sebti, S.M. The farnesyltransferase inhibitor, FTI-2153, inhibits bipolar spindle formation during mitosis independently of transformation and Ras and p53 mutation status. *Cell Death Differ.* **2002**, *9*, 702–709. [CrossRef] [PubMed]
331. Hussein, D.; Taylor, S.S. Farnesylation of Cenp-F is required for G2/M progression and degradation after mitosis. *J. Cell Sci.* **2002**, *115*, 3403–3414. [PubMed]
332. Brown, K.D.; Coulson, R.M.; Yen, T.J.; Cleveland, D.W. Cyclin-like accumulation and loss of the putative kinetochore motor CENP-E results from coupling continuous synthesis with specific degradation at the end of mitosis. *J. Cell Biol.* **1994**, *125*, 1303–1312. [CrossRef] [PubMed]
333. Gurden, M.D.J.; Holland, A.J.; van Zon, W.; Tighe, A.; Vergnolle, M.A.; Andres, D.A.; Spielmann, H.P.; Malumbres, M.; Wolthuis, R.M.F.; Cleveland, D.W.; et al. Cdc20 is required for the post-anaphase, KEN-dependent degradation of centromere protein F. *J. Cell Sci.* **2010**, *123*, 321–330. [CrossRef] [PubMed]

334. Holland, A.J.; Reis, R.M.; Niessen, S.; Pereira, C.; Andres, D.A.; Spielmann, H.P.; Cleveland, D.W.; Desai, A.; Gassmann, R. Preventing farnesylation of the dynein adaptor Spindly contributes to the mitotic defects caused by farnesyltransferase inhibitors. *Mol. Biol. Cell* **2015**, *26*, 1845–1856. [CrossRef] [PubMed]
335. Moudgil, D.K.; Westcott, N.; Famulski, J.K.; Patel, K.; Macdonald, D.; Hang, H.; Chan, G.K.T. A novel role of farnesylation in targeting a mitotic checkpoint protein, human Spindly, to kinetochores. *J. Cell Biol.* **2015**, *208*, 881–896. [CrossRef] [PubMed]
336. Hewitt, L.; Tighe, A.; Santaguida, S.; White, A.M.; Jones, C.D.; Musacchio, A.; Green, S.; Taylor, S.S. Sustained Mps1 activity is required in mitosis to recruit O-Mad2 to the Mad1-C-Mad2 core complex. *J. Cell Biol.* **2010**, *190*, 25–34. [CrossRef] [PubMed]
337. Jelluma, N.; Brenkman, A.B.; van den Broek, N.J.F.; Cruijsen, C.W.A.; van Osch, M.H.J.; Lens, S.M.A.; Medema, R.H.; Kops, G.J.P.L. Mps1 phosphorylates Borealin to control Aurora B activity and chromosome alignment. *Cell* **2008**, *132*, 233–246. [CrossRef] [PubMed]
338. Maure, J.-F.; Kitamura, E.; Tanaka, T.U. Mps1 kinase promotes sister-kinetochore bi-orientation by a tension-dependent mechanism. *Curr. Biol.* **2007**, *17*, 2175–2182. [CrossRef] [PubMed]
339. Wang, X.; Yu, H.; Xu, L.; Zhu, T.; Zheng, F.; Fu, C.; Wang, Z.; Dou, Z. Dynamic autophosphorylation of mps1 kinase is required for faithful mitotic progression. *PLoS ONE* **2014**, *9*, e104723. [CrossRef] [PubMed]
340. Maciejowski, J.; George, K.A.; Terret, M.-E.; Zhang, C.; Shokat, K.M.; Jallepalli, P.V. Mps1 directs the assembly of Cdc20 inhibitory complexes during interphase and mitosis to control M phase timing and spindle checkpoint signaling. *J. Cell Biol.* **2010**, *190*, 89–100. [CrossRef] [PubMed]
341. Santaguida, S.; Tighe, A.; D'Alise, A.M.; Taylor, S.S.; Musacchio, A. Dissecting the role of MPS1 in chromosome biorientation and the spindle checkpoint through the small molecule inhibitor reversine. *J. Cell Biol.* **2010**, *190*, 73–87. [CrossRef] [PubMed]
342. Abrieu, A.; Magnaghi-Jaulin, L.; Kahana, J.A.; Peter, M.; Castro, A.; Vigneron, S.; Lorca, T.; Cleveland, D.W.; Labbe, J.C. Mps1 is a kinetochore-associated kinase essential for the vertebrate mitotic checkpoint. *Cell* **2001**, *106*, 83–93. [CrossRef]
343. Verhey, K.J.; Gaertig, J. The tubulin code. *Cell Cycle* **2007**, *6*, 2152–2160. [CrossRef] [PubMed]
344. Janke, C. The tubulin code: Molecular components, readout mechanisms, and functions. *J. Cell Biol.* **2014**, *206*, 461–472. [CrossRef] [PubMed]
345. Barisic, M.; Silva e Sousa, R.; Tripathy, S.K.; Magiera, M.M.; Zaytsev, A.V.; Pereira, A.L.; Janke, C.; Grishchuk, E.L.; Maiato, H. Mitosis. Microtubule detyrosination guides chromosomes during mitosis. *Science* **2015**, *348*, 799–803. [CrossRef] [PubMed]
346. Barisic, M.; Maiato, H. The Tubulin Code: A Navigation System for Chromosomes during Mitosis. *Trends Cell Biol.* **2016**, *26*, 766–775. [CrossRef] [PubMed]
347. Park, I.Y.; Powell, R.T.; Tripathi, D.N.; Dere, R.; Ho, T.H.; Blasius, T.L.; Chiang, Y.C.; Davis, I.J.; Fahey, C.C.; Hacker, K.E.; et al. Dual Chromatin and Cytoskeletal Remodeling by SETD2. *Cell* **2016**, *166*, 950–962. [CrossRef] [PubMed]
348. Hammond, J.W.; Huang, C.F.; Kaech, S.; Jacobson, C.; Banker, G.; Verhey, K.J. Posttranslational modifications of tubulin and the polarized transport of kinesin-1 in neurons. *Mol. Biol. Cell* **2010**, *21*, 572–583. [CrossRef] [PubMed]
349. Konishi, Y.; Setou, M. Tubulin tyrosination navigates the kinesin-1 motor domain to axons. *Nat. Neurosci.* **2009**, *12*, 559–567. [CrossRef] [PubMed]
350. Maas, C.; Belgardt, D.; Lee, H.K.; Heisler, F.F.; Lappe-Siefke, C.; Magiera, M.M.; van Dijk, J.; Hausrat, T.J.; Janke, C.; Kneussel, M. Synaptic activation modifies microtubules underlying transport of postsynaptic cargo. *Proc. Natl. Acad. Sci. USA* **2009**, *106*, 8731–8736. [CrossRef] [PubMed]
351. Reed, N.A.; Cai, D.; Blasius, T.L.; Jih, G.T.; Meyhofer, E.; Gaertig, J.; Verhey, K.J. Microtubule acetylation promotes kinesin-1 binding and transport. *Curr. Biol.* **2006**, *16*, 2166–2172. [CrossRef] [PubMed]
352. Kaul, N.; Soppina, V.; Verhey, K.J. Effects of alpha-tubulin K40 acetylation and detyrosination on kinesin-1 motility in a purified system. *Biophys. J.* **2014**, *106*, 2636–2643. [CrossRef] [PubMed]
353. Sirajuddin, M.; Rice, L.M.; Vale, R.D. Regulation of microtubule motors by tubulin isotypes and post-translational modifications. *Nat. Cell Biol.* **2014**, *16*, 335–344. [CrossRef] [PubMed]
354. Bobinnec, Y.; Moudjou, M.; Fouquet, J.P.; Desbruyeres, E.; Edde, B.; Bornens, M. Glutamylation of centriole and cytoplasmic tubulin in proliferating non-neuronal cells. *Cell Motil. Cytoskeleton* **1998**, *39*, 223–232. [CrossRef]

355. Gundersen, G.G.; Bulinski, J.C. Distribution of tyrosinated and nontyrosinated α-tubulin during mitosis. *J. Cell Biol.* **1986**, *102*, 1118–1126. [CrossRef] [PubMed]
356. Gundersen, G.G.; Kalnoski, M.H.; Bulinski, J.C. Distinct populations of microtubules: Tyrosinated and nontyrosinated alpha tubulin are distributed differently in vivo. *Cell* **1984**, *38*, 779–789. [CrossRef]
357. Wilson, P.J.; Forer, A. Effects of nanomolar taxol on crane-fly spermatocyte spindles indicate that acetylation of kinetochore microtubules can be used as a marker of poleward tubulin flux. *Cell Motil. Cytoskeleton* **1997**, *37*, 20–32. [CrossRef]
358. McKenney, R.J.; Huynh, W.; Vale, R.D.; Sirajuddin, M. Tyrosination of alpha-tubulin controls the initiation of processive dynein-dynactin motility. *EMBO J.* **2016**, *35*, 1175–1185. [CrossRef] [PubMed]
359. Nirschl, J.J.; Magiera, M.M.; Lazarus, J.E.; Janke, C.; Holzbaur, E.L. alpha-Tubulin Tyrosination and CLIP-170 Phosphorylation Regulate the Initiation of Dynein-Driven Transport in Neurons. *Cell Rep.* **2016**, *14*, 2637–2652. [CrossRef] [PubMed]
360. Peris, L.; Thery, M.; Faure, J.; Saoudi, Y.; Lafanechere, L.; Chilton, J.K.; Gordon-Weeks, P.; Galjart, N.; Bornens, M.; Wordeman, L.; et al. Tubulin tyrosination is a major factor affecting the recruitment of CAP-Gly proteins at microtubule plus ends. *J. Cell Biol.* **2006**, *174*, 839–849. [CrossRef] [PubMed]
361. Kubo, T.; Yanagisawa, H.A.; Yagi, T.; Hirono, M.; Kamiya, R. Tubulin polyglutamylation regulates axonemal motility by modulating activities of inner-arm dyneins. *Curr. Biol.* **2010**, *20*, 441–445. [CrossRef] [PubMed]
362. Wang, Z.; Sheetz, M.P. The C-terminus of tubulin increases cytoplasmic dynein and kinesin processivity. *Biophys. J.* **2000**, *78*, 1955–1964. [CrossRef]
363. Alper, J.D.; Decker, F.; Agana, B.; Howard, J. The motility of axonemal dynein is regulated by the tubulin code. *Biophys. J.* **2014**, *107*, 2872–2880. [CrossRef] [PubMed]
364. Sardar, H.S.; Gilbert, S.P. Microtubule capture by mitotic kinesin centromere protein E (CENP-E). *J. Biol. Chem.* **2012**, *287*, 24894–24904. [CrossRef] [PubMed]
365. Belyy, V.; Schlager, M.A.; Foster, H.; Reimer, A.E.; Carter, A.P.; Yildiz, A. The mammalian dynein-dynactin complex is a strong opponent to kinesin in a tug-of-war competition. *Nat. Cell Biol.* **2016**, *18*, 1018–1024. [CrossRef] [PubMed]
366. Iniguez, A.; Allard, J. Spatial pattern formation in microtubule post-translational modifications and the tight localization of motor-driven cargo. *J. Math. Biol.* **2016**. [CrossRef] [PubMed]
367. Skiniotis, G.; Cochran, J.C.; Muller, J.; Mandelkow, E.; Gilbert, S.P.; Hoenger, A. Modulation of kinesin binding by the C-termini of tubulin. *EMBO J.* **2004**, *23*, 989–999. [CrossRef] [PubMed]
368. Neumann, E.; Garcia-Saez, I.; DeBonis, S.; Wade, R.H.; Kozielski, F.; Conway, J.F. Human kinetochore-associated kinesin CENP-E visualized at 17 A resolution bound to microtubules. *J. Mol. Biol.* **2006**, *362*, 203–211. [CrossRef] [PubMed]
369. Garcia-Saez, I.; Yen, T.; Wade, R.H.; Kozielski, F. Crystal structure of the motor domain of the human kinetochore protein CENP-E. *J. Mol. Biol.* **2004**, *340*, 1107–1116. [CrossRef] [PubMed]
370. Musinipally, V.; Howes, S.; Alushin, G.M.; Nogales, E. The microtubule binding properties of CENP-E's C-terminus and CENP-F. *J. Mol. Biol.* **2013**, *425*, 4427–4441. [CrossRef] [PubMed]
371. Wang, Q.; Crevenna, A.H.; Kunze, I.; Mizuno, N. Structural basis for the extended CAP-Gly domains of p150(glued) binding to microtubules and the implication for tubulin dynamics. *Proc. Natl. Acad. Sci. USA* **2014**, *111*, 11347–11352. [CrossRef] [PubMed]
372. Yan, S.; Guo, C.; Hou, G.; Zhang, H.; Lu, X.; Williams, J.C.; Polenova, T. Atomic-resolution structure of the CAP-Gly domain of dynactin on polymeric microtubules determined by magic angle spinning NMR spectroscopy. *Proc. Natl. Acad. Sci. USA* **2015**, *112*, 14611–14616. [CrossRef] [PubMed]
373. Pereira, A.J.; Matos, I.; Lince-Faria, M.; Maiato, H. Dissecting mitosis with laser microsurgery and RNAi in Drosophila cells. *Methods Mol. Biol.* **2009**, *545*, 145–164. [PubMed]
374. McIntosh, J.R.; Hepler, P.K.; Van Wie, D.G. Model for mitosis. *Nature* **1969**, *224*, 659–663. [CrossRef]
375. Goode, D. Microtubule turnover as a mechanism of mitosis and its possible evolution. *Biosystems* **1981**, *14*, 271–287. [CrossRef]
376. Margolis, R.L.; Wilson, L.; Keifer, B.I. Mitotic mechanism based on intrinsic microtubule behaviour. *Nature* **1978**, *272*, 450–452. [CrossRef] [PubMed]
377. Nicklas, R.B.; Kubai, D.F.; Hays, T.S. Spindle microtubules and their mechanical associations after micromanipulation in anaphase. *J. Cell Biol.* **1982**, *95*, 91–104. [CrossRef] [PubMed]

378. Mastronarde, D.N.; McDonald, K.L.; Ding, R.; McIntosh, J.R. Interpolar spindle microtubules in PTK cells. *J. Cell Biol.* **1993**, *123*, 1475–1489. [CrossRef] [PubMed]
379. Shimamoto, Y.; Maeda, Y.T.; Ishiwata, S.; Libchaber, A.J.; Kapoor, T.M. Insights into the micromechanical properties of the metaphase spindle. *Cell* **2011**, *145*, 1062–1074. [CrossRef] [PubMed]
380. Vladimirou, E.; McHedlishvili, N.; Gasic, I.; Armond, J.W.; Samora, C.P.; Meraldi, P.; McAinsh, A.D. Nonautonomous Movement of Chromosomes in Mitosis. *Dev. Cell* **2013**, *27*, 60–71. [CrossRef] [PubMed]
381. Pereira, A.J.; Maiato, H. Maturation of the kinetochore-microtubule interface and the meaning of metaphase. *Chromosome Res.* **2012**, *20*, 563–577. [CrossRef] [PubMed]
382. Cross, R.A.; McAinsh, A. Prime movers: The mechanochemistry of mitotic kinesins. *Nat. Rev. Mol. Cell Biol.* **2014**, *15*, 257–271. [CrossRef] [PubMed]
383. Royle, S.J. The role of clathrin in mitotic spindle organisation. *J. Cell Sci.* **2012**, *125*, 19–28. [CrossRef] [PubMed]
384. Maiato, H.; Sampaio, P.; Sunkel, C.E. Microtubule-associated proteins and their essential roles during mitosis. *Int. Rev. Cytol.* **2004**, *241*, 53–153. [PubMed]
385. Hoffman, D.B.; Pearson, C.G.; Yen, T.J.; Howell, B.J.; Salmon, E.D. Microtubule-dependent changes in assembly of microtubule motor proteins and mitotic spindle checkpoint proteins at PtK1 kinetochores. *Mol. Biol. Cell* **2001**, *12*, 1995–2009. [CrossRef] [PubMed]
386. Ohashi, A.; Ohori, M.; Iwai, K.; Nambu, T.; Miyamoto, M.; Kawamoto, T.; Okaniwa, M. A Novel Time-Dependent CENP-E Inhibitor with Potent Antitumor Activity. *PLoS ONE* **2015**, *10*, e0144675. [CrossRef] [PubMed]
387. Gorbsky, G.J. Cohesion fatigue. *Curr. Biol.* **2013**, *23*, R986–R988. [CrossRef] [PubMed]
388. Daum, J.R.; Potapova, T.A.; Sivakumar, S.; Daniel, J.J.; Flynn, J.N.; Rankin, S.; Gorbsky, G.J. Cohesion fatigue induces chromatid separation in cells delayed at metaphase. *Curr. Biol.* **2011**, *21*, 1018–1024. [CrossRef] [PubMed]
389. Bannigan, A.; Lizotte-Waniewski, M.; Riley, M.; Baskin, T.I. Emerging molecular mechanisms that power and regulate the anastral mitotic spindle of flowering plants. *Cell Motil. Cytoskelet.* **2008**, *65*, 1–11. [CrossRef] [PubMed]
390. Yamada, M.; Goshima, G. Mitotic spindle assembly in land plants: Molecules and mechanisms. *Biology* **2017**, *6*, 6. [CrossRef] [PubMed]
391. Khodjakov, A.; Cole, R.W.; Bajer, A.S.; Rieder, C.L. The force for poleward chromosome motion in Haemanthus cells acts along the length of the chromosome during metaphase but only at the kinetochore during anaphase. *J. Cell Biol.* **1996**, *132*, 1093–1104. [CrossRef] [PubMed]
392. Schuh, M.; Ellenberg, J. Self-organization of MTOCs replaces centrosome function during acentrosomal spindle assembly in live mouse oocytes. *Cell* **2007**, *130*, 484–498. [CrossRef] [PubMed]
393. Dumont, J.; Desai, A. Acentrosomal spindle assembly and chromosome segregation during oocyte meiosis. *Trends Cell Biol.* **2012**, *22*, 241–249. [CrossRef] [PubMed]
394. Bennabi, I.; Terret, M.E.; Verlhac, M.H. Meiotic spindle assembly and chromosome segregation in oocytes. *J. Cell Biol.* **2016**, *215*, 611–619. [CrossRef] [PubMed]
395. Kitajima, T.S.; Ohsugi, M.; Ellenberg, J. Complete kinetochore tracking reveals error-prone homologous chromosome biorientation in mammalian oocytes. *Cell* **2011**, *146*, 568–581. [CrossRef] [PubMed]
396. Ohsugi, M.; Adachi, K.; Horai, R.; Kakuta, S.; Sudo, K.; Kotaki, H.; Tokai-Nishizumi, N.; Sagara, H.; Iwakura, Y.; Yamamoto, T. Kid-mediated chromosome compaction ensures proper nuclear envelope formation. *Cell* **2008**, *132*, 771–782. [CrossRef] [PubMed]
397. Gui, L.; Homer, H. Spindle assembly checkpoint signalling is uncoupled from chromosomal position in mouse oocytes. *Development* **2012**, *139*, 1941–1946. [CrossRef] [PubMed]
398. Radford, S.J.; Hoang, T.L.; Gluszek, A.A.; Ohkura, H.; McKim, K.S. Lateral and End-On Kinetochore Attachments Are Coordinated to Achieve Bi-orientation in Drosophila Oocytes. *PLoS Genet* **2015**, *11*, e1005605. [CrossRef] [PubMed]
399. Wignall, S.M.; Villeneuve, A.M. Lateral microtubule bundles promote chromosome alignment during acentrosomal oocyte meiosis. *Nat. Cell Biol.* **2009**, *11*, 839–844. [CrossRef] [PubMed]
400. Dumont, J.; Oegema, K.; Desai, A. A kinetochore-independent mechanism drives anaphase chromosome separation during acentrosomal meiosis. *Nat. Cell Biol.* **2010**, *12*, 894–901. [CrossRef] [PubMed]
401. Wickstead, B.; Gull, K. Dyneins across eukaryotes: A comparative genomic analysis. *Traffic* **2007**, *8*, 1708–1721. [CrossRef] [PubMed]

402. ten Hoopen, R.; Schleker, T.; Manteuffel, R.; Schubert, I. Transient CENP-E-like kinetochore proteins in plants. *Chromosome Res.* **2002**, *10*, 561–570. [CrossRef] [PubMed]
403. Naito, H.; Goshima, G. NACK kinesin is required for metaphase chromosome alignment and cytokinesis in the moss Physcomitrella patens. *Cell Struct. Funct.* **2015**, *40*, 31–41. [CrossRef] [PubMed]
404. Moutinho-Pereira, S.; Stuurman, N.; Afonso, O.; Hornsveld, M.; Aguiar, P.; Goshima, G.; Vale, R.D.; Maiato, H. Genes involved in centrosome-independent mitotic spindle assembly in Drosophila S2 cells. *Proc. Natl. Acad. Sci. USA* **2013**, *110*, 19808–19813. [CrossRef] [PubMed]
405. Thompson, S.L.; Bakhoum, S.F.; Compton, D.A. Mechanisms of chromosomal instability. *Curr. Biol.* **2010**, *20*, R285–R295. [CrossRef] [PubMed]
406. Carter, S.L.; Eklund, A.C.; Kohane, I.S.; Harris, L.N.; Szallasi, Z. A signature of chromosomal instability inferred from gene expression profiles predicts clinical outcome in multiple human cancers. *Nat. Genet.* **2006**, *38*, 1043–1048. [CrossRef] [PubMed]
407. Choi, C.-M.; Seo, K.W.; Jang, S.J.; Oh, Y.-M.; Shim, T.-S.; Kim, W.S.; Lee, D.-S.; Lee, S.-D. Chromosomal instability is a risk factor for poor prognosis of adenocarcinoma of the lung: Fluorescence in situ hybridization analysis of paraffin-embedded tissue from Korean patients. *Lung Cancer* **2009**, *64*, 66–70. [CrossRef] [PubMed]
408. Lee, A.J.X.; Endesfelder, D.; Rowan, A.J.; Walther, A.; Birkbak, N.J.; Futreal, P.A.; Downward, J.; Szallasi, Z.; Tomlinson, I.P.M.; Howell, M.; et al. Chromosomal instability confers intrinsic multidrug resistance. *Cancer Res.* **2011**, *71*, 1858–1870. [CrossRef] [PubMed]
409. McClelland, S.E.; Burrell, R.A.; Swanton, C. Chromosomal instability: A composite phenotype that influences sensitivity to chemotherapy. *Cell Cycle* **2009**, *8*, 3262–3266. [CrossRef] [PubMed]
410. Swanton, C.; Nicke, B.; Schuett, M.; Eklund, A.C.; Ng, C.; Li, Q.; Hardcastle, T.; Lee, A.; Roy, R.; East, P.; et al. Chromosomal instability determines taxane response. *Proc. Natl. Acad. Sci. USA* **2009**, *106*, 8671–8676. [CrossRef] [PubMed]
411. Birkbak, N.J.; Eklund, A.C.; Li, Q.; McClelland, S.E.; Endesfelder, D.; Tan, P.; Tan, I.B.; Richardson, A.L.; Szallasi, Z.; Swanton, C. Paradoxical relationship between chromosomal instability and survival outcome in cancer. *Cancer Res.* **2011**, *71*, 3447–3452. [CrossRef] [PubMed]
412. Burrell, R.A.; Juul, N.; Johnston, S.R.; Reis-Filho, J.S.; Szallasi, Z.; Swanton, C. Targeting chromosomal instability and tumour heterogeneity in HER2-positive breast cancer. *J. Cell. Biochem.* **2010**, *111*, 782–790. [CrossRef] [PubMed]
413. Roschke, A.V.; Kirsch, I.R. Targeting cancer cells by exploiting karyotypic complexity and chromosomal instability. *Cell Cycle* **2005**, *4*, 679–682. [CrossRef] [PubMed]
414. Gordon, D.J.; Resio, B.; Pellman, D. Causes and consequences of aneuploidy in cancer. *Nat. Rev. Genet.* **2012**, *13*, 189–203. [CrossRef] [PubMed]
415. Maiato, H.; Logarinho, E. Mitotic spindle multipolarity without centrosome amplification. *Nat. Cell Biol.* **2014**, *16*, 386–394. [CrossRef] [PubMed]
416. Weaver, B.A.A.; Bonday, Z.Q.; Putkey, F.R.; Kops, G.J.P.L.; Silk, A.D.; Cleveland, D.W. Centromere-associated protein-E is essential for the mammalian mitotic checkpoint to prevent aneuploidy due to single chromosome loss. *J. Cell Biol.* **2003**, *162*, 551–563. [CrossRef] [PubMed]
417. Silk, A.D.; Zasadil, L.M.; Holland, A.J.; Vitre, B.; Cleveland, D.W.; Weaver, B.A. Chromosome missegregation rate predicts whether aneuploidy will promote or suppress tumors. *Proc. Natl. Acad. Sci. USA* **2013**, *110*, E4134–E4141. [CrossRef] [PubMed]
418. Weaver, B.A.A.; Silk, A.D.; Montagna, C.; Verdier-Pinard, P.; Cleveland, D.W. Aneuploidy acts both oncogenically and as a tumor suppressor. *Cancer Cell* **2007**, *11*, 25–36. [CrossRef] [PubMed]
419. Zasadil, L.M.; Britigan, E.M.C.; Ryan, S.D.; Kaur, C.; Guckenberger, D.J.; Beebe, D.J.; Moser, A.R.; Weaver, B.A. High rates of chromosome missegregation suppress tumor progression but do not inhibit tumor initiation. *Mol. Biol. Cell* **2016**, *27*, 1981–1989. [CrossRef] [PubMed]
420. Morais da Silva, S.; Moutinho-Santos, T.; Sunkel, C.E. A tumor suppressor role of the Bub3 spindle checkpoint protein after apoptosis inhibition. *J. Cell Biol.* **2013**, *201*, 385–393. [CrossRef] [PubMed]
421. Clemente-Ruiz, M.; Muzzopappa, M.; Milán, M. Tumor suppressor roles of CENP-E and Nsl1 in Drosophila epithelial tissues. *Cell Cycle* **2014**, *13*, 1450–1455. [CrossRef] [PubMed]
422. Kullmann, F.; Judex, M.; Ballhorn, W.; Jüsten, H.P.; Wessinghage, D.; Welsh, J.; Yen, T.J.; Lang, B.; Hittle, J.C.; McClelland, M.; et al. Kinesin-like protein CENP-E is upregulated in rheumatoid synovial fibroblasts. *Arthritis Res.* **1999**, *1*, 71–80. [CrossRef] [PubMed]

423. Kung, P.-P.; Martinez, R.; Zhu, Z.; Zager, M.; Blasina, A.; Rymer, I.; Hallin, J.; Xu, M.; Carroll, C.; Chionis, J.; et al. Chemogenetic evaluation of the mitotic kinesin CENP-E reveals a critical role in triple-negative breast cancer. *Mol. Cancer Ther.* **2014**, *13*, 2104–2115. [CrossRef] [PubMed]
424. Liu, Z.; Ling, K.; Wu, X.; Cao, J.; Liu, B.; Li, S.; Si, Q.; Cai, Y.; Yan, C.; Zhang, Y.; et al. Reduced expression of cenp-e in human hepatocellular carcinoma. *J. Exp. Clin. Cancer Res.* **2009**, *28*, 156. [CrossRef] [PubMed]
425. Kumar, A.; Purohit, R. Computational screening and molecular dynamics simulation of disease associated nsSNPs in CENP-E. *Mutat. Res.* **2012**, *738–773*, 28–37. [CrossRef] [PubMed]
426. Mirzaa, G.M.; Vitre, B.; Carpenter, G.; Abramowicz, I.; Gleeson, J.G.; Paciorkowski, A.R.; Cleveland, D.W.; Dobyns, W.B.; O'Driscoll, M. Mutations in CENPE define a novel kinetochore-centromeric mechanism for microcephalic primordial dwarfism. *Hum. Genet.* **2014**, *133*, 1023–1039. [CrossRef] [PubMed]
427. Nagahara, M.; Nishida, N.; Iwatsuki, M.; Ishimaru, S.; Mimori, K.; Tanaka, F.; Nakagawa, T.; Sato, T.; Sugihara, K.; Hoon, D.S.B.; et al. Kinesin 18A expression: Clinical relevance to colorectal cancer progression. *Int. J. Cancer* **2011**, *129*, 2543–2552. [CrossRef] [PubMed]
428. Zhang, C.; Zhu, C.; Chen, H.; Li, L.; Guo, L.; Jiang, W.; Lu, S.H. Kif18A is involved in human breast carcinogenesis. *Carcinogenesis* **2010**, *31*, 1676–1684. [CrossRef] [PubMed]
429. Rucksaken, R.; Khoontawad, J.; Roytrakul, S.; Pinlaor, P.; Hiraku, Y.; Wongkham, C.; Pairojkul, C.; Boonmars, T.; Pinlaor, S. Proteomic analysis to identify plasma orosomucoid 2 and kinesin 18A as potential biomarkers of cholangiocarcinoma. *Cancer Biomark.* **2012**, *12*, 81–95. [CrossRef] [PubMed]
430. Tooker, B.C.; Newman, L.S.; Bowler, R.P.; Karjalainen, A.; Oksa, P.; Vainio, H.; Pukkala, E.; Brandt-Rauf, P.W. Proteomic detection of cancer in asbestosis patients using SELDI-TOF discovered serum protein biomarkers. *Biomarkers* **2011**, *16*, 181–191. [CrossRef] [PubMed]
431. Zhu, H.; Xu, W.; Zhang, H.; Liu, J.; Xu, H.; Lu, S.; Dang, S.; Kuang, Y.; Jin, X.; Wang, Z. Targeted deletion of Kif18a protects from colitis-associated colorectal (CAC) tumors in mice through impairing Akt phosphorylation. *Biochem. Biophys. Res. Commun.* **2013**, *438*, 97–102. [CrossRef] [PubMed]
432. Kurasawa, Y.; Earnshaw, W.C.; Mochizuki, Y.; Dohmae, N.; Todokoro, K. Essential roles of KIF4 and its binding partner PRC1 in organized central spindle midzone formation. *EMBO J.* **2004**, *23*, 3237–3248. [CrossRef] [PubMed]
433. Mazumdar, M.; Lee, J.-H.; Sengupta, K.; Ried, T.; Rane, S.; Misteli, T. Tumor formation via loss of a molecular motor protein. *Curr. Biol.* **2006**, *16*, 1559–1564. [CrossRef] [PubMed]
434. Gao, J.; Sai, N.; Wang, C.; Sheng, X.; Shao, Q.; Zhou, C.; Shi, Y.; Sun, S.; Qu, X.; Zhu, C. Overexpression of chromokinesin KIF4 inhibits proliferation of human gastric carcinoma cells both in vitro and in vivo. *Tumour Biol.* **2011**, *32*, 53–61. [CrossRef] [PubMed]
435. Narayan, G.; Bourdon, V.; Chaganti, S.; Arias-Pulido, H.; Nandula, S.V.; Rao, P.H.; Gissmann, L.; Dürst, M.; Schneider, A.; Pothuri, B.; et al. Gene dosage alterations revealed by cDNA microarray analysis in cervical cancer: Identification of candidate amplified and overexpressed genes. *Genes Chromosom. Cancer* **2007**, *46*, 373–384. [CrossRef] [PubMed]
436. Taniwaki, M.; Takano, A.; Ishikawa, N.; Yasui, W.; Inai, K.; Nishimura, H.; Tsuchiya, E.; Kohno, N.; Nakamura, Y.; Daigo, Y. Activation of KIF4A as a prognostic biomarker and therapeutic target for lung cancer. *Clin. Cancer Res.* **2007**, *13*, 6624–6631. [CrossRef] [PubMed]
437. Jordan, M.A.; Wilson, L. Microtubules as a target for anticancer drugs. *Nat. Rev. Cancer* **2004**, *4*, 253–265. [CrossRef] [PubMed]
438. Gotaskie, G.E.; Andreassi, B.F. Paclitaxel: A new antimitotic chemotherapeutic agent. *Cancer Pract.* **1994**, *2*, 27–33. [PubMed]
439. Kavallaris, M. Microtubules and resistance to tubulin-binding agents. *Nat. Rev. Cancer* **2010**, *10*, 194–204. [CrossRef] [PubMed]
440. Zhou, X.J.; Rahmani, R. Preclinical and clinical pharmacology of vinca alkaloids. *Drugs* **1992**, *44*, 1–16. [CrossRef] [PubMed]
441. Jordan, M.A.; Toso, R.J.; Thrower, D.; Wilson, L. Mechanism of mitotic block and inhibition of cell proliferation by taxol at low concentrations. *Proc. Natl. Acad. Sci. USA* **1993**, *90*, 9552–9556. [CrossRef] [PubMed]
442. Jordan, M.A.; Wendell, K.; Gardiner, S.; Derry, W.B.; Copp, H.; Wilson, L. Mitotic block induced in HeLa cells by low concentrations of paclitaxel (Taxol) results in abnormal mitotic exit and apoptotic cell death. *Cancer Res.* **1996**, *56*, 816–825. [PubMed]

443. Jordan, M.A.; Thrower, D.; Wilson, L. Effects of vinblastine, podophyllotoxin and nocodazole on mitotic spindles. Implications for the role of microtubule dynamics in mitosis. *J. Cell Sci.* **1992**, *102*, 401–416. [PubMed]
444. Yang, Z.; Kenny, A.E.; Brito, D.A.; Rieder, C.L. Cells satisfy the mitotic checkpoint in Taxol, and do so faster in concentrations that stabilize syntelic attachments. *J. Cell Biol.* **2009**, *186*, 675–684. [CrossRef] [PubMed]
445. Zasadil, L.M.; Andersen, K.A.; Yeum, D.; Rocque, G.B.; Wilke, L.G.; Tevaarwerk, A.J.; Raines, R.T.; Burkard, M.E.; Weaver, B.A. Cytotoxicity of paclitaxel in breast cancer is due to chromosome missegregation on multipolar spindles. *Sci. Transl. Med.* **2014**, *6*, 229ra43. [CrossRef] [PubMed]
446. Manchado, E.; Guillamot, M.; Malumbres, M. Killing cells by targeting mitosis. *Cell Death Differ.* **2012**, *19*, 369–377. [CrossRef] [PubMed]
447. Miglarese, M.R.; Carlson, R.O. Development of new cancer therapeutic agents targeting mitosis. *Expert Opin. Investig. Drugs* **2006**, *15*, 1411–1425. [CrossRef] [PubMed]
448. Jackson, J.R.; Patrick, D.R.; Dar, M.M.; Huang, P.S. Targeted anti-mitotic therapies: can we improve on tubulin agents? *Nat. Rev. Cancer* **2007**, *7*, 107–117. [CrossRef] [PubMed]
449. Wood, K.W.; Lad, L.; Luo, L.; Qian, X.; Knight, S.D.; Nevins, N.; Brejc, K.; Sutton, D.; Gilmartin, A.G.; Chua, P.R.; et al. Antitumor activity of an allosteric inhibitor of centromere-associated protein-E. *Proc. Natl. Acad. Sci. USA* **2010**, *107*, 5839–5844. [CrossRef] [PubMed]
450. Qian, X.; McDonald, A.; Zhou, H.-J.; Adams, N.D.; Parrish, C.A.; Duffy, K.J.; Fitch, D.M.; Tedesco, R.; Ashcraft, L.W.; Yao, B.; et al. Discovery of the First Potent and Selective Inhibitor of Centromere-Associated Protein E: GSK923295. *ACS Med. Chem. Lett.* **2010**, *1*, 30–34. [CrossRef] [PubMed]
451. Balamuth, N.J.; Wood, A.; Wang, Q.; Jagannathan, J.; Mayes, P.; Zhang, Z.; Chen, Z.; Rappaport, E.; Courtright, J.; Pawel, B.; et al. Serial transcriptome analysis and cross-species integration identifies centromere-associated protein E as a novel neuroblastoma target. *Cancer Res.* **2010**, *70*, 2749–2758. [CrossRef] [PubMed]
452. Hu, Z.; Kuo, W.-l.; Das, D.; Ziyad, S.; Gu, S.; Bhattacharya, S.; Wyrobek, A.; Wang, N.; Feiler, H.; Wooster, R.; et al. Abstract #5572: Small molecular inhibitor of centromere-associated protein E (CENP-E), GSK923295A inhibits cell growth in breast cancer cells. *Cancer Res.* **2009**, *69*, 5572.
453. Lock, R.B.; Carol, H.; Morton, C.L.; Keir, S.T.; Reynolds, C.P.; Kang, M.H.; Maris, J.M.; Wozniak, A.W.; Gorlick, R.; Kolb, E.A.; et al. Initial testing of the CENP-E inhibitor GSK923295A by the pediatric preclinical testing program. *Pediatr. Blood Cancer* **2012**, *58*, 916–923. [CrossRef] [PubMed]
454. Mayes, P.A.; Degenhardt, Y.Y.; Wood, A.; Toporovskya, Y.; Diskin, S.J.; Haglund, E.; Moy, C.; Wooster, R.; Maris, J.M. Mitogen-activated protein kinase (MEK/ERK) inhibition sensitizes cancer cells to centromere-associated protein E inhibition. *Int. J. Cancer* **2013**, *132*, E149–E157. [CrossRef] [PubMed]
455. Sutton, D.; Gilmartin, A.; Kusnierz, A.; Sung, C.-M.; Luo, L.; Carson, J.; Laquerre, S.; Cornwell, W.; King, A.; Knight, S.; et al. A potent and selective inhibitor of the mitotic kinesin CENP-E (GSK923295A), demonstrates a novel mechanism of inhibiting tumor cell proliferation and shows activity against a broad panel of human tumor cell lines in vitro. *Am. Assoc. Cancer Res.* **2007**, *6*, A111.
456. Bennett, A.; Bechi, B.; Tighe, A.; Thompson, S.; Procter, D.J.; Taylor, S.S. Cenp-E inhibitor GSK923295: Novel synthetic route and use as a tool to generate aneuploidy. *Oncotarget* **2015**, *6*, 20921–20932. [CrossRef] [PubMed]
457. Tcherniuk, S.O.; Oleinikov, A.V. Pgp efflux pump decreases the cytostatic effect of CENP-E inhibitor GSK923295. *Cancer Lett.* **2015**, *361*, 97–103. [CrossRef] [PubMed]
458. Ohashi, A.; Ohori, M.; Iwai, K.; Nakayama, Y.; Nambu, T.; Morishita, D.; Kawamoto, T.; Miyamoto, M.; Hirayama, T.; Okaniwa, M.; et al. Aneuploidy generates proteotoxic stress and DNA damage concurrently with p53-mediated post-mitotic apoptosis in SAC-impaired cells. *Nat. Commun.* **2015**, *6*, 7668. [CrossRef] [PubMed]
459. Kim, J.-H.; Lee, H.-S.; Lee, N.C.O.; Goncharov, N.V.; Kumeiko, V.; Masumoto, H.; Earnshaw, W.C.; Kouprina, N.; Larionov, V. Development of a novel HAC-based "gain of signal" quantitative assay for measuring chromosome instability (CIN) in cancer cells. *Oncotarget* **2016**, *7*, 14841–14856. [PubMed]
460. Henderson, M.C.; Shaw, Y.-J.Y.; Wang, H.; Han, H.; Hurley, L.H.; Flynn, G.; Dorr, R.T.; Von Hoff, D.D. UA62784, a novel inhibitor of centromere protein E kinesin-like protein. *Mol. Cancer Ther.* **2009**, *8*, 36–44. [CrossRef] [PubMed]
461. Tcherniuk, S.; Deshayes, S.; Sarli, V.; Divita, G.; Abrieu, A. UA62784 Is a cytotoxic inhibitor of microtubules, not CENP-E. *Chem. Biol.* **2011**, *18*, 631–641. [CrossRef] [PubMed]

462. Maiato, H.; Logarinho, E. Motor-dependent and -independent roles of CENP-E at kinetochores: The cautionary tale of UA62784. *Chem. Biol.* **2011**, *18*, 679–680. [CrossRef] [PubMed]
463. Shaw, A.Y.; Henderson, M.C.; Flynn, G.; Samulitis, B.; Han, H.; Stratton, S.P.; Chow, H.H.S.; Hurley, L.H.; Dorr, R.T. Characterization of novel diaryl oxazole-based compounds as potential agents to treat pancreatic cancer. *J. Pharmacol. Exp. Ther.* **2009**, *331*, 636–647. [CrossRef] [PubMed]
464. Ding, X.; Yan, F.; Yao, P.; Yang, Z.; Wan, W.; Wang, X.; Liu, J.; Gao, X.; Abrieu, A.; Zhu, T.; et al. Probing CENP-E function in chromosome dynamics using small molecule inhibitor syntelin. *Cell Res.* **2010**, *20*, 1386–1389. [CrossRef] [PubMed]
465. Chung, V.; Heath, E.I.; Schelman, W.R.; Johnson, B.M.; Kirby, L.C.; Lynch, K.M.; Botbyl, J.D.; Lampkin, T.A.; Holen, K.D. First-time-in-human study of GSK923295, a novel antimitotic inhibitor of centromere-associated protein E (CENP-E), in patients with refractory cancer. *Cancer Chemother. Pharmacol.* **2012**, *69*, 733–741. [CrossRef] [PubMed]
466. Capell, B.C.; Erdos, M.R.; Madigan, J.P.; Fiordalisi, J.J.; Varga, R.; Conneely, K.N.; Gordon, L.B.; Der, C.J.; Cox, A.D.; Collins, F.S. Inhibiting farnesylation of progerin prevents the characteristic nuclear blebbing of Hutchinson-Gilford progeria syndrome. *Proc. Natl. Acad. Sci. USA* **2005**, *102*, 12879–12884. [CrossRef] [PubMed]
467. Buckner, F.S.; Eastman, R.T.; Yokoyama, K.; Gelb, M.H.; Van Voorhis, W.C. Protein farnesyl transferase inhibitors for the treatment of malaria and African trypanosomiasis. *Curr. Opin. Investig. Drugs* **2005**, *6*, 791–797. [PubMed]
468. Gordon, L.B.; Kleinman, M.E.; Miller, D.T.; Neuberg, D.S.; Giobbie-Hurder, A.; Gerhard-Herman, M.; Smoot, L.B.; Gordon, C.M.; Cleveland, R.; Snyder, B.D.; et al. Clinical trial of a farnesyltransferase inhibitor in children with Hutchinson-Gilford progeria syndrome. *Proc. Natl. Acad. Sci. USA* **2012**, *109*, 16666–16671. [CrossRef] [PubMed]
469. Nallan, L.; Bauer, K.D.; Bendale, P.; Rivas, K.; Yokoyama, K.; Horney, C.P.; Pendyala, P.R.; Floyd, D.; Lombardo, L.J.; Williams, D.K.; et al. Protein farnesyltransferase inhibitors exhibit potent antimalarial activity. *J. Med. Chem.* **2005**, *48*, 3704–3713. [CrossRef] [PubMed]
470. Wiesner, J.; Kettler, K.; Sakowski, J.; Ortmann, R.; Katzin, A.M.; Kimura, E.A.; Silber, K.; Klebe, G.; Jomaa, H.; Schlitzer, M. Farnesyltransferase inhibitors inhibit the growth of malaria parasites in vitro and in vivo. *Angew. Chem. Int. Ed. Engl.* **2004**, *43*, 251–254. [CrossRef] [PubMed]
471. Shen, M.; Pan, P.; Li, Y.; Li, D.; Yu, H.; Hou, T. Farnesyltransferase and geranylgeranyltransferase I: Structures, mechanism, inhibitors and molecular modeling. *Drug Discov. Today* **2015**, *20*, 267–276. [CrossRef] [PubMed]
472. Moorthy, N.S.; Sousa, S.F.; Ramos, M.J.; Fernandes, P.A. Farnesyltransferase inhibitors: A comprehensive review based on quantitative structural analysis. *Curr. Med. Chem.* **2013**, *20*, 4888–4923. [CrossRef] [PubMed]
473. Agrawal, A.G.; Somani, R.R. Farnesyltransferase inhibitor as anticancer agent. *Mini Rev. Med. Chem.* **2009**, *9*, 638–652. [CrossRef] [PubMed]
474. Kho, Y.; Kim, S.C.; Jiang, C.; Barma, D.; Kwon, S.W.; Cheng, J.; Jaunbergs, J.; Weinbaum, C.; Tamanoi, F.; Falck, J.; et al. A tagging-via-substrate technology for detection and proteomics of farnesylated proteins. *Proc. Natl. Acad. Sci. USA* **2004**, *101*, 12479–12484. [CrossRef] [PubMed]
475. Clark, G.J.; Kinch, M.S.; Rogers-Graham, K.; Sebti, S.M.; Hamilton, A.D.; Der, C.J. The Ras-related protein Rheb is farnesylated and antagonizes Ras signaling and transformation. *J. Biol. Chem.* **1997**, *272*, 10608–10615. [PubMed]
476. Holstein, S.A.; Hohl, R.J. Is there a future for prenyltransferase inhibitors in cancer therapy? *Curr. Opin. Pharmacol.* **2012**, *12*, 704–709. [CrossRef] [PubMed]
477. Kohl, N.E.; Mosser, S.D.; deSolms, S.J.; Giuliani, E.A.; Pompliano, D.L.; Graham, S.L.; Smith, R.L.; Scolnick, E.M.; Oliff, A.; Gibbs, J.B. Selective inhibition of ras-dependent transformation by a farnesyltransferase inhibitor. *Science* **1993**, *260*, 1934–1937. [CrossRef] [PubMed]
478. Lee, K.H.; Koh, M.; Moon, A. Farnesyl transferase inhibitor FTI-277 inhibits breast cell invasion and migration by blocking H-Ras activation. *Oncol. Lett.* **2016**, *12*, 2222–2226. [CrossRef] [PubMed]
479. Cox, A.D.; Der, C.J. Farnesyltransferase inhibitors and cancer treatment: Targeting simply Ras? *Biochim. Biophys. Acta* **1997**, *1333*, F51–F71. [CrossRef]
480. Sepp-Lorenzino, L.; Ma, Z.; Rands, E.; Kohl, N.E.; Gibbs, J.B.; Oliff, A.; Rosen, N. A peptidomimetic inhibitor of farnesyl:protein transferase blocks the anchorage-dependent and -independent growth of human tumor cell lines. *Cancer Res.* **1995**, *55*, 5302–5309. [PubMed]

481. Karp, J.E.; Lancet, J.E.; Kaufmann, S.H.; End, D.W.; Wright, J.J.; Bol, K.; Horak, I.; Tidwell, M.L.; Liesveld, J.; Kottke, T.J.; et al. Clinical and biologic activity of the farnesyltransferase inhibitor R115777 in adults with refractory and relapsed acute leukemias: A phase 1 clinical-laboratory correlative trial. *Blood* **2001**, *97*, 3361–3369. [CrossRef] [PubMed]
482. Rolland, D.; Camara-Clayette, V.; Barbarat, A.; Salles, G.; Coiffier, B.; Ribrag, V.; Thieblemont, C. Farnesyltransferase inhibitor R115777 inhibits cell growth and induces apoptosis in mantle cell lymphoma. *Cancer Chemother. Pharmacol.* **2008**, *61*, 855–863. [CrossRef] [PubMed]
483. Adjei, A.A.; Davis, J.N.; Bruzek, L.M.; Erlichman, C.; Kaufmann, S.H. Synergy of the protein farnesyltransferase inhibitor SCH66336 and cisplatin in human cancer cell lines. *Clin. Cancer Res.* **2001**, *7*, 1438–1445. [PubMed]
484. Russo, P.; Malacarne, D.; Falugi, C.; Trombino, S.; O'Connor, P.M. RPR-115135, a farnesyltransferase inhibitor, increases 5-FU- cytotoxicity in ten human colon cancer cell lines: role of p53. *Int. J. Cancer* **2002**, *100*, 266–275. [CrossRef] [PubMed]
485. Brassard, D.L.; English, J.M.; Malkowski, M.; Kirschmeier, P.; Nagabhushan, T.L.; Bishop, W.R. Inhibitors of farnesyl protein transferase and MEK1,2 induce apoptosis in fibroblasts transformed with farnesylated but not geranylgeranylated H-Ras. *Exp. Cell Res.* **2002**, *273*, 138–146. [CrossRef] [PubMed]
486. Edamatsu, H.; Gau, C.L.; Nemoto, T.; Guo, L.; Tamanoi, F. Cdk inhibitors, roscovitine and olomoucine, synergize with farnesyltransferase inhibitor (FTI) to induce efficient apoptosis of human cancer cell lines. *Oncogene* **2000**, *19*, 3059–3068. [CrossRef] [PubMed]
487. Nagai, T.; Ohmine, K.; Fujiwara, S.; Uesawa, M.; Sakurai, C.; Ozawa, K. Combination of tipifarnib and rapamycin synergistically inhibits the growth of leukemia cells and overcomes resistance to tipifarnib via alteration of cellular signaling pathways. *Leuk. Res.* **2010**, *34*, 1057–1063. [CrossRef] [PubMed]
488. Moasser, M.M.; Sepp-Lorenzino, L.; Kohl, N.E.; Oliff, A.; Balog, A.; Su, D.S.; Danishefsky, S.J.; Rosen, N. Farnesyl transferase inhibitors cause enhanced mitotic sensitivity to taxol and epothilones. *Proc. Natl. Acad. Sci. USA* **1998**, *95*, 1369–1374. [CrossRef] [PubMed]
489. Karp, J.E.; Kaufmann, S.H.; Adjei, A.A.; Lancet, J.E.; Wright, J.J.; End, D.W. Current status of clinical trials of farnesyltransferase inhibitors. *Curr. Opin. Oncol.* **2001**, *13*, 470–476. [CrossRef]
490. Santos, E.S.; Rosenblatt, J.D.; Goodman, M. Role of farnesyltransferase inhibitors in hematologic malignancies. *Expert Rev. Anticancer Ther.* **2004**, *4*, 843–856. [CrossRef] [PubMed]
491. Sebti, S.M.; Adjei, A.A. Farnesyltransferase inhibitors. *Semin. Oncol.* **2004**, *31*, 28–39. [CrossRef] [PubMed]
492. Karp, J.E.; Lancet, J.E. Targeting the process of farynesylation for therapy of hematologic malignancies. *Curr. Mol. Med.* **2005**, *5*, 643–652. [CrossRef] [PubMed]
493. Epling-Burnette, P.K.; Loughran, T.P., Jr. Suppression of farnesyltransferase activity in acute myeloid leukemia and myelodysplastic syndrome: Current understanding and recommended use of tipifarnib. *Expert Opin. Investig. Drugs* **2010**, *19*, 689–698. [CrossRef] [PubMed]
494. Rao, S.; Cunningham, D.; de Gramont, A.; Scheithauer, W.; Smakal, M.; Humblet, Y.; Kourteva, G.; Iveson, T.; Andre, T.; Dostalova, J.; et al. Phase III double-blind placebo-controlled study of farnesyl transferase inhibitor R115777 in patients with refractory advanced colorectal cancer. *J. Clin. Oncol.* **2004**, *22*, 3950–3957. [CrossRef] [PubMed]
495. Van Cutsem, E.; van de Velde, H.; Karasek, P.; Oettle, H.; Vervenne, W.L.; Szawlowski, A.; Schoffski, P.; Post, S.; Verslype, C.; Neumann, H.; et al. Phase III trial of gemcitabine plus tipifarnib compared with gemcitabine plus placebo in advanced pancreatic cancer. *J. Clin. Oncol.* **2004**, *22*, 1430–1438. [CrossRef] [PubMed]
496. Macdonald, J.S.; McCoy, S.; Whitehead, R.P.; Iqbal, S.; Wade, J.L., 3rd; Giguere, J.K.; Abbruzzese, J.L. A phase II study of farnesyl transferase inhibitor R115777 in pancreatic cancer: A Southwest oncology group (SWOG 9924) study. *Investig. New Drugs* **2005**, *23*, 485–487. [CrossRef] [PubMed]
497. Harousseau, J.L.; Martinelli, G.; Jedrzejczak, W.W.; Brandwein, J.M.; Bordessoule, D.; Masszi, T.; Ossenkoppele, G.J.; Alexeeva, J.A.; Beutel, G.; Maertens, J.; et al. A randomized phase 3 study of tipifarnib compared with best supportive care, including hydroxyurea, in the treatment of newly diagnosed acute myeloid leukemia in patients 70 years or older. *Blood* **2009**, *114*, 1166–1173. [CrossRef] [PubMed]
498. Stieglitz, E.; Ward, A.F.; Gerbing, R.B.; Alonzo, T.A.; Arceci, R.J.; Liu, Y.L.; Emanuel, P.D.; Widemann, B.C.; Cheng, J.W.; Jayaprakash, N.; et al. Phase II/III trial of a pre-transplant farnesyl transferase inhibitor in juvenile myelomonocytic leukemia: A report from the Children's Oncology Group. *Pediatr. Blood Cancer* **2015**, *62*, 629–636. [CrossRef] [PubMed]

499. Gajewski, T.F.; Salama, A.K.; Niedzwiecki, D.; Johnson, J.; Linette, G.; Bucher, C.; Blaskovich, M.A.; Sebti, S.M.; Haluska, F.; Cancer and Leukemia Group B; et al. Phase II study of the farnesyltransferase inhibitor R115777 in advanced melanoma (CALGB 500104). *J. Transl. Med.* **2012**, *10*, 246. [CrossRef] [PubMed]
500. Burnett, A.K.; Russell, N.H.; Culligan, D.; Cavanagh, J.; Kell, J.; Wheatley, K.; Virchis, A.; Hills, R.K.; Milligan, D.; AML Working Group of the UK National Cancer Research Institute. The addition of the farnesyl transferase inhibitor, tipifarnib, to low dose cytarabine does not improve outcome for older patients with AML. *Br. J. Haematol.* **2012**, *158*, 519–522. [CrossRef] [PubMed]
501. Meier, W.; du Bois, A.; Rau, J.; Gropp-Meier, M.; Baumann, K.; Huober, J.; Wollschlaeger, K.; Kreienberg, R.; Canzler, U.; Schmalfeldt, B.; et al. Randomized phase II trial of carboplatin and paclitaxel with or without lonafarnib in first-line treatment of epithelial ovarian cancer stage IIB-IV. *Gynecol. Oncol.* **2012**, *126*, 236–240. [CrossRef] [PubMed]
502. Adjei, A.A.; Croghan, G.A.; Erlichman, C.; Marks, R.S.; Reid, J.M.; Sloan, J.A.; Pitot, H.C.; Alberts, S.R.; Goldberg, R.M.; Hanson, L.J.; et al. A Phase I trial of the farnesyl protein transferase inhibitor R115777 in combination with gemcitabine and cisplatin in patients with advanced cancer. *Clin. Cancer Res.* **2003**, *9*, 2520–2526. [PubMed]
503. Siegel-Lakhai, W.S.; Crul, M.; Zhang, S.; Sparidans, R.W.; Pluim, D.; Howes, A.; Solanki, B.; Beijnen, J.H.; Schellens, J.H. Phase I and pharmacological study of the farnesyltransferase inhibitor tipifarnib (Zarnestra, R115777) in combination with gemcitabine and cisplatin in patients with advanced solid tumours. *Br. J. Cancer* **2005**, *93*, 1222–1229. [CrossRef] [PubMed]
504. Sparano, J.A.; Moulder, S.; Kazi, A.; Vahdat, L.; Li, T.; Pellegrino, C.; Munster, P.; Malafa, M.; Lee, D.; Hoschander, S.; et al. Targeted inhibition of farnesyltransferase in locally advanced breast cancer: A phase I and II trial of tipifarnib plus dose-dense doxorubicin and cyclophosphamide. *J. Clin. Oncol.* **2006**, *24*, 3013–3018. [CrossRef] [PubMed]
505. Medeiros, B.C.; Landau, H.J.; Morrow, M.; Lockerbie, R.O.; Pitts, T.; Eckhardt, S.G. The farnesyl transferase inhibitor, tipifarnib, is a potent inhibitor of the MDR1 gene product, P-glycoprotein, and demonstrates significant cytotoxic synergism against human leukemia cell lines. *Leukemia* **2007**, *21*, 739–746. [CrossRef] [PubMed]
506. Jabbour, E.; Kantarjian, H.; Ravandi, F.; Garcia-Manero, G.; Estrov, Z.; Verstovsek, S.; O'Brien, S.; Faderl, S.; Thomas, D.A.; Wright, J.J.; et al. A phase 1–2 study of a farnesyltransferase inhibitor, tipifarnib, combined with idarubicin and cytarabine for patients with newly diagnosed acute myeloid leukemia and high-risk myelodysplastic syndrome. *Cancer* **2011**, *117*, 1236–1244. [CrossRef] [PubMed]
507. Li, T.; Guo, M.; Gradishar, W.J.; Sparano, J.A.; Perez, E.A.; Wang, M.; Sledge, G.W. A phase II trial of capecitabine in combination with the farnesyltransferase inhibitor tipifarnib in patients with anthracycline-treated and taxane-resistant metastatic breast cancer: An Eastern Cooperative Oncology Group Study (E1103). *Breast Cancer Res. Treat.* **2012**, *134*, 345–352. [CrossRef] [PubMed]
508. Kim, E.S.; Kies, M.S.; Fossella, F.V.; Glisson, B.S.; Zaknoen, S.; Statkevich, P.; Munden, R.F.; Summey, C.; Pisters, K.M.; Papadimitrakopoulou, V.; et al. Phase II study of the farnesyltransferase inhibitor lonafarnib with paclitaxel in patients with taxane-refractory/resistant nonsmall cell lung carcinoma. *Cancer* **2005**, *104*, 561–569. [CrossRef] [PubMed]
509. Karp, J.E.; Smith, B.D.; Gojo, I.; Lancet, J.E.; Greer, J.; Klein, M.; Morris, L.; Levis, M.J.; Gore, S.D.; Wright, J.J.; et al. Phase II trial of tipifarnib as maintenance therapy in first complete remission in adults with acute myelogenous leukemia and poor-risk features. *Clin. Cancer Res.* **2008**, *14*, 3077–3082. [CrossRef] [PubMed]
510. Castro-Castro, A.; Janke, C.; Montagnac, G.; Paul-Gilloteaux, P.; Chavrier, P. ATAT1/MEC-17 acetyltransferase and HDAC6 deacetylase control a balance of acetylation of alpha-tubulin and cortactin and regulate MT1-MMP trafficking and breast tumor cell invasion. *Eur. J. Cell Biol.* **2012**, *91*, 950–960. [CrossRef] [PubMed]
511. Boggs, A.E.; Vitolo, M.I.; Whipple, R.A.; Charpentier, M.S.; Goloubeva, O.G.; Ioffe, O.B.; Tuttle, K.C.; Slovic, J.; Lu, Y.; Mills, G.B.; et al. α-Tubulin acetylation elevated in metastatic and basal-like breast cancer cells promotes microtentacle formation, adhesion, and invasive migration. *Cancer Res.* **2015**, *75*, 203–215. [CrossRef] [PubMed]

512. Kashiwaya, K.; Nakagawa, H.; Hosokawa, M.; Mochizuki, Y.; Ueda, K.; Piao, L.; Chung, S.; Hamamoto, R.; Eguchi, H.; Ohigashi, H.; et al. Involvement of the tubulin tyrosine ligase-like family member 4 polyglutamylase in PELP1 polyglutamylation and chromatin remodeling in pancreatic cancer cells. *Cancer Res.* **2010**, *70*, 4024–4033. [CrossRef] [PubMed]
513. Wasylyk, C.; Zambrano, A.; Zhao, C.; Brants, J.; Abecassis, J.; Schalken, J.A.; Rogatsch, H.; Schaefer, G.; Pycha, A.; Klocker, H.; et al. Tubulin tyrosine ligase like 12 links to prostate cancer through tubulin posttranslational modification and chromosome ploidy. *Int. J. Cancer* **2010**, *127*, 2542–2553. [CrossRef] [PubMed]
514. Brants, J.; Semenchenko, K.; Wasylyk, C.; Robert, A.; Carles, A.; Zambrano, A.; Pradeau-Aubreton, K.; Birck, C.; Schalken, J.A.; Poch, O.; et al. Tubulin tyrosine ligase like 12, a TTLL family member with SET- and TTL-like domains and roles in histone and tubulin modifications and mitosis. *PLoS ONE* **2012**, *7*, e51258. [CrossRef] [PubMed]
515. Rocha, C.; Papon, L.; Cacheux, W.; Marques Sousa, P.; Lascano, V.; Tort, O.; Giordano, T.; Vacher, S.; Lemmers, B.; Mariani, P.; et al. Tubulin glycylases are required for primary cilia, control of cell proliferation and tumor development in colon. *EMBO J.* **2014**, *33*, 2247–2260. [CrossRef] [PubMed]
516. Lafanechere, L.; Courtay-Cahen, C.; Kawakami, T.; Jacrot, M.; Rudiger, M.; Wehland, J.; Job, D.; Margolis, R.L. Suppression of tubulin tyrosine ligase during tumor growth. *J. Cell Sci.* **1998**, *111*, 171–181. [PubMed]
517. Mialhe, A.; Lafanechere, L.; Treilleux, I.; Peloux, N.; Dumontet, C.; Bremond, A.; Panh, M.H.; Payan, R.; Wehland, J.; Margolis, R.L.; et al. Tubulin detyrosination is a frequent occurrence in breast cancers of poor prognosis. *Cancer Res.* **2001**, *61*, 5024–5027.
518. Kato, C.; Miyazaki, K.; Nakagawa, A.; Ohira, M.; Nakamura, Y.; Ozaki, T.; Imai, T.; Nakagawara, A. Low expression of human tubulin tyrosine ligase and suppressed tubulin tyrosination/detyrosination cycle are associated with impaired neuronal differentiation in neuroblastomas with poor prognosis. *Int. J. Cancer* **2004**, *112*, 365–375. [CrossRef] [PubMed]
519. Soucek, K.; Kamaid, A.; Phung, A.D.; Kubala, L.; Bulinski, J.C.; Harper, R.W.; Eiserich, J.P. Normal and prostate cancer cells display distinct molecular profiles of alpha-tubulin posttranslational modifications. *Prostate* **2006**, *66*, 954–965. [CrossRef] [PubMed]
520. Kuroda, H.; Saito, K.; Kuroda, M.; Suzuki, Y. Differential expression of glu-tubulin in relation to mammary gland disease. *Virchows Arch.* **2010**, *457*, 477–482. [CrossRef] [PubMed]
521. Whipple, R.A.; Matrone, M.A.; Cho, E.H.; Balzer, E.M.; Vitolo, M.I.; Yoon, J.R.; Ioffe, O.B.; Tuttle, K.C.; Yang, J.; Martin, S.S. Epithelial-to-mesenchymal transition promotes tubulin detyrosination and microtentacles that enhance endothelial engagement. *Cancer Res.* **2010**, *70*, 8127–8137. [CrossRef] [PubMed]
522. Kreuger, M.R.; Grootjans, S.; Biavatti, M.W.; Vandenabeele, P.; D'Herde, K. Sesquiterpene lactones as drugs with multiple targets in cancer treatment: Focus on parthenolide. *Anticancer Drugs* **2012**, *23*, 883–896. [PubMed]
523. Curry, E.A., 3rd; Murry, D.J.; Yoder, C.; Fife, K.; Armstrong, V.; Nakshatri, H.; O'Connell, M.; Sweeney, C.J. Phase I dose escalation trial of feverfew with standardized doses of parthenolide in patients with cancer. *Investig. New Drugs* **2004**, *22*, 299–305. [CrossRef] [PubMed]
524. Ghantous, A.; Sinjab, A.; Herceg, Z.; Darwiche, N. Parthenolide: From plant shoots to cancer roots. *Drug Discov. Today* **2013**, *18*, 894–905. [CrossRef] [PubMed]
525. Bork, P.M.; Schmitz, M.L.; Kuhnt, M.; Escher, C.; Heinrich, M. Sesquiterpene lactone containing Mexican Indian medicinal plants and pure sesquiterpene lactones as potent inhibitors of transcription factor NF-κB. *FEBS Lett.* **1997**, *402*, 85–90. [CrossRef]
526. Hehner, S.P.; Hofmann, T.G.; Droge, W.; Schmitz, M.L. The antiinflammatory sesquiterpene lactone parthenolide inhibits NF-κB by targeting the IκB kinase complex. *J. Immunol.* **1999**, *163*, 5617–5623. [PubMed]
527. Kwok, B.H.; Koh, B.; Ndubuisi, M.I.; Elofsson, M.; Crews, C.M. The anti-inflammatory natural product parthenolide from the medicinal herb Feverfew directly binds to and inhibits IκB kinase. *Chem. Biol.* **2001**, *8*, 759–766. [CrossRef]
528. Garcia-Pineres, A.J.; Castro, V.; Mora, G.; Schmidt, T.J.; Strunck, E.; Pahl, H.L.; Merfort, I. Cysteine 38 in p65/NF-κB plays a crucial role in DNA binding inhibition by sesquiterpene lactones. *J. Biol. Chem.* **2001**, *276*, 39713–39720. [CrossRef] [PubMed]
529. Fonrose, X.; Ausseil, F.; Soleilhac, E.; Masson, V.; David, B.; Pouny, I.; Cintrat, J.C.; Rousseau, B.; Barette, C.; Massiot, G.; et al. Parthenolide inhibits tubulin carboxypeptidase activity. *Cancer Res.* **2007**, *67*, 3371–3378. [CrossRef] [PubMed]

530. Whipple, R.A.; Vitolo, M.I.; Boggs, A.E.; Charpentier, M.S.; Thompson, K.; Martin, S.S. Parthenolide and costunolide reduce microtentacles and tumor cell attachment by selectively targeting detyrosinated tubulin independent from NF-κB inhibition. *Breast Cancer Res.* **2013**, *15*, R83. [CrossRef] [PubMed]
531. Shanmugam, R.; Kusumanchi, P.; Appaiah, H.; Cheng, L.; Crooks, P.; Neelakantan, S.; Peat, T.; Klaunig, J.; Matthews, W.; Nakshatri, H.; et al. A water soluble parthenolide analog suppresses in vivo tumor growth of two tobacco-associated cancers, lung and bladder cancer, by targeting NF-κB and generating reactive oxygen species. *Int. J. Cancer* **2011**, *128*, 2481–2494. [CrossRef] [PubMed]
532. Shanmugam, R.; Kusumanchi, P.; Appaiah, H.; Cheng, L.; Crooks, P.; Neelakantan, S.; Peat, T.; Klaunig, J.; Matthews, W.; Nakshatri, H.; et al. Naturally occurring asteriscunolide A induces apoptosis and activation of mitogen-activated protein kinase pathway in human tumor cell lines. *Mol. Carcinog.* **2010**, *49*, 488–499.
533. Rozenblat, S.; Grossman, S.; Bergman, M.; Gottlieb, H.; Cohen, Y.; Dovrat, S. Induction of G2/M arrest and apoptosis by sesquiterpene lactones in human melanoma cell lines. *Biochem. Pharmacol.* **2008**, *75*, 369–382. [CrossRef] [PubMed]
534. Guzman, M.L.; Rossi, R.M.; Karnischky, L.; Li, X.; Peterson, D.R.; Howard, D.S.; Jordan, C.T. The sesquiterpene lactone parthenolide induces apoptosis of human acute myelogenous leukemia stem and progenitor cells. *Blood* **2005**, *105*, 4163–4169. [CrossRef] [PubMed]
535. Carnero, A.; Garcia-Mayea, Y.; Mir, C.; Lorente, J.; Rubio, I.T.; ME, L.L. The cancer stem-cell signaling network and resistance to therapy. *Cancer Treat. Rev.* **2016**, *49*, 25–36. [CrossRef] [PubMed]
536. Valent, P.; Bonnet, D.; De Maria, R.; Lapidot, T.; Copland, M.; Melo, J.V.; Chomienne, C.; Ishikawa, F.; Schuringa, J.J.; Stassi, G.; et al. Cancer stem cell definitions and terminology: The devil is in the details. *Nat. Rev. Cancer* **2012**, *12*, 767–775. [CrossRef] [PubMed]
537. Kawasaki, B.T.; Hurt, E.M.; Kalathur, M.; Duhagon, M.A.; Milner, J.A.; Kim, Y.S.; Farrar, W.L. Effects of the sesquiterpene lactone parthenolide on prostate tumor-initiating cells: An integrated molecular profiling approach. *Prostate* **2009**, *69*, 827–837. [CrossRef] [PubMed]
538. Guzman, M.L.; Rossi, R.M.; Neelakantan, S.; Li, X.; Corbett, C.A.; Hassane, D.C.; Becker, M.W.; Bennett, J.M.; Sullivan, E.; Lachowicz, J.L.; et al. An orally bioavailable parthenolide analog selectively eradicates acute myelogenous leukemia stem and progenitor cells. *Blood* **2007**, *110*, 4427–4435. [CrossRef] [PubMed]
539. Carlisi, D.; Buttitta, G.; Di Fiore, R.; Scerri, C.; Drago-Ferrante, R.; Vento, R.; Tesoriere, G. Parthenolide and DMAPT exert cytotoxic effects on breast cancer stem-like cells by inducing oxidative stress, mitochondrial dysfunction and necrosis. *Cell Death Dis.* **2016**, *7*, e2194. [CrossRef] [PubMed]
540. Shanmugam, R.; Kusumanchi, P.; Cheng, L.; Crooks, P.; Neelakantan, S.; Matthews, W.; Nakshatri, H.; Sweeney, C.J. A water-soluble parthenolide analogue suppresses in vivo prostate cancer growth by targeting NFkappaB and generating reactive oxygen species. *Prostate* **2010**, *70*, 1074–1086. [CrossRef] [PubMed]
541. Sweeney, C.J.; Mehrotra, S.; Sadaria, M.R.; Kumar, S.; Shortle, N.H.; Roman, Y.; Sheridan, C.; Campbell, R.A.; Murry, D.J.; Badve, S.; et al. The sesquiterpene lactone parthenolide in combination with docetaxel reduces metastasis and improves survival in a xenograft model of breast cancer. *Mol. Cancer Ther.* **2005**, *4*, 1004–1012. [CrossRef] [PubMed]
542. Zhang, D.; Qiu, L.; Jin, X.; Guo, Z.; Guo, C. Nuclear factor-kappaB inhibition by parthenolide potentiates the efficacy of Taxol in non-small cell lung cancer in vitro and in vivo. *Mol. Cancer Res.* **2009**, *7*, 1139–1149. [CrossRef] [PubMed]
543. Liu, Y.; Lu, W.L.; Guo, J.; Du, J.; Li, T.; Wu, J.W.; Wang, G.L.; Wang, J.C.; Zhang, X.; Zhang, Q. A potential target associated with both cancer and cancer stem cells: A combination therapy for eradication of breast cancer using vinorelbine stealthy liposomes plus parthenolide stealthy liposomes. *J. Control. Release* **2008**, *129*, 18–25. [CrossRef] [PubMed]

biology MDPI

Review

A Cell Biological Perspective on Past, Present and Future Investigations of the Spindle Assembly Checkpoint

Ajit P. Joglekar

Cell & Developmental Biology, University of Michigan Medical School, Ann Arbor, MI 48109, USA; ajitj@umich.edu; Tel.: +1-734-764-2474; Fax: +1-734-615-8500

Academic Editor: J. Richard McIntosh
Received: 4 October 2016; Accepted: 14 November 2016; Published: 19 November 2016

Abstract: The spindle assembly checkpoint (SAC) is a quality control mechanism that ensures accurate chromosome segregation during cell division. It consists of a mechanochemical signal transduction mechanism that senses the attachment of chromosomes to the spindle, and a signaling cascade that inhibits cell division if one or more chromosomes are not attached. Extensive investigations of both these component systems of the SAC have synthesized a comprehensive understanding of the underlying molecular mechanisms. This review recounts the milestone results that elucidated the SAC, compiles a simple model of the complex molecular machinery underlying the SAC, and highlights poorly understood facets of the biochemical design and cell biological operation of the SAC that will drive research forward in the near future.

Keywords: mitosis; spindle assembly checkpoint; signal transduction; aneuploidy

1. Introduction

The primary objective of mitosis is to create two cells with identical genomes. To achieve this, the dividing cell must commence the process of cell division only after every chromosome is stably attached to spindle microtubules emanating from opposite spindle poles. If cell division occurs in the presence of unattached kinetochores, then the result is either chromosome missegregation or loss, and the creation of genetically abnormal, aneuploid cells. To avoid this fate, the dividing cell enforces the requirement for stable kinetochore-microtubule attachment using a cell cycle control known as the spindle assembly checkpoint (SAC). The SAC is a mechanosensitive signaling cascade that ties the progress of the cell cycle machinery with the mechanics of kinetochore biorientation. It is activated by unattached kinetochores, which recruit many different SAC proteins and generate the inhibitory "wait-anaphase" signal. Once the last unattached kinetochore attaches to spindle microtubules, the "wait-anaphase" signal rapidly dissipates, and anaphase ensues. This simple "on-off" operation of the SAC belies an intricate interplay between complex signal transduction machinery embedded in the kinetochore and an equally complex signaling cascade that involves kinases, phosphatases, and numerous SAC signaling proteins. A tight coupling between the signal transduction machinery and signaling cascade of the SAC is essential to minimize chromosome loss and maintain genome stability during cell division.

Understanding the elegant design of the SAC requires an in-depth understanding of both of its component systems: the kinetochore-based mechanochemical signal transduction mechanism that senses the absence of microtubule attachment and the signaling cascade that amplifies and spreads the anaphase-inhibitory signal through the entire cell. This understanding is necessary, because the misregulation of either system can have dire consequences on the genetic stability and health of both daughter cells. Aberrant expression of kinetochore and signaling proteins involved in the SAC are strongly correlated with tumorigenesis and cancer. However, whether and how the aberrant expression directly leads to tumorigenesis is not known. A mechanistic understanding of the kinetochore-based machinery and a quantitative understanding of the SAC signaling cascade can elucidate the causal links that likely connect aberrant SAC function, chromosome missegregation, and tumorigenesis.

This review considers the molecular mechanisms underlying the SAC and their operation from the perspective of cell biology. Extensive molecular, structural, and biochemical investigations of the SAC over the last two decades have achieved a nearly complete description of its signaling cascade, and they elucidate how the kinetochore controls this cascade. Therefore, the following goals were set for this review. The first goal is to briefly summarize the conceptual leaps achieved in understanding the SAC. This summary will highlight studies that deeply influenced the mitosis field, and which continue to guide investigations of the SAC today. The summary uses logical rather than chronological linkage. The second goal is to synthesize a succinct working model for the operation of the SAC. Many expert reviews that delve into the structural details of the SAC signaling proteins and the biochemistry of the SAC were recently published [1,2]. Therefore, this knowledge will be organized in the context of cell biology, so that it is easy to grasp even for readers outside the field of cell division. The final goal for this review is to discuss the major gaps in our understanding of the SAC, and pose four broad questions that are likely to drive future investigations into the SAC.

2. Early Hints of a Pathway that Monitors Chromosome Alignment and Controls Anaphase Onset

The foundation for our current understanding of the SAC was established by cell biological investigations conducted almost sixty years ago. For this, the adoption of cine-microscopy proved to be the enabling development. With cine-microscopy, the complete sequence of mitotic events could be documented in real-time for the first time. These observations revealed that the alignment of chromosomes at the metaphase plate is important for the timely onset of anaphase. In one study in particular, Bajer and Mole-Bajer describe the importance of chromosome alignment quite succinctly [3]: "In some cases anaphase does not begin, but 'waits' for the chromosome to move to the plate, beginning a few minutes after this has reached it." An even stronger correlation between chromosome misalignment and delayed anaphase emerged from observations of the first meiotic division in mantid spermatocytes [4]. Mantid spermatocytes contain three sex chromosomes: X1, X2, and Y. Normally, the two X chromosomes pair with the Y to form the sex trivalent. However, if this pairing of sex chromosomes is not successful, the sex chromosomes are unable to align at the metaphase plate. The presence of such misaligned sex chromosomes block cell division for long periods of time. The causative link between chromosome alignment and anaphase onset was made apparent by Zirkle's experiments [5]. Zirkle used a UV laser micro-beam to irradiate the cytoplasm in the vicinity of metaphase spindles. Laser irradiation damaged the spindle, and dislodged chromosomes from the metaphase plate. Strikingly, the dividing cell containing the misaligned chromosomes remained in mitosis, and initiated anaphase only after these chromosomes realigned at the metaphase plate. Although these results did not implicate unattached kinetochores as the reason for the cell cycle block, they established that the metazoan cell waits until all chromosomes are aligned at the spindle equator before initiating anaphase.

In this context, it is necessary to discuss the truly innovative experiments conducted by Nicklas [6]. Although Nicklas' experiments were designed to investigate chromosome movement and alignment, their findings deeply influenced our conceptualization of the mechanism of SAC signaling. Using a glass microneedle, Nicklas directly pushed, pulled, and prodded aligned chromosomes during the first meiosis in grasshopper spermatocytes to observe the establishment of bipolar attachments (he referred to this process as "chromosome reorientation"). He noted that the spermatocytes never entered anaphase in the presence of unattached chromosomes. More importantly, he demonstrated that kinetochore-microtubule attachments are stabilized by the application of an opposing mechanical force. The finding that mechanical forces arising from kinetochore interactions with the spindle alter the biochemistry within the kinetochore would provide the basis for the hypothesis that the mechanical force generated by kinetochore-microtubule attachment silences the SAC [7].

Such careful observations of dividing cells derived from diverse organisms provided ample and strong evidence of a system that monitors the alignment of chromosomes at the metaphase plate, and that delays anaphase onset if this alignment is not achieved. However, the significance of these observations remained unclear for nearly three decades, because the biochemical basis of the cell cycle and the concept of cell cycle control were not yet fully understood.

3. Discovery of the Spindle Assembly Checkpoint

Until the early 1970s, the cell cycle was viewed as a prescribed sequence of activities and events that proliferating eukaryotic cells progress through [8,9]. According to this view, the completion of each step in the cell cycle sequence was required for and followed by the next step; feedback mechanisms that enforce quality control were not envisioned. Several studies in the 1970s and 1980s challenged this view. Genetic studies discovered many genes, which when mutated, arrested cells in specific stages of the cell cycle [10,11]. Clearly, specific functions encoded by specific genes were essential for the completion of each cell cycle stage. At the same time, biochemical investigations of the synchronous cell divisions that occur during early embryonic development revealed that specific biochemical activities had to be stimulated and then silenced to drive the cell cycle [12]. These findings forced a reconsideration of the nature of the cell cycle. Whether the cell cycle is a set of sequential processes, or if control mechanisms monitor the completion of each process and prevent further progress in case the previous process is not satisfactorily accomplished, became a fundamental question that needed to be addressed.

The existence of feedback mechanisms controlling the progression of the cell cycle was first confirmed by the ground-breaking study by Weinert and Hartwell [13]. The authors screened for genes that are necessary for the cell cycle arrest induced by damaged DNA. They discovered *RAD9* (wild-type gene names appear in upper case italics, mutant genes appear in lower case italics, protein names appear in Roman letters), a gene that is dispensable for the normal cell cycle progression but essential for the cell cycle arrest induced by DNA damage. The discovery of *RAD9* revealed that not only does the eukaryotic cell encode genes that drive cell cycle progression, but also genes that actively monitor the cell cycle, and prevent progress if a crucial step is not satisfactorily completed. This seminal discovery demonstrated that the cell employs feedback controls or checkpoints that monitor the progress of at least one cell cycle phase in order to maintain the quality of the genome [13,14].

The discovery of the DNA damage checkpoint prompted a reconsideration of the early cell biological observations of cell division arrest induced by unaligned and/or unattached chromosomes. Biophysical and biochemical studies of tubulin and microtubules had discovered a number of small molecules that depolymerize microtubules, and thus act as spindle poisons. It was also known that treatment of eukaryotic cells with spindle poisons not only destroys the spindle structure, but also arrests the cells in mitosis [15–17]. Furthermore, experimental disruption of kinetochore assembly created chromosomes that could not stably attach to the spindle, and cells containing such unattached or weakly attached chromosomes delayed anaphase for several hours [18]. Could this cell cycle arrest seen in cells with spindle damage also be instituted by another checkpoint? In 1991, two seminal studies discovered two sets of genes that are necessary for arresting cells in mitosis in the presence of microtubule poisons [19,20]. Following the conceptual framework established by the DNA damage checkpoint, these studies hypothesized that specific genes are necessary to institute the metaphase arrest triggered by spindle damage, and therefore, the mutation of these genes will allow cells to enter anaphase even though the spindle is damaged and chromosome segregation is impaired. By screening mutant budding yeast cells that cannot arrest in the presence of spindle damage, they discovered two classes of aptly named genes: mitotic arrest deficient (*MAD*), and budding uninhibited by benzimidazole (*BUB*) genes [19,20]. Because this checkpoint appeared to respond to damage to the mitotic spindle, it was termed the spindle assembly checkpoint. This discovery confirmed that the eukaryotic cell uses a surveillance mechanism to regulate the metaphase to anaphase transition.

4. Elucidation of the Design and Operation of the SAC

The discovery of the SAC unleashed a decade-long search for genes and proteins involved in the implementing it. In retrospect, these efforts resemble an exciting game of solving a complex jigsaw puzzle. Important pieces of this puzzle, in the form of major SAC activities, were already in hand. It was known that the SAC detects spindle damage and then blocks the onset of anaphase in response (Figure 1). To be able to accomplish these functions, the SAC must perform at least three activities: (1) detect damage to the spindle and/or unaligned chromosomes in the spindle, (2) inhibit anaphase onset, and (3) prevent sister chromatid separation. With this knowledge, the race to solve the SAC jigsaw puzzle was underway.

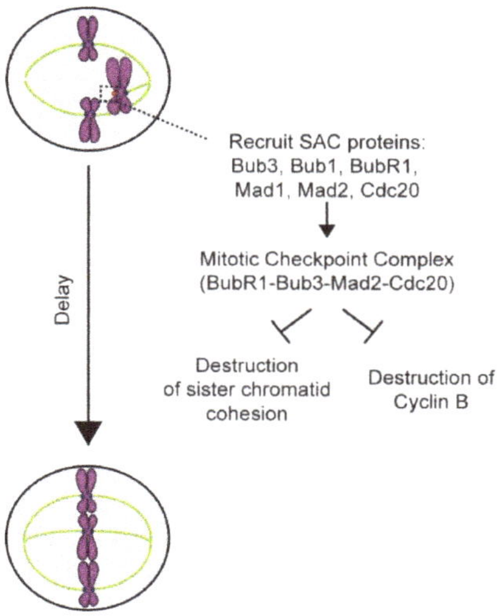

Figure 1. Schematic of the overall design of the spindle assembly checkpoint (SAC).

As discussed earlier, the significance of chromosome attachment to the spindle for timely anaphase onset was clear from early cytological studies. Therefore, cell biologists suspected that the mitotic arrest observed upon spindle damage was a response to the creation of unattached chromosomes rather than damage to the spindle. To explain why cells with damaged spindles arrest in mitosis, McIntosh presented a clear, mechanistic hypothesis [7]. He proposed that the centromeric region of unattached chromosomes generates a "wait-anaphase" signal in order to prevent anaphase onset. Influenced by Nicklas' vivid demonstration of the ability of mechanical forces to influence the biochemistry of kinetochore-microtubule attachment, McIntosh also proposed that the tension in the centromeric region of each chromosome, which is generated by the opposing forces generated by sister kinetochores, plays a critical role: it stops the production of the "wait-anaphase" signal, and thus silences the SAC. Evidence in support of McIntosh's hypotheses accumulated quickly through cell biological experimentation. For example, mutations in the DNA sequence of the genetically defined point centromere found in budding yeast significantly delayed mitosis [21]. This result independently confirmed the correlation between centromere function and cell cycle progression that had been established by the Earnshaw group [22]. Rieder's classic experiment involving laser ablation provided unequivocal evidence for the activating role of unattached kinetochores in SAC [23,24]. Rieder showed that cells containing even one unattached kinetochore were blocked in metaphase. Importantly, he showed that the ablation of this single unattached kinetochore by a focused laser beam was sufficient to remove the metaphase block, and allow the cell to enter anaphase within minutes. This result confirmed that the kinetochore detects a lack of microtubule attachment, and transduces this information into a biochemical signal to prevent anaphase onset.

How does the unattached kinetochore transduce information regarding the lack of microtubule attachment and convey it to the biochemical machinery that drives the cell cycle? Elucidation of the biochemical activities involved in this signal transduction process took place at a rapid pace, because the genes involved in SAC signaling were already known. The discovery of *MAD2* led to the characterization of the function of its product, Mad2, in human cells and in *Xenopus* extracts [25,26]. This work confirmed that Mad2 is necessary for activating the SAC, and more importantly, demonstrated that Mad2 localizes exclusively at unattached kinetochores. These findings revealed that unattached kinetochores are also the site of the biochemical activity that generates the "wait-anaphase" signal. Characterization of the Mad2 protein also led to the discovery of its binding partner, the protein Mad1 [27]. Subsequent studies localized other SAC proteins to unattached kinetochores as well, and found that this localization is highly dynamic [28,29]. The dynamic nature of SAC protein localization lent credence to the notion that kinetochores assemble the "wait-anaphase" signal, which then spreads throughout the cell volume to inhibit anaphase onset.

After unattached kinetochores were established as the site of SAC signal generation, the focus of research turned to the mechanism by which the SAC signal prevents anaphase onset. Vital clues to this puzzle were already in hand: that cyclin-dependent kinase 1 regulates mitosis, that its activator, cyclin B, is degraded in anaphase, and that this degradation occurs via a ubiquitin-mediated pathway [30–32]. Clever genetic screens and biochemical experiments based on these clues led researchers to subunits of the anaphase promoting complex/cyclosome (APC/C) [33,34]. With the discovery of the APC/C, researchers focused their attention on how the cell inhibits APC/C prior to anaphase. Using genetic screens of mutant alleles of genes implicated in the cell division cycle (*CDC* genes), two studies discovered the activating subunits of the APC/C: Cdc20 and Cdh1 [35,36]. The discovery of Cdc20 shifted research focus to the biochemical nature of the kinetochore generated "wait-anaphase" signal. These investigations found that the SAC proteins Bub3, Mad2, Mad3/BubR1, and Cdc20 interact with one another, and that this interaction is necessary to sequester Cdc20 [37,38]. The complex of these four proteins came to be known as the Mitotic Checkpoint Complex (MCC). Careful biochemical characterization revealed that formation of the MCC depletes Cdc20 from the cytosol, and thus deprives APC/C of its activating subunit. Thus it was finally clear that unattached kinetochores activate the SAC by recruiting SAC proteins, and enabling them to bind to and sequester Cdc20. Sequestration of Cdc20 keeps APC/C inactive and inhibits anaphase onset. As discussed later, a very recent study demonstrates that the MCC uses yet another mechanism in order to act as a potent inhibitor of the APC/C [39].

In addition to preventing anaphase onset, the SAC must also protect the cohesion between sister chromatids during the cell cycle arrest. This notion was supported by the observation that cells carrying mutations in the APC/C genes not only arrested in mitosis, but also failed to separate sister chromatids [40,41]. A genetic screen based on this observation yielded the precocious dissociation of sister chromatids gene (*PDS1*), and revealed that mutations in *PDS1* led to the premature separation of sister chromatids prior to anaphase. Furthermore, this observation suggested that the APC/C might target a protein involved in sister chromatid cohesion for degradation. In fact, earlier studies had shown that complete degradation of cyclin B, the main APC/C target known at the time, is not necessary for anaphase onset [42]. Using the discovery of *PDS1* as a toe-hold, researchers designed yet another genetic screen that yielded subunits of the Cohesin complex, a remarkable protein clamp that hold sister chromatids together, and which must be broken apart by APC/C-directed proteolysis to allow sister chromatid separation [43,44]. Discoveries of *PDS1* and the Cohesin complex clarified the role of Pds1 as the inhibitor of Cohesin destruction [45,46]. With this information, the protease that cleaves Cohesin, known as separase, was also discovered [47,48]. These discoveries outlined the third process that is necessary for the effective operation of the SAC.

In this manner, the major pieces of the jigsaw puzzle of the SAC were set in place in less than 10 years after its discovery. This knowledge led to the discovery of new proteins and activities critical to SAC function. Of note, the kinetochore proteins Ndc80 and the Spc105 were identified [49]. These proteins were later revealed to be critical components of the SAC activation machinery. The involvement of Mps1 kinase in SAC signaling was also revealed [50]. Thus, a firm foundation for defining the molecular components and biochemical activities of the SAC was established.

5. Molecular Mechanisms Underlying SAC Activation and Inactivation

The operation of the SAC during cell division is deceptively simple. Each unattached kinetochore generates a biochemical signal to inhibit APC/C activity, and thereby preventing the degradation of mitotic proteins and sister chromatid cohesion. Once the last unattached kinetochore forms a stable attachment to the mitotic spindle, the APC/C is unleashed and anaphase ensues. This seemingly simple sequence of events requires the interlocked operation of three distinct processes, each of which employs the coordinated activity of many different proteins. These processes are: (1) detection of the lack of attachment by the kinetochores, (2) recruitment of SAC proteins to the unattached kinetochore and production of the wait-anaphase signal, and (3) rapid inactivation of the "wait-anaphase" signal after the last kinetochore forms stable attachment to the spindle. The following discussion presents a concise description of the current understanding of the mechanisms underlying these individual processes.

5.1. Detection of the Lack of "End-On" Microtubule Attachment to the Kinetochore

During cell division, the eukaryotic kinetochore grabs onto an approximately 40 nm section of the plus-ends of one or more microtubules even as these plus-ends grow and shrink [51,52]. The kinetochore is exquisitely sensitive to such "end-on" attachment: it is able to distinguish this type of attachment from the absence of microtubule attachment as well as from the so-called "lateral" attachment to the microtubule lattice. Moreover, the kinetochore responds to a change in its end-on attachment state almost instantaneously as inferred from the loss or recruitment of SAC proteins by the kinetochore upon the gain or loss of end-on attachment [53,54]. How does the kinetochore detect changes in its attachment state and then transduce this information into a biochemical signal?

The kinetochore relies on two properties to control the SAC: the biochemical activities of a trio of protein components and their nanoscale organization in the kinetochore (Figure 2a). The three protein components are the kinetochore protein complexes Ndc80 and Spc105/KNL1-ZWINT1 and the Mps1 kinase. The function of each protein is clear. The Ndc80 complex is the essential recruitment site for Mps1 in the kinetochore, while Spc105/KNL1 is the primary target of Mps1 kinase activity [55–58]. Mps1 phosphorylates Spc105/KNL1 at several phosphorylation sites to start a biochemical cascade that recruits all the SAC proteins, including Mad2, to the kinetochore. Importantly, the phosphorylation of Spc105/KNL1 by Mps1 is exquisitely sensitive to end-on microtubule attachment. The sensitivity stems from the ingenious design of the end-on kinetochore-microtubule attachment. End-on attachment is established by the Calponin Homology domains of the Ndc80 complex, the same domains that recruit Mps1 [59,60]. Moreover, the microtubule-binding and Mps1-binding surfaces in the Calponin Homology domains partially overlap. Consequently, Mps1 directly competes with the microtubule tip for binding the Calponin Homology domains. It robustly binds unattached kinetochores, but gets dislodged during the formation of end-on microtubule attachment (Figure 2b top). This prevents the phosphorylation of Spc105/KNL1, and thus disrupts SAC signaling.

The competition between Mps1 and the microtubule tip for binding to the Calponin Homology domain removes a large fraction of Mps1, but not all of it. Moreover, under certain conditions, kinetochores with end-on attachments retain Mps1 and recruit SAC proteins, but they do not delay anaphase onset [61,62]. Finally, Mps1 kinase activity is present within the metaphase kinetochore possessing end-on microtubule attachment, because it activates the SAC if Mad1 is artificially tethered to the kinetochore [63–65]. Why does the residual Mps1 in the metaphase kinetochore not phosphorylate Spc105/KNL1 nor activate the SAC?

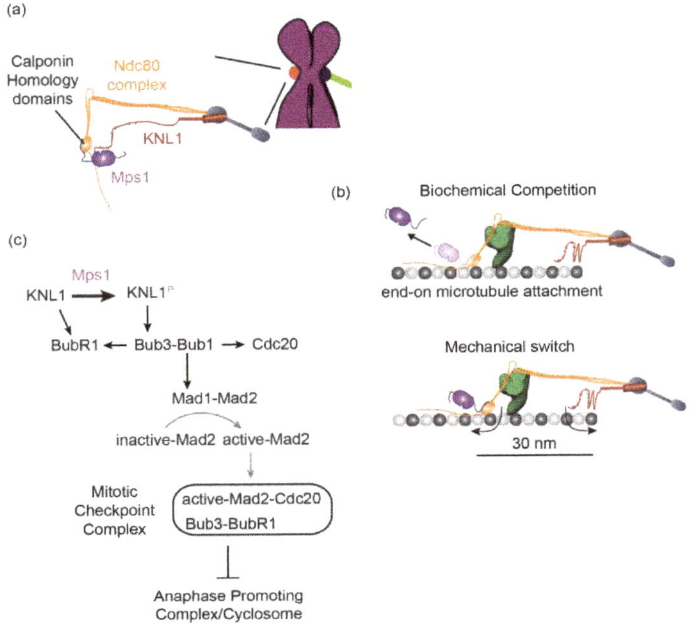

Figure 2. (**a**) Cartoon depicting the three-component microtubule-sensing mechanism in a kinetochore that lacks end-on microtubule attachment. (**b**) Microtubule attachment disrupts SAC signaling via two distinct mechanisms. The cartoons depict 1D visualization of the budding yeast kinetochore [66]. The green shape represents the Dam1 ring found in budding yeast; the blue rod like molecule is the Mtw1/Mis12 complex. Top: end-on attachment dislodges a large fraction of the Mps1 from the kinetochore. Bottom: End-on attachment separates the Calponin-Homology domains and the phosphodomain of Spc105/KNL1 from each other. (**c**) A simplified schematic of the biochemical interaction network that recruits SAC proteins to the unattached kinetochore. Black arrows represent binding to proteins localized in the kinetochore. Gray arrow indicates a conformational change that converts inactive Mad2 (also known as "Open" Mad2) into its active form ("Closed" Mad2).

The answer to this question is likely to lie in the architecture of the kinetochore-microtubule attachment [66]. High-resolution colocalization of fluorescently labeled kinetochore proteins in budding yeast, *Drosophila*, and human kinetochores revealed that each protein occupies a distinct average position along the long axis of the end-on kinetochore-microtubule attachment [67–69]. Importantly, the Calponin Homology domains are separated from the phosphodomain of Spc105/KNL1 by a distance of ~30 nm. This means that the Mps1 bound to the Calponin Homology domains will also be 30 nm away from Spc105/KNL1, its phosphorylation target. This small separation can be crucial for the SAC as demonstrated by the analysis of the budding yeast kinetochore [66]. Like human kinetochores, budding yeast kinetochores retain a fraction of Mps1, even after forming end-on microtubule attachment [55,66]. In budding yeast, the only reason why this residual Mps1 cannot activate the SAC is that the 30 nm separation between the Mps1 binding site in the kinetochore and Spc105/KNL1 in the kinetochore-microtubule attachment prevents Mps1 from phosphorylating Spc105/KNL1. In fact, experimentally bridging the 30 nm gap between the kinase and its substrate, either by moving Mps1 closer to Spc105/KNL1, or vice versa, is sufficient to re-activate the SAC even on kinetochores with stable end-on microtubule attachments. These findings suggest that the yeast kinetochore functions as a mechanical toggle-switch comprising the Calponin Homology domains of the Ndc80 complex and the phosphodomain of Spc105/KNL1. Bringing the two terminals of this toggle-switch close together turns the SAC on, whereas their separation turns the SAC off. Only stable end-on microtubule attachment can reliably and persistently separate the two terminals from one-another (Figure 2b, bottom). Structural properties of the Ndc80 complex, most notably its flexible hinge, likely facilitate the microtubule in pulling the Calponin Homology domain away from the phosphodomain of Spc105/KNL1 [70]. Similarly, the phosphodomain of Spc105/KNL1 also binds to the microtubule, which probably prevents it from approaching the Calponin Homology domains [71]. Thus, the nanoscale architecture of the yeast kinetochore plays an essential role in its ability to detect end-on microtubule attachment.

These investigations elucidate how the eukaryotic kinetochore senses end-on microtubule attachment. Even though the role of kinetochore architecture in sensing microtubule attachment was revealed using the budding yeast kinetochore, this role is likely to pertain to a wide range of organisms because of the remarkable conservation of the architecture of the end-on kinetochore-microtubule attachment. This role also highlights the functional significance of such attachments during mitosis. End-on kinetochore-microtubule attachments have been found in mitotic cells of nearly every eukaryotic organism that has been studied so far [52]. However, such attachments are not necessary for chromosome congression; achievement and maintenance of chromosome congression reflects a balance of opposing forces acting on each chromosome. Indeed, the holocentric chromosomes in *C. elegans* congress to the spindle equator using lateral kinetochore-microtubule attachments during meiosis [72]. Even in HeLa cells, chromosome congression to the metaphase plate can be achieved using lateral attachments, if the force generation mechanisms in the mitotic spindle are suitably manipulated to facilitate this process [73]. Yet, end-on attachment is highly conserved throughout eukaryotic evolution. An obvious reason for this conservation may be that end-on attachment seamlessly integrates the feedback control mechanism and the force generation machinery in the kinetochore [74,75].

5.2. Generation of the Mitotic Checkpoint Complex

Unattached kinetochores recruit SAC signaling proteins from the cytosol to generate the "wait-anaphase" signal in the form of the Mitotic Checkpoint Complex (Figure 2c). This recruitment is achieved by a cascade of biochemical interactions that recruit Bub3, Bub1, BubR1, Cdc20, as well as Mad1 and Mad2. This cascade is initiated when Mps1 phosphorylates Spc105/KNL1 [76–78]. Mps1 targets several sites within Spc105/KNL1 with the consensus amino acid sequence "Met-Glu-Lys-Thr", commonly referred to as MELT repeats. Each phosphorylated MELT repeat can bind one molecule of the Bub3-Bub1 protein complex by making contact with residues in both Bub3 and Bub1 [79]. The recruitment of the Bub3-Bub1 complex is the key event, because Bub1 provides a binding interface for BubR1, Cdc20, and the Mad1-Mad2 complex [80,81]. BubR1 also recruits Cdc20 [81]. Finally, the metazoan Spc105/KNL1 protein contains two related sequences known as KI motifs because of their amino acid sequence, each of which binds directly to Bub1 and BubR1 respectively in an Mps1-independent manner [82–84]. Bub1 molecules recruited to the kinetochore as part of the Bub3-Bub1 complex are also phosphorylated by Mps1, which enables them to interact with and recruit the heterotetrameric Mad1-Mad2 complex to the kinetochore [80,85]. These regulated biochemical interactions together ensure that all components of the MCC are localized within an unattached kinetochore, and pave the way for MCC formation. However, the Mad1-Mad2 complex recruited to the kinetochore in this manner does not become a part of the MCC. Instead it participates in catalyzing the conversion of "inactive" conformations of cytosolic Mad2 into an "active" conformation, which then gets incorporated into the MCC [86].

The molecular mechanism by which Mad2 switches between active and inactive conformations is yet another fascinating process in the cell cycle, and its molecular details have been the subject of several studies [1]. Suffice it to say here that the Mad2 molecule complexed with Mad1 assumes the active conformation, and it forms a conformational heterodimer with cytosolic Mad2 molecules that are in the inactive conformation, thus recruiting these molecules to the kinetochore [87]. The Mad1-Mad2 complex is then thought to act as a template that converts inactive Mad2 molecules into their active form [86]. Intriguingly, Mps1 kinase activity is also required for the conversion of the inactive Mad2 conformation into the active conformation [88,89]. The functional effect of the Mps1-mediated phosphorylation in the conformational change is unknown. However, the involvement of Mps1 in every step of the SAC cascade suggests that it functions as a "licensing kinase" that firmly tethers the entire signaling cascade to unattached kinetochores.

In addition to SAC protein recruitment, several molecular interactions add to the complexity of the central SAC signaling cascade and ensure robust SAC signaling. The Aurora B kinase, which is involved in the error correction pathway, enhances SAC signaling by promoting Mps1 recruitment [90]. Polo-like kinase 1, which licenses centrosome duplication, also phosphorylates Cdc20 molecules within the kinetochore to prevent them from activating the APC/C [80,91]. The ultimate goal of this network of biochemical interactions is to generate the MCC, which then inhibits APC/C. It is important to note that this simple description does not explain the remarkable potency with which the MCC inhibits the APC/C: one or a few unattached kinetochores generate enough MCC to affect mitotic progression within ~5 minutes [54]. Recently published biochemical experimentation explains why the APC/C is highly sensitive to MCC [39]. As discussed earlier, the MCC reduces APC/C activation by sequestering Cdc20. Additionally, it also binds to a second molecule of Cdc20 that is already complexed with APC/C, and by doing so, inhibits the APC/C that has been activated. Finally, APC/C itself also contributes to the maintenance of the SAC by targeting Cdc20 for degradation [92].

From the perspective of cell biology, the complex biochemical interactions that generate MCC must meet two demands. First, they must ensure that a single unattached kinetochore can produce a sufficiently large quantity of MCC so that anaphase is delayed and chromosome missegregation is averted. Second, they must also ensure that the generation of MCC does not scale linearly with the number of unattached kinetochores in the cell [93]. A dividing cell contains a large number of unattached kinetochores in prophase. If these kinetochores produce a proportionately large quantity of MCC, then the result could be the accumulation of a vast excess of MCC, and consequently, unnecessarily delay in anaphase onset even after all chromosomes attach to the spindle. Meeting these contrasting demands using a biochemical signaling cascade is challenging. In fact, cell biological experimentation suggests that the kinetics of MCC generation and the steady-state MCC concentration can fall short of the target necessary for complete APC/C inhibition. If this happens and APC/C activity is not fully inhibited, the residual activity steadily degrades cyclin B, and perhaps other mitotic proteins, until cyclin B levels fall below the threshold necessary to maintain the biochemical state of the cell corresponding to mitosis [94]. As a result, the cell enters anaphase even as it contains unattached kinetochores [95,96].

The significance of the kinetics of MCC generation and steady-state concentration of MCC was demonstrated by a set of three elegant studies. One of these studies, which was discussed earlier, created different numbers of unattached kinetochores by destroying their attachment to the spindle in metaphase cells, and asked whether these cells activated the SAC and arrested in mitosis [54]. This study found that the unattached kinetochores do not always inhibit anaphase. Despite recruiting normal levels of SAC proteins, these kinetochores cannot fully suppress APC/C activity presumably because they cannot produce a sufficient quantity of MCC. This observation implies that the signaling cascade of the SAC must be calibrated such that a single unattached kinetochore produces a sufficiently large signal at a high rate. Two other studies, one in fission yeast and the other in human cells, demonstrated that the duration of the SAC-mediated metaphase arrest inversely correlates with the rate of cyclin B degradation. Therefore, the steady-state MCC concentration also directly correlates with the duration of the mitotic arrest achieved [97,98]. The functional significance of the kinetics of MCC generation and steady-state MCC concentration in the dividing cell is also apparent from the results of two unrelated studies. In human cells, MCC generation begins in interphase via interactions of the

Mad1-Mad2 complex with the Mps1 kinase at the nuclear envelop [99]. Although the exact mechanism of MCC generation in this case is unclear, this pool of MCC is required for accurate chromosome segregation. This surprising result suggests that the MCC generated during interphase inhibits APC/C activity during early mitosis to slow down mitotic progression, while the kinetochore-based MCC generation cascade is ramping up. Thus, the interphase MCC generation effectively acts as a buffering mechanism to minimize APC/C activity during early mitosis. Finally, the significance of the MCC generation capacity of the kinetochore was also demonstrated by observations of SAC signaling in *Xenopus* egg extracts. These experiments revealed that kinetochores expand their signaling capacity in order to bolster the steady-state MCC concentration [100].

5.3. Inactivation of the SAC

SAC inactivation is mediated by two distinct processes. The first process involves the silencing of the SAC signaling events within a kinetochore. The kinetochore loses SAC proteins within a couple of minutes after establishing end-on microtubule attachments. Critical to this process is the disruption phosphorylation of Spc105/KNL1, and potentially Bub1 and Mad1, by end-on microtubule attachment as discussed earlier. This event enables Protein Phosphatase 1 (PP1) to remove the phosphorylation on Spc105/KNL1. In fact, KNL1/Spc105 uses a conserved PP1 recruitment motif within its phosphodomain to recruit PP1 to the kinetochore [101,102]. Moreover, recent work shows that Protein Phosphatase 2A (PP2-B56) recruited by the kinetochore-bound BubR1 also dephosphorylates Spc105/KNL1 [103]. Finally, a microtubule-binding component of the metazoan kinetochore, the spindle and kinetochore-associated (SKA) complex, was also shown to recruit PP1 and promote SAC silencing [104]. Additionally, metazoan kinetochores employ dynein motors, which strip kinetochore-bound Mad1-Mad2 complexes and carry them to the spindle poles along the kinetochore-attached microtubules [53]. These processes together ensure that kinetochores with end-on microtubule attachments do not signal.

The second process of SAC inactivation rapidly dissipates the checkpoint signal in the cytosol. Metazoan cells employ two mechanisms to achieve this. A protein known as p31comet uses a particularly fascinating mechanism. p31comet structurally mimics the active conformation of Mad2 [105]. This allows it to bind to MCC, and extract Cdc20 from it. Additionally, a specialized ATPase called Thyroid receptor hormone interacting protein (TRIP13) converts the active form of Mad2 into the inactive form [106–108]. Both mechanisms are operational throughout mitosis, not just in anaphase. They ensure that anaphase ensues without delay after the last unattached kinetochore forms stable attachment.

The process of SAC silencing is usually rapid, as evidenced by live-cell observations that show that anaphase onset takes place within 15 minutes after the last unattached kinetochore forms end-on attachments [24]. Unnecessary delay in anaphase onset, even after all chromosomes have established bipolar attachment, is unlikely to have any positive outcomes. In fact, prolonged mitotic arrest is often deleterious to the cell [95]. A large fraction of the cells that arrest in mitosis undergo apoptosis. They also suffer from "cohesion fatigue" due to the degradation of sister chromatid cohesion over time [109–112]. Prolonged mitosis can also alter cell fate in the tissue context [113]. The mechanisms discussed above are likely crucial for avoiding these negative outcomes.

6. Directions for Future Investigations of the SAC

The extensive research spanning over two decades affords us a deep, mechanistic understanding of many facets of the SAC. This understanding provides a solid foundation to attack areas of the SAC that are not well-understood. The discussion below highlights four such areas.

6.1. What Does the Kinetochore Respond to—End-On Attachment to the Kinetochore, an Architectural Change within the Kinetochore Induced by Such Attachment, or Both?

This topic has been the subject of debate for many years [114,115]. As discussed in the previous section, two different mechanisms have been shown to disrupt SAC signaling at the kinetochore: biochemical competition and attachment-induced separation of two protein domains. Although it is clear that the physical separation of the Mps1 kinase bound to the Calponin-Homology domains from the Spc105/LNL1 phosphodomain is essential for SAC silencing in budding yeast, whether this mechanism is also important for SAC silencing in metazoan kinetochores must be addressed. In comparison to the budding yeast kinetochore, kinetochores in most other eukaryotes offer an additional challenge to SAC silencing. These kinetochores typically bind the plus-ends of many microtubules unlike the budding yeast kinetochore, which binds just one microtubule [52]. Furthermore, these microtubule attachments are dynamic: old attachments are lost and new attachments form even in metaphase. This means that the metazoan kinetochore will contain a number of unbound Calponin Homology domains. Why don't these domains recruit Mps1 and activate the SAC? Neither the biochemical competition model nor the mechanical switch model offers a satisfactory explanation. It is likely that additional mechanisms suppress signaling activity from metaphase kinetochores. Alternatively, it is possible that metaphase kinetochores may harbor trace SAC signaling activity that is not detectable by conventional methods.

It is also important to note here that the attachment-induced separation of two kinetochore proteins from one-another has sometimes been construed as "intra-kinetochore stretch", and hence considered a tension-based mechanism [116–118]. However, physical separation of two mechanically unlinked protein domains does not necessarily require a large force. The Ndc80 complex is linked to Spc105/KNL1 on the centromeric ends of the respective molecules. Crucially, however, the Calponin Homology domains and the unstructured phosphodomain of Spc105/KNL1 are not linked to each other, and as such they are likely to be free to move within a certain radius about the linked, centromeric ends of the two protein molecules. Therefore, sustained separation of their free ends can be achieved by the maintenance of the architecture of the end-on kinetochore-microtubule attachment. If this hypothesis is true, then it creates the possibility that end-on attachment is necessary, but not sufficient, to silence the SAC.

6.2. How Does the Kinetochore Generate a Sufficiently Large Quantity of MCC at a High Rate?

The kinetochore contains a rather small number of molecules of Ndc80 and Spc105/KNL1. For example, a human kinetochore contains approximately 250 molecules, whereas the kinetochore in the much smaller budding yeast contains only 8 molecules of Ndc80 and Spc105/KNL1 [119–121]. Yet, this small number of molecules is capable of generating a sufficiently large quantity of MCC, and delay cell division. In this regard, it is interesting that each Spc105/KNL1 molecule contains a large number of MELT motifs: ~19 in human cells [78]. Since each MELT motif can bind one Bub3-Bub1 complex, 19 MELT motifs, in principle, should be able to bind 19 Bub3-Bub1 molecules, and generate a 19-fold higher SAC signal. However, careful analysis of Bub3 recruitment reveals that on average only 30% of the MELT motifs bind Bub3-Bub1 [93,122]. In fact, engineered KNL1 molecules with just a single MELT motif and the KI repeats suffice to activate the SAC when cells are treated with nocodazole [82]. Why Spc105/KNL1 contains many MELT motifs but uses only a small fraction of these motifs, and how it succeeds in generating a strong wait-anaphase signal, are fundamental questions at the heart of the SAC. A significant challenge in addressing these questions is that the biochemical reactions leading up to the generation of the MCC take place in the nanoscopic structure of the kinetochore. The crowded environment of the kinetochore makes it extremely difficult to measure the biochemical rates of individual reactions. Whether and how the localization of SAC proteins within unattached kinetochores alters or enhances the rate of MCC generation is also a key question that needs to be addressed.

6.3. Is the SAC a Switch or a Rheostat?

This complex question does not have a simple answer. The operation of the SAC during cell division gives the distinct impression of a switch-like behavior [4,23,24]. The response of the kinetochore to end-on microtubule attachment is also switch-like. On the other hand, the SAC can have different strengths specified by different the steady-state level of MCC in the dividing cell. Consequently, an active SAC can produce different lengths of delay in anaphase onset [97,123]. This SAC behavior is analogous to that of a rheostat. It is also worth noting that a kinetochore can be actively generating MCC, but unable to inhibit anaphase [54]. Therefore, analysis of the SAC that takes into account the two separate systems that underlie its operation: the kinetochore-based SAC activation system and the cytoplasmic SAC signaling cascade, is needed to define the operation of the SAC in its entirety.

6.4. Can a Defective SAC Cause Aneuploidy?

The majority of tumors contain aneuploid cells, and exhibit high rates of chromosome missegregation. Because of the critical role that the SAC plays in ensuring accurate chromosome segregation, aberrant SAC signaling is likely to be an essential aspect of cancer cell biology. In fact, cancer cells quite frequently misregulate the expression of one or more critical SAC protein including Mad2, Bub3, Bub1, and BubR1 [124]. The strongest data implicating aberrant SAC come from studies of mouse models [125]. However, whether aberrant expression of SAC proteins is directly responsible for generating aneuploidy that leads to tumorigenesis and cancer is not clear. This is because many cancer cell lines appear to have a functional SAC, even if they express SAC proteins aberrantly [126]. Therefore, whether an aberrant SAC is the causative factor of aneuploidy and tumorigenesis, or if it is a consequence of aneuploidy arising from other factors, needs to be determined.

The uncertainty regarding the role of the SAC in cancer cell biology likely stems from the fact that the SAC is a biochemical approximation of a toggle-switch. Despite its switch like operation, the strength of the SAC depends on the steady-state level of MCC generation and the rate of MCC generation by individual kinetochores [54,97,98]. This means that the conventional methodology for assessing SAC function, which is to depolymerize the spindle and quantify the duration of cell cycle arrest, suffers from a key limitation. This method quantitates only the maximum strength of the SAC. It cannot determine whether and how subtle changes in SAC strength, i.e., the potency of MCC generation, caused by aberrant expression of one or more SAC proteins affect chromosome segregation accuracy. This is because conventional assays based on spindle depolymerization generate a large number of unattached kinetochores, and thus mimic the prophase, when the dividing cell contains a large number of unattached kinetochores. As mitosis progresses, unattached kinetochores attach to spindle microtubules and cease to signal, and finally just one unattached signaling kinetochore is left. This is when optimal strength of the SAC is the most critical. Only if the last unattached kinetochore reliably delays cell division, chromosome missegregation will be averted. The misregulation of SAC genes can impact the ability of this kinetochore to delay cell division, and hence increase the rate of chromosomal instability. Future studies of the SAC will require new techniques to quantify subtle changes in the SAC signaling cascade, and then study whether such changes elevate the rate of chromosome missegregation during cell division.

7. Conclusions

Research spanning over two decades has revealed the elegant biochemical design, the molecular complexity, and the efficient cell biological operation of the SAC. On-going innovative research is adding new dimensions to the SAC field. For example, a truly fascinating field is the evolutionary biology SAC genes and proteins [127,128]. Scaling of SAC strength with changing cell size during development is another topic that merits attention [129]. These investigations will add to this knowledge, and fully define the molecular mechanisms and design of the SAC.

A pressing need for enabling a complete understanding the SAC is the technical capability to experimentally control it in vivo, and then quantify the individual biochemical reactions in the SAC signaling cascade. Mathematical models to simulate the operation of SAC in space and time are also necessary. Only such models can account for how the operation of the SAC changes in the context of a number of parameters, biochemical (concentrations of SAC proteins), physical (kinetochore size and the volume of the dividing cell), and physiological (species specific duration of the cell cycle, number of chromosomes, etc.), that characterize the dividing cell [96,130,131]. Integration of quantitative data with mathematical modeling will likely elucidate the biochemical design and cell biological operation of the SAC.

Acknowledgments: The author would like to thank Alex Kukreja, Mara Duncan, and Yukiko Yamashita for their critical reading of the manuscript. This work was supported by R01-GM-112992.

Conflicts of Interest: The author declares no conflict of interest.

References

1. Musacchio, A. The molecular biology of spindle assembly checkpoint signaling dynamics. *Curr. Biol.* **2015**, *25*, R1002–R1018. [CrossRef] [PubMed]
2. Wieser, S.; Pines, J. The biochemistry of mitosis. *Cold Spring Harb. Perspect. Biol.* **2015**, *7*. [CrossRef] [PubMed]
3. Bajer, A.; Molè-Bajer, J. Cine-micrographic studies on mitosis in endosperm. *Chromosoma* **1955**, *7*, 558–607. [CrossRef]
4. Callan, H.G.; Jacobs, P.A. The meiotic process in *Mantis religiosa* L. males. *J. Genet.* **1957**, *55*, 200–217. [CrossRef]
5. Zirkle, R.E. Ultraviolet-microbeam irradiation of newt-cell cytoplasm: Spindle destruction, false anaphase, and delay of true anaphase. *Radiat. Res.* **1970**, *41*, 516–537. [CrossRef] [PubMed]
6. Nicklas, R.B. Chromosome micromanipulation. II. Induced reorientation and the experimental control of segregation in meiosis. *Chromosoma* **1967**, *21*, 17–50. [CrossRef] [PubMed]
7. McIntosh, J.R. Structural and mechanical control of mitotic progression. *Cold Spring Harb. Symp. Quant. Biol.* **1991**, *56*, 613–619. [CrossRef] [PubMed]
8. Neskovic, B.A. Developmental phases in intermitosis and the preparation for mitosis of mammalian cells in vitro. *Int. Rev. Cytol.* **1968**, *24*, 71–97. [PubMed]
9. Mazia, D. Mitosis and the physiology of cell division. In *The Cell*; Brachet, J., Mirsky, A.E., Eds.; Academic Press: Cambridge, MA, USA, 1961; Volume III, pp. 77–413.
10. Pringle, J.R. The use of conditional lethal cell cycle mutants for temporal and functional sequence mapping of cell cycle events. *J. Cell. Physiol.* **1978**, *95*, 393–405. [CrossRef] [PubMed]
11. Simchen, G. Cell cycle mutants. *Annu. Rev. Genet.* **1978**, *12*, 161–191. [CrossRef] [PubMed]
12. Evans, T.; Rosenthal, E.T.; Youngblom, J.; Distel, D.; Hunt, T. Cyclin: A protein specified by maternal mRNA in sea urchin eggs that is destroyed at each cleavage division. *Cell* **1983**, *33*, 389–396. [CrossRef]
13. Weinert, T.A.; Hartwell, L.H. The RAD9 gene controls the cell cycle response to DNA damage in Saccharomyces cerevisiae. *Science* **1988**, *241*, 317–322. [CrossRef] [PubMed]
14. Hartwell, L.H.; Weinert, T.A. Checkpoints: Controls that ensure the order of cell cycle events. *Science* **1989**, *246*, 629–634. [CrossRef] [PubMed]
15. Zieve, G.W.; Turnbull, D.; Mullins, J.M.; McIntosh, J.R. Production of large numbers of mitotic mammalian cells by use of the reversible microtubule inhibitor nocodazole. Nocodazole accumulated mitotic cells. *Exp. Cell Res.* **1980**, *126*, 397–405. [CrossRef]
16. Umesono, K.; Toda, T.; Hayashi, S.; Yanagida, M. Cell division cycle genes NDA2 and NDA3 of the fission yeast Schizosaccharomyces pombe control microtubular organization and sensitivity to anti-mitotic benzimidazole compounds. *J. Mol. Biol.* **1983**, *168*, 271–284. [CrossRef]
17. Jacobs, C.W.; Adams, A.E.; Szaniszlo, P.J.; Pringle, J.R. Functions of microtubules in the *Saccharomyces cerevisiae* cell cycle. *J. Cell Biol.* **1988**, *107*, 1409–1426. [CrossRef] [PubMed]
18. Bernat, R.L.; Borisy, G.G.; Rothfield, N.F.; Earnshaw, W.C. Injection of anticentromere antibodies in interphase disrupts events required for chromosome movement at mitosis. *J. Cell Biol.* **1990**, *111*, 1519–1533. [CrossRef] [PubMed]

19. Li, R.; Murray, A.W. Feedback control of mitosis in budding yeast. *Cell* **1991**, *66*, 519–531. [CrossRef]
20. Hoyt, M.A.; Totis, L.; Roberts, B.T. *S. cerevisiae* genes required for cell cycle arrest in response to loss of microtubule function. *Cell* **1991**, *66*, 507–517. [CrossRef]
21. Spencer, F.; Hieter, P. Centromere DNA mutations induce a mitotic delay in *Saccharomyces cerevisiae*. *Proc. Natl. Acad. Sci. USA* **1992**, *89*, 8908–8912. [CrossRef] [PubMed]
22. Bernat, R.L.; Delannoy, M.R.; Rothfield, N.F.; Earnshaw, W.C. Disruption of centromere assembly during interphase inhibits kinetochore morphogenesis and function in mitosis. *Cell* **1991**, *66*, 1229–1238. [CrossRef]
23. Rieder, C.L.; Schultz, A.; Cole, R.; Sluder, G. Anaphase onset in vertebrate somatic cells is controlled by a checkpoint that monitors sister kinetochore attachment to the spindle. *J. Cell Biol.* **1994**, *127*, 1301–1310. [CrossRef] [PubMed]
24. Rieder, C.L.; Cole, R.W.; Khodjakov, A.; Sluder, G. The checkpoint delaying anaphase in response to chromosome monoorientation is mediated by an inhibitory signal produced by unattached kinetochores. *J. Cell Biol.* **1995**, *130*, 941–948. [CrossRef] [PubMed]
25. Li, Y.; Benezra, R. Identification of a human mitotic checkpoint gene: hsMAD2. *Science* **1996**, *274*, 246–248. [CrossRef] [PubMed]
26. Chen, R.H.; Waters, J.C.; Salmon, E.D.; Murray, A.W. Association of spindle assembly checkpoint component XMAD2 with unattached kinetochores. *Science* **1996**, *274*, 242–246. [CrossRef] [PubMed]
27. Hardwick, K.G.; Murray, A.W. Mad1p, a phosphoprotein component of the spindle assembly checkpoint in budding yeast. *J. Cell Biol.* **1995**, *131*, 709–720. [CrossRef] [PubMed]
28. Howell, B.J.; Hoffman, D.B.; Fang, G.; Murray, A.W.; Salmon, E.D. Visualization of Mad2 dynamics at kinetochores, along spindle fibers, and at spindle poles in living cells. *J. Cell Biol.* **2000**, *150*, 1233–1250. [CrossRef] [PubMed]
29. Howell, B.J.; Moree, B.; Farrar, E.M.; Stewart, S.; Fang, G.; Salmon, E.D. Spindle checkpoint protein dynamics at kinetochores in living cells. *Curr. Biol.* **2004**, *14*, 953–964. [CrossRef] [PubMed]
30. Holloway, S.L.; Glotzer, M.; King, R.W.; Murray, A.W. Anaphase is initiated by proteolysis rather than by the inactivation of maturation-promoting factor. *Cell* **1993**, *73*, 1393–1402. [CrossRef]
31. Glotzer, M.; Murray, A.W.; Kirschner, M.W. Cyclin is degraded by the ubiquitin pathway. *Nature* **1991**, *349*, 132–138. [CrossRef] [PubMed]
32. Surana, U.; Robitsch, H.; Price, C.; Schuster, T.; Fitch, I.; Futcher, A.B.; Nasmyth, K. The role of CDC28 and cyclins during mitosis in the budding yeast *S. cerevisiae*. *Cell* **1991**, *65*, 145–161. [CrossRef]
33. Sudakin, V.; Ganoth, D.; Dahan, A.; Heller, H.; Hershko, J.; Luca, F.C.; Ruderman, J.V.; Hershko, A. The cyclosome, a large complex containing cyclin-selective ubiquitin ligase activity, targets cyclins for destruction at the end of mitosis. *Mol. Biol. Cell* **1995**, *6*, 185–197. [CrossRef] [PubMed]
34. Irniger, S.; Piatti, S.; Michaelis, C.; Nasmyth, K. Genes involved in sister chromatid separation are needed for B-type cyclin proteolysis in budding yeast. *Cell* **1995**, *81*, 269–278. [CrossRef]
35. Schwab, M.; Lutum, A.S.; Seufert, W. Yeast Hct1 is a regulator of Clb2 cyclin proteolysis. *Cell* **1997**, *90*, 683–693. [CrossRef]
36. Hwang, L.H.; Lau, L.F.; Smith, D.L.; Mistrot, C.A.; Hardwick, K.G.; Hwang, E.S.; Amon, A.; Murray, A.W. Budding yeast Cdc20: A target of the spindle checkpoint. *Science* **1998**, *279*, 1041–1044. [CrossRef] [PubMed]
37. Hardwick, K.G.; Johnston, R.C.; Smith, D.L.; Murray, A.W. MAD3 encodes a novel component of the spindle checkpoint which interacts with Bub3p, Cdc20p, and Mad2p. *J. Cell Biol.* **2000**, *148*, 871–882. [CrossRef] [PubMed]
38. Sudakin, V.; Chan, G.K.; Yen, T.J. Checkpoint inhibition of the APC/C in HeLa cells is mediated by a complex of BUBR1, BUB3, CDC20, and MAD2. *J. Cell Biol.* **2001**, *154*, 925–936. [CrossRef] [PubMed]
39. Izawa, D.; Pines, J. The mitotic checkpoint complex binds a second CDC20 to inhibit active APC/C. *Nature* **2015**, *517*, 631–634. [CrossRef] [PubMed]
40. Yamamoto, A.; Guacci, V.; Koshland, D. Pds1p is required for faithful execution of anaphase in the yeast, *Saccharomyces cerevisiae*. *J. Cell Biol.* **1996**, *133*, 85–97. [CrossRef] [PubMed]
41. Yamamoto, A.; Guacci, V.; Koshland, D. Pds1p, an inhibitor of anaphase in budding yeast, plays a critical role in the APC and checkpoint pathway(s). *J. Cell Biol.* **1996**, *133*, 99–110. [CrossRef] [PubMed]
42. Surana, U.; Amon, A.; Dowzer, C.; McGrew, J.; Byers, B.; Nasmyth, K. Destruction of the CDC28/CLB mitotic kinase is not required for the metaphase to anaphase transition in budding yeast. *EMBO J.* **1993**, *12*, 1969–1978. [PubMed]

43. Michaelis, C.; Ciosk, R.; Nasmyth, K. Cohesins: Chromosomal proteins that prevent premature separation of sister chromatids. *Cell* **1997**, *91*, 35–45. [CrossRef]
44. Guacci, V.; Koshland, D.; Strunnikov, A. A direct link between sister chromatid cohesion and chromosome condensation revealed through the analysis of MCD1 in *S. cerevisiae*. *Cell* **1997**, *91*, 47–57. [CrossRef]
45. Zou, H.; McGarry, T.J.; Bernal, T.; Kirschner, M.W. Identification of a vertebrate sister-chromatid separation inhibitor involved in transformation and tumorigenesis. *Science* **1999**, *285*, 418–422. [CrossRef] [PubMed]
46. Uhlmann, F.; Lottspeich, F.; Nasmyth, K. Sister-chromatid separation at anaphase onset is promoted by cleavage of the cohesin subunit Scc1. *Nature* **1999**, *400*, 37–42. [PubMed]
47. Uhlmann, F.; Wernic, D.; Poupart, M.A.; Koonin, E.V.; Nasmyth, K. Cleavage of cohesin by the CD clan protease separin triggers anaphase in yeast. *Cell* **2000**, *103*, 375–386. [CrossRef]
48. Ciosk, R.; Zachariae, W.; Michaelis, C.; Shevchenko, A.; Mann, M.; Nasmyth, K. An ESP1/PDS1 complex regulates loss of sister chromatid cohesion at the metaphase to anaphase transition in yeast. *Cell* **1998**, *93*, 1067–1076. [CrossRef]
49. Wigge, P.A.; Jensen, O.N.; Holmes, S.; Soues, S.; Mann, M.; Kilmartin, J.V. Analysis of the Saccharomyces spindle pole by matrix-assisted laser desorption/ionization (MALDI) mass spectrometry. *J. Cell Biol.* **1998**, *141*, 967–977. [CrossRef] [PubMed]
50. Hardwick, K.G.; Weiss, E.; Luca, F.C.; Winey, M.; Murray, A.W. Activation of the budding yeast spindle assembly checkpoint without mitotic spindle disruption. *Science* **1996**, *273*, 953–956. [CrossRef] [PubMed]
51. Rieder, C.L. The Formation, structure, and composition of the mammalian kinetochore and kinetochore fiber. *Int. Rev. Cytol.* **1982**, *79*, 1–58. [PubMed]
52. McIntosh, J.R.; O'Toole, E.; Zhudenkov, K.; Morphew, M.; Schwartz, C.; Ataullakhanov, F.I.; Grishchuk, E.L. Conserved and divergent features of kinetochores and spindle microtubule ends from five species. *J. Cell Biol.* **2013**, *200*, 459–474. [CrossRef] [PubMed]
53. Howell, B.J.; McEwen, B.F.; Canman, J.C.; Hoffman, D.B.; Farrar, E.M.; Rieder, C.L.; Salmon, E.D. Cytoplasmic dynein/dynactin drives kinetochore protein transport to the spindle poles and has a role in mitotic spindle checkpoint inactivation. *J. Cell Biol.* **2001**, *155*, 1159–1172. [CrossRef] [PubMed]
54. Dick, A.E.; Gerlich, D.W. Kinetic framework of spindle assembly checkpoint signalling. *Nat. Cell Biol.* **2013**, *15*, 1370–1377. [CrossRef] [PubMed]
55. Hiruma, Y.; Sacristan, C.; Pachis, S.T.; Adamopoulos, A.; Kuijt, T.; Ubbink, M.; von Castelmur, E.; Perrakis, A.; Kops, G.J. CELL DIVISION CYCLE. Competition between MPS1 and microtubules at kinetochores regulates spindle checkpoint signaling. *Science* **2015**, *348*, 1264–1267. [CrossRef] [PubMed]
56. Ji, Z.; Gao, H.; Yu, H. CELL DIVISION CYCLE. Kinetochore attachment sensed by competitive Mps1 and microtubule binding to Ndc80C. *Science* **2015**, *348*, 1260–1264. [CrossRef] [PubMed]
57. Dou, Z.; Liu, X.; Wang, W.; Zhu, T.; Wang, X.; Xu, L.; Abrieu, A.; Fu, C.; Hill, D.L.; Yao, X. Dynamic localization of Mps1 kinase to kinetochores is essential for accurate spindle microtubule attachment. *Proc. Natl. Acad. Sci. USA* **2015**, *112*, E4546–E4555. [CrossRef] [PubMed]
58. Kemmler, S.; Stach, M.; Knapp, M.; Ortiz, J.; Pfannstiel, J.; Ruppert, T.; Lechner, J. Mimicking Ndc80 phosphorylation triggers spindle assembly checkpoint signalling. *EMBO J.* **2009**, *28*, 1099–1110. [CrossRef] [PubMed]
59. DeLuca, J.G.; Moree, B.; Hickey, J.M.; Kilmartin, J.V.; Salmon, E.D. hNuf2 inhibition blocks stable kinetochore-microtubule attachment and induces mitotic cell death in HeLa cells. *J. Cell Biol.* **2002**, *159*, 549–555. [CrossRef] [PubMed]
60. DeLuca, J.G.; Howell, B.J.; Canman, J.C.; Hickey, J.M.; Fang, G.; Salmon, E.D. Nuf2 and Hec1 are required for retention of the checkpoint proteins Mad1 and Mad2 to kinetochores. *Curr. Biol.* **2003**, *13*, 2103–2109. [CrossRef] [PubMed]
61. Vázquez-Novelle, M.D.; Petronczki, M. Relocation of the Chromosomal Passenger Complex Prevents Mitotic Checkpoint Engagement at Anaphase. *Curr. Biol.* **2010**, *20*, 1402–1407. [CrossRef] [PubMed]
62. Oliveira, R.A.; Hamilton, R.S.; Pauli, A.; Davis, I.; Nasmyth, K. Cohesin cleavage and Cdk inhibition trigger formation of daughter nuclei. *Nat. Cell Biol.* **2010**, *12*, 185–192. [CrossRef] [PubMed]
63. Kuijt, T.E.; Omerzu, M.; Saurin, A.T.; Kops, G.J. Conditional targeting of MAD1 to kinetochores is sufficient to reactivate the spindle assembly checkpoint in metaphase. *Chromosoma* **2014**, *123*, 471–480. [CrossRef] [PubMed]

64. Ballister, E.R.; Riegman, M.; Lampson, M.A. Recruitment of Mad1 to metaphase kinetochores is sufficient to reactivate the mitotic checkpoint. *J. Cell Biol.* **2014**, *204*, 901–908. [CrossRef] [PubMed]
65. Maldonado, M.; Kapoor, T.M. Constitutive Mad1 targeting to kinetochores uncouples checkpoint signalling from chromosome biorientation. *Nat. Cell Biol.* **2011**, *13*, 475–482. [CrossRef] [PubMed]
66. Aravamudhan, P.; Goldfarb, A.A.; Joglekar, A.P. The kinetochore encodes a mechanical switch to disrupt spindle assembly checkpoint signalling. *Nat. Cell Biol.* **2015**, *17*, 868–879. [CrossRef] [PubMed]
67. Joglekar, A.P.; Bloom, K.; Salmon, E.D. In vivo protein architecture of the eukaryotic kinetochore with nanometer scale accuracy. *Curr. Biol.* **2009**, *19*, 694–699. [CrossRef] [PubMed]
68. Wan, X.; O'Quinn, R.P.; Pierce, H.L.; Joglekar, A.P.; Gall, W.E.; DeLuca, J.G.; Carroll, C.W.; Liu, S.T.; Yen, T.J.; McEwen, B.F.; et al. Protein architecture of the human kinetochore microtubule attachment site. *Cell* **2009**, *137*, 672–684. [CrossRef] [PubMed]
69. Schittenhelm, R.B.; Chaleckis, R.; Lehner, C.F. Intrakinetochore localization and essential functional domains of Drosophila Spc105. *EMBO J.* **2009**, *28*, 2374–2386. [CrossRef] [PubMed]
70. Wang, H.W.; Long, S.; Ciferri, C.; Westermann, S.; Drubin, D.; Barnes, G.; Nogales, E. Architecture and flexibility of the yeast Ndc80 kinetochore complex. *J. Mol. Biol.* **2008**, *383*, 894–903. [CrossRef] [PubMed]
71. Espeut, J.; Cheerambathur, D.K.; Krenning, L.; Oegema, K.; Desai, A. Microtubule binding by KNL-1 contributes to spindle checkpoint silencing at the kinetochore. *J. Cell Biol.* **2012**, *196*, 469–482. [CrossRef] [PubMed]
72. Wignall, S.M.; Villeneuve, A.M. Lateral microtubule bundles promote chromosome alignment during acentrosomal oocyte meiosis. *Nat. Cell Biol.* **2009**, *11*, 839–844. [CrossRef] [PubMed]
73. Cai, S.; O'Connell, C.B.; Khodjakov, A.; Walczak, C.E. Chromosome congression in the absence of kinetochore fibres. *Nat. Cell Biol.* **2009**, *11*, 832–838. [CrossRef] [PubMed]
74. Joglekar, A.P.; Aravamudhan, P. How the kinetochore switches off the spindle assembly checkpoint. *Cell Cycle* **2016**, *15*, 7–8. [CrossRef] [PubMed]
75. Aravamudhan, P.; Felzer-Kim, I.; Gurunathan, K.; Joglekar, A.P. Assembling the protein architecture of the budding yeast kinetochore-microtubule attachment using FRET. *Curr. Biol.* **2014**, *24*, 1437–1446. [CrossRef] [PubMed]
76. Shepperd, L.A.; Meadows, J.C.; Sochaj, A.M.; Lancaster, T.C.; Zou, J.; Buttrick, G.J.; Rappsilber, J.; Hardwick, K.G.; Millar, J.B. Phosphodependent recruitment of Bub1 and Bub3 to Spc7/KNL1 by Mph1 kinase maintains the spindle checkpoint. *Curr. Biol.* **2012**, *22*, 891–899. [CrossRef] [PubMed]
77. London, N.; Ceto, S.; Ranish, J.A.; Biggins, S. Phosphoregulation of Spc105 by Mps1 and PP1 regulates Bub1 localization to kinetochores. *Curr. Biol.* **2012**, *22*, 900–906. [CrossRef] [PubMed]
78. Vleugel, M.; Tromer, E.; Omerzu, M.; Groenewold, V.; Nijenhuis, W.; Snel, B.; Kops, G.J. Arrayed BUB recruitment modules in the kinetochore scaffold KNL1 promote accurate chromosome segregation. *J. Cell Biol.* **2013**, *203*, 943–955. [CrossRef] [PubMed]
79. Primorac, I.; Weir, J.R.; Chiroli, E.; Gross, F.; Hoffmann, I.; van Gerwen, S.; Ciliberto, A.; Musacchio, A. Bub3 reads phosphorylated MELT repeats to promote spindle assembly checkpoint signaling. *Elife* **2013**, *2*. [CrossRef] [PubMed]
80. Jia, L.; Li, B.; Yu, H. The Bub1-Plk1 kinase complex promotes spindle checkpoint signalling through Cdc20 phosphorylation. *Nat. Commun.* **2016**, *7*. [CrossRef] [PubMed]
81. Di Fiore, B.; Davey, N.E.; Hagting, A.; Izawa, D.; Mansfeld, J.; Gibson, T.J.; Pines, J. The ABBA Motif Binds APC/C Activators and Is Shared by APC/C Substrates and Regulators. *Dev. Cell* **2015**, *32*, 358–372. [CrossRef] [PubMed]
82. Krenn, V.; Overlack, K.; Primorac, I.; van Gerwen, S.; Musacchio, A. KI motifs of human Knl1 enhance assembly of comprehensive spindle checkpoint complexes around MELT repeats. *Curr. Biol.* **2014**, *24*, 29–39. [CrossRef] [PubMed]
83. Kiyomitsu, T.; Murakami, H.; Yanagida, M. Protein interaction domain mapping of human kinetochore protein Blinkin reveals a consensus motif for binding of spindle assembly checkpoint proteins Bub1 and BubR1. *Mol. Cell Biol.* **2011**, *31*, 998–1011. [CrossRef] [PubMed]
84. Krenn, V.; Wehenkel, A.; Li, X.; Santaguida, S.; Musacchio, A. Structural analysis reveals features of the spindle checkpoint kinase Bub1-kinetochore subunit Knl1 interaction. *J. Cell Biol.* **2012**, *196*, 451–467. [CrossRef] [PubMed]

85. London, N.; Biggins, S. Mad1 kinetochore recruitment by Mps1-mediated phosphorylation of Bub1 signals the spindle checkpoint. *Genes Dev.* **2014**, *28*, 140–152. [CrossRef] [PubMed]
86. De Antoni, A.; Pearson, C.G.; Cimini, D.; Canman, J.C.; Sala, V.; Nezi, L.; Mapelli, M.; Sironi, L.; Faretta, M.; Salmon, E.D.; Musacchio, A. The Mad1/Mad2 complex as a template for Mad2 activation in the spindle assembly checkpoint. *Curr. Biol.* **2005**, *15*, 214–225. [CrossRef] [PubMed]
87. Mapelli, M.; Massimiliano, L.; Santaguida, S.; Musacchio, A. The MAD2 conformational dimer: Structure and implications for the Spindle Assembly Checkpoint. *Cell* **2007**, *131*, 730–743. [CrossRef] [PubMed]
88. Tipton, A.R.; Ji, W.; Sturt-Gillespie, B.; Bekier, M.E., 2nd; Wang, K.; Taylor, W.R.; Liu, S.T. Monopolar spindle 1 (MPS1) kinase promotes production of closed MAD2 (C-MAD2) conformer and assembly of the mitotic checkpoint complex. *J. Biol. Chem.* **2013**, *288*, 35149–35158. [CrossRef] [PubMed]
89. Hewitt, L.; Tighe, A.; Santaguida, S.; White, A.M.; Jones, C.D.; Musacchio, A.; Green, S.; Taylor, S.S. Sustained Mps1 activity is required in mitosis to recruit O-Mad2 to the Mad1-C-Mad2 core complex. *J. Cell Biol.* **2010**, *190*, 25–34. [CrossRef] [PubMed]
90. Krenn, V.; Musacchio, A. The Aurora B Kinase in Chromosome Bi-Orientation and Spindle Checkpoint Signaling. *Front. Oncol.* **2015**, *5*. [CrossRef] [PubMed]
91. O'Connor, A.; Maffini, S.; Rainey, M.D.; Kaczmarczyk, A.; Gaboriau, D.; Musacchio, A.; Santocanale, C. Requirement for PLK1 kinase activity in the maintenance of a robust spindle assembly checkpoint. *Biol. Open* **2015**, *5*, 11–19. [CrossRef] [PubMed]
92. Nilsson, J.; Yekezare, M.; Minshull, J.; Pines, J. The APC/C maintains the spindle assembly checkpoint by targeting Cdc20 for destruction. *Nat. Cell Biol.* **2008**, *10*, 1411–1420. [CrossRef] [PubMed]
93. Aravamudhan, P.; Chen, R.; Roy, B.; Sim, J.; Joglekar, A.P. Dual mechanisms regulate the recruitment of spindle assembly checkpoint proteins to the budding yeast kinetochore. *Mol. Biol. Cell* **2016**, *27*, 3405–3417. [CrossRef] [PubMed]
94. Brito, D.A.; Rieder, C.L. Mitotic checkpoint slippage in humans occurs via cyclin B destruction in the presence of an active checkpoint. *Curr. Biol.* **2006**, *16*, 1194–1200. [CrossRef] [PubMed]
95. Rieder, C.L.; Maiato, H. Stuck in division or passing through: What happens when cells cannot satisfy the spindle assembly checkpoint. *Dev. Cell* **2004**, *7*, 637–651. [CrossRef] [PubMed]
96. He, E.; Kapuy, O.; Oliveira, R.A.; Uhlmann, F.; Tyson, J.J.; Novák, B. System-level feedbacks make the anaphase switch irreversible. *Proc. Natl. Acad. Sci. USA* **2011**, *108*, 10016–10021. [CrossRef] [PubMed]
97. Collin, P.; Nashchekina, O.; Walker, R.; Pines, J. The spindle assembly checkpoint works like a rheostat rather than a toggle switch. *Nat. Cell Biol.* **2013**, *15*, 1378–1385. [CrossRef] [PubMed]
98. Heinrich, S.; Geissen, E.M.; Kamenz, J.; Trautmann, S.; Widmer, C.; Drewe, P.; Knop, M.; Radde, N.; Hasenauer, J.; Hauf, S. Determinants of robustness in spindle assembly checkpoint signalling. *Nat. Cell Biol.* **2013**, *15*, 1328–1339. [CrossRef] [PubMed]
99. Rodriguez-Bravo, V.; Maciejowski, J.; Corona, J.; Buch, H.K.; Collin, P.; Kanemaki, M.T.; Shah, J.V.; Jallepalli, P.V. Nuclear pores protect genome integrity by assembling a premitotic and Mad1-dependent anaphase inhibitor. *Cell* **2014**, *156*, 1017–1031. [CrossRef] [PubMed]
100. Wynne, D.J.; Funabiki, H. Kinetochore function is controlled by a phospho-dependent coexpansion of inner and outer components. *J. Cell Biol.* **2015**, *210*, 899–916. [CrossRef] [PubMed]
101. Liu, D.; Vleugel, M.; Backer, C.B.; Hori, T.; Fukagawa, T.; Cheeseman, I.M.; Lampson, M.A. Regulated targeting of protein phosphatase 1 to the outer kinetochore by KNL1 opposes Aurora B kinase. *J. Cell Biol.* **2010**, *188*, 809–820. [CrossRef] [PubMed]
102. Rosenberg, J.S.; Cross, F.R.; Funabiki, H. KNL1/Spc105 recruits PP1 to silence the spindle assembly checkpoint. *Curr. Biol.* **2011**, *21*, 942–947. [CrossRef] [PubMed]
103. Espert, A.; Uluocak, P.; Bastos, R.N.; Mangat, D.; Graab, P.; Gruneberg, U. PP2A-B56 opposes Mps1 phosphorylation of Knl1 and thereby promotes spindle assembly checkpoint silencing. *J. Cell Biol.* **2014**, *206*, 833–842. [CrossRef] [PubMed]
104. Sivakumar, S.; Janczyk, P.L.; Qu, Q.; Brautigam, C.A.; Stukenberg, P.T.; Yu, H.; Gorbsky, G.J. The human SKA complex drives the metaphase-anaphase cell cycle transition by recruiting protein phosphatase 1 to kinetochores. *Elife* **2016**, *5*. [CrossRef] [PubMed]
105. Yang, M.; Li, B.; Tomchick, D.R.; Machius, M.; Rizo, J.; Yu, H.; Luo, X. p31comet blocks Mad2 activation through structural mimicry. *Cell* **2007**, *131*, 744–755. [CrossRef] [PubMed]

106. Eytan, E.; Wang, K.; Miniowitz-Shemtov, S.; Sitry-Shevah, D.; Kaisari, S.; Yen, T.J.; Liu, S.T.; Hershko, A. Disassembly of mitotic checkpoint complexes by the joint action of the AAA-ATPase TRIP13 and p31(comet). *Proc. Natl. Acad. Sci. USA* **2014**, *111*, 12019–12024. [CrossRef] [PubMed]
107. Ye, Q.; Rosenberg, S.C.; Moeller, A.; Speir, J.A.; Su, T.Y.; Corbett, K.D. TRIP13 is a protein-remodeling AAA+ATPase that catalyzes MAD2 conformation switching. *Elife* **2015**, *4*. [CrossRef] [PubMed]
108. Miniowitz-Shemtov, S.; Eytan, E.; Kaisari, S.; Sitry-Shevah, D.; Hershko, A. Mode of interaction of TRIP13 AAA-ATPase with the Mad2-binding protein p31comet and with mitotic checkpoint complexes. *Proc. Natl. Acad. Sci. USA* **2015**, *112*, 11536–11540. [CrossRef] [PubMed]
109. Gorbsky, G.J. Cohesion fatigue. *Curr. Biol.* **2013**, *23*, R986–R988. [CrossRef] [PubMed]
110. Daum, J.R.; Potapova, T.A.; Sivakumar, S.; Daniel, J.J.; Flynn, J.N.; Rankin, S.; Gorbsky, G.J. Cohesion fatigue induces chromatid separation in cells delayed at metaphase. *Curr. Biol.* **2011**, *21*, 1018–1024. [CrossRef] [PubMed]
111. Meitinger, F.; Anzola, J.V.; Kaulich, M.; Richardson, A.; Stender, J.D.; Benner, C.; Glass, C.K.; Dowdy, S.F.; Desai, A.; Shiau, A.K.; Oegema, K. 53BP1 and USP28 mediate p53 activation and G1 arrest after centrosome loss or extended mitotic duration. *J. Cell Biol.* **2016**, *214*, 155–166. [CrossRef] [PubMed]
112. Stevens, D.; Gassmann, R.; Oegema, K.; Desai, A. Uncoordinated loss of chromatid cohesion is a common outcome of extended metaphase arrest. *PLoS ONE* **2011**, *6*, e22969. [CrossRef] [PubMed]
113. Pilaz, L.J.; McMahon, J.J.; Miller, E.E.; Lennox, A.L.; Suzuki, A.; Salmon, E.; Silver, D.L. Prolonged Mitosis of Neural Progenitors Alters Cell Fate in the Developing Brain. *Neuron* **2016**, *89*, 83–99. [CrossRef] [PubMed]
114. Shah, J.V.; Cleveland, D.W. Waiting for anaphase: Mad2 and the spindle assembly checkpoint. *Cell* **2000**, *103*, 997–1000. [CrossRef]
115. Etemad, B.; Kops, G.J. Attachment issues: Kinetochore transformations and spindle checkpoint silencing. *Curr. Opin. Cell Biol.* **2016**, *39*, 101–108. [CrossRef] [PubMed]
116. Magidson, V.; He, J.; Ault, J.G.; O'Connell, C.B.; Yang, N.; Tikhonenko, I.; McEwen, B.F.; Sui, H.; Khodjakov, A. Unattached kinetochores rather than intrakinetochore tension arrest mitosis in taxol-treated cells. *J. Cell Biol.* **2016**, *212*, 307–319. [CrossRef] [PubMed]
117. Tauchman, E.C.; Boehm, F.J.; DeLuca, J.G. Stable kinetochore-microtubule attachment is sufficient to silence the spindle assembly checkpoint in human cells. *Nat. Commun.* **2015**, *6*. [CrossRef] [PubMed]
118. Etemad, B.; Kuijt, T.E.; Kops, G.J. Kinetochore-microtubule attachment is sufficient to satisfy the human spindle assembly checkpoint. *Nat. Commun.* **2015**, *6*. [CrossRef] [PubMed]
119. Suzuki, A.; Badger, B.L.; Salmon, E.D. A quantitative description of Ndc80 complex linkage to human kinetochores. *Nat. Commun.* **2015**, *6*. [CrossRef] [PubMed]
120. Aravamudhan, P.; Felzer-Kim, I.; Joglekar, A.P. The budding yeast point centromere associates with two Cse4 molecules during mitosis. *Curr. Biol.* **2013**, *23*, 770–774. [CrossRef] [PubMed]
121. Joglekar, A.P.; Bouck, D.C.; Molk, J.N.; Bloom, K.S.; Salmon, E.D. Molecular architecture of a kinetochore-microtubule attachment site. *Nat. Cell Biol.* **2006**, *8*, 581–585. [CrossRef] [PubMed]
122. Vleugel, M.; Omerzu, M.; Groenewold, V.; Hadders, M.A.; Lens, S.M.; Kops, G.J. Sequential multisite phospho-regulation of KNL1-BUB3 interfaces at mitotic kinetochores. *Mol. Cell* **2015**, *57*, 824–835. [CrossRef] [PubMed]
123. Westhorpe, F.G.; Tighe, A.; Lara-Gonzalez, P.; Taylor, S.S. p31comet-mediated extraction of Mad2 from the MCC promotes efficient mitotic exit. *J. Cell Sci.* **2011**, *124*, 3905–3916. [CrossRef] [PubMed]
124. Kops, G.J.; Weaver, B.A.; Cleveland, D.W. On the road to cancer: Aneuploidy and the mitotic checkpoint. *Nat. Rev. Cancer* **2005**, *5*, 773–785. [CrossRef] [PubMed]
125. Schvartzman, J.M.; Sotillo, R.; Benezra, R. Mitotic chromosomal instability and cancer: Mouse modelling of the human disease. *Nat. Rev. Cancer* **2010**, *10*, 102–115. [CrossRef] [PubMed]
126. Tighe, A.; Johnson, V.L.; Albertella, M.; Taylor, S.S. Aneuploid colon cancer cells have a robust spindle checkpoint. *EMBO Rep.* **2001**, *2*, 609–614. [CrossRef] [PubMed]
127. Vleugel, M.; Hoogendoorn, E.; Snel, B.; Kops, G.J. Evolution and function of the mitotic checkpoint. *Dev. Cell* **2012**, *23*, 239–250. [CrossRef] [PubMed]
128. Tromer, E.; Snel, B.; Kops, G.J. Widespread Recurrent Patterns of Rapid Repeat Evolution in the Kinetochore Scaffold KNL1. *Genome Biol. Evol.* **2015**, *7*, 2383–2393. [CrossRef] [PubMed]
129. Galli, M.; Morgan, D.O. Cell Size Determines the Strength of the Spindle Assembly Checkpoint during Embryonic Development. *Dev. Cell* **2016**, *36*, 344–352. [CrossRef] [PubMed]

130. Sear, R.P.; Howard, M. Modeling dual pathways for the metazoan spindle assembly checkpoint. *Proc. Natl. Acad. Sci. USA* **2006**, *103*, 16758–16763. [CrossRef] [PubMed]
131. Doncic, A.; Ben-Jacob, E.; Barkai, N. Evaluating putative mechanisms of the mitotic spindle checkpoint. *Proc. Natl. Acad. Sci. USA* **2005**, *102*, 6332–6337. [CrossRef] [PubMed]

biology

MDPI

Review

Mechanisms to Avoid and Correct Erroneous Kinetochore-Microtubule Attachments

Michael A. Lampson [1,*] and Ekaterina L. Grishchuk [2,*]

1 Department of Biology, University of Pennsylvania, Philadelphia, PA 19104, USA
2 Department of Physiology, Perelman School of Medicine, University of Pennsylvania, Philadelphia, PA 19104, USA
* Correspondences: lampson@sas.upenn.edu (M.A.L.); gekate@mail.med.upenn.edu (E.L.G.)

Academic Editor: J. Richard McIntosh
Received: 3 November 2016; Accepted: 28 December 2016; Published: 5 January 2017

Abstract: In dividing vertebrate cells multiple microtubules must connect to mitotic kinetochores in a highly stereotypical manner, with each sister kinetochore forming microtubule attachments to only one spindle pole. The exact sequence of events by which this goal is achieved varies considerably from cell to cell because of the variable locations of kinetochores and spindle poles, and randomness of initial microtubule attachments. These chance encounters with the kinetochores nonetheless ultimately lead to the desired outcome with high fidelity and in a limited time frame, providing one of the most startling examples of biological self-organization. This chapter discusses mechanisms that contribute to accurate chromosome segregation by helping dividing cells to avoid and resolve improper microtubule attachments.

Keywords: Aurora B kinase; kinetochore geometry; microtubule turnover; tension-dependent regulation

1. Introduction

Segregating chromosomes equally is a non-trivial task, which cells face every time they divide. Healthy human cells have 46 chromosomes, all of which must be duplicated and then segregated equally [1,2]. If segregation fails, the daughter cells may acquire an inappropriate number of chromosomes (aneuploidy), which is associated with severe developmental abnormalities and diseases, including cancer [3–6]. The correct outcome is achieved reproducibly and with high accuracy despite a large stochasticity and variability in many parameters: number and size of the chromosomes, their initial locations, and stochastic behavior of spindle microtubules (MTs), to name a few. MTs originating from the pole can be captured by the kinetochore from almost any angle [7]. Subsequent kinetochore motions, translational and rotational, bring it in contact with more MTs. A human kinetochore ultimately binds 20 MTs on average [8,9], but the real task is to ensure that all these MTs are to one pole, while the MTs at the sister kinetochore are connected to the opposite pole, referred to as the amphitelic configuration. If MTs from two poles attach to sister kinetochores completely randomly, the probability of finding such a configuration is less than 10^{-10}. The accuracy of segregation is orders of magnitude better, although the exact frequency of chromosome loss in somatic human cells in vivo is not well known. In normal cells, the mis-segregation rate during one division lies in the range 10^{-4}–10^{-3} per chromosome [10–12], corresponding to one mis-segregating chromosome in 10^2–10^3 dividing cells. Immortalized human cell lines, which are typically derived from different tumors, are often aneuploid, and the mis-segregation rate per chromosome is ~10 times higher [13]. Even at this rate, the segregation outcome is impressive and indicates robust self-organization to avoid or resolve MT attachment errors. In addition, a surveillance system buys time if normal mitotic progression is perturbed, as considered in another chapter in this issue by A. Joglekar. Below, we focus on principal mechanisms that resolve MT attachment errors and collectively ensure accurate chromosome segregation.

2. Literature Review Sections

2.1. Error Correction of Syntelic Attachments

The conceptual framework for error correction was laid out by B. Nicklas, whose work focused on syntelic attachment errors, in which kinetochores that should be attached to opposite poles are instead attached to the same spindle pole, in meiosis I (Figure 1). Error correction for this type of attachment results from Darwinian selection of the amphitelic kinetochore MT (KMT) configuration (reviewed in [7]). The basis for this mechanism is that KMT attachments are inherently unstable, so errors can be corrected via trial and error, as MTs are released from kinetochores and new ones are captured (called KMT turnover). Nicklas suggested that correction of syntelic attachments involves repeated cycles of complete detachment and reattachment of KMTs until the correct, amphitelic configuration is encountered. As for any selection process, a feedback is required to reinforce the desired outcome. Nicklas's work established the dominant idea in the field that this feedback is provided by tension exerted by spindle MTs pulling sister kinetochores in opposite directions. Because of the spindle geometry, tension is higher for amphitelic attachments than for incorrect attachments, so these attachments can be discriminated from each other. Classic experiments in grasshopper spermatocytes [14] provided evidence supporting this idea: syntelic attachments are unstable unless tension is applied with a micromanipulation needle by pulling the chromosomes away from the attached pole (Figure 1).

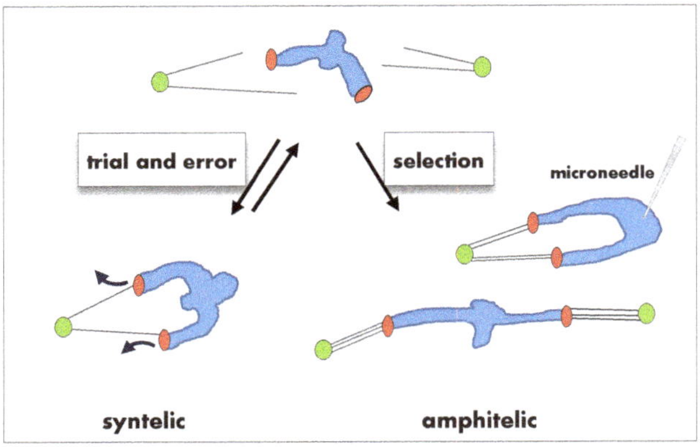

Figure 1. Tension-dependent error correction for syntelic kinetochore microtubule (KMT) attachments. Kinetochores (red) of a meiotic bivalent (blue) repeatedly bind MTs from different poles (green), but KMT attachments in the syntelic configuration are short-lived. Tension arises when the amphitelic configuration is encountered, inducing stable KMT attachments. Syntelic attachments can be stabilized artificially by applying tension with a microneedle. In meiosis I, as shown here, kinetochores of homologous chromosomes attach to opposite spindle poles in the amphitelic orientation. In mitotic cells, kinetochores of sister chromatids would attach to opposite poles.

Analogous manipulations in vertebrate tissue culture cells have proven difficult [15], so a direct demonstration of tension-induced stabilization of KMT attachments during somatic cell division is lacking. In a molecular variation of Nicklas's experiment, tension was applied in mitotic Drosophila cells by overexpression of a chromokinesin, a plus-end-directed MT motor that localizes to chromosome arms [16]. Increasing chromokinesin levels leads to increased polar ejection forces acting on chromosomes during mitosis. In cells with monopolar chromosomes, these forces increase tension on syntelic attachments and stabilize them. These complementary approaches and other correlative observations provide compelling evidence for tension-dependent stabilization of kinetochore-MT attachments both in mitosis and in meiosis I. At a molecular level this "catch bond" behavior is unusual because increasing tension usually breaks a bond, increasing its dissociation under increasing force. The catch bond will also eventually break if the force is too large, but at intermediate tension this bond is strengthened. Interestingly, Nicklas thought that tension was likely to stabilize the anchorage of MTs at the pole [7]. Current thinking for mitosis is that the tension-modulated bonds are located at the kinetochore, as discussed later in this chapter, although the exact mechanism could vary in different organisms.

2.2. Error Correction of Merotelic Attachments

In mammalian cells, multiple MTs interact with each kinetochore, and it is difficult to visualize these individual attachments in real time. Although quantitative information is generally lacking, syntelic attachments seem to form rarely during a typical mitosis in human cells [17]. A more common error is merotelic attachment: connection of one sister kinetochore to MTs from both poles, (Figure 2). Both syntelic and merotelic attachments can lead to chromosome segregation errors. Interestingly, cells typically delay anaphase in the presence of syntelic attachments, but merotelic attachments are more dangerous because anaphase can start prior to correcting them [17]. As a result, MTs attached to the wrong pole impede chromosome motion toward the correct pole, leading to lagging chromosomes in anaphase [18,19]. Such chromosomes have been suggested to constitute the most common pathway to create aneuploidy in cancer cells [13,20–22]. It is counter-intuitive that the more frequent MT attachment errors are not monitored by a checkpoint, which suggests that normal cell division mechanisms are sufficient for coping with these errors. Little is known about these mechanisms, however, because merotelic attachments are difficult to study without introducing large perturbations.

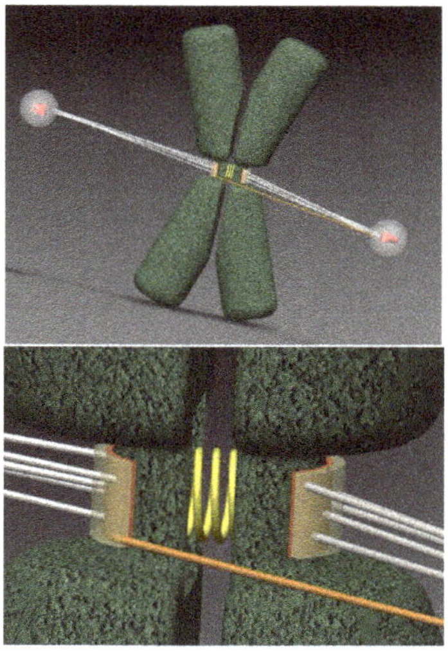

Figure 2. Merotelic KMT configuration for a congressed chromosome. Top image is a three-dimensional representation of a mammalian chromosome (green) positioned midway between two spindle poles. Sister chromatids (green) are connected by a stretchable centromere, represented as a spring (yellow). Bottom image is an enlargement of sister kinetochores, depicted as semi-transparent layers. Most of the attached MTs are in the proper amphitelic configuration (grey), extending from opposite spindle poles. However, although the chromosome has congressed and its sister kinetochores are positioned back-to-back, they can still bind improper merotelic MTs (one such MT is shown in orange). Computer-generated images are snapshots from video material in [23].

The most straightforward hypothesis for merotelic attachments is that they are corrected based on the same principle mechanism as syntelic attachments, but direct application of the trial-and-error mechanism to merotelic attachment is problematic. Indeed, for syntelic error correction, the correct configuration can be selected if the response to high and low tension is simply "on" and "off" (Figure 1), so the kinetochore cycles between states with attached MTs and with complete detachment, until finding the configuration that generates tension. Merotelic attachments may also experience a dramatic loss of tension if the number of incorrect MTs is relatively large, in which case a complete KMT detachment may take place. However, cycles of complete kinetochore detachment are not a common feature of somatic mitosis, so merotelic errors appear to improve gradually over the course of prometaphase and metaphase [24]. It is still possible, of course, to view this process as tension-dependent Darwinian selection of the correct KMT configuration by assuming that tension gradually increases with more amphitelic KMTs, leading to slower detachment of MTs from amphitelic kinetochores and faster from merotelic (Figure 3A). This assumption is reasonable if all MTs, correct or incorrect, generate pulling force. Because these MTs pull in different directions, tension between sister kinetochores can potentially serve as a readout for the number of properly attached KMTs. Since the number of amphitelic KMTs improves gradually during a typical mitosis, tension should also modulate KMT attachment stability gradually, guiding the evolution of KMT configurations. In this framework, tension-dependent regulation at a single kinetochore pair is indiscriminate, i.e., it stabilizes all KMTs to the same extent, regardless of whether they are correct or incorrect.

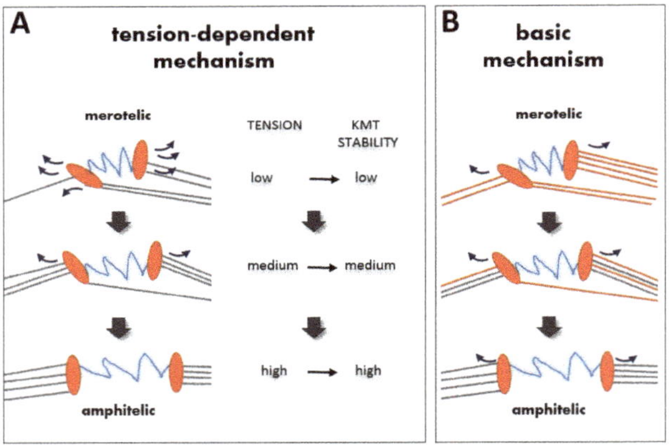

Figure 3. Models for correction of merotelic KMT attachment errors. Correction of merotelic attachments in vertebrate cells proceeds gradually during prometaphase and metaphase without necessarily losing all KMT attachments. (**A**) In the tension-dependent error-correction model, stretching between sister kinetochores (red ovals connected with blue springs) is the primary signal that modulates the turnover of KMT attachments (depicted with curved arrows). As kinetochores bind and lose KMTs, probing different configurations, the configurations that produce higher tension are assumed to induce higher KMT stability. Because their KMTs detach less frequently, such configurations last longer. This gradual evolution would lead eventually to the completely amphitelic configuration, which generates maximal tension. In the non-discriminate version of this model (as shown), all KMTs are affected similarly. Alternatively, in the selective version of this model, lifetime of merotelic vs. amphitelic KMTs is assumed to be regulated differently (see text for details). (**B**) In the basic mechanism, the gradual correction of merotelic attachments proceeds with no change in the turnover of KMT attachments. The rate of KMT turnover in this model does not depend on tension and is not selective, so all old KMTs (orange) eventually detach, whether correct and incorrect, and are replaced with new KMTs (grey). New KMTs preferentially attach correctly, favored by the back-to-back geometry of sister kinetochores.

This general concept is appealing but raises many interesting and still unanswered questions. For example, are two sister kinetochores affected by this mechanism equally? As noted above, if the tension between two sisters is the main regulator, the degree of destabilization of KMT attachments should be the same for both kinetochores, even though one of them may have already acquired the correct set and destabilizing this successful outcome would be counterproductive. Also counterproductive would be destabilization of amphitelically attached kinetochores with a reduced number of KMTs, which is normally quite variable [9]. The required sensitivity of such a tension-dependent mechanism also raises some concerns. For example, is it feasible for such gradual regulation to discriminate and correct a single merotelic MT? One improperly attached KMT out of 20 is expected to reduce tension at the kinetochore by only 5%, which is less than the normal variation in inter-kinetochore tension associated with directional chromosome instability during metaphase oscillations. The overall speed of this regulatory mechanism, which is inherently limited by the rate of KMT turnover, may also become an issue, because during normal mitotic duration the kinetochores would have to sort through a large number of wrong KMT configurations, the essence of the Darwinian selection mechanism. In general, designing a robust regulatory mechanism that senses a small change (1 KMT) for a large range of input signals (from 1 to 20 KMTs), while quickly inducing a strong (stabilizing) response after all KMTs have attached properly, is not trivial, so future experiments and theoretical calculations should address the feasibility of the Darwinian regulatory mechanism for merotelic errors.

Finally, it is not clear how these views of error correction can be reconciled with the concept of force balance at metaphase kinetochores [25,26]. Because the tension-dependent stabilization model is concerned only with the magnitude of inter-kinetochore tension, it is often overlooked that this selection mechanism could be accompanied by a large and variable imbalance in the kinetochore pulling forces. Unbalanced forces acting on sister kinetochores result in directed chromosome motion, but large chromosomal excursions are rare during metaphase. Polar ejection forces and kinetochore-localized regulators of KMT dynamics are thought to dampen chromosome oscillations (e.g., [27]). These forces can also affect inter-kinetochore tension, however, so why they do not interfere with the tension-driven error-correction mechanism is unclear.

An interesting possibility is that mammalian cells need not make 100% accurate, amphitelic attachments. Perhaps some degree of merotely can be tolerated because it would lead to a fairly small reduction of the pulling force in anaphase. In the ensuing tug-of-war between properly attached kinetochore MTs and a few MTs pulling in the wrong direction, the main KMT bundle can apparently win [28], leading to normal segregation of a merotelic kinetochore. This strategy seems dangerous, however, because chromosome velocity in anaphase slows down while the wrong attachments are being resolved, thereby delaying arrival of the affected chromosome at the site of nuclear envelope reformation. The late arriving chromosome may end up excluded from the main nucleus, forming its own micronucleus [29]. Normal somatic cells arrest the cell cycle in the presence of such micronuclei, but in cells with p53 mutation, disastrous rearrangement of DNA in a micronucleus may take place, contributing to tumorigenesis [30,31]. Nonetheless, this back-up mechanism for correcting merotelic attachments during anaphase segregation may play an important role when the number of merotelic KMTs is small, relieving the difficult requirement for high sensitivity of the error-correction mechanism. It is also interesting in this respect that the rate of chromosome mis-segregation reported for mitotic human cells is higher than in yeast: ~10-fold when adjusted for chromosome number [12]. The reasons for this difference are unclear but could indicate more relaxed requirements for fidelity of chromosome segregation in human cells.

2.3. Tension-Independent Error-Correction Mechanism

Although tension-dependent MT stabilization traditionally receives major attention for its role in error correction, other factors are known to play a significant role. In particular, the specific geometry of vertebrate kinetochores is an important contributor to the accuracy of mitotic chromosome segregation [32]. It has long been recognized that the back-to-back arrangement of sister kinetochores creates geometric constraints which favor sister kinetochore attachment to opposite spindle poles [33]. Little is known, however, about the stringency of these constraints and the exact role they play in mitosis. In the traditional view of Darwinian selection via tension-dependent stabilization, this geometry does not help to correct the already formed merotelic MTs but rather reduces the number of initial wrong attachments [7].

The alternative view, which is not incompatible with the first one, is that geometric constraints constitute an integral part of a simplified error-correction mechanism that does not rely on tension-dependent stabilization, but incorporates KMT turnover in addition to restrictive kinetochore geometry (Figure 3B). This mechanism can be understood intuitively assuming that improper kinetochore MTs are acquired during early stages of mitosis, and that their subsequent release is the rate-limiting step for successful bi-orientation [34]. Indeed, after the kinetochores become positioned favorably (i.e. midway between spindle poles), all bound KMTs will be gradually released due to turnover. Stringent geometric constraints, however, will favor their replacement only with proper, amphitelic MTs. This mechanism is feasible because the turnover time for KMTs in human cells is 2–6 min [35–37]. With this release rate, all KMTs that attached to kinetochores earlier in mitosis will be replaced during the normal duration of metaphase (10–20 min). Interestingly, prolonging mitosis appears to elicit a relatively small improvement in error correction (2.7-fold decrease in lagging chromosomes) [24]. This finding suggests that the normal mitotic clock closely matches the kinetics of KMT turnover. A combination of these matching kinetics, the back-up mechanism to resolve small numbers of merotelic KMTs in anaphase and the overall segregation accuracy that can be tolerated, may explain why vertebrate cells do not utilize a checkpoint to buy more time to achieve perfect amphitelic KMT attachments.

The positive effect from this tension-independent error-correction mechanism is maximal when the kinetochore pair is positioned favorably. Thus, it should benefit strongly from expedient chromosome congression, because achieving a position midway between the poles marks the start of a productive time toward the steady-state configuration with a minimal number of merotelic KMTs. Consistent with this view, chromosomes can congress via different mechanisms and even without bi-orientation [38], enabling them to assume a midway position quickly. Interestingly, for successful operation of this error-correction mechanism, it is not necessary to assume that the incorrect KMTs are less stable than the correct ones, so in this sense regulation of KMT stability is indiscriminate [39]. Calculations suggest that if small numbers of merotelic MTs can be tolerated, a combination of indiscriminate KMT turnover and geometric constraints can enable normal segregation of 45 out of 46 chromosomes during cell division [23]. This fidelity is less than physiological but is a vast improvement over completely random attachments, so it is appropriate to call the underlying processes the "basic" error-correction mechanism. In the next two sections we discuss in more detail the specific roles played by kinetochore geometry and KMT turnover in ensuring accurate chromosome segregation within the framework of the basic mechanism.

2.3.1. How Strongly Should Kinetochores Shield Themselves from Wrong MTs?

The stringency of geometric constraints appears to be different at different mitotic stages. Since the basic mechanism is most productive after the chromosomes have congressed, shielding them from wrong MTs at this stage is critical. In dividing mammalian cells, however, the binding of inappropriate MTs to metaphase kinetochores is still possible [24]. Various factors could explain this observation. For example, the spindle structure is not linear, so not all kinetochores can become aligned perfectly along its axis. It is also not completely static: the entire spindle moves within the cell, the poles move relative to one another, and chromosomes are subjected to various forces, including thermal. The centromeric material is relatively elastic, allowing sister kinetochores to deviate from perfect back-to-back orientation. Together, the resulting motions could permit capture of inappropriate MTs even on aligned and oriented sister kinetochores. Calculations show that the frequency of such attachments will directly affect the number of merotelic KMTs at anaphase onset (Figure 4A). Thus, any mechanical or structural feature that stabilizes the co-alignment of the kinetochore pair and spindle axes would help to optimize operation of the basic mechanism and improve the accuracy of segregation. In this sense, the entire spindle and its mechanical properties constitute an important factor in mitotic error correction, explaining why segregation accuracy can be reduced via so many different molecular perturbations.

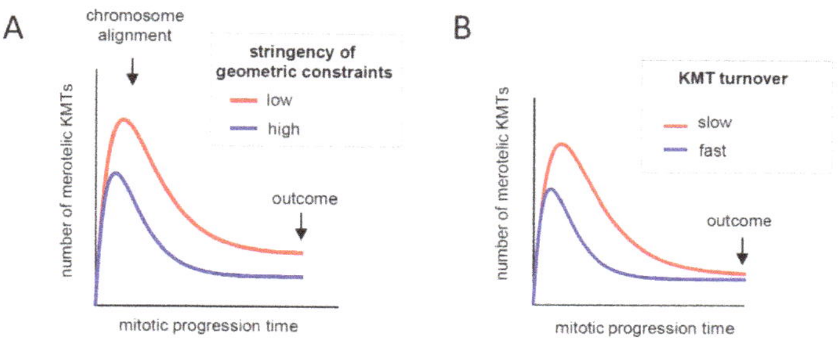

Figure 4. Quantitative analysis of the evolution of KMT configurations during mitotic progression. Graphs illustrate changes in the number of merotelic KMTs during mitotic progression, as predicted by the basic error-correction model [23]. After mitosis starts, the number of attached MTs, some of which are merotelic, increases sharply. As chromosomes become aligned midway on the spindle, the number of merotelic KMTs begins to decline. However, merotelic KMTs are not eliminated completely by this mechanism because new erroneous attachments continue to form even on congressed chromosomes, albeit at lower frequency. Thus, the system tends toward a steady state, in which the rates of forming wrong attachments and eliminating them through non-discriminatory turnover are balanced. (**A**) With more stringent geometric constraints, the final outcome for the KMT configuration at anaphase onset is improved because the rate of capturing wrong KMTs at steady-state is reduced. (**B**) The rate of KMT turnover does not affect the final outcome (black arrow). However, if anaphase starts before the steady state is reached, cells with slower KMT turnover will have more merotelic KMTs.

Geometric constraints during earlier mitotic stages appear to be much less stringent. In mammalian cells, ~10% of chromosomes have sister kinetochores in a side-by-side orientation [40], which is highly permissive to the formation of multiple merotelic attachments. During prometaphase the relaxed geometric constraints may be advantageous because they increase the probability of MT capture. Calculations show that the traditional "search and capture" mechanism with interphase MT dynamics is inefficient [41]. Mitotic cells speed up this search by increasing MT dynamics, but some studies suggest that capture is still problematic in a crowded environment with a large number of chromosomes [42,43]. If kinetochore geometry during prometaphase was as restrictive as is desirable for metaphase kinetochores, the capture of MTs early in mitosis would be greatly delayed [23], so the frequency of MT capture by kinetochores is likely to be a limiting factor for mitotic progression. It appears that prometaphase kinetochores overcome this limitation by enlarging their coronas and increasing curvature, despite the elevated risk of binding wrong MTs [44–46]. Little is known, however, about the underlying molecular mechanisms and the processes by which kinetochores shed coronas and become more compact during metaphase, when the capture of inappropriate MTs must be avoided.

2.3.2. What Determines the Rate of KMT Turnover during Mitosis?

KMT turnover is central for error correction in both tension-dependent and tension-independent models, but the degree of KMT stabilization at amphitelic kinetochores is often overstated. The KMT half-life time in many human cells increases from 2–3 min in prometaphase to 4–6 min in metaphase, indicating not more than 2–3-fold stabilization [35–37]. The tension-dependent model explains this effect by evolution of KMT configurations from merotelic to amphitelic. Interestingly, however, careful measurements of KMT turnover in perturbed mitotic cells suggest that turnover decreases later in mitosis owing to a timing mechanism, which is not dependent on achieving the amphitelic configuration [36].

Error correction via the basic mechanism does not require that KMT turnover slows down in metaphase, but other aspects of mitotic physiology could explain the constrained kinetics. As discussed earlier in this review, a lower limit on the KMT turnover rate is likely to be imposed by the normal duration of metaphase, because if anaphase starts before all KMTs have been replaced with new ones, some wrong KMTs that could have been removed will remain and lead to lagging chromosomes [19] (Figure 4B). This logic suggests that faster turnover should increase the rate of error correction and promote higher fidelity of chromosome segregation. Such correlation is indeed observed in different cell lines [21]. Moreover, moderate destabilization of the overly stable kinetochore MTs in cancer cell lines improves the accuracy of segregation (reviewed in [47]). Consistently, normal cells do not tolerate induced MT stabilization, as it leads to chromosome segregation errors.

These results make sense in the framework of the basic mechanism because merotelic errors are not detected and metaphase duration is limited, so cells with slower turnover should have more lagging chromosomes when anaphase starts. But why then do normal dividing cells slow down their rate of KMT turnover in metaphase, opposite to what this logic suggests? With no such retardation, anaphase could start earlier without compromising the accuracy of this mechanism (Figure 5, red curve). The answer may lie in the physiological requirement to form a robust kinetochore fiber, a process that depends strongly on KMT turnover and may impose an upper limit on its rate. Calculations show that a fully sized kinetochore fiber cannot form if KMTs exchange too quickly, regardless of how long the cell waits [23]. Building a kinetochore fiber with 20-25 KMTs, is problematic if the KMT half-life is 2–3 min, as in prometaphase (Figure 5, blue curve). To acquire this set, the half-life of KMTs should be increased to 4–6 min. Thus, increasing the overall stability of KMT attachments in metaphase may be a compromise reflecting the multiple roles played by KMT turnover during mitosis and the need to balance accuracy of chromosome segregation, speed of mitosis and acquisition of fully sized KMT fibers [23].

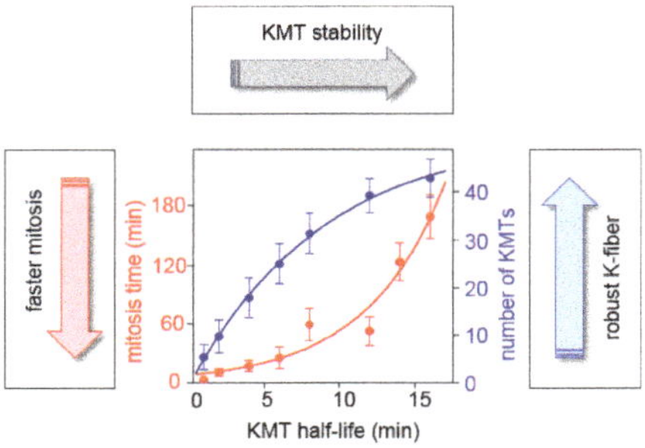

Figure 5. Competing constraints on KMT turnover during mitotic progression. Graph illustrates how KMT turnover (represented as KMT half-life, horizontal axis) affects the overall speed of mitosis and the total number of acquired KMTs (based on the error-correction model in [23]). The speed of mitotic progression (left axis) is evaluated based on the time required to achieve the steady-state KMT configuration. At steady state, the number of attached KMTs (right axis) and the fraction of merotelic KMTs (not shown) have stopped changing, and increasing the duration of mitosis does not generate more KMTs or improve accuracy. When KMT turnover is slow, i.e., KMT half-life is longer, attachments are more stable because they have low release rate. The time required to release all old KMTs increases quickly with increasing KMT stability (red curve), so shorter KMT lifetime is required for speedy mitotic progression. However, the number of KMTs in a K-fiber is lower for shorter KMT lifetime (blue curve). Thus, the limited time of mitotic progression and the acquisition of a full set of KMTs represent the competing constraints on the rate of KMT turnover during mitosis.

2.4. *Molecular Mechanisms of Tension-Dependent Feedback for KMT Stabilization*

2.4.1. Direct Regulation of MT Dynamics

As described above, the basic error-correction mechanism can potentially provide vast improvement in accuracy over random MT attachments, but the expected mis-segregation rate for this mechanism is still high: 10^{-1}–10^{-2}. This theoretical prediction is close to the rate in some cancer cell lines, but the mis-segregation rate measured in human RPE-1 cells is much lower: ~10^{-4} [13]. Therefore, additional error-correction mechanisms in normal cells should provide more than 100-fold improvement in the rate of chromosome mis-segregation on top of the basic mechanism. It is likely that such fidelity is achieved by a combination of the basic mechanism and tension-dependent stabilization of proper attachments. While many questions remain about the operation of such a combined mechanism, molecular details of how tension regulates KMT stability, a long-standing goal in the field, are beginning to emerge.

Tension has been suggested to directly regulate dynamics of kinetochore-bound MT ends because pulling force applied to the depolymerizing MT end slows depolymerization and promotes MT rescue [48,49]. Furthermore, tension of a few piconewtons applied via an optical trap to isolated yeast kinetochores increases the lifetime of MT attachment, although the stability decreases with further increase in force [50]. This bell-shaped dependency suggests the long-sought catch-bond, but apparently it does not result from a specific molecular linkage. Rather, this behavior reflects a property of the MT to switch into polymerization under pulling tension. Since yeast kinetochores bind much more strongly to growing MT plus-ends than to shortening ones, force increases the lifetime of attachment to dynamic MT ends by promoting a longer duration for the MT polymerization state. If this mechanism contributes to tension-sensitive stabilization in vivo, then KMT plus-ends should spend more time in the polymerizing state when they are in an amphitelic configuration than when they have incorrect attachments. This prediction has not been directly tested, but in budding yeast, in which bi-oriented kinetochores oscillate in the absence of poleward MT flux, MT polymerization at one kinetochore is balanced by depolymerization at the sister, so both would not be in the strongly attached state simultaneously. In some organisms poleward MT flux can increase the time spent in the polymerizing state for both sisters, but the velocity of flux in mammalian cells is significantly lower than the rate of chromosome oscillations, so at metaphase the kinetochore-bound MT ends spend approximately equal time in polymerizing and depolymerizing states. The alternating states are confirmed by visualization of an EB (end-binding) protein [51,52], which binds to polymerizing MT ends and shows intermittent localization at oscillating kinetochores. It is not clear if this EB localization pattern changes as merotelic attachments are being corrected on congressed chromosomes, as would be predicted if the direct mechanism played a significant role in error correction.

2.4.2. Aurora B-Dependent Mechanisms

Several other proposed models feature the mitotic kinase Aurora B as a key regulator of kinetochore-MT interactions. Work in yeast and in vertebrate cells shows that Aurora B phosphorylates kinetochore substrates to regulate MT binding, promotes turnover of KMT attachments, and plays an essential role in accurate chromosome segregation [53–61]. Furthermore, phosphorylation of Aurora B substrates at kinetochores is generally inversely proportional to tension, although the MT-binding protein Ndc80 is not fully rephosphorylated after KMT attachments that have already formed are disrupted with nocodazole [62–67]. These observations and others (recently reviewed in [68,69]) are consistent with the hypothesis that Aurora B-dependent phosphorylation of kinetochore substrates destabilizes incorrect KMT attachments in the absence of tension.

These results also imply that Aurora B could be a mediator of tension, but how does it affect KMT stability? Aurora B can destabilize attachments in two ways. The first is by directly promoting detachment of MTs from the kinetochore, as suggested by experiments in vitro [63,70–73]. Consistent with the in vitro findings, kinetochores in mitotic cells cannot maintain stable attachment to MTs when

Aurora B activity is increased by targeting the kinase to kinetochores [62]. Phosphorylation of Aurora B substrates at kinetochores, such as the Ndc80 complex, reduces MT binding affinity (reviewed in [68]), and expression of phosphomimetic mutants of Ndc80 in cells leads to fewer KMTs [74]. Perturbing other Aurora B substrates, such as the MT-binding Dam1 and Ska1 complexes, also destabilizes kinetochore-MT attachments [68]. Second, Aurora B can promote catastrophe and depolymerization of KMTs, as shown in vitro using phosphomimetic mutants of Ndc80 and Dam1 [72,73]. Observations in live cells indicate that at syntelic attachments Aurora B promotes MT depolymerization, rather than immediate detachment. In this case, the syntelic kinetochores are first pulled towards the spindle pole, and detachment occurs subsequently as a result of activity of the related Aurora A kinase, which is enriched at the pole [59,75,76].

2.4.3. Spatial Separation Model

Why does Aurora B activity at kinetochores change in response to interkinetochore tension? The spatial separation model, first suggested based on experiments in budding yeast [56], proposes that tension sensing depends on the changing distance between Aurora B, which is enriched at the inner centromere, and its MT-binding substrates at the outer kinetochore. When KMT attachments are amphitelic, tension pulls bi-oriented sister kinetochores away from the inner centromere, separating the kinase from its substrates, thereby reducing phosphorylation. This model is supported by the observations described above, showing that phosphorylation decreases with tension. Furthermore, repositioning Aurora B so that it localizes in close proximity to kinetochore substrates leads to increased phosphorylation and destabilization of KMT attachments [62].

Consistent with this model, multiple observations indicate that perturbing centromere localization of Aurora B disrupts normal regulation of kinetochore-MT attachments. As part of the chromosome passenger complex (CPC), Aurora B is targeted to the inner centromere through binding of the CPC to phosphorylation marks on histones H3 and H2A [77–80]. Mutation of these phosphorylation sites in fission yeast prevents CPC targeting and leads to severe bi-orientation defects. Moreover, these defects are largely rescued by restoring CPC targeting, e.g., by fusing a CPC component, Survivin, to a chromodomain that binds pericentromeric heterochromatin [80]. In human cells, either mutation of Survivin or inhibition of the Haspin kinase, which is responsible for the H3 phosphorylation mark, prevents normal centromere localization of Aurora B, and leads to defects in the correction of attachment errors [79,81]. The same mutation in chicken Survivin did not affect cell growth, but the error-correction process was not explicitly tested in these cells [82]. Furthermore, preventing CPC targeting to centromeres by mutation of Cdk phosphorylation sites also leads to chromosome bi-orientation defects in both fission yeast and human cells. Again, these defects are rescued by restoring localization of the CPC using a chromodomain fused to Survivin [83]. Together these observations provide strong support for the importance of the inner centromere pool of Aurora B for error correction.

The spatial separation model explains the importance of the centromeric pool of Aurora B by suggesting that a continuous gradient of Aurora B activity extends from the sites of its enrichment at the inner centromere, while kinetochore substrates change their positions within this gradient depending on tension. Indeed, position changes of as little as 30–50 nm are associated with different levels of phosphorylation of both endogenous and exogenous Aurora B substrates at mammalian kinetochores [63,67]. How can such a steep gradient of kinase activity be established and maintained within the kinetochore, given that it is located hundreds of nanometers away from sites of Aurora B enrichment at the inner centromere? One idea links properties of the spatial activity gradient with the ability of Aurora B to activate itself via autophosphorylation, and to become conversely inactivated by a phosphatase [84–88]. Autoactivation in the context of a coupled kinase-phosphatase system leads to nonlinear, bistable behavior of kinase activity both in vitro and in cells, suggesting a physically plausible model for the formation of a steep kinase activity gradient [89]. In this model, Aurora B activates itself at the sites of highest concentration (i.e., inner centromere), overcoming inhibition by phosphatase. This activity then spreads throughout the chromatin thanks to a reaction-diffusion mechanism, in which active kinase molecules phosphorylate and activate additional kinase molecules that are either chromatin-bound or diffusing in the cytosol. As the Aurora B concentration decreases away from the inner centromere, phosphatase activity switches Aurora B to the inactive state in a highly nonlinear manner. Calculations suggest that this mechanism can potentially establish a steep gradient of kinase activity specifically in the kinetochore region, far away from the sites of highest Aurora B concentration [89].

This hypothesis is additionally interesting because it implies that the centromeric chromatin and the kinetochore should be viewed as a continuous mechano-chemical medium. Stretching of this medium, which takes place when sister kinetochores are amphitelic, changes its biochemical properties due to changes in local Aurora B concentration everywhere within this medium (Figure 6). Because the coupled kinase-phosphatase system linked to this stretchable mechanical matrix is bistable, tension can exert strong and accurate control over the position and steepness of the Aurora B activity gradient at the kinetochore. This and other possible models for long-range effects of the centromeric Aurora B pool await critical examination.

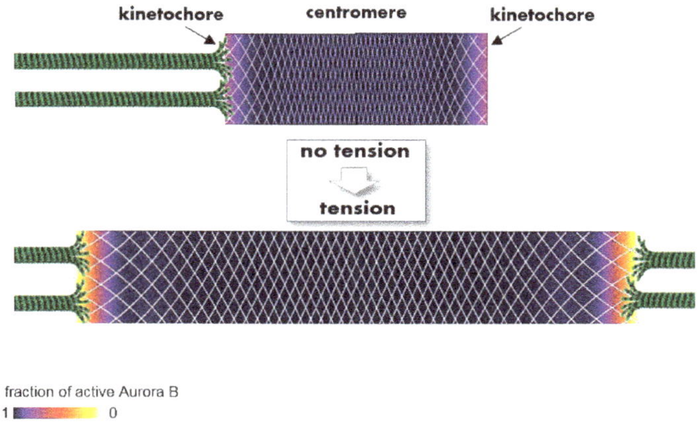

Figure 6. A molecular model to explain how tension regulates phosphorylation of kinetochore proteins that bind KMTs and regulate their attachment lifetime. Color-coded plots show the spatial distribution of active Aurora B kinase within a continuous flexible matrix (white mesh), encompassing the centromeric chromatin and two sister kinetochores (based on the theoretical model in [89]). Aurora B kinase is enriched strongly in the middle of the centromere (not shown), where it becomes highly active (purple and blue colors) due to trans-molecular auto-phosphorylation. With no tension this activity propagates from the centromere throughout the entire matrix, as active kinase "ignites" the nearby kinase, overcoming opposing phosphatases. As a result, kinase activity at the kinetochores is high with no tension, reducing KMT lifetime. When amphitelic KMTs stretch the connecting matrix, local Aurora B concentration is reduced everywhere (shown by increased spacing of the white mesh). However, the local concentration of active kinase does not decrease proportionally owing to the highly nonlinear, bistable nature of the underlying kinase-phosphatase switch. Aurora B kinase activity remains high within centromeric heterochromatin but drops sharply at the outer kinetochore (orange/red colors), forming phosphorylation gradients within the kinetochores. While this model provides a biophysical explanation for tension-dependent, long-range regulation of Aurora B kinase activity, the physiological significance of the kinetochore activity gradient seen in mammalian cells remains to be understood.

2.4.4. Alternative Models

Although there is strong evidence that Aurora B function in regulating kinetochore-MT interactions depends on its targeting to the inner centromere in both fission yeast and mammalian cells, the situation in budding yeast is different. Removal of the normal centromere pool of Aurora B through mutation of its binding partner, INCENP, does not have major consequences for chromosome segregation or mitotic progression in budding yeast [90]. The reasons for this discrepancy are unclear and may reflect differences in Aurora B function between budding yeast and other systems that have been examined. For example, budding yeast is unusual in that each kinetochore binds only a single MT, and therefore there is no need to correct merotelic attachment errors, which may reduce the burden on Aurora B in destabilizing such attachments.

Alternatives to the spatial separation model, which could be applied to all organisms, not just yeast, assign no specific regulatory role to the centromeric Aurora B pool, and explain its importance for chromosome segregation by some other function [90]. For tension-dependent error correction, some models invoke a specialized pool of the kinase localized at kinetochores, in close proximity to its outer kinetochore targets. In one such model, stretching within the kinetochore itself separates Aurora B from its substrates, and an elongated INCENP subunit of the CPC may act as a flexible arm that determines how far the kinase can reach from its kinetochore binding site [68,91]. This model could explain how attachments are selectively stabilized, but only if stretching within the kinetochore correlates with tension on bi-oriented sister kinetochores, which does not appear to be the case [92–94]. Another model proposes a tension-sensitive binding site for Aurora B at the kinetochore [90]. In this case, increased tension would lead to a loss of kinase binding and stabilization of KMT attachments. It has also been proposed that Aurora B substrates are somehow inaccessible to the kinase when tension is high [69]. Yet another possibility is that phosphatase activity rather than kinase activity is regulated in response to tension, as a mechanism to control phosphorylation status of Aurora B substrates. Aurora B activity reduces kinetochore recruitment of protein phosphatase 1 (PP1) [95,96], but there is no evidence that phosphatase activity is directly regulated by tension. Because neither kinetochore binding sites for Aurora B nor tension-sensitive substrates have yet been identified, these different models are difficult to evaluate, and testing them is a challenging future goal.

2.4.5. Selectivity of Aurora B-Dependent Error-Correction Mechanisms

Thus far, we have considered tension-dependent and tension-independent error-correction models in the frame of the simplest hypothesis that the underlying regulatory mechanisms are indiscriminate, i.e., at one kinetochore they effect the stability of correct and incorrect KMTs similarly. However, current views on Aurora B-dependent error correction often assume that modulation of KMT stability is selective, and the mechanism applies only to incorrectly attached MTs, while properly attached MTs at the same kinetochore are not regulated or are regulated oppositely [97]. It is not clear how this discrimination would work molecularly, and the mechanisms could be different for syntelic and merotelic errors.

Correction of syntelic attachments in yeast has been proposed to require that Aurora B specifically destabilizes the end-on attached syntelic MTs, while lateral MT attachments are resistant to its activity [98]. In this way Aurora B could promote repeated trial-and-error cycles until the tension-producing configuration is found and stabilized (Figure 1). Because amphitelic yeast KMTs are highly stable and do not turn over [99], Aurora B-dependent destabilization at this stage may not be needed.

With merotelic attachments that are common in cells with multiple KMTs, selectivity of a tension-dependent mechanism is difficult to explain because tension is shared by all MTs attached at the kinetochore, so it should affect them equally. One possibility is that regulation is dependent on the direction of tension, not just its magnitude. If merotelic KMTs are oriented more laterally to the kinetochore plane than amphitelic KMTs, their relative stabilities could be different [100]. Another model suggests that merotelic KMTs extend between the sister kinetochores; in this case, a centromere-localized kinesin-13 could selectively destabilize erroneous KMTs [101]. Yet another model builds on the observation that kinetochores with a bundle of merotelic KMTs are stretched such that part of the merotelic sister kinetochore extends towards the inner centromere, where Aurora

B is enriched. In this case, kinetochore proteins bound to the plus-ends of incorrect KMTs may be closer to Aurora B and subject to increased kinase activity, while the ends of correct MTs would be unaffected [24,28,35,97]. Such strong kinetochore deformations are rarely seen, and it has been difficult to directly test this and other models because of the challenge of correlating differences in phosphorylation, MT end location and stability within a single kinetochore.

Another model for how Aurora B destabilizes incorrect attachments is through regulation of kinase levels at the inner centromere. Aurora B is recruited to centromeres of chromosomes that are not properly aligned at the metaphase plate in multiple human cell lines [66,102], and at merotelic attachments in *Xenopus* cells [103]. This mechanism could act globally, i.e., affecting all KMTs at a merotelic kinetochore. For selective destabilization of merotelic KMTs it would have to rely on some signal to distinguish them from correct KMTs, but how such selectivity could be achieved remains unclear. Interestingly, a kinetochore configuration with merotelic KMTs has been found to persist longer than expected based on the measured metaphase KMT turnover rate [28,97]. This is in contrast with the idea of selective destabilization, which predicts that erroneous KMTs should be replaced faster. The continued presence of merotelic KMTs, however, could be explained by the basic error-correction mechanism, in which force from the merotelic KMTs prevents normal orientation of sister kinetochores [23]. On kinetochores that are not well oriented, the probability of binding wrong MTs is higher, so the merotelic configuration can persist. While future work may reveal a mechanism to selectively target incorrect vs. correct MTs at the same kinetochore, it is also possible that tension-dependent error-correction mechanisms regulate KMT turnover in a non-discriminating manner.

3. Conclusions

In cells with large number of kinetochore microtubules (KMTs), a surprisingly good accuracy of spindle microtubule (MT) attachments can potentially be achieved by the combined effects of back-to-back kinetochore geometry and indiscriminate KMT turnover. Optimal operation of this tension-independent error-correction mechanism requires expedient chromosome congression and moderate modulation of kinetochore geometry and turnover rate. These crucial properties are constrained by competing physiological requirements, such as formation of fully sized kinetochore fibers, and the speed and accuracy of segregation. The basic correction mechanism operates in combination with tension-induced stabilization of the amphitelic KMT configuration. Compelling evidence supports the essential role of the tension-dependent mechanism, especially for correcting syntelic attachments. While many questions remain about operation of this mechanism for correcting merotelic attachments, several molecular components that link tension and KMT stability, such as Aurora B kinase, have been identified and are under investigation.

Acknowledgments: This work is supported in part by National Institutes of Health grant GM083988 to Michael A. Lampson and Ekaterina L. Grishchuk, and by Research Scholar Grant RGS-14-018—01-CCG from the American Cancer Society to Ekaterina L. Grishchuk. We thank A. Zaytsev for preparing figure graphs and excellent technical assistance.

Conflicts of Interest: The authors declare no conflict of interest.

References

1. Rieder, C.L.; Salmon, E.D. The vertebrate cell kinetochore and its roles during mitosis. *Trends Cell Biol.* **1998**, *8*, 310–318. [CrossRef]
2. Walczak, C.E.; Cai, S.; Khodjakov, A. Mechanisms of chromosome behaviour during mitosis. *Nat. Rev. Mol. Cell Biol.* **2010**, *11*, 91–102. [CrossRef] [PubMed]
3. Weaver, B.A.; Cleveland, D.W. Does aneuploidy cause cancer? *Curr. Opin. Cell. Biol.* **2006**, *18*, 658–667. [CrossRef] [PubMed]
4. Bakhoum, S.F.; Compton, D.A. Chromosomal instability and cancer: A complex relationship with therapeutic potential. *J. Clin. Investig.* **2012**, *122*, 1138–1143. [CrossRef] [PubMed]

5. Nicholson, J.M.; Cimini, D. Link between aneuploidy and chromosome instability. *Int. Rev. Cell Mol. Biol.* **2015**, *315*, 299–317. [PubMed]
6. Silk, A.D.; Zasadil, L.M.; Holland, A.J.; Vitre, B.; Cleveland, D.W.; Weaver, B.A. Chromosome missegregation rate predicts whether aneuploidy will promote or suppress tumors. *Proc. Natl. Acad. Sci. USA* **2013**, *110*, E4134–E4141. [CrossRef] [PubMed]
7. Nicklas, R.B. How cells get the right chromosomes. *Science* **1997**, *275*, 632–637. [CrossRef] [PubMed]
8. Wendell, K.L.; Wilson, L.; Jordan, M.A. Mitotic block in HeLa cells by vinblastine: Ultrastructural changes in kinetochore-microtubule attachment and in centrosomes. *J. Cell Sci.* **1993**, *104*, 261–274. [PubMed]
9. McEwen, B.F.; Chan, G.K.; Zubrowski, B.; Savoian, M.S.; Sauer, M.T.; Yen, T.J. CENP-E is essential for reliable bioriented spindle attachment, but chromosome alignment can be achieved via redundant mechanisms in mammalian cells. *Mol. Biol. Cell* **2001**, *12*, 2776–2789. [CrossRef] [PubMed]
10. Knouse, K.A.; Wu, J.; Whittaker, C.A.; Amon, A. Single cell sequencing reveals low levels of aneuploidy across mammalian tissues. *Proc. Natl. Acad. Sci. USA* **2014**, *111*, 13409–13414. [CrossRef] [PubMed]
11. Van den Bos, H.; Spierings, D.C.J.; Taudt, A.S.; Bakker, B.; Porubský, D.; Falconer, E.; Novoa, C.; Halsema, N.; Kazemier, H.G.; Hoekstra-Wakker, K.; et al. Single-cell whole genome sequencing reveals no evidence for common aneuploidy in normal and Alzheimer's disease neurons. *Genome Biol.* **2016**. [CrossRef]
12. Santaguida, S.; Amon, A. Short- and long-term effects of chromosome mis-segregation and aneuploidy. *Nat. Rev. Mol. Cell Biol.* **2015**, *16*, 473–485. [CrossRef] [PubMed]
13. Thompson, S.L.; Compton, D.A. Examining the link between chromosomal instability and aneuploidy in human cells. *J. Cell Biol.* **2008**, *180*, 665–672. [CrossRef] [PubMed]
14. Nicklas, R.B.; Koch, C.A. Chromosome micromanipulation. 3. Spindle fiber tension and the reorientation of mal-oriented chromosomes. *J. Cell Biol.* **1969**, *43*, 40–50. [CrossRef] [PubMed]
15. Skibbens, R.V.; Salmon, E.D. Micromanipulation of chromosomes in mitotic vertebrate tissue cells: Tension controls the state of kinetochore movement. *Exp. Cell Res.* **1997**, *235*, 314–324. [CrossRef] [PubMed]
16. Cane, S.; Ye, A.A.; Luks-Morgan, S.J.; Maresca, T.J. Elevated polar ejection forces stabilize kinetochore-microtubule attachments. *J. Cell Biol.* **2013**, *200*, 203–218. [CrossRef] [PubMed]
17. Cimini, D. Merotelic kinetochore orientation, aneuploidy, and cancer. *Biochim. Biophys. Acta* **2008**, *1786*, 32–40. [CrossRef] [PubMed]
18. Cimini, D.; Fioravanti, D.; Salmon, E.D.; Degrassi, F. Merotelic kinetochore orientation versus chromosome mono-orientation in the origin of lagging chromosomes in human primary cells. *J. Cell Sci.* **2002**, *115*, 507–515. [PubMed]
19. Bakhoum, S.F.; Compton, D.A. Kinetochores and disease: Keeping microtubule dynamics in check! *Curr. Opin. Cell Biol.* **2012**, *24*, 64–70. [CrossRef] [PubMed]
20. Cimini, D.; Howell, B.; Maddox, P.; Khodjakov, A.; Degrassi, F.; Salmon, E.D. Merotelic kinetochore orientation is a major mechanism of aneuploidy in mitotic mammalian tissue cells. *J. Cell Biol.* **2001**, *153*, 517–527. [CrossRef] [PubMed]
21. Bakhoum, S.F.; Genovese, G.; Compton, D.A. Deviant kinetochore microtubule dynamics underlie chromosomal instability. *Curr. Biol.* **2009**, *19*, 1937–1942. [CrossRef] [PubMed]
22. Bakhoum, S.F.; Silkworth, W.T.; Nardi, I.K.; Nicholson, J.M.; Compton, D.A.; Cimini, D. The mitotic origin of chromosomal instability. *Curr. Biol.* **2014**, *24*, R148–R149. [CrossRef] [PubMed]
23. Zaytsev, A.V.; Grishchuk, E.L. Basic mechanism for biorientation of mitotic chromosomes is provided by the kinetochore geometry and indiscriminate turnover of kinetochore microtubules. *Mol. Biol. Cell* **2015**. [CrossRef] [PubMed]
24. Cimini, D.; Moree, B.; Canman, J.C.; Salmon, E.D. Merotelic kinetochore orientation occurs frequently during early mitosis in mammalian tissue cells and error correction is achieved by two different mechanisms. *J. Cell Sci.* **2003**, *116*, 4213–4225. [CrossRef] [PubMed]
25. Pereira, A.J.; Maiato, H. Maturation of the kinetochore-microtubule interface and the meaning of metaphase. *Chromosome Res.* **2012**, *20*, 563–577. [CrossRef] [PubMed]
26. Civelekoglu-Scholey, G.; Scholey, J.M. Mitotic force generators and chromosome segregation. *Cell. Mol. Life Sci.* **2010**, *67*, 2231–2250. [CrossRef] [PubMed]
27. Stumpff, J.; Wagenbach, M.; Franck, A.; Asbury, C.L.; Wordeman, L. Kif18A and chromokinesins confine centromere movements via microtubule growth suppression and spatial control of kinetochore tension. *Dev. Cell* **2012**, *22*, 1017–1029. [CrossRef] [PubMed]

28. Cimini, D.; Cameron, L.A.; Salmon, E.D. Anaphase spindle mechanics prevent mis-segregation of merotelically oriented chromosomes. *Curr. Biol.* **2004**, *14*, 2149–2155. [CrossRef] [PubMed]
29. Thompson, S.L.; Compton, D.A. Chromosome missegregation in human cells arises through specific types of kinetochore-microtubule attachment errors. *Proc. Natl. Acad. Sci. USA* **2011**, *108*, 17974–17978. [CrossRef] [PubMed]
30. Leibowitz, M.L.; Zhang, C.-Z.; Pellman, D. Chromothripsis: A New Mechanism for Rapid Karyotype Evolution. *Annu. Rev. Genet.* **2015**, *49*, 183–211. [CrossRef] [PubMed]
31. Hinchcliffe, E.H.; Day, C.A.; Karanjeet, K.B.; Fadness, S.; Langfald, A.; Vaughan, K.T.; Dong, Z. Chromosome missegregation during anaphase triggers p53 cell cycle arrest through histone H3.3 Ser31 phosphorylation. *Nat. Cell Biol.* **2016**, *18*, 668–675. [CrossRef] [PubMed]
32. Journey, L.; Whaley, A. Kinetochore Ultrastructure in Vincristine-Treated Mammalian Cells. *J. Cell Sci.* **1970**, *7*, 49–54.
33. Östergren, G. The mechanism of co-orientation in bivalents and multivalents. *Hereditas* **1951**, *37*, 85–156. [CrossRef]
34. Nicklas, R.B.; Ward, S.C. Elements of error correction in mitosis: Microtubule capture, release, and tension. *J. Cell Biol.* **1994**, *126*, 1241–1253. [CrossRef] [PubMed]
35. Cimini, D.; Wan, X.; Hirel, C.B.; Salmon, E.D. Aurora kinase promotes turnover of kinetochore microtubules to reduce chromosome segregation errors. *Curr. Biol.* **2006**, *16*, 1711–1718. [CrossRef] [PubMed]
36. Kabeche, L.; Compton, D.A. Cyclin A regulates kinetochore microtubules to promote faithful chromosome segregation. *Nature* **2013**, *502*, 110–113. [CrossRef] [PubMed]
37. Zhai, Y.; Kronebusch, P.J.; Borisy, G.G. Kinetochore microtubule dynamics and the metaphase-anaphase transition. *J. Cell Biol.* **1995**, *131*, 721–734. [CrossRef] [PubMed]
38. Kapoor, T.M.; Lampson, M.A.; Hergert, P.; Cameron, L.; Cimini, D.; Salmon, E.D.; McEwen, B.F.; Khodjakov, A. Chromosomes can congress to the metaphase plate before biorientation. *Science* **2006**, *311*, 388–391. [CrossRef] [PubMed]
39. Bakhoum, S.F.; Thompson, S.L.; Manning, A.L.; Compton, D.A. Genome stability is ensured by temporal control of kinetochore-microtubule dynamics. *Nat. Cell Biol.* **2009**, *11*, 27–35. [CrossRef] [PubMed]
40. Loncarek, J.; Kisurina-Evgenieva, O.; Vinogradova, T.; Hergert, P.; La Terra, S.; Kapoor, T.M.; Khodjakov, A. The centromere geometry essential for keeping mitosis error free is controlled by spindle forces. *Nature* **2007**, *450*, 745–749. [CrossRef] [PubMed]
41. Holy, T.E.; Leibler, S. Dynamic instability of microtubules as an efficient way to search in space. *Proc. Natl. Acad. Sci. USA* **1994**, *91*, 5682–5685. [CrossRef] [PubMed]
42. Wollman, R.; Cytrynbaum, E.N.; Jones, J.T.; Meyer, T.; Scholey, J.M.; Mogilner, A. Efficient chromosome capture requires a bias in the "search-and-capture" process during mitotic-spindle assembly. *Curr. Biol.* **2005**, *15*, 828–832. [CrossRef] [PubMed]
43. Paul, R.; Wollman, R.; Silkworth, W.T.; Nardi, I.K.; Cimini, D.; Mogilner, A. Computer simulations predict that chromosome movements and rotations accelerate mitotic spindle assembly without compromising accuracy. *Proc. Natl. Acad. Sci. USA* **2009**, *106*, 15708–15713. [CrossRef] [PubMed]
44. Hoffman, D.B.; Pearson, C.G.; Yen, T.J.; Howell, B.J.; Salmon, E.D. Microtubule-dependent changes in assembly of microtubule motor proteins and mitotic spindle checkpoint proteins at PtK1 kinetochores. *Mol. Biol. Cell* **2001**, *12*, 1995–2009. [CrossRef] [PubMed]
45. Magidson, V.; Paul, R.; Yang, N.; Ault, J.G.; O'Connell, C.B.; Tikhonenko, I.; McEwen, B.F.; Mogilner, A.; Khodjakov, A. Adaptive changes in the kinetochore architecture facilitate proper spindle assembly. *Nat. Cell Biol.* **2015**, *17*, 1134–1144. [CrossRef] [PubMed]
46. Wynne, D.J.; Funabiki, H. Kinetochore function is controlled by a phospho-dependent coexpansion of inner and outer components. *J. Cell Biol.* **2015**, *210*, 899–916. [CrossRef] [PubMed]
47. Godek, K.M.; Kabeche, L.; Compton, D.A. Regulation of kinetochore-microtubule attachments through homeostatic control during mitosis. *Nat. Rev. Mol. Cell Biol.* **2014**, *16*, 57–64. [CrossRef] [PubMed]
48. Grishchuk, E.L.; Molodtsov, M.I.; Ataullakhanov, F.I.; McIntosh, J.R. Force production by disassembling microtubules. *Nature* **2005**, *438*, 384–388. [CrossRef] [PubMed]
49. Franck, A.D.; Powers, A.F.; Gestaut, D.R.; Gonen, T.; Davis, T.N.; Asbury, C.L. Tension applied through the Dam1 complex promotes microtubule elongation providing a direct mechanism for length control in mitosis. *Nat. Cell Biol.* **2007**, *9*, 832–837. [CrossRef] [PubMed]

50. Akiyoshi, B.; Sarangapani, K.K.; Powers, A.F.; Nelson, C.R.; Reichow, S.L.; Arellano-Santoyo, H.; Gonen, T.; Ranish, J.A.; Asbury, C.L.; Biggins, S. Tension directly stabilizes reconstituted kinetochore-microtubule attachments. *Nature* **2010**, *468*, 576–579. [CrossRef] [PubMed]
51. Armond, J.W.; Vladimirou, E.; Erent, M.; McAinsh, A.D.; Burroughs, N.J. Probing microtubule polymerisation state at single kinetochores during metaphase chromosome motion. *J. Cell Sci.* **2015**, *3*, 1991–2001. [CrossRef] [PubMed]
52. Tirnauer, J.S.; Canman, J.C.; Salmon, E.D.; Mitchison, T.J. EB1 targets to kinetochores with attached, polymerizing microtubules. *Mol. Biol. Cell* **2002**, *13*, 4308–4316. [CrossRef] [PubMed]
53. Biggins, S.; Severin, F.F.; Bhalla, N.; Sassoon, I.; Hyman, A.A.; Murray, A.W. The conserved protein kinase Ipl1 regulates microtubule binding to kinetochores in budding yeast. *Genes Dev.* **1999**, *13*, 532–544. [CrossRef] [PubMed]
54. Cheeseman, I.M.; Anderson, S.; Jwa, M.; Green, E.M.; Kang, J.; Yates, J.R., 3rd; Chan, C.S.; Drubin, D.G.; Barnes, G. Phospho-regulation of kinetochore-microtubule attachments by the Aurora kinase Ipl1p. *Cell* **2002**, *111*, 163–172. [CrossRef]
55. Francisco, L.; Wang, W.; Chan, C.S. Type 1 protein phosphatase acts in opposition to IpL1 protein kinase in regulating yeast chromosome segregation. *Mol. Cell. Biol.* **1994**, *14*, 4731–4740. [CrossRef] [PubMed]
56. Tanaka, T.U.; Rachidi, N.; Janke, C.; Pereira, G.; Galova, M.; Schiebel, E.; Stark, M.J.; Nasmyth, K. Evidence that the Ipl1-Sli15 (Aurora kinase-INCENP) complex promotes chromosome bi-orientation by altering kinetochore-spindle pole connections. *Cell* **2002**, *108*, 317–329. [CrossRef]
57. Hauf, S.; Cole, R.W.; LaTerra, S.; Zimmer, C.; Schnapp, G.; Walter, R.; Heckel, A.; van Meel, J.; Rieder, C.L.; Peters, J.M. The small molecule Hesperadin reveals a role for Aurora B in correcting kinetochore-microtubule attachment and in maintaining the spindle assembly checkpoint. *J. Cell Biol.* **2003**, *161*, 281–294. [CrossRef] [PubMed]
58. Ditchfield, C.; Johnson, V.L.; Tighe, A.; Ellston, R.; Haworth, C.; Johnson, T.; Mortlock, A.; Keen, N.; Taylor, S.S. Aurora B couples chromosome alignment with anaphase by targeting BubR1, Mad2, and Cenp-E to kinetochores. *J. Cell Biol.* **2003**, *161*, 267–280. [CrossRef] [PubMed]
59. Lampson, M.A.; Renduchitala, K.; Khodjakov, A.; Kapoor, T.M. Correcting improper chromosome-spindle attachments during cell division. *Nat. Cell Biol.* **2004**, *6*, 232–237. [CrossRef] [PubMed]
60. Kallio, M.J.; McCleland, M.L.; Stukenberg, P.T.; Gorbsky, G.J. Inhibition of aurora B kinase blocks chromosome segregation, overrides the spindle checkpoint, and perturbs microtubule dynamics in mitosis. *Curr. Biol.* **2002**, *12*, 900–905. [CrossRef]
61. Chan, C.S.; Botstein, D. Isolation and characterization of chromosome-gain and increase-in-ploidy mutants in yeast. *Genetics* **1993**, *135*, 677–691. [PubMed]
62. Liu, D.; Vader, G.; Vromans, M.J.M.; Lampson, M.A.; Lens, S.M.A. Sensing chromosome bi-orientation by spatial separation of aurora B kinase from kinetochore substrates. *Science* **2009**, *323*, 1350–1353. [CrossRef] [PubMed]
63. Welburn, J.P.; Vleugel, M.; Liu, D.; Yates, J.R., 3rd; Lampson, M.A.; Fukagawa, T.; Cheeseman, I.M. Aurora B phosphorylates spatially distinct targets to differentially regulate the kinetochore-microtubule interface. *Mol. Cell* **2010**, *38*, 383–392. [CrossRef] [PubMed]
64. Keating, P.; Rachidi, N.; Tanaka, T.U.; Stark, M.J. Ipl1-dependent phosphorylation of Dam1 is reduced by tension applied on kinetochores. *J. Cell Sci.* **2009**, *122*, 4375–4382. [CrossRef] [PubMed]
65. DeLuca, K.F.; Lens, S.M.; DeLuca, J.G. Temporal changes in Hec1 phosphorylation control kinetochore-microtubule attachment stability during mitosis. *J. Cell Sci.* **2011**, *124*, 622–634. [CrossRef] [PubMed]
66. Salimian, K.J.; Ballister, E.R.; Smoak, E.M.; Wood, S.; Panchenko, T.; Lampson, M.A.; Black, B.E. Feedback control in sensing chromosome biorientation by the Aurora B kinase. *Curr. Biol.* **2011**, *21*, 1158–1165. [CrossRef] [PubMed]
67. Suzuki, A.; Badger, B.L.; Wan, X.; DeLuca, J.G.; Salmon, E.D. The Architecture of CCAN Proteins Creates a Structural Integrity to Resist Spindle Forces and Achieve Proper Intrakinetochore Stretch. *Dev. Cell* **2014**, *30*, 717–730. [CrossRef] [PubMed]
68. Krenn, V.; Musacchio, A. The Aurora B Kinase in Chromosome Bi-Orientation and Spindle Checkpoint Signaling. *Front. Oncol.* **2015**, *5*, 225. [CrossRef] [PubMed]

69. Sarangapani, K.K.; Asbury, C.L. Catch and release: How do kinetochores hook the right microtubules during mitosis? *Trends Genet.* **2014**, *30*, 150–159. [CrossRef] [PubMed]
70. Cheeseman, I.M.; Chappie, J.S.; Wilson-Kubalek, E.M.; Desai, A. The conserved KMN network constitutes the core microtubule-binding site of the kinetochore. *Cell* **2006**, *127*, 983–997. [CrossRef] [PubMed]
71. Gestaut, D.R.; Graczyk, B.; Cooper, J.; Widlund, P.O.; Zelter, A.; Wordeman, L.; Asbury, C.L.; Davis, T.N. Phosphoregulation and depolymerization-driven movement of the Dam1 complex do not require ring formation. *Nat. Cell Biol.* **2008**, *10*, 407–414. [CrossRef] [PubMed]
72. Sarangapani, K.K.; Akiyoshi, B.; Duggan, N.M.; Biggins, S.; Asbury, C.L. Phosphoregulation promotes release of kinetochores from dynamic microtubules via multiple mechanisms. *Proc. Natl. Acad. Sci. USA* **2013**, *110*, 7282–7287. [CrossRef] [PubMed]
73. Umbreit, N.T.; Gestaut, D.R.; Tien, J.F.; Vollmar, B.S.; Gonen, T.; Asbury, C.L.; Davis, T.N. The Ndc80 kinetochore complex directly modulates microtubule dynamics. *Proc. Natl. Acad. Sci. USA* **2012**, *109*, 16113–16118. [CrossRef] [PubMed]
74. Zaytsev, A.V.; Sundin, L.J.R.; DeLuca, K.F.; Grishchuk, E.L.; DeLuca, J.G. Accurate phosphoregulation of kinetochore-microtubule affinity requires unconstrained molecular interactions. *J. Cell Biol.* **2014**, *206*, 45–59. [CrossRef] [PubMed]
75. Ye, A.A.; Deretic, J.; Hoel, C.M.; Hinman, A.W.; Cimini, D.; Welburn, J.P.; Maresca, T.J. Aurora A Kinase Contributes to a Pole-Based Error Correction Pathway. *Curr. Biol.* **2015**, *25*, 1842–1851. [CrossRef] [PubMed]
76. Chmátal, L.; Yang, K.; Schultz, R.M.; Lampson, M.A. Spatial Regulation of Kinetochore Microtubule Attachments by Destabilization at Spindle Poles in Meiosis I. *Curr. Biol.* **2015**, *25*, 1835–1841. [CrossRef] [PubMed]
77. Kawashima, S.A.; Yamagishi, Y.; Honda, T.; Ishiguro, K.; Watanabe, Y. Phosphorylation of H2A by Bub1 prevents chromosomal instability through localizing shugoshin. *Science* **2010**, *327*, 172–177. [CrossRef] [PubMed]
78. Kelly, A.E.; Ghenoiu, C.; Xue, J.Z.; Zierhut, C.; Kimura, H.; Funabiki, H. Survivin Reads Phosphorylated Histone H3 Threonine 3 to Activate the Mitotic Kinase Aurora B. *Science* **2010**. [CrossRef] [PubMed]
79. Wang, F.; Dai, J.; Daum, J.R.; Niedzialkowska, E.; Banerjee, B.; Stukenberg, P.T.; Gorbsky, G.J.; Higgins, J.M. Histone H3 Thr-3 phosphorylation by Haspin positions Aurora B at centromeres in mitosis. *Science* **2010**, *330*, 231–235. [CrossRef] [PubMed]
80. Yamagishi, Y.; Honda, T.; Tanno, Y.; Watanabe, Y. Two histone marks establish the inner centromere and chromosome bi-orientation. *Science* **2010**, *330*, 239–243. [CrossRef] [PubMed]
81. Wang, F.; Ulyanova, N.P.; Daum, J.R.; Patnaik, D.; Kateneva, A.V.; Gorbsky, G.J.; Higgins, J.M. Haspin inhibitors reveal centromeric functions of Aurora B in chromosome segregation. *J. Cell Biol.* **2012**, *199*, 251–268. [CrossRef] [PubMed]
82. Yue, Z.; Carvalho, A.; Xu, Z.; Yuan, X.; Cardinale, S.; Ribeiro, S.; Lai, F.; Ogawa, H.; Gudmundsdottir, E.; Gassmann, R.; et al. Deconstructing Survivin: Comprehensive genetic analysis of Survivin function by conditional knockout in a vertebrate cell line. *J. Cell Biol.* **2008**, *183*, 279–296. [CrossRef] [PubMed]
83. Tsukahara, T.; Tanno, Y.; Watanabe, Y. Phosphorylation of the CPC by Cdk1 promotes chromosome bi-orientation. *Nature* **2010**. [CrossRef] [PubMed]
84. Bishop, J.D.; Schumacher, J.M. Phosphorylation of the carboxyl terminus of inner centromere protein (INCENP) by the Aurora B Kinase stimulates Aurora B kinase activity. *J. Biol. Chem.* **2002**, *277*, 27577–27580. [CrossRef] [PubMed]
85. Honda, R.; Korner, R.; Nigg, E.A. Exploring the functional interactions between Aurora B, INCENP, and survivin in mitosis. *Mol. Biol. Cell* **2003**, *14*, 3325–3341. [CrossRef] [PubMed]
86. Yasui, Y.; Urano, T.; Kawajiri, A.; Nagata, K.; Tatsuka, M.; Saya, H.; Furukawa, K.; Takahashi, T.; Izawa, I.; Inagaki, M. Autophosphorylation of a newly identified site of Aurora-B is indispensable for cytokinesis. *J. Biol. Chem.* **2004**, *279*, 12997–13003. [CrossRef] [PubMed]
87. Sessa, F.; Mapelli, M.; Ciferri, C.; Tarricone, C.; Areces, L.B.; Schneider, T.R.; Stukenberg, P.T.; Musacchio, A. Mechanism of Aurora B activation by INCENP and inhibition by hesperadin. *Mol. Cell* **2005**, *18*, 379–391. [CrossRef] [PubMed]
88. Kelly, A.E.; Sampath, S.C.; Maniar, T.A.; Woo, E.M.; Chait, B.T.; Funabiki, H. Chromosomal enrichment and activation of the aurora B pathway are coupled to spatially regulate spindle assembly. *Dev. Cell* **2007**, *12*, 31–43. [CrossRef] [PubMed]

89. Zaytsev, A.V.; Segura-Pena, D.; Godzi, M.; Calderon, A.; Ballister, E.R.; Stamatov, R.; Mayo, A.M.; Peterson, L.; Black, B.E.; Ataullakhanov, F.I.; et al. Bistability of a coupled aurora B kinase-phosphatase system in cell division. *eLife* **2016**, *5*, e10644. [CrossRef] [PubMed]
90. Campbell, C.S.; Desai, A. Tension sensing by Aurora B kinase is independent of survivin-based centromere localization. *Nature* **2013**, *497*, 118–121. [CrossRef] [PubMed]
91. Samejima, K.; Platani, M.; Wolny, M.; Ogawa, H.; Vargiu, G.; Knight, P.J.; Peckham, M.; Earnshaw, W.C. The inner centromere protein (INCENP) coil is a single α-helix (SAH) domain that binds directly to microtubules and is important for chromosome passenger complex (CPC) localization and function in mitosis. *J. Biol. Chem.* **2015**, *290*, 21460–21472. [CrossRef] [PubMed]
92. Uchida, K.S.K.; Takagaki, K.; Kumada, K.; Hirayama, Y.; Noda, T.; Hirota, T. Kinetochore stretching inactivates the spindle assembly checkpoint. *J. Cell Biol.* **2009**, *184*, 383–390. [CrossRef] [PubMed]
93. Dumont, S.; Salmon, E.D.; Mitchison, T.J. Deformations within moving kinetochores reveal different sites of active and passive force generation. *Science* **2012**, *337*, 355–358. [CrossRef] [PubMed]
94. Khodjakov, A.; Pines, J. Centromere tension: A divisive issue. *Nat. Cell Biol.* **2010**, *12*, 919–923. [CrossRef] [PubMed]
95. Liu, D.; Vleugel, M.; Backer, C.B.; Hori, T.; Fukagawa, T.; Cheeseman, I.M.; Lampson, M.A. Regulated targeting of protein phosphatase 1 to the outer kinetochore by KNL1 opposes Aurora B kinase. *J. Cell Biol.* **2010**, *188*, 809–820. [CrossRef] [PubMed]
96. Rosenberg, J.S.; Cross, F.R.; Funabiki, H. KNL1/Spc105 recruits PP1 to silence the spindle assembly checkpoint. *Curr. Biol.* **2011**, *21*, 942–947. [CrossRef] [PubMed]
97. Gregan, J.; Polakova, S.; Zhang, L.; Tolić-Nørrelykke, I.M.; Cimini, D. Merotelic kinetochore attachment: Causes and effects. *Trends Cell Biol.* **2011**, *21*, 374–381. [CrossRef] [PubMed]
98. Kalantzaki, M.; Kitamura, E.; Zhang, T.; Mino, A.; Novák, B.; Tanaka, T.U. Kinetochore-microtubule error correction is driven by differentially regulated interaction modes. *Nat. Cell Biol.* **2015**. [CrossRef]
99. Pearson, C.G.; Yeh, E.; Gardner, M.; Odde, D.; Salmon, E.D.; Bloom, K. Stable kinetochore-microtubule attachment constrains centromere positioning in metaphase. *Curr. Biol.* **2004**, *14*, 1962–1967. [CrossRef] [PubMed]
100. McIntosh, J.R.; Hering, G.E. Spindle fiber action and chromosome movement. *Annu. Rev. Cell Biol.* **1991**, *7*, 403–426. [CrossRef] [PubMed]
101. Kline-Smith, S.L.; Khodjakov, A.; Hergert, P.; Walczak, C.E. Depletion of centromeric MCAK leads to chromosome congression and segregation defects due to improper kinetochore attachments. *Mol. Biol. Cell* **2004**, *15*, 1146–1159. [CrossRef] [PubMed]
102. Tanno, Y.; Susumu, H.; Kawamura, M.; Sugimura, H.; Honda, T.; Watanabe, Y. The inner centromere-shugoshin network prevents chromosomal instability. *Science* **2015**, *349*, 1237–1240. [CrossRef] [PubMed]
103. Knowlton, A.L.; Lan, W.; Stukenberg, P.T. Aurora B is enriched at merotelic attachment sites, where it regulates MCAK. *Curr. Biol.* **2006**, *16*, 1705–1710. [CrossRef] [PubMed]

biology

MDPI

Review

Mitotic Spindle Assembly in Land Plants: Molecules and Mechanisms

Moé Yamada and Gohta Goshima *

Graduate School of Science, Division of Biological Science, Nagoya University, Furo-cho, Chikusa-ku, Nagoya 464-8602, Japan; yamada.moe@a.mbox.nagoya-u.ac.jp
* Correspondence: goshima@bio.nagoya-u.ac.jp; Tel.: +81-52-788-6175; Fax: +81-52-788-6174

Academic Editor: J. Richard McIntosh
Received: 1 October 2016; Accepted: 8 January 2017; Published: 25 January 2017

Abstract: In textbooks, the mitotic spindles of plants are often described separately from those of animals. How do they differ at the molecular and mechanistic levels? In this chapter, we first outline the process of mitotic spindle assembly in animals and land plants. We next discuss the conservation of spindle assembly factors based on database searches. Searches of >100 animal spindle assembly factors showed that the genes involved in this process are well conserved in plants, with the exception of two major missing elements: centrosomal components and subunits/regulators of the cytoplasmic dynein complex. We then describe the spindle and phragmoplast assembly mechanisms based on the data obtained from robust gene loss-of-function analyses using RNA interference (RNAi) or mutant plants. Finally, we discuss future research prospects of plant spindles.

Keywords: mitosis; kinetochore; centrosome; dynein; kinesin; augmin; gamma-tubulin; *Arabidopsis thaliana*; *Physcomitrella patens*; *Haemanthus*

---

1. Microscopic Overview of the Spindle Assembly

Mitotic spindle formation involves several key events, such as microtubule (MT) generation, bipolarity establishment, pole focusing, length control, and chromosome capture/alignment. Since the early days of spindle research, plant spindles have often been described separately from those of animal cells, perhaps owing to the apparent differences in their overall structure [1,2]. Most noticeably, land plants lack centrosomes, the dominant MT nucleating and organising centre in animal somatic cells; the metaphase spindle is generally barrel-shaped without a single focusing point at the pole (Figure 1). In the later stages of mitosis, plant cells uniquely assemble phragmoplasts that are MT arrays for cell plate material deposition. How, then, do land plant and animal spindles differ at the molecular level? In this chapter, we first outline the process of mitotic spindle assembly in animals and plants before discussing the molecular factors involved in this process.

1.1. *Mitotic Spindle Assembly in Animals*

The start of mitosis in animal somatic cells is characterised by the maturation of centrosomes during prophase. Centrosomes serve as the dominant MT generation sites as well as MT organising centres during spindle assembly [3]. After nuclear envelope breakdown (NEBD), two additional mechanisms operate to produce more MTs [4,5]. One is chromosome-mediated nucleation, in which a chromosome-associated protein activates the MT nucleation/stabilisation machinery around the chromosomes [6]. The other is MT-dependent MT nucleation, where new MTs are nucleated in a branching fashion from the existing MTs, such as those nucleated via the centrosomal or chromosomal pathway [7]. In some cell types, these three pathways act in concert, whereas in others, one or two pathways do not play major roles [8]. Regardless of their source, MTs are oriented in a bipolar manner by the action of MT-based motor proteins and through stable bipolar association with kinetochore MTs [9]. Two poles of the metaphase spindle are well focused as spindle MTs remain associated with the centrosome at the pole or crosslinked with each other by motors and MAPs [10,11]. Regulators of MT dynamics also play critical roles in spindle morphogenesis; alteration of MT dynamics affects the length and pole organisation of the metaphase spindle [12]. During prometaphase and metaphase, sister kinetochores are bioriented and attach to the plus end of MTs; completion of this process is essential for equal segregation of sister chromatids into two daughter cells. In addition, a defect in the kinetochore-MT attachment causes force imbalance in the spindle that affects spindle length. During anaphase, kinetochore MTs are depolymerised, whereas interpolar MTs elongate. In addition, MTs are de novo generated by the actions of three MT generating pathways that also operate during pre-anaphase [13]. Motor- and MAP-dependent crosslinking of those MTs at the midzone leads to the appearance of the characteristic central spindle structure during anaphase that is required for subsequent cytokinesis [14].

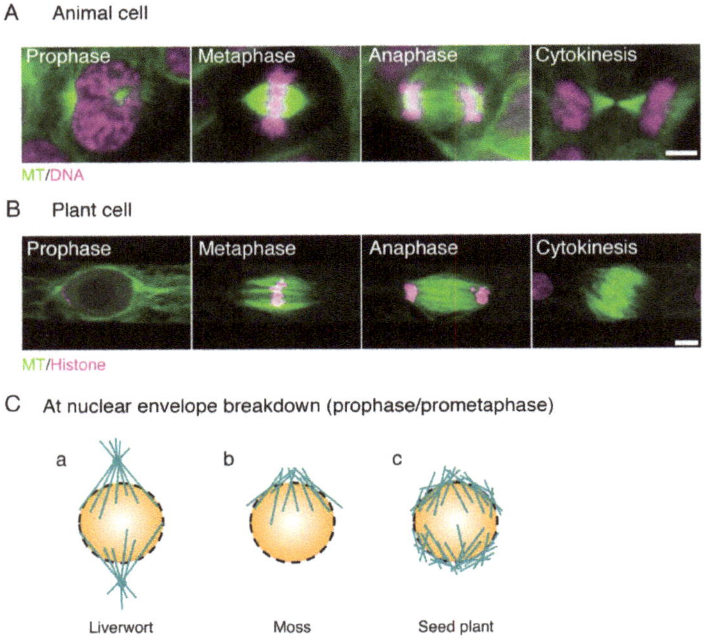

Figure 1. Animal and plant spindles. (**A**) Human HCT116 cells stained with anti-α-tubulin antibody (green) and DAPI (purple); (**B**) Spindles in the moss *Physcomitrella. patens* highlighted by GFP-tubulin (green) and histone H2B-RFP (purple). The two main differences between animal and plant spindles are (1) the presence of centrosomes and well-developed astral MTs in animal spindles, and (2) the morphology of the anaphase spindle (the 'phragmoplast' in plants). Bars, 5 μm; (**C**) distinct types of MT formation at the beginning of prometaphase (at NEBD) in liverwort, moss, and angiosperm. (**a**) In liverwort, polar organisers (POs) are assembled and act as MTOCs (microtubule-organising centres). Unlike the centrosome, however, the PO is merged into the spindle and cannot be observed as a distinct structure in metaphase [15,16]. The genes required for PO formation are unknown; (**b**) In moss protonemata, MTs are asymmetrically accumulated around the nucleus and are more abundant on the apical side [17]. An RNAi study indicated that their formation depends on γ-tubulin, but not augmin; (**c**) In most angiosperm cell types, two loosely organised MT structures known as 'polar caps' are detected around the nucleus [18]. γ-Tubulin is localised at this region and MTs are actively generated [19].

1.2. Mitotic Spindle Assembly in Seed Plants

Researchers have elucidated the mechanism of acentrosomal spindle formation in land plants through microscopic observation and have revealed the processes common to, and different from, animal somatic cells [20,21]. One of the best-characterised cell types with regard to mitotic spindle assembly is the endosperm of African blood lily *Haemanthus*. In the absence of centrosomes, abundant MTs are detected around the nuclear envelope during prophase [22]. Immunofluorescence microscopy identified MT converging centres within the MT cloud, which was consistent with the idea that they are the major MT nucleation sites at this stage [23,24]. MTs around the nucleus are gradually organised into a spindle-like structure, called the 'prophase spindle' (or 'prospindle'). The prophase spindle has either a bipolar fusiform or multipolar structure [23]. After NEBD, MTs emanating from the converging centres associate with kinetochores to form kinetochore MTs [24]. MTs are also likely nucleated near the chromosome/kinetochore independent of prophase spindles during the prometaphase as an MT depolymerisation/regrowth assay detected chromosome-proximal MT formation [25]. Those MTs are then organised into an overall bipolar configuration. Electron microscopy showed that the majority of the MTs are oriented in such a way that plus ends are pointed to the chromosome/kinetochore, similar to animal spindles [26]. However, the metaphase spindle is barrel-shaped rather than fusiform, as the pole is not tightly focused at one point; multiple kinetochore and non-kinetochore MTs are converged or cross-linked locally and, thus, multiple mini-poles are observed [23]. Immunostaining of MTs also identified 'fir tree' structures within the spindle, in which many MTs branched off from kinetochore MTs [27]. With the start of anaphase, sister chromatids are separated and then segregated to the pole by kinetochore MT depolymerisation, analogous to animal spindles. During telophase, the phragmoplast forms and is followed by centrifugal expansion towards the cell cortex [28].

Arabidopsis thaliana is currently the most frequently used plant organism for genetic studies, and the mitotic spindle assembly process has been observed in several *Arabidopsis* tissues and suspension cells [19,29,30]. The tobacco BY-2 cell line is another popular system for mitosis imaging [31]. In these cells, MTs accumulate at the nuclear envelope and form prophase spindles (also called 'polar caps'), as occurs in *Haemanthus* endosperm. Upon NEBD, MTs emanating from polar caps become a source of spindle MTs. Thus, the initial spindle assembly process in prometaphase is similar to that observed in the *Haemanthus* endosperm. The processes of metaphase, anaphase, and telophase are also analogous to those described for the endosperm. Unlike the endosperm, however, most seed-plant tissues have the preprophase band (PPB) that is a structure consisting of parallel MT arrays beneath the cell cortex that appears prior to mitosis and marks the future division plate. The PPB ensures the bipolarity of prophase spindles. While this structure is critical for division plane determination and polar cap bipolarity [32], we will not discuss this structure further in this review as they generally disappear or degenerate during the prophase and are dispensable for bipolar metaphase spindle assembly per se (see [33,34] as recent reviews on PPBs).

1.3. Mitotic Spindle Assembly in Bryophytes

Bryophytes have also been the subject of microscopic analysis of the mitotic spindle formation process. This process, particularly during the prophase, is somewhat different from that observed in seed plants (Figure 1C). In liverwort, prior to the appearance of prophase spindles, centrosome-like MT organising centres (MTOCs), called polar organisers (PO), appear in the cytoplasm [35,36]. The PO is similar to centrosomes in that they produce astral MTs. However, the PO is a transient structure that does not have centriole core, and during spindle formation, is no longer identified as a discrete structure. To the best of our knowledge, the roles of POs have not been experimentally demonstrated. It is, however, plausible that they function as an MT nucleation centre, as well as ensuring spindle bipolarity because they are stained well with antibodies against γ-tubulin, the major MT nucleator in eukaryotes [37,38]. In moss, conversely, PO-like structures have not been observed; instead, MTs are enriched around the nuclear envelope in prophase. These MTs emanating from the nuclear envelope represent the major source of prometaphase spindles, similar to *Haemanthus* [15,17]. In the hornwort,

MTOCs are associated with plastids [15,39]. Despite the apparent differences in the earliest phase of spindle assembly, the morphology of the metaphase spindle of bryophytes is similar to that of *Haemanthus* endosperm, suggesting that a similar molecular factor is involved in the spindle assembly process during the prometaphase [15].

2. Conservation of Spindle Assembly Factors

The molecular factors in yeast and animal spindles have been extensively surveyed using genetics (including RNAi screening) and biochemistry (such as mass spectrometric protein identification). It is believed that most of the key factors have been identified [5,40–43]. However, since the experimental system is cumbersome and the genes are highly redundant, few genes required for spindle assembly in plants have been identified using these techniques. To characterise the putative molecular factors involved in plant spindle assembly, we performed an extensive database search (including BLAST sequence homology searching) on 131 known animal and yeast spindle factors (Table 1). As the targets, we selected *Arabidopsis* and the moss *Physcomitrella patens*, for which complete genome sequences are available [44,45], and molecular dissection of the spindle is arguably most advanced. For some genes, similar or more extensive homologue lists have previously been generated by other researchers (e.g., [46,47]); we double-checked the conservation/non-conservation of those genes with our procedure and included the references in the table.

In our search, homologues for most of the animal proteins were identified. All *Arabidopsis* genes are conserved in the moss *P. patens*; this suggests that they are likely to be found in a vast majority of land plant species. Nevertheless, our search failed to identify many components of three functional modules, namely, centrosomes, the cytoplasmic dynein complex, and kinetochores.

Table 1. List of homologues of spindle assembly factors.

	Generic Name	*H. sapiens*	*D. melanogaster*	*S. pombe*	*P. patens*	*A. thaliana* (* *identified with BLAST*)	*A. thaliana* Gene Acession #	References
Kinase/ phosphatase/ signalling	Cdk1	■	■	■	■	■ *	AT3G48750	
	Aurora kinase	■	■	■	■	■ *	AT4G32830 etc.	[48]
	Plk1	■	■	■				[49]
	Haspin	■	■	■	■	■ *	AT1G09450	
	Ran	■	■	■	■	■ *	AT5G20010 etc.	
	RCC1	■	■	■	■	■ *	AT5G63860 etc.	
	RanGAP	■	■	■	■	■ *	AT3G63130, AT5G19320	
	PP2A	■	■	■	■	■ *	AT1G69960 etc.	[50]
	Endosulfine	■	■	■	■	■	AT1G69510	[51]
	PP1	■	■	■	■	■	AT2G29400	[50]
	PP6	■	■	■	■	■ *	AT3G19980	
Centriole	Plk4	■	■					[52]
	Sas4	■	■		■			
	Sas5/Ana2/STIL	■	■					
	Sas6	■	■		■			[52]
	Spd2/CEP192	■	■					
	Ana1/CEP295	■	■					
	Ana3/Rotatin	■	■					
Motor/MAPs	Kin4/chromokinesin	■	■		■	■ *	AT5G60930 etc.	[53,54]
	Kin5	■	■	■	■	■ *	AT2G28620 etc.	
	Kin6	■	■					
	Kin7/CENP-E	■	■		■	■ *	AT3G10180 etc.	
	Kin8	■	■	■	■	■ *	AT1G18550, AT3G49650	
	Kin12/KIF15	■			■	■ *	AT3G19050 etc.	
	Kin13	■	■		■	■ *	AT3G16060, AT3G16630	
	Kin14	■	■	■	■	■ *	AT4G21270 etc.	

Table 1. *Cont.*

	Generic Name	*H. sapiens*	*D. melanogaster*	*S. pombe*	*P. patens*	*A. thaliana* (* identified with BLAST)	*A. thaliana* Gene Acession #	References
	DHC	■	■	■				
	DIC	■	■	■	■			[55]
	DLC (LC8)	■	■	■	■	■ *	AT4G15930 etc.	
	Dynactin p50	■	■	■				
	Dynactin p150	■	■	■	■			
	Dynactin ARP1	■	■	■				[56]
	PRC1/MAP65/Ase1	■	■	■	■	■ *	AT3G60840 etc.	
	Katanin (p60)	■	■		■	■ *	AT1G80350	
	HURP	■	■					
	TACC	■	■	■				
	TPX2	■	■		■	■ *	AT1G03780 etc.	[57]
Nucleation	γ-Tubulin	■	■	■	■	■ *	AT3G61650, AT5G05620	[58]
	GCP2/3	■	■	■	■	■ *	AT5G17410, AT5G06680	[46]
	GCP4/5/6	■	■	■	■	■ *	At3g53760 etc.	[46,59]
	NEDD1	■	■		■	■ *	AT5G05970	[60]
	Mzt1	■	■	■	■	■ *	AT1G73790, AT4G09550	[61–63]
	Mzt2	■						[64]
	Augmin (8 subunits)	■	■		■	■ *	At5g40740 etc.	[17,46,65]
	Pericentrin/D-plp	■	■					[64]
	AKAP9	■	■					[64,66]
	SPC110/Pcp1			■				
	CDK5RAP2/Cnn	■	■	■				[64]
	Myomegalin	■						
Microtubule plus end	ch-TOG/XMAP215	■	■	■	■	■ *	AT2G35630	[67]
	EB1	■	■	■	■	■ *	AT5G62500 etc.	
	SLAIN/Sentin	■	■					
	CLIP170	■	■	■				
	CLASP	■	■	■	■	■ *	AT2G20190	
	SKAP	■						
	Astrin	■						

Table 1. *Cont.*

	Generic Name	*H. sapiens*	*D. melanogaster*	*S. pombe*	*P. patens*	*A. thaliana* (* *identified with BLAST*)	*A. thaliana* Gene Acession #	References
Microtubule minus end	CAMSAP	■	■					[68]
	Msd1/SSX2IP	■		■	■	■ *	AT5G57410 etc.	[69]
	ASPM	■	■		■	■ *	AT4G21820	
	CaM	■	■	■	■	■ *	AT2G27030 etc.	
	NuMA	■	■					
	Microspherule	■	■		■	■ *	AT3G54350 etc.	
Chromosome	CAP-D2	■	■	■	■	■ *	AT3G57060	[70]
	SMC2	■	■	■	■	■ *	AT3G47460, AT5G62410	[70,71]
	CAP-H	■	■	■	■	■ *	AT2G32590	[70,72]
	SMC4	■	■	■	■	■ *	AT5G48600	[70–73]
	CAP-G	■	■	■	■	■ *	AT5G37630	[70]
	Topo II	■	■	■	■	■ *	AT3G23890	
	Rad21	■	■	■	■	■ *	AT5G16270 etc.	[74,75]
	SCC3	■	■	■	■	■ *	AT2G47980	[75]
	SMC1	■	■	■	■	■ *	AT3G54670	
	SMC3	■	■	■	■	■ *	AT2G27170	
	SCC2	■	■	■	■	■ *	AT5G15540	[76]
	SCC4	■	■	■	■	■ *	AT5G51340	
	Eco1	■	■	■	■	■ *	AT4G31400	[77]
	Sororin	■	■					[78]
	Wapl	■	■	■	■	■ *	AT1G11060	
	PDS5	■	■	■	■	■ *	AT5G47690 etc.	
	HP1	■	■	■	■	■ *	AT5G17690	
	Sgo1	■	■	■	■	■	AT3G10440, AT5G04320	[79]
	Borealin	■	■	■	■	■ *	AT4g39630	
	INCENP	■	■	■	■	■	AT5g55820	[80]
	Survivin	■	■	■				
	CENP-B	■		■				
	Mis18	■		■				
	Mis18BP1	■		■	■	■	At5g02520	[81]
	HJURP	■		■				

Table 1. *Cont.*

	Generic Name	*H. sapiens*	*D. melanogaster*	*S. pombe*	*P. patens*	*A. thaliana* (* *identified with BLAST*)	*A. thaliana* Gene Acession #	References
Kinetochore/ centromere	Cal1		■					
	CENP-A	■	■	■	■	■ *	AT1G01370	[82]
	CENP-C	■	■	■	■	■ *	AT1G15660	[83]
	CENP-S	■		■	■	■ *	AT5G50930	[84]
	CENP-X	■		■	■	■	AT1G78790	[85]
	CENP-T	■		■				
	CENP-W	■		■				
	CENP-L	■		■				
	CENP-N	■		■				
	CENP-H	■		■				
	CENP-I	■		■				
	CENP-K	■		■				
	CENP-M	■						
	CENP-O	■		■	■	■ *	AT5G10710	
	CENP-P	■		■				
	CENP-Q	■		■				
	CENP-U	■		■				
	CENP-R	■						
	Mis12	■	■	■	■	■	AT5G35520	[86]
	Dsn1/Mis13	■		■	■	■	AT3G27520	
	Nnf1	■	■	■	■	■	AT4G19350	
	Nsl1/Mis14	■	■	■				
	KNL1	■	■	■	■	■	AT2G04235	[87]
	Ndc80	■	■	■	■	■ *	AT3G54630	
	Nuf2	■	■	■	■	■ *	AT1G61000	
	Spc24	■	■	■	■	■	AT3G08880, AT5G01570	
	Spc25	■	■	■	■	■ *	AT3G48210	
	Ska1	■			■	■ *	AT3G60660	
	Ska2	■			■	■	AT2G24970	
	Ska3	■			■	■	AT5G06590	
	Dam1			■				
	CENP-F	■						

Table 1. *Cont.*

	Generic Name	*H. sapiens*	*D. melanogaster*	*S. pombe*	*P. patens*	*A. thaliana* (* *identified with BLAST*)	*A. thaliana* Gene Acession #	References
Spindle assembly checkpoint (SAC)	Mad1	■	■	■	■	■ *	AT5G49880	
	Mad2	■	■	■	■	■ *	AT3G25980	[88]
	Mad3 /BubR1	■	■	■	■	■ *	AT2G33560, AT5G05510	[88,89]
	Bub1	■	■	■	■	■ *	AT2G20635	
	Bub3	■	■	■	■	■ *	AT3G19590, AT1G49910	
	Mps1	■	■	■	■	■ *	AT1G77720	[90]
	Tpr	■	■	■	■	■ *	AT1G79280	[91]
	Cdc20	■	■	■	■	■ *	AT4G33270 etc.	
	Spindly	■	■					
	Rod	■	■					
	Zwilch	■	■					
	Zw10	■	■		■	■ *	AT2G32900	[92]

Black boxes indicate that homologous genes are present. Boxes are left blank if no clear homologues are present. The *Arabidopsis* genes identified with the BLAST search are marked with asterisks (*). Gene accession numbers for *Arabidopsis* genes were presented; however, just one or two numbers were provided when more homologues were identified or for protein complexes (augmin and GCPs). The identities of other subunits are found in the references presented in a separate column.

The homologous genes were sought as follows:

1. Animal and yeast genes required for spindle assembly were found in the literature [40,52,64,93];
2. The amino acid sequences of the *Homo sapiens* proteins were retrieved from the NCBI database. When multiple isoforms were identified, only one randomly selected isoform was used;
3. *Drosophila melanogaster* (fruit fly) and *Schizosaccharomyces. pombe* (fission yeast) homologues were sought in the NCBI 'Homologene' or 'Gene' search. When clear homologues were not identified, the BLAST search was performed;
4. Homologous genes of *Arabidopsis thaliana* and *Physcomitrella. patens* were sought using BLAST (query: human or yeast protein);
5. If no clear homologues could be identified, the databases for individual species were searched (PomBase, fly base, PHYSCObase, or TAIR). For the query, human (or, in some instances, fly) gene names or keywords (e.g., 'centromere', 'kinetochore', or 'CENP') were used;
6. If homologous genes were still not identified, the name was searched using Google Scholar and PubMed;
7. The sequences of plant Dsn1/Nnf1/Spc24/Ska3 and Msd1 were provided by Dr. Geert Kops (Utrecht University, The Netherlands) and Dr. Takashi Hashimoto (Nara Institute of Science and Technology, Japan), respectively.

2.1. Centrosome Proteins

As expected, animal centriole proteins, such as the cartwheel component Sas6, are mostly missing from the *Arabidopsis* genome. In contrast, pericentriolar proteins, like the subunits of the γ-tubulin ring complex (γ-TuRC) are more conserved. This makes sense because γ-TuRC functions at other places besides the centrosome, including the spindle MTs [15,17,58,94,95]. However, the regulators of γ-TuRC at the animal centrosome, such as the localisation factor/activator CDK5RAP2/Cnn, Plk1 kinase (polo-like kinase), and pericentrin, are largely unidentified in plants. In animals and yeasts, γ-TuRC alone lacks potent MT nucleating activity. Plants, therefore, might possess some plant-specific γ-TuRC activation factors. However, since centrosomal components are often difficult to identify by BLAST because of low sequence identity, their homologues may be present in the genome but have not yet been identified [96]. Several centriole components are found in the moss genome; these genes are likely to be required for the formation of the basal body that is used for flagella assembly in sperm [97].

2.2. Dynein Complex and Its Localisation Factors

Cytoplasmic dynein forms a large complex with several associated subunits and is a major MT-based motor protein in animals and fungi [98]. It moves towards the minus-end of MTs, delivers various cargoes, and generates force on the MT. In mitosis, cargoes include mitotic checkpoint proteins, chromosomes, and free cytoplasmic MTs. As previously shown [99], almost the entire dynein complex is absent from the *Arabidopsis* genome (except for the LC8 subunit that binds to other proteins such as myosin [100]). Furthermore, dynein adaptor proteins at the kinetochore or cell cortex, such as Rod, Zwilch, Spindly, or NuMA, are also missing. Thus, almost the entire dynein functional module has been lost in *Arabidopsis*. Since dynein plays various important roles in animals, plants must have developed an alternative force-generating system. One candidate is kinesin-14, which, like dynein, has minus-end-directed motility [101–103] and plays a partially redundant role in spindle pole organisation in animal cells [9,104]. Moss does not have cytoplasmic dynein but has axonemal dynein that is likely used for sperm motility [105]. Several dynein accessory subunits found in moss may be associated with the axonemal dynein heavy chain.

2.3. Kinetochore Components

Factors required for high-ordered chromosome organisation, such as condensin and cohesin complexes, and core components of the mitotic checkpoint [106] are highly conserved. However, many components of the kinetochore (the MT attachment site during mitosis) could not be identified. They might be present but could not be identified via BLAST, as kinetochore protein sequences are, in general, highly divergent among species, even within the metazoans (e.g., *Drosophila melanogaster* and humans) [107]. Current biochemical research has elucidated kinetochore subcomplexes [93]. We have identified at least one component per subcomplex in plants, with the exception of the CENP-H/I/K/M subcomplex. This suggests that other components with low sequence similarity are also present. It is also possible, however, that plants have either lost certain subunits or acquired plant-specific components. The former case is seen in *Drosophila*, in which most of the CENP components were lost during evolution. Systematic studies, such as those involving proteomics, are necessary to identify the complete set of kinetochore components in plants.

3. Molecular Mechanisms of Spindle Assembly in Land Plants

In this section, we draw a current molecular model showing how spindles and phragmoplasts are assembled in plant cells (Figure 2). The diagram described here is based on experimental results obtained using reliable methodology such as mutant or RNAi analysis. The knowledge was derived mainly from *P. patens*, which is a system that allows rapid loss-of-function analysis and high-resolution time-lapse microscopy [108], and *Arabidopsis*, which has a rich history of mutant collection [29]. However, since genes are well conserved across land plant species, the basic mechanism could be conserved in other plant cell types. We do not include information obtained solely for animal orthologues; however, it is possible that the uncharacterised plant homologues of animal spindle proteins have identical molecular activities and functions.

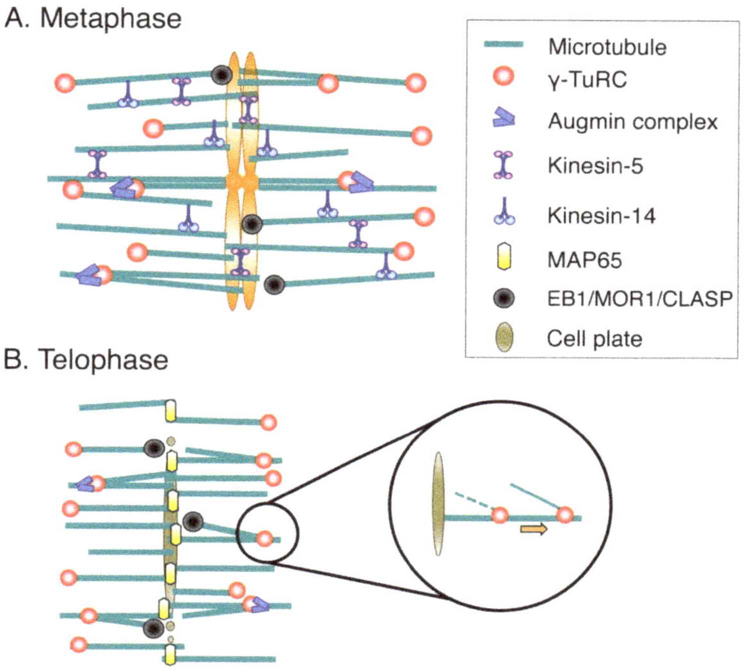

Figure 2. Molecular factors for spindle/phragmoplast assembly. (**A**) During prometaphase and metaphase, MTs are nucleated mainly by the γ-tubulin ring complex (γ-TuRC) and its recruitment/activation factor, the augmin complex. Multiple cross-linking proteins, including kinesin-5 and kinesin-14, shape the spindle. Spindle length is regulated by conserved MT plus-end-regulating proteins (EB1/MOR1/CLASP); (**B**) The MT-based arrays assembled after sister chromatid separation are called phragmoplasts. The central factors for MT generation in the phragmoplast are γ-TuRC and augmin, whereas MAP65 is an essential MT cross-linker that ensures phragmoplast bipolarity. MT plus ends are regulated by the same set of proteins as those acting during metaphase. Newly nucleated MTs are transported poleward via an unknown molecular mechanism.

3.1. Spindle Assembly

Genetic analyses of γ-tubulin and their associated subunits have clarified the pivotal role of the γ-TuRC in spindle MT generation [59,60,109,110]. Recently, a mechanism underlying γ-TuRC activation was also uncovered. Studies on moss and *Arabidopsis* indicated that the eight-subunit complex augmin is a key factor in increasing spindle MTs during prometaphase via γ-TuRC localisation and activation [17,65] (Figure 2A). Augmin was originally identified in *Drosophila* cells as a protein complex that drives MT-dependent MT generation by recruiting γ-tubulin onto existing spindle MTs [111]. In moss, RNAi knockdown of augmin subunits reduced MTs to ~50%, suggesting that at least half of the spindle MTs were generated via augmin-dependent, branching nucleation during prometaphase [17]. Since RNAi knockdown left behind residual augmin proteins, it is likely that 50% is an underestimate. The fir-tree structure observed in *Haemanthus* endosperm spindles might represent augmin-dependent MTs [27]. Prior to the discovery of augmin, MT-dependent MT generation was described in detail for the cortical MT arrays in tobacco and *Arabidopsis* cells [95]. Recent studies in *Arabidopsis* demonstrated that this branching nucleation is an augmin-dependent process [112]. In contrast, the origin of augmin-independent spindle MTs after augmin RNAi is unknown. It is possible that, analogous to animal cells [8], prophase MTs, and chromatin-mediated nucleation in the prometaphase (which depends on RanGTP or aurora kinase) play a role in producing these MTs.

The molecular mechanism by which plant cells achieve bipolar arrangement without centrosomes is not well understood. In animals, the key molecules in the bipolar arrangement are kinesin-5 and kinesin-12, which cross-link and slide apart anti-parallel MT overlaps in the spindle midzone [5]. This model may apply to certain plant tissues since an *Arabidopsis* mutant of kinesin-5 exhibits monopolar spindle formation in roots [113]. However, in moss protonemata, RNAi knockdown of kinesin-5 did not show monopolarisation. Moreover, GFP-tagged kinesin-5 is scarcely detected at the midzone [114]. Genes encoding kinesin-12 are amplified in plants. Therefore, it is possible that this motor redundantly plays a major role in bipolarity establishment and maintenance in some plant cell types. A comprehensive functional analysis of the kinesin-12 subfamily is required to test this hypothesis.

In animals, spindle coalescence is mediated by the partially redundant functions of kinesin-14 and cytoplasmic dynein [9,104]. Kinesin-14 has a second MT binding site in its tail domain and works as an MT cross-linker. In plants, two closely related kinesin-14 proteins, ATK1 and ATK5, have been shown to play a similar role [30,115–117]. When ATK5 was absent in root cells, spindles were less focused than they were in control cells [30]. However, whether the MT converging centres observed at the pole [23] are solely organised by the kinesin-14 motor remains unclear; in fly cells, this local crosslinking was dependent on an additional factor, ASPM/Asp [11,118], which is conserved, but uncharacterised, in plants.

Spindle length appears to be controlled by MT dynamics at plus ends, similar to animal cells. XMAP215/Dis1 family protein is an established MT polymerase, and the *mor1* (XMAP215/Dis1 orthologue) mutant has shorter spindles in *Arabidopsis* [67,119]. EB1 is also a critical regulator of MT plus ends, with shorter spindle formation reported for the *Arabidopsis eb1c* mutant [120]. The cytoplasmic linker-associated protein (CLASP) is an essential factor for MT polymerisation at the kinetochore in animals [121], and mutations in this gene in *Arabidopsis* resulted in significantly shorter spindles [122]. In the animal spindle, CLASP-dependent MT polymerisation and motor-dependent, poleward MT transport/sliding are coupled with minus end depolymerisation by the kinesin-13 depolymerase to maintain spindle length at the steady state ('MT flux'; [123]). MT flux has been observed in plant spindles, suggesting that MT minus ends are also regulated by a depolymerising factor [124].

3.2. Phragmoplast Assembly

The phragmoplast begins to assemble upon sister chromatid separation (Figure 2B). The overall structural similarity, namely, bipolar MT array with anti-parallel MT interdigitation in the middle, have raised the notion that the phragmoplast is analogous to the central spindle or midbody in animal cells [125]. Recently reported data on MT generation further support this idea. In both structures, some MTs are constantly generated de novo in an augmin- and γ-tubulin-dependent manner [13,17]. When augmin is depleted in moss, MTs are diminished and phragmoplasts eventually disappear before they reach the cell cortex. A plausible explanation for this phenotype is that augmin utilises existing MTs, such as those carried over from metaphase, as templates for new MT nucleation. About 50% of the MTs in the central spindle of animal cells are generated in an augmin-dependent manner [13]. However, animal cells seem to have additional MT generation pathways during anaphase: at that time, hepatoma up-regulated proteins (HURP) are involved in chromosome-proximal MT generation [13]. HURP-like proteins have not been identified in the plant genome.

The key factor that maintains phragmoplast bipolarity is MAP65 (PRC1/Ase1 orthologue). This is an anti-parallel MT cross-linking protein whose activity is conserved in both yeasts and animals [126].

When three paralogous MAP65 genes are simultaneously knocked down in moss, MT bipolarity is lost and cytokinesis fails [127]. In *Arabidopsis*, bipolarity is maintained in known MAP65 mutants [128,129] but it is possible that multiple MAP65s work redundantly, and cross-linking activity persists in the mutant [130]. Other proteins, such as kinesins, might also constitute a redundant cross-linking mechanism [131].

The signalling pathway underlying phragmoplast MT regulation differs between animals and plants. In animals, the key kinases required for proper central spindle assembly are Plk1 and Aurora-B. They are concentrated at the midzone and phosphorylate multiple MT-regulating proteins including the MAP65 orthologue PRC1 [132,133]. In contrast, plants do not have Plk1. The aurora kinase constitutes a signalling pathway during cytokinesis, but the mutant exhibits a defect in orientation, but not assembly per se of the phragmoplast [134]. In plants, the MAP kinase cascade also constitutes the late mitotic signalling pathway (called the NACK-PQR pathway) [135]. One of the downstream factors in plants is also MAP65. The phosphorylation of MAP65 down-regulates its MT-bundling activity which, in turn, stimulates the progression of cytokinesis [136]. Thus, in plants, the development of MAPK signalling might have compensated for the loss of Plk1 kinase.

The phragmoplast length is regulated by MT-associated proteins; similar to metaphase spindles, shorter phragmoplasts are observed in the mutants of MOR1, EB1, and CLASP [119,120,122]. In addition, katanin-mediated severing may affect MT length in some cell types [137]. MT flux is also observed within the phragmoplast, but the molecules responsible remain to be identified [138].

4. Conclusions and Future Perspectives on Spindle Research in Plants

At first glance, plant and animal spindles look quite different. However, the database search suggests that only a few of the mitotic elements present in animals are missing from the plant genome. Our queries using animals could not elucidate the genes that evolved uniquely in plants. Therefore, it is possible that some plant-specific genes for spindle assembly have yet to be identified. Nevertheless, most of the gene repertoire is probably common to both animals and plants.

However, the mechanism of plant spindle assembly is formally not yet well understood. First, although gene conservation predicts that homologous proteins possess similar biochemical activity, are found in similar locations, and execute similar functions, each of these *must* be tested experimentally. Indeed, recent 'repeat' experiments using the plant orthologues of well-characterised animal genes have revealed unexpected functions, such as the role of kinesin-5 in chromosome alignment or cytokinesis [114] and a γ-TuRC-interacting protein in centromere integrity [61]. Regarding localisation dynamics, a comprehensive study in moss showed that 42 out of 43 mitotic kinesins were localised at a site not observed in animal studies [114]. Until recently, 'repeat' experiments intended to confirm animal study results were very time-consuming due to the lack of a model cell system for rapid investigation. The recent development of quick and robust loss-of-function tools such as conditional RNAi and CRISPR/Cas9-based genome editing technology in moss and liverwort, as well as advances in live microscopy, have provided an opportunity to delve into *putative* spindle assembly factors in plants [17,139–142].

Second, it is not yet known how plants compensate for the lack of two major components, centrosomes and dynein. In animal and yeast, these two components are critical for spindle positioning, which is a crucial process in determining the cell division axis and symmetry/asymmetry [143]. It has been shown that genetic perturbation of the PPB causes division axis abnormalities in seed plants [33,34]. Yet little is known about the molecular mechanism underlying the PPB-dependent determination of the spindle axis. It is remarkable that plants evolved a unique PPB-based mechanism to substitute for centrosome function [144]. The mechanism by which plant spindles are oriented in the proper direction remains a fascinating, and as yet unanswered, question.

Acknowledgments: We thank Geert Kops (Utrecht University, The Netherlands) and Takashi Hashimoto (Nara Institute of Science and Technology, Japan) for providing information about some putative plant spindle/chromosome proteins and Ami Ito (Nagoya University) for furnishing human spindle images. The plant research conducted in our laboratory is supported by the TORAY Science Foundation and by JSPS KAKNHI (15K14540 and 15H01227). Moé Yamada is a recipient of the JSPS pre-doctoral fellowship.

Conflicts of Interest: The authors declare no conflict of interest.

References

1. Inoue, S.; Sato, H. Cell motility by labile association of molecules. The nature of mitotic spindle fibers and their role in chromosome movement. *J. Gen. Physiol.* **1967**, *50*, S259–S292. [CrossRef]
2. Karp, G. *Cell and Molecular Biology: Concepts and Experiments*, 6th ed.; John Wiley: Hoboken, NJ, USA, 2010.
3. Conduit, P.T.; Richens, J.H.; Wainman, A.; Holder, J.; Vicente, C.C.; Pratt, M.B.; Dix, C.I.; Novak, Z.A.; Dobbie, I.M.; Schermelleh, L.; et al. A molecular mechanism of mitotic centrosome assembly in drosophila. *eLife* **2014**, *3*, e03399. [CrossRef] [PubMed]
4. Duncan, T.; Wakefield, J.G. 50 ways to build a spindle: The complexity of microtubule generation during mitosis. *Chromosome Res.* **2011**, *19*, 321–333. [CrossRef] [PubMed]
5. Reber, S.; Hyman, A.A. Emergent properties of the metaphase spindle. *Cold Spring Harb. Perspect. Biol.* **2015**, *7*, a015784. [CrossRef] [PubMed]
6. Walczak, C.E.; Heald, R. Mechanisms of mitotic spindle assembly and function. *Int. Rew. Cytol.* **2008**, *265*, 111–158.
7. Goshima, G.; Kimura, A. New look inside the spindle: Microtubule-Dependent microtubule generation within the spindle. *Curr. Opin. Cell Biol.* **2010**, *22*, 44–49. [CrossRef] [PubMed]
8. Hayward, D.; Metz, J.; Pellacani, C.; Wakefield, J.G. Synergy between multiple microtubule-generating pathways confers robustness to centrosome-driven mitotic spindle formation. *Dev. Cell* **2014**, *28*, 81–93. [CrossRef] [PubMed]
9. Walczak, C.E.; Vernos, I.; Mitchison, T.J.; Karsenti, E.; Heald, R. A model for the proposed roles of different microtubule-based motor proteins in establishing spindle bipolarity. *Curr. Biol.* **1998**, *8*, 903–913. [CrossRef]
10. Hatsumi, M.; Endow, S.A. Mutants of the microtubule motor protein, nonclaret disjunctional, affect spindle structure and chromosome movement in meiosis and mitosis. *J. Cell Sci.* **1992**, *101*, 547–559. [PubMed]
11. Wakefield, J.G.; Bonaccorsi, S.; Gatti, M. The drosophila protein asp is involved in microtubule organization during spindle formation and cytokinesis. *J. Cell Biol.* **2001**, *153*, 637–648. [CrossRef] [PubMed]
12. Goshima, G.; Scholey, J.M. Control of mitotic spindle length. *Annu. Rev. Cell Dev. Biol.* **2010**, *26*, 21–57. [CrossRef] [PubMed]
13. Uehara, R.; Goshima, G. Functional central spindle assembly requires de novo microtubule generation in the interchromosomal region during anaphase. *J. Cell Biol.* **2010**, *191*, 259–267. [CrossRef] [PubMed]
14. Glotzer, M. The 3ms of central spindle assembly: Microtubules, motors and maps. *Nat. Rev. Mol. Cell Biol.* **2009**, *10*, 9–20. [CrossRef] [PubMed]
15. Brown, R.C.; Lemmon, B.E. Dividing without centrioles: Innovative plant microtubule organizing centres organize mitotic spindles in bryophytes, the earliest extant lineages of land plants. *AoB Plants* **2011**, *2011*. [CrossRef] [PubMed]
16. Buschmann, H.; Holtmannspotter, M.; Borchers, A.; O'Donoghue, M.T.; Zachgo, S. Microtubule dynamics of the centrosome-like polar organizers from the basal land plant marchantia polymorpha. *New Phytol.* **2016**, *209*, 999–1013. [CrossRef] [PubMed]
17. Nakaoka, Y.; Miki, T.; Fujioka, R.; Uehara, R.; Tomioka, A.; Obuse, C.; Kubo, M.; Hiwatashi, Y.; Goshima, G. An inducible rna interference system in physcomitrella patens reveals a dominant role of augmin in phragmoplast microtubule generation. *Plant Cell* **2012**, *24*, 1478–1493. [CrossRef] [PubMed]
18. Lloyd, C.; Chan, J. Not so divided: The common basis of plant and animal cell division. *Nat. Rev. Mol. Cell Biol.* **2006**, *7*, 147–152. [CrossRef] [PubMed]
19. Chan, J.; Calder, G.; Fox, S.; Lloyd, C. Localization of the microtubule end binding protein eb1 reveals alternative pathways of spindle development in arabidopsis suspension cells. *Plant Cell* **2005**, *17*, 1737–1748. [CrossRef] [PubMed]
20. Baskin, T.I.; Cande, W.Z. The structure and function of the mitotic spindle in flowering plants. *Ann. Rev. Plant Physiol. Plant Mol. Biol.* **1990**, *41*, 277–315. [CrossRef]
21. Smirnova, E.A.; Bajer, A.S. Spindle poles in higher plant mitosis. *Cell Motil Cytoskelet.* **1992**, *23*, 1–7. [CrossRef] [PubMed]

22. De Mey, J.; Lambert, A.M.; Bajer, A.S.; Moeremans, M.; De Brabander, M. Visualization of microtubules in interphase and mitotic plant cells of haemanthus endosperm with the immuno-gold staining method. *Proc. Natl. Acad. Sci. USA* **1982**, *79*, 1898–1902. [CrossRef] [PubMed]
23. Smirnova, E.A.; Bajer, A.S. Microtubule converging centers and reorganization of the interphase cytoskeleton and the mitotic spindle in higher plant haemanthus. *Cell Motil Cytoskelet.* **1994**, *27*, 219–233. [CrossRef] [PubMed]
24. Smirnova, E.A.; Bajer, A.S. Early stages of spindle formation and independence of chromosome and microtubule cycles in haemanthus endosperm. *Cell Motil Cytoskelet.* **1998**, *40*, 22–37. [CrossRef]
25. Falconer, M.M.; Donaldson, G.; Seagull, R.W. Mtocs in higher-plant cells—An immunofluorescent study of microtubule assembly sites following depolymerization by apm. *Protoplasma* **1988**, *144*, 46–55. [CrossRef]
26. Euteneuer, U.; Jackson, W.T.; McIntosh, J.R. Polarity of spindle microtubules in haemanthus endosperm. *J. Cell Biol.* **1982**, *94*, 644–653. [CrossRef] [PubMed]
27. Bajer, A.S.; Mole-Bajer, J. Reorganization of microtubules in endosperm cells and cell fragments of the higher plant haemanthus in vivo. *J. Cell Biol.* **1986**, *102*, 263–281. [CrossRef] [PubMed]
28. Lambert, A.M.; Bajer, A.S. Dynamics of spindle fibers and microtubules during anaphase and phragmoplast formation. *Chromosoma* **1972**, *39*, 101–144. [CrossRef]
29. Liu, B.; Ho, C.M.; Lee, Y.R. Microtubule reorganization during mitosis and cytokinesis: Lessons learned from developing microgametophytes in arabidopsis thaliana. *Front. Plant Sci.* **2011**. [CrossRef] [PubMed]
30. Ambrose, J.C.; Cyr, R. The kinesin atk5 functions in early spindle assembly in arabidopsis. *Plant Cell.* **2007**, *19*, 226–236. [CrossRef] [PubMed]
31. Hayashi, T.; Sano, T.; Kutsuna, N.; Kumagai-Sano, F.; Hasezawa, S. Contribution of anaphase b to chromosome separation in higher plant cells estimated by image processing. *Plant Cell Physiol.* **2007**, *48*, 1509–1513. [CrossRef] [PubMed]
32. Ambrose, J.C.; Cyr, R. Mitotic spindle organization by the preprophase band. *Mol. Plant* **2008**, *1*, 950–960. [CrossRef] [PubMed]
33. Muller, S.; Wright, A.J.; Smith, L.G. Division plane control in plants: New players in the band. *Trends Cell Biol.* **2009**, *19*, 180–188. [CrossRef] [PubMed]
34. Rasmussen, C.G.; Humphries, J.A.; Smith, L.G. Determination of symmetric and asymmetric division planes in plant cells. *Annu. Rev. Plant Biol.* **2011**, *62*, 387–409. [CrossRef] [PubMed]
35. Fowke, L.C.; Pickett-Heaps, J.D. Electron microscope study of vegetative cell division in two species of marchantia. *Can. J. Bot.* **1978**, *56*, 467–475. [CrossRef]
36. Brown, R.C.; Lemmon, B.E. Polar organizers mark division axis prior to preprophase band formation in mitosis of the hepaticreboulia hemisphaerica (bryophyta). *Protoplasma* **1990**, *156*, 74–81. [CrossRef]
37. Shimamura, M.; Brown, R.C.; Lemmon, B.E.; Akashi, T.; Mizuno, K.; Nishihara, N.; Tomizawa, K.; Yoshimoto, K.; Deguchi, H.; Hosoya, H.; et al. Gamma-Tubulin in basal land plants: Characterization, localization, and implication in the evolution of acentriolar microtubule organizing centers. *Plant Cell* **2004**, *16*, 45–59. [CrossRef] [PubMed]
38. Brown, R.C.; Lemmon, B.E.; Horio, T. Gamma-Tubulin localization changes from discrete polar organizers to anastral spindles and phragmoplasts in mitosis of marchantia polymorpha l. *Protoplasma* **2004**, *224*, 187–193. [CrossRef] [PubMed]
39. Brown, R.C.; Lemmon, B.E. Preprophase microtubule systems and development of the mitotic spindle in hornworts (bryophyta). *Protoplasma* **1988**, *143*, 11–21. [CrossRef]
40. Goshima, G.; Wollman, R.; Goodwin, N.; Zhang, J.M.; Scholey, J.M.; Vale, R.D.; Stuurman, N. Genes required for mitotic spindle assembly in drosophila s2 cells. *Science* **2007**, *316*, 417–421. [CrossRef] [PubMed]
41. Sonnichsen, B.; Koski, L.B.; Walsh, A.; Marschall, P.; Neumann, B.; Brehm, M.; Alleaume, A.M.; Artelt, J.; Bettencourt, P.; Cassin, E.; et al. Full-Genome rnai profiling of early embryogenesis in caenorhabditis elegans. *Nature* **2005**, *434*, 462–469. [CrossRef] [PubMed]
42. Hutchins, J.R.; Toyoda, Y.; Hegemann, B.; Poser, I.; Heriche, J.K.; Sykora, M.M.; Augsburg, M.; Hudecz, O.; Buschhorn, B.A.; Bulkescher, J.; et al. Systematic analysis of human protein complexes identifies chromosome segregation proteins. *Science* **2010**, *328*, 593–599. [CrossRef] [PubMed]
43. Hoyt, M.A.; Geiser, J.R. Genetic analysis of the mitotic spindle. *Annu. Rev. Genet.* **1996**, *30*, 7–33. [CrossRef] [PubMed]

44. Rensing, S.A.; Lang, D.; Zimmer, A.D.; Terry, A.; Salamov, A.; Shapiro, H.; Nishiyama, T.; Perroud, P.F.; Lindquist, E.A.; Kamisugi, Y.; et al. The physcomitrella genome reveals evolutionary insights into the conquest of land by plants. *Science* **2008**, *319*, 64–69. [CrossRef] [PubMed]
45. Arabidopsis Genome, I. Analysis of the genome sequence of the flowering plant arabidopsis thaliana. *Nature* **2000**, *408*, 796–815. [CrossRef] [PubMed]
46. Hamada, T. Microtubule organization and microtubule-associated proteins in plant cells. *Int. Rev. Cell Mol. Biol.* **2014**, *312*, 1–52. [PubMed]
47. Gardiner, J. The evolution and diversification of plant microtubule-associated proteins. *Plant J.* **2013**, *75*, 219–229. [CrossRef] [PubMed]
48. Demidov, D.; Van Damme, D.; Geelen, D.; Blattner, F.R.; Houben, A. Identification and dynamics of two classes of aurora-like kinases in arabidopsis and other plants. *Plant Cell* **2005**, *17*, 836–848. [CrossRef] [PubMed]
49. De Carcer, G.; Manning, G.; Malumbres, M. From plk1 to plk5: Functional evolution of polo-like kinases. *Cell Cycle* **2011**, *10*, 2255–2262. [CrossRef] [PubMed]
50. Moorhead, G.B.; De Wever, V.; Templeton, G.; Kerk, D. Evolution of protein phosphatases in plants and animals. *Biochem. J.* **2009**, *417*, 401–409. [CrossRef] [PubMed]
51. Labandera, A.M.; Vahab, A.R.; Chaudhuri, S.; Kerk, D.; Moorhead, G.B. The mitotic pp2a regulator ensa/arpp-19 is remarkably conserved across plants and most eukaryotes. *Biochem. Biophys. Res. Commun.* **2015**, *458*, 739–744. [CrossRef] [PubMed]
52. Carvalho-Santos, Z.; Machado, P.; Branco, P.; Tavares-Cadete, F.; Rodrigues-Martins, A.; Pereira-Leal, J.B.; Bettencourt-Dias, M. Stepwise evolution of the centriole-assembly pathway. *J. Cell Sci.* **2010**, *123*, 1414–1426. [CrossRef] [PubMed]
53. Zhu, C.; Dixit, R. Functions of the arabidopsis kinesin superfamily of microtubule-based motor proteins. *Protoplasma* **2012**, *249*, 887–899. [CrossRef] [PubMed]
54. Vanstraelen, M.; Inze, D.; Geelen, D. Mitosis-specific kinesins in arabidopsis. *Trends Plant Sci.* **2006**, *11*, 167–175. [CrossRef] [PubMed]
55. Wickstead, B.; Gull, K. Dyneins across eukaryotes: A comparative genomic analysis. *Traffic* **2007**, *8*, 1708–1721. [CrossRef] [PubMed]
56. Kandasamy, M.K.; Deal, R.B.; McKinney, E.C.; Meagher, R.B. Plant actin-related proteins. *Trends Plant Sci.* **2004**, *9*, 196–202. [CrossRef] [PubMed]
57. Vos, J.W.; Pieuchot, L.; Evrard, J.L.; Janski, N.; Bergdoll, M.; de Ronde, D.; Perez, L.H.; Sardon, T.; Vernos, I.; Schmit, A.C. The plant tpx2 protein regulates prospindle assembly before nuclear envelope breakdown. *Plant Cell* **2008**, *20*, 2783–2797. [CrossRef] [PubMed]
58. Liu, B.; Joshi, H.C.; Wilson, T.J.; Silflow, C.D.; Palevitz, B.A.; Snustad, D.P. Gamma-tubulin in arabidopsis: Gene sequence, immunoblot, and immunofluorescence studies. *Plant Cell* **1994**, *6*, 303–314. [CrossRef] [PubMed]
59. Kong, Z.; Hotta, T.; Lee, Y.R.; Horio, T.; Liu, B. The {gamma}-tubulin complex protein gcp4 is required for organizing functional microtubule arrays in arabidopsis thaliana. *Plant Cell* **2010**, *22*, 191–204. [CrossRef] [PubMed]
60. Zeng, C.J.; Lee, Y.R.; Liu, B. The wd40 repeat protein nedd1 functions in microtubule organization during cell division in arabidopsis thaliana. *Plant Cell* **2009**, *21*, 1129–1140. [CrossRef] [PubMed]
61. Batzenschlager, M.; Lermontova, I.; Schubert, V.; Fuchs, J.; Berr, A.; Koini, M.A.; Houlne, G.; Herzog, E.; Rutten, T.; Alioua, A.; et al. Arabidopsis mzt1 homologs gip1 and gip2 are essential for centromere architecture. *Proc. Natl. Acad. Sci. USA* **2015**, *112*, 8656–8660. [CrossRef] [PubMed]
62. Nakamura, M.; Yagi, N.; Kato, T.; Fujita, S.; Kawashima, N.; Ehrhardt, D.W.; Hashimoto, T. Arabidopsis gcp3-interacting protein 1/mozart 1 is an integral component of the gamma-tubulin-containing microtubule nucleating complex. *Plant J.* **2012**, *71*, 216–225. [CrossRef] [PubMed]
63. Janski, N.; Masoud, K.; Batzenschlager, M.; Herzog, E.; Evrard, J.L.; Houlne, G.; Bourge, M.; Chaboute, M.E.; Schmit, A.C. The gcp3-interacting proteins gip1 and gip2 are required for gamma-tubulin complex protein localization, spindle integrity, and chromosomal stability. *Plant Cell* **2012**, *24*, 1171–1187. [CrossRef] [PubMed]
64. Lin, T.C.; Neuner, A.; Schiebel, E. Targeting of gamma-tubulin complexes to microtubule organizing centers: Conservation and divergence. *Trends Cell Biol.* **2015**, *25*, 296–307. [CrossRef] [PubMed]

65. Hotta, T.; Kong, Z.; Ho, C.M.; Zeng, C.J.; Horio, T.; Fong, S.; Vuong, T.; Lee, Y.R.; Liu, B. Characterization of the arabidopsis augmin complex uncovers its critical function in the assembly of the acentrosomal spindle and phragmoplast microtubule arrays. *Plant Cell* **2012**, *24*, 1494–1509. [CrossRef] [PubMed]
66. Kawaguchi, S.; Zheng, Y. Characterization of a drosophila centrosome protein cp309 that shares homology with kendrin and cg-nap. *Mol. Biol. Cell* **2004**, *15*, 37–45. [CrossRef] [PubMed]
67. Kawamura, E.; Wasteneys, G.O. Mor1, the arabidopsis thaliana homologue of xenopus map215, promotes rapid growth and shrinkage, and suppresses the pausing of microtubules in vivo. *J. Cell Sci.* **2008**, *121*, 4114–4123. [CrossRef] [PubMed]
68. Jiang, K.; Hua, S.; Mohan, R.; Grigoriev, I.; Yau, K.W.; Liu, Q.; Katrukha, E.A.; Altelaar, A.F.; Heck, A.J.; Hoogenraad, C.C.; et al. Microtubule minus-end stabilization by polymerization-driven camsap deposition. *Dev. Cell* **2014**, *28*, 295–309. [CrossRef] [PubMed]
69. Hamada, T.; Nagasaki-Takeuchi, N.; Kato, T.; Fujiwara, M.; Sonobe, S.; Fukao, Y.; Hashimoto, T. Purification and characterization of novel microtubule-associated proteins from arabidopsis cell suspension cultures. *Plant Physiol.* **2013**, *163*, 1804–1816. [CrossRef] [PubMed]
70. Hirano, T. Condensins: Organizing and segregating the genome. *Curr. Biol.* **2005**, *15*, R265–R275. [CrossRef] [PubMed]
71. Siddiqui, N.U. Mutations in arabidopsis condensin genes disrupt embryogenesis, meristem organization and segregation of homologous chromosomes during meiosis. *Development* **2003**, *130*, 3283–3295. [CrossRef] [PubMed]
72. Fujimoto, S.; Yonemura, M.; Matsunaga, S.; Nakagawa, T.; Uchiyama, S.; Fukui, K. Characterization and dynamic analysis of arabidopsis condensin subunits, atcap-h and atcap-h2. *Planta* **2005**, *222*, 293–300. [CrossRef] [PubMed]
73. Siddiqui, N.U.; Rusyniak, S.; Hasenkampf, C.A.; Riggs, C.D. Disruption of the arabidopsis smc4 gene, atcap-c, compromises gametogenesis and embryogenesis. *Planta* **2006**, *223*, 990–997. [CrossRef] [PubMed]
74. Bhatt, A.M.; Lister, C.; Page, T.; Fransz, P.; Findlay, K.; Jones, G.H.; Dickinson, H.G.; Dean, C. The dif1 gene of arabidopsis is required for meioticchromosome segregation and belongs to the rec8/rad21cohesin gene family. *Plant J.* **1999**, *19*, 463–472. [CrossRef] [PubMed]
75. Schubert, V.; Weissleder, A.; Ali, H.; Fuchs, J.; Lermontova, I.; Meister, A.; Schubert, I. Cohesin gene defects may impair sister chromatid alignment and genome stability in arabidopsis thaliana. *Chromosoma* **2009**, *118*, 591–605. [CrossRef] [PubMed]
76. Sebastian, J.; Ravi, M.; Andreuzza, S.; Panoli, A.P.; Marimuthu, M.P.; Siddiqi, I. The plant adherin atscc2 is required for embryogenesis and sister-chromatid cohesion during meiosis in arabidopsis. *Plant J.* **2009**, *59*, 1–13. [CrossRef] [PubMed]
77. Singh, D.K.; Andreuzza, S.; Panoli, A.P.; Siddiqi, I. Atctf7 is required for establishment of sister chromatid cohesion and association of cohesin with chromatin during meiosis in arabidopsis. *BMC Plant Biol.* **2013**. [CrossRef] [PubMed]
78. Ladurner, R.; Kreidl, E.; Ivanov, M.P.; Ekker, H.; Idarraga-Amado, M.H.; Busslinger, G.A.; Wutz, G.; Cisneros, D.A.; Peters, J.M. Sororin actively maintains sister chromatid cohesion. *EMBO J.* **2016**, *35*, 635–653. [CrossRef] [PubMed]
79. Zamariola, L.; De Storme, N.; Tiang, C.L.; Armstrong, S.J.; Franklin, F.C.; Geelen, D. Sgo1 but not sgo2 is required for maintenance of centromere cohesion in arabidopsis thaliana meiosis. *Plant Reprod.* **2013**, *26*, 197–208. [CrossRef] [PubMed]
80. Kirioukhova, O.; Johnston, A.J.; Kleen, D.; Kagi, C.; Baskar, R.; Moore, J.M.; Baumlein, H.; Gross-Hardt, R.; Grossniklaus, U. Female gametophytic cell specification and seed development require the function of the putative arabidopsis incenp ortholog wyrd. *Development* **2011**, *138*, 3409–3420. [CrossRef] [PubMed]
81. Lermontova, I.; Kuhlmann, M.; Friedel, S.; Rutten, T.; Heckmann, S.; Sandmann, M.; Demidov, D.; Schubert, V.; Schubert, I. Arabidopsis kinetochore null2 is an upstream component for centromeric histone h3 variant cenh3 deposition at centromeres. *Plant Cell* **2013**, *25*, 3389–3404. [CrossRef] [PubMed]
82. Talbert, P.B.; Masuelli, R.; Tyagi, A.P.; Comai, L.; Henikoff, S. Centromeric localization and adaptive evolution of an arabidopsis histone h3 variant. *Plant Cell* **2002**, *14*, 1053–1066. [CrossRef] [PubMed]
83. Ogura, Y.; Shibata, F.; Sato, H.; Murata, M. Characterization of a cenp-c homolog in arabidopsis thaliana. *Genes Genet. Syst.* **2004**, *79*, 139–144. [CrossRef] [PubMed]

84. Dangel, N.J.; Knoll, A.; Puchta, H. Mhf1 plays fanconi anaemia complementation group M protein (fancm)-dependent and fancm-independent roles in DNA repair and homologous recombination in plants. *Plant J.* **2014**, *78*, 822–833. [CrossRef] [PubMed]
85. Girard, C.; Crismani, W.; Froger, N.; Mazel, J.; Lemhemdi, A.; Horlow, C.; Mercier, R. Fancm-associated proteins mhf1 and mhf2, but not the other fanconi anemia factors, limit meiotic crossovers. *Nucleic Acids Res.* **2014**, *42*, 9087–9095. [CrossRef] [PubMed]
86. Sato, H.; Shibata, F.; Murata, M. Characterization of a mis12 homologue in arabidopsis thaliana. *Chromosome Res.* **2005**, *13*, 827–834. [CrossRef] [PubMed]
87. Tromer, E.; Snel, B.; Kops, G.J. Widespread recurrent patterns of rapid repeat evolution in the kinetochore scaffold knl1. *Genome Biol. Evol.* **2015**, *7*, 2383–2393. [CrossRef] [PubMed]
88. Caillaud, M.C.; Paganelli, L.; Lecomte, P.; Deslandes, L.; Quentin, M.; Pecrix, Y.; Le Bris, M.; Marfaing, N.; Abad, P.; Favery, B. Spindle assembly checkpoint protein dynamics reveal conserved and unsuspected roles in plant cell division. *PLoS ONE* **2009**, *4*, e6757. [CrossRef] [PubMed]
89. Wang, M.; Tang, D.; Luo, Q.; Jin, Y.; Shen, Y.; Wang, K.; Cheng, Z. Brk1, a bub1-related kinase, is essential for generating proper tension between homologous kinetochores at metaphase i of rice meiosis. *Plant Cell* **2012**, *24*, 4961–4973. [CrossRef] [PubMed]
90. de Oliveira, E.A.; Romeiro, N.C.; Ribeiro Eda, S.; Santa-Catarina, C.; Oliveira, A.E.; Silveira, V.; de Souza Filho, G.A.; Venancio, T.M.; Cruz, M.A. Structural and functional characterization of the protein kinase mps1 in arabidopsis thaliana. *PLoS ONE* **2012**, *7*, e45707. [CrossRef] [PubMed]
91. Xu, X.M.; Rose, A.; Muthuswamy, S.; Jeong, S.Y.; Venkatakrishnan, S.; Zhao, Q.; Meier, I. Nuclear pore anchor, the arabidopsis homolog of tpr/mlp1/mlp2/megator, is involved in mrna export and sumo homeostasis and affects diverse aspects of plant development. *Plant Cell* **2007**, *19*, 1537–1548. [CrossRef] [PubMed]
92. Starr, D.A.; Williams, B.C.; Li, Z.; Etemad-Moghadam, B.; Dawe, R.K.; Goldberg, M.L. Conservation of the centromere/kinetochore protein zw10. *J. Cell Biol.* **1997**, *138*, 1289–1301. [CrossRef] [PubMed]
93. McKinley, K.L.; Cheeseman, I.M. The molecular basis for centromere identity and function. *Nat. Rev. Mol. Cell Biol.* **2016**, *17*, 16–29. [CrossRef] [PubMed]
94. Drykova, D.; Cenklova, V.; Sulimenko, V.; Volc, J.; Draber, P.; Binarova, P. Plant gamma-tubulin interacts with alphabeta-tubulin dimers and forms membrane-associated complexes. *Plant Cell* **2003**, *15*, 465–480. [CrossRef] [PubMed]
95. Murata, T.; Sonobe, S.; Baskin, T.I.; Hyodo, S.; Hasezawa, S.; Nagata, T.; Horio, T.; Hasebe, M. Microtubule-dependent microtubule nucleation based on recruitment of gamma-tubulin in higher plants. *Nat. Cell Biol.* **2005**, *7*, 961–968. [CrossRef] [PubMed]
96. Azimzadeh, J.; Nacry, P.; Christodoulidou, A.; Drevensek, S.; Camilleri, C.; Amiour, N.; Parcy, F.; Pastuglia, M.; Bouchez, D. Arabidopsis tonneau1 proteins are essential for preprophase band formation and interact with centrin. *Plant Cell* **2008**, *20*, 2146–2159. [CrossRef] [PubMed]
97. Hodges, M.E.; Wickstead, B.; Gull, K.; Langdale, J.A. The evolution of land plant cilia. *New Phytol.* **2012**, *195*, 526–540. [CrossRef] [PubMed]
98. Kardon, J.R.; Vale, R.D. Regulators of the cytoplasmic dynein motor. *Nat. Rev. Mol. Cell Biol.* **2009**, *10*, 854–865. [CrossRef] [PubMed]
99. Lawrence, C.J.; Morris, N.R.; Meagher, R.B.; Dawe, R.K. Dyneins have run their course in plant lineage. *Traffic* **2001**, *2*, 362–363. [CrossRef] [PubMed]
100. Rapali, P.; Szenes, A.; Radnai, L.; Bakos, A.; Pal, G.; Nyitray, L. Dynll/lc8: A light chain subunit of the dynein motor complex and beyond. *FEBS J.* **2011**, *278*, 2980–2996. [CrossRef] [PubMed]
101. Endow, S.A. Determinants of molecular motor directionality. *Nat. Cell Biol.* **1999**, *1*, E163–E167. [CrossRef] [PubMed]
102. Jonsson, E.; Yamada, M.; Vale, R.D.; Goshima, G. Clustering of a kinesin-14 motor enables processive retrograde microtubule-based transport in plants. *Nat. Plants* **2015**, *1*. [CrossRef] [PubMed]
103. Furuta, K.; Furuta, A.; Toyoshima, Y.Y.; Amino, M.; Oiwa, K.; Kojima, H. Measuring collective transport by defined numbers of processive and nonprocessive kinesin motors. *Proc. Natl. Acad. Sci. USA* **2013**, *110*, 501–506. [CrossRef] [PubMed]
104. Goshima, G.; Nedelec, F.; Vale, R.D. Mechanisms for focusing mitotic spindle poles by minus end-directed motor proteins. *J. Cell Biol.* **2005**, *171*, 229–240. [CrossRef] [PubMed]
105. Hyams, J.S.; Campbell, C.J. Widespread absence of outer dynein arms in the spermatozoids of lower plants. *Cell Biol. Int. Rep.* **1985**, *9*, 841–848. [CrossRef]

106. Komaki, S.; Schnittger, A. The spindle checkpoint in plants-a green variation over a conserved theme? *Curr. Opin. Plant Biol.* **2016**, *34*, 84–91. [CrossRef] [PubMed]
107. Przewloka, M.R.; Zhang, W.; Costa, P.; Archambault, V.; D'Avino, P.P.; Lilley, K.S.; Laue, E.D.; McAinsh, A.D.; Glover, D.M. Molecular analysis of core kinetochore composition and assembly in drosophila melanogaster. *PLoS ONE* **2007**, *2*, e478. [CrossRef] [PubMed]
108. Miki, T.; Nishina, M.; Goshima, G. Rnai screening identifies the armadillo repeat-containing kinesins responsible for microtubule-dependent nuclear positioning in physcomitrella patens. *Plant Cell Physiol.* **2015**, *56*, 737–749. [CrossRef] [PubMed]
109. Binarova, P.; Cenklova, V.; Prochazkova, J.; Doskocilova, A.; Volc, J.; Vrlik, M.; Bogre, L. Gamma-Tubulin is essential for acentrosomal microtubule nucleation and coordination of late mitotic events in arabidopsis. *Plant Cell* **2006**, *18*, 1199–1212. [CrossRef] [PubMed]
110. Pastuglia, M.; Azimzadeh, J.; Goussot, M.; Camilleri, C.; Belcram, K.; Evrard, J.L.; Schmit, A.C.; Guerche, P.; Bouchez, D. Gamma-Tubulin is essential for microtubule organization and development in arabidopsis. *Plant Cell* **2006**, *18*, 1412–1425. [CrossRef] [PubMed]
111. Goshima, G.; Mayer, M.; Zhang, N.; Stuurman, N.; Vale, R.D. Augmin: A protein complex required for centrosome-independent microtubule generation within the spindle. *J. Cell Biol.* **2008**, *181*, 421–429. [CrossRef] [PubMed]
112. Liu, T.; Tian, J.; Wang, G.; Yu, Y.; Wang, C.; Ma, Y.; Zhang, X.; Xia, G.; Liu, B.; Kong, Z. Augmin triggers microtubule-dependent microtubule nucleation in interphase plant cells. *Curr. Biol.* **2014**, *24*, 2708–2713. [CrossRef] [PubMed]
113. Bannigan, A.; Scheible, W.R.; Lukowitz, W.; Fagerstrom, C.; Wadsworth, P.; Somerville, C.; Baskin, T.I. A conserved role for kinesin-5 in plant mitosis. *J. Cell Sci.* **2007**, *120*, 2819–2827. [CrossRef] [PubMed]
114. Miki, T.; Naito, H.; Nishina, M.; Goshima, G. Endogenous localizome identifies 43 mitotic kinesins in a plant cell. *Proc. Natl. Acad. Sci. USA* **2014**, *111*, E1053–E1061. [CrossRef] [PubMed]
115. Chen, C.; Marcus, A.; Li, W.; Hu, Y.; Calzada, J.P.; Grossniklaus, U.; Cyr, R.J.; Ma, H. The arabidopsis atk1 gene is required for spindle morphogenesis in male meiosis. *Development* **2002**, *129*, 2401–2409. [PubMed]
116. Ambrose, J.C.; Li, W.; Marcus, A.; Ma, H.; Cyr, R. A minus-end-directed kinesin with plus-end tracking protein activity is involved in spindle morphogenesis. *Mol. Biol. Cell* **2005**, *16*, 1584–1592. [CrossRef] [PubMed]
117. Marcus, A.I.; Li, W.; Ma, H.; Cyr, R.J. A kinesin mutant with an atypical bipolar spindle undergoes normal mitosis. *Mol. Biol. Cell* **2003**, *14*, 1717–1726. [CrossRef] [PubMed]
118. Ito, A.; Goshima, G. Microcephaly protein asp focuses the minus ends of spindle microtubules at the pole and within the spindle. *J. Cell Biol.* **2015**, *211*, 999–1009. [CrossRef] [PubMed]
119. Kawamura, E.; Himmelspach, R.; Rashbrooke, M.C.; Whittington, A.T.; Gale, K.R.; Collings, D.A.; Wasteneys, G.O. Microtubule organization 1 regulates structure and function of microtubule arrays during mitosis and cytokinesis in the arabidopsis root. *Plant Physiol.* **2006**, *140*, 102–114. [CrossRef] [PubMed]
120. Komaki, S.; Abe, T.; Coutuer, S.; Inze, D.; Russinova, E.; Hashimoto, T. Nuclear-localized subtype of end-binding 1 protein regulates spindle organization in arabidopsis. *J. Cell Sci.* **2010**, *123*, 451–459. [CrossRef] [PubMed]
121. Maiato, H.; Khodjakov, A.; Rieder, C.L. Drosophila clasp is required for the incorporation of microtubule subunits into fluxing kinetochore fibres. *Nat. Cell Biol.* **2005**, *7*, 42–47. [CrossRef] [PubMed]
122. Ambrose, J.C.; Shoji, T.; Kotzer, A.M.; Pighin, J.A.; Wasteneys, G.O. The arabidopsis clasp gene encodes a microtubule-associated protein involved in cell expansion and division. *Plant Cell* **2007**, *19*, 2763–2775. [CrossRef] [PubMed]
123. Rogers, G.C.; Rogers, S.L.; Sharp, D.J. Spindle microtubules in flux. *J. Cell Sci.* **2005**, *118*, 1105–1116. [CrossRef] [PubMed]
124. Dhonukshe, P.; Vischer, N.; Gadella, T.W., Jr. Contribution of microtubule growth polarity and flux to spindle assembly and functioning in plant cells. *J. Cell Sci.* **2006**, *119*, 3193–3205. [CrossRef] [PubMed]
125. Otegui, M.S.; Verbrugghe, K.J.; Skop, A.R. Midbodies and phragmoplasts: Analogous structures involved in cytokinesis. *Trends Cell Biol.* **2005**, *15*, 404–413. [CrossRef] [PubMed]
126. Duellberg, C.; Fourniol, F.J.; Maurer, S.P.; Roostalu, J.; Surrey, T. End-Binding proteins and ase1/prc1 define local functionality of structurally distinct parts of the microtubule cytoskeleton. *Trends Cell. Biol* **2013**, *23*, 54–63. [CrossRef] [PubMed]

127. Kosetsu, K.; de Keijzer, J.; Janson, M.E.; Goshima, G. Microtubule-associated protein65 is essential for maintenance of phragmoplast bipolarity and formation of the cell plate in physcomitrella patens. *Plant Cell* **2013**, *25*, 4479–4492. [CrossRef] [PubMed]
128. Ho, C.M.; Hotta, T.; Guo, F.; Roberson, R.W.; Lee, Y.R.; Liu, B. Interaction of antiparallel microtubules in the phragmoplast is mediated by the microtubule-associated protein map65–3 in arabidopsis. *Plant Cell* **2011**, *23*, 2909–2923. [CrossRef] [PubMed]
129. Muller, S.; Smertenko, A.; Wagner, V.; Heinrich, M.; Hussey, P.J.; Hauser, M.T. The plant microtubule-associated protein atmap65–3/ple is essential for cytokinetic phragmoplast function. *Curr. Biol.* **2004**, *14*, 412–417. [CrossRef] [PubMed]
130. Sasabe, M.; Kosetsu, K.; Hidaka, M.; Murase, A.; Machida, Y. Arabidopsis thaliana map65–1 and map65–2 function redundantly with map65–3/pleiade in cytokinesis downstream of mpk4. *Plant Signal Behav.* **2011**, *6*, 743–747. [CrossRef] [PubMed]
131. Hiwatashi, Y.; Obara, M.; Sato, Y.; Fujita, T.; Murata, T.; Hasebe, M. Kinesins are indispensable for interdigitation of phragmoplast microtubules in the moss physcomitrella patens. *Plant Cell* **2008**, *20*, 3094–3106. [CrossRef] [PubMed]
132. Archambault, V.; Carmena, M. Polo-Like kinase-activating kinases: Aurora a, aurora b and what else? *Cell Cycle* **2012**, *11*, 1490–1495. [CrossRef] [PubMed]
133. Hu, C.K.; Ozlu, N.; Coughlin, M.; Steen, J.J.; Mitchison, T.J. Plk1 negatively regulates prc1 to prevent premature midzone formation before cytokinesis. *Mol. Biol. Cell* **2012**, *23*, 2702–2711. [CrossRef] [PubMed]
134. Van Damme, D.; De Rybel, B.; Gudesblat, G.; Demidov, D.; Grunewald, W.; De Smet, I.; Houben, A.; Beeckman, T.; Russinova, E. Arabidopsis alpha aurora kinases function in formative cell division plane orientation. *Plant Cell* **2011**, *23*, 4013–4024. [CrossRef] [PubMed]
135. Sasabe, M.; Machida, Y. Regulation of organization and function of microtubules by the mitogen-activated protein kinase cascade during plant cytokinesis. *Cytoskeleton* **2012**, *69*, 913–918. [CrossRef] [PubMed]
136. Sasabe, M.; Soyano, T.; Takahashi, Y.; Sonobe, S.; Igarashi, H.; Itoh, T.J.; Hidaka, M.; Machida, Y. Phosphorylation of ntmap65–1 by a map kinase down-regulates its activity of microtubule bundling and stimulates progression of cytokinesis of tobacco cells. *Genes Dev.* **2006**, *20*, 1004–1014. [CrossRef] [PubMed]
137. Panteris, E.; Adamakis, I.D.; Voulgari, G.; Papadopoulou, G. A role for katanin in plant cell division: Microtubule organization in dividing root cells of fra2 and lue1arabidopsis thaliana mutants. *Cytoskeleton* **2011**, *68*, 401–413. [CrossRef] [PubMed]
138. Murata, T.; Sano, T.; Sasabe, M.; Nonaka, S.; Higashiyama, T.; Hasezawa, S.; Machida, Y.; Hasebe, M. Mechanism of microtubule array expansion in the cytokinetic phragmoplast. *Nat. Commun.* **2013**. [CrossRef] [PubMed]
139. Collonnier, C.; Epert, A.; Mara, K.; Maclot, F.; Guyon-Debast, A.; Charlot, F.; White, C.; Schaefer, D.G.; Nogue, F. Crispr-cas9-mediated efficient directed mutagenesis and rad51-dependent and rad51-independent gene targeting in the moss physcomitrella patens. *Plant Biotechnol. J.* **2016**. [CrossRef]
140. Sugano, S.S.; Shirakawa, M.; Takagi, J.; Matsuda, Y.; Shimada, T.; Hara-Nishimura, I.; Kohchi, T. Crispr/cas9-mediated targeted mutagenesis in the liverwort marchantia polymorpha l. *Plant Cell. Physiol.* **2014**, *55*, 475–481. [CrossRef] [PubMed]
141. Ishizaki, K.; Nishihama, R.; Yamato, K.T.; Kohchi, T. Molecular genetic tools and techniques for marchantia polymorpha research. *Plant Cell Physiol.* **2016**, *57*, 262–270. [CrossRef] [PubMed]
142. Nakaoka, Y.; Kimura, A.; Tani, T.; Goshima, G. Cytoplasmic nucleation and atypical branching nucleation generate endoplasmic microtubules in physcomitrella patens. *Plant Cell* **2015**, *27*, 228–242. [CrossRef] [PubMed]
143. Kiyomitsu, T. Mechanisms of daughter cell-size control during cell division. *Trends Cell Biol.* **2015**, *25*, 286–295. [CrossRef] [PubMed]
144. Buschmann, H.; Zachgo, S. The evolution of cell division: From streptophyte algae to land plants. *Trends Plant Sci.* **2016**, *21*, 872–883. [CrossRef] [PubMed]

biology

MDPI

Review

Anaphase A: Disassembling Microtubules Move Chromosomes toward Spindle Poles

Charles L. Asbury

Department of Physiology & Biophysics, University of Washington, Seattle, WA 98195, USA; casbury@uw.edu

Academic Editor: J. Richard McIntosh
Received: 30 December 2016; Accepted: 10 February 2017; Published: 17 February 2017

Abstract: The separation of sister chromatids during anaphase is the culmination of mitosis and one of the most strikingly beautiful examples of cellular movement. It consists of two distinct processes: Anaphase A, the movement of chromosomes toward spindle poles via shortening of the connecting fibers, and anaphase B, separation of the two poles from one another via spindle elongation. I focus here on anaphase A chromosome-to-pole movement. The chapter begins by summarizing classical observations of chromosome movements, which support the current understanding of anaphase mechanisms. Live cell fluorescence microscopy studies showed that poleward chromosome movement is associated with disassembly of the kinetochore-attached microtubule fibers that link chromosomes to poles. Microtubule-marking techniques established that kinetochore-fiber disassembly often occurs through loss of tubulin subunits from the kinetochore-attached plus ends. In addition, kinetochore-fiber disassembly in many cells occurs partly through 'flux', where the microtubules flow continuously toward the poles and tubulin subunits are lost from minus ends. Molecular mechanistic models for how load-bearing attachments are maintained to disassembling microtubule ends, and how the forces are generated to drive these disassembly-coupled movements, are discussed.

Keywords: anaphase A; kinetochore; chromosome-to-pole motion; microtubule poleward flux; conformational wave; biased diffusion

---

1. Introduction and Distinction between Anaphase "A" and "B"

In his classic 1961 volume on cell division, Daniel Mazia referred to anaphase as the act of chromosome movement that gives mitosis its meaning [1] (p. 95). The term, anaphase, was originally coined over 130 years ago [2]. By Mazia's time it had come to refer—as it still does today—to the phase of mitosis when sister chromatids are moving apart from one another toward opposite sides of the cell. The onset of anaphase is one of the most abrupt events of mitosis, making it cytologically useful as a reference for the timing of other mitotic events. It is also one of the most strikingly beautiful examples of cellular movement.

Anaphase consists of at least two distinct processes, traditionally referred to as "anaphase A" and "anaphase B". Anaphase A is the movement of chromosomes toward the spindle poles via shortening of the connecting fibers; it is the focus of this chapter (Figure 1). Anaphase B, which is covered in the subsequent chapter by Scholey et al. [3], is the separation of the two poles from one another via elongation of the spindle. The distinction between anaphase A and B is more than a mere descriptive convenience. The two processes occur simultaneously in many cell types; but they are mechanistically distinct, a fact that has been appreciated since well before the underlying mechanisms were understood [4]. Anaphase A can be further divided into at least two mechanistically distinct sub-processes, as discussed below.

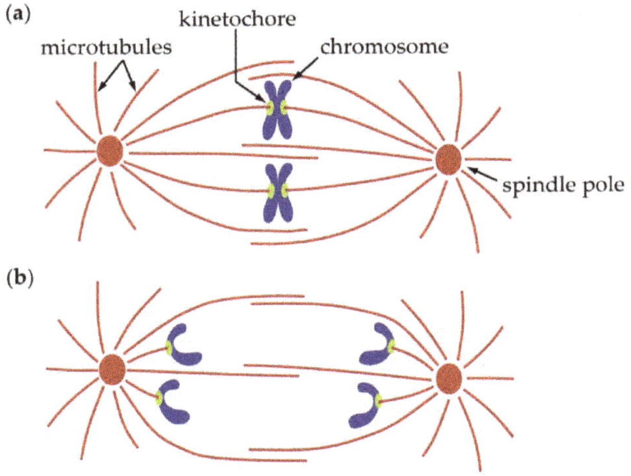

Figure 1. Schematic diagram of a spindle in metaphase (**a**) and anaphase (**b**). Only the chromosome-to-pole, "anaphase A" motion is depicted here; it is the focus of this chapter. Separation of the two spindle poles from one another via elongation of the spindle, "anaphase B", is discussed in the subsequent chapter by Scholey et al. [3].

This chapter begins with a description of chromosome movements during anaphase, which have been studied for over a century. Nevertheless, it is worthwhile to summarize the classical findings that support our current understanding and are sometimes taken for granted. Next is a description of microtubule dynamics within the spindle, another pillar of our modern view of anaphase. The remainder of the chapter is devoted to a discussion of force generation, which occurs also in earlier phases of mitosis but is most obvious during anaphase. Where and how are the forces that drive anaphase A generated? What roles are played by microtubule-based motor proteins and by the microtubules themselves? Evidence that the microtubules convert chemical energy into mechanical work is presented. Mechanistic concepts are emphasized, rather than specific molecules, with the hope that the discussion will be accessible and interesting, even for readers less familiar with mitosis.

2. Centromeres and Kinetochores Usually Lead Anaphase Movements While Chromosome Arms Follow

The idea that chromosomes are moved, during anaphase A and other phases as well, by forces exerted on them at kinetochores is so well established that the observations on which it rests are scarcely mentioned anymore. Condensed mitotic chromosomes are visible by brightfield microscopy, particularly when phase or differential interference contrast is used. Thus, as summarized in Chapter 1 of this volume [5], they have been observed for over a century. In certain cell types, the mitotic chromosomes are relatively long and their primary constrictions—their centromeres—are also discernable. Because these centromeric constrictions usually lead during mitotic chromosome movements (Figure 1), it is clear that they are major sites where force is transmitted to the chromosomes. Indeed, this is why they were given the name, kinetochores ("movement places") [6]. Kinetochores in anaphase tend to move in straight paths toward the spindle poles, while the chromosome arms, following the kinetochores, swing and trace out more complex paths. Reflecting on these 'rag-doll' like movements, Mazia famously compared the role of chromosomes in mitosis to "that of a corpse at a funeral: they provide the reason for the proceedings but do not take an active part in them" [1] (p. 212).

The general rule that kinetochores lead while chromosome arms follow applies in many cell types including vertebrates [7] and yeasts [8], but there are exceptions, such as in plant endosperm [9], and in crane-fly spermatocytes [10], where arms sometimes lead. These alternative cases remind us that forces are also exerted directly on chromosome arms, although the primary motive forces for anaphase are commonly exerted at kinetochores. The chromosome arms in certain well-studied mitotic cell types (e.g., Newt lung [7]) are pushed continually away from the spindle poles. These antipoleward forces have been dubbed the "polar winds" (or "polar ejection forces" [11]). They must be overcome by the kinetochores to drive anaphase poleward movement, and they explain why the chromosome arms usually point away from the poles in these cells. For other cell types, in which the chromosome arms sometimes lead the motion, the polar winds can blow toward, rather than away from the spindle poles [9,10]. Plant endosperm is an interesting case where chromosome arms first experience poleward forces prior to metaphase and then later, after anaphase onset, the situation reverses and arms experience away-from-the-pole forces (Figure 2) [9]. In crane-fly spermatocytes, however, poleward forces are apparently exerted on chromosome arms even during anaphase, providing an additional force that assists rather than opposes the kinetochores [10].

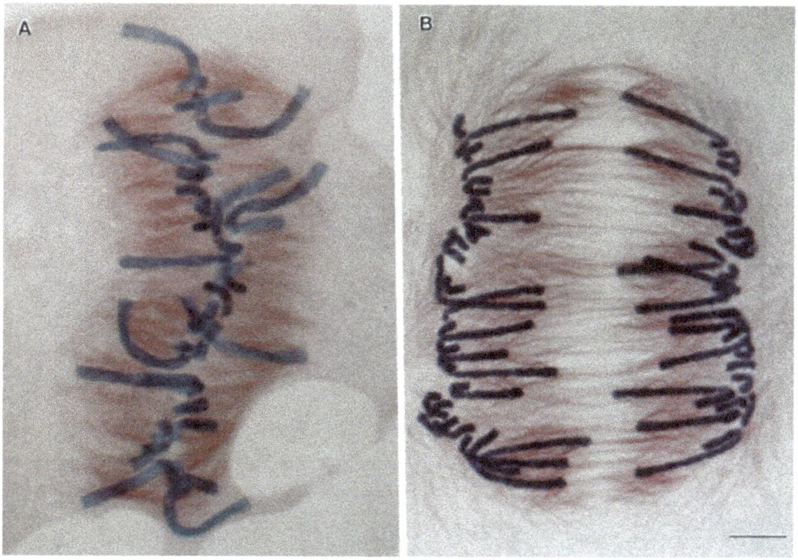

Figure 2. Light micrographs of metaphase (**a**) and late anaphase (**b**) plant endosperm (*Haemanthus*) spindles. During metaphase in these plant cells the chromosome arms are bent in the direction of the spindle poles. This behavior differs from what is seen in animal somatic cells, where chromosome arms are pushed continually away from spindle poles [11]. These *Haemanthus* images are reprinted from [9], and are displayed under the terms of a Creative Commons License (Attribution-Noncommerical-Share Alike 3.0 Unported license, as described at http://creativecommons.org/licenses/by-nc-sa/3.0/). Scale bar, 10 μm.

3. Poleward Movement during Anaphase A Is Mostly but Not Entirely Unidirectional

The poleward movement of kinetochores in anaphase is mostly unidirectional, but not always. Reversals in direction, similar to the oscillations seen earlier in prometaphase and metaphase, can continue in anaphase, but a poleward bias is generally maintained [12] (Figure 3). This bi-directional, back-and-forth movement has been named 'directional instability'. It bears a striking resemblance to the intrinsic 'dynamic instability' of microtubule filaments, which stochastically switch between periods of shortening and growth [13,14], and suggests an intimate coupling between chromosome movements and microtubule dynamics, as discussed below. Although anaphase begins abruptly, anaphase chromosome movements are not perfectly synchronous. A kinetochore moving poleward in anaphase can reverse direction, transiently moving anti-poleward while its peers continue their poleward march. Neighboring chromosomes within a cell can also move closely past one another in opposite directions, e.g., when anaphase occurs prematurely, prior to formation of a proper metaphase plate (e.g., see [1] (p. 288) and [15]). A chromosome can also become erroneously attached to the spindle, with one of its kinetochores attached simultaneously to microtubules emanating from both poles. These "merotelically" attached chromosomes lag behind their properly ("amphitelically") attached peers during anaphase [16]. Together these observations demonstrate that kinetochores are moved individually, rather than as a group. (Likewise, the mitotic error correction machinery acts at the individual kinetochore level, as described in the chapter in this volume by Grishchuk and Lampson [17].)

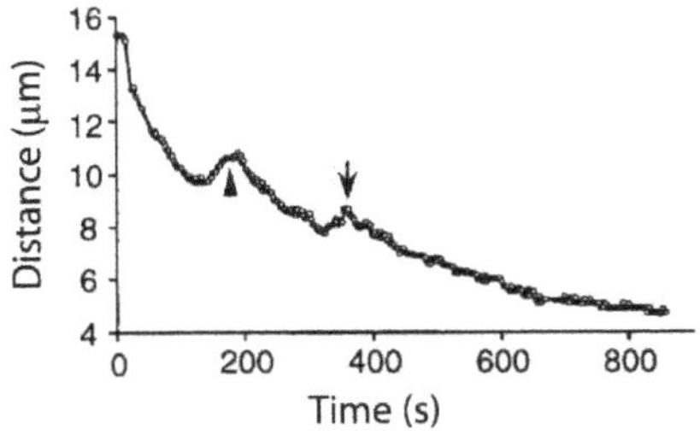

Figure 3. Example of kinetochore directional instability during anaphase A in a newt lung cell. Anaphase A chromosome-to-pole movement of the kinetochore is interrupted by transient reversals in directionality. This graph is reprinted from [12], and is displayed under the terms of a Creative Commons License (Attribution-Noncommerical-Share Alike 3.0 Unported license, as described at http://creativecommons.org/licenses/by-nc-sa/3.0/).

Perhaps the most direct evidence supporting the primacy of kinetochores for moving chromosomes comes from UV ablation studies, which began as early as the 1950s [18]. If the kinetochores of a single chromosome are damaged by UV irradiation, the remaining chromosome arms drift rather than following their un-irradiated peers [7,18,19]. In contrast, a chromosome whose arm has been ablated follows the normal patterns of movement.

4. Poleward Chromosome Movement Is Coupled to Shortening of the Connecting Microtubules

Modern theories about chromosome movement began to emerge with the structural understanding of spindle architecture afforded by electron microscopy. Several distinct categories of microtubule filaments exist, with well-defined polarities (as discussed thoroughly in Chapter 1 of this volume [5]). The most important for anaphase A are the kinetochore-attached microtubules, which have one end, their fast-growing 'plus' end, located at a kinetochore, while their 'minus' ends project poleward. In medium-sized and larger spindles, many microtubules terminate together at each kinetochore and these are bundled together to form a kinetochore fiber. Some but not necessarily all the microtubules in a kinetochore fiber extend all the way to a spindle pole [20]. In the tiny spindles of budding yeast, the situation is simpler, with just one microtubule linking each kinetochore to a pole [21].

Advances in tubulin biochemistry and live-cell fluorescence microscopy have provided a fascinating view of the dynamics of microtubules in living spindles [22–25]. Time-lapse movies of large mammalian cells with fluorescent-tags on their kinetochores and their microtubules show that movement of a kinetochore is coupled to growth or shortening of the microtubule fibers to which it is attached [25]. During anaphase A, kinetochore-associated fibers shorten, without becoming noticeably thicker. This shortening of kinetochore fibers seems to draw the chromosomes poleward. In many cell types, microtubule-marking techniques (fluorescence photobleaching, photoactivation, and speckle microscopy) have shown that kinetochore fiber shortening occurs partly via loss of tubulin subunits from the kinetochore-attached plus ends (Figure 4a,b). How a kinetochore can maintain a persistent, load-bearing attachment to a microtubule tip that is disassembling under its grip is only poorly understood. Some models are discussed below.

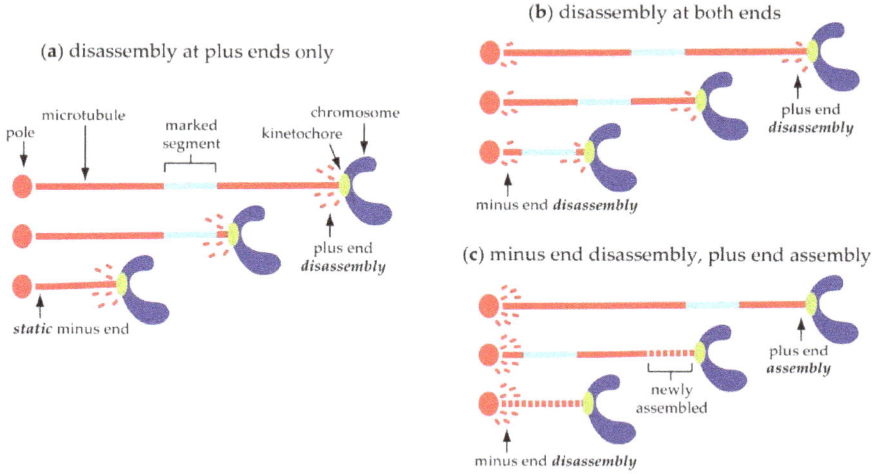

Figure 4. Chromosome-to-pole motion during anaphase A is coupled to microtubule disassembly. (**a**) Simple mechanism with disassembly occurring only at microtubule plus ends, as seen in yeasts, where minus end attachments to the poles are static and no flux occurs [26–29]. (**b**) Dual mechanism, as in cultured mitotic human cells, where chromosome-to-pole motion is a superposition of a kinetochore's movement relative to the microtubules, which is coupled to plus end disassembly, and the microtubules' flux relative to the poles, which is coupled to minus end disassembly [30]. (**c**) Mechanism observed for autosomal half-bivalents in meiotic crane-fly spermatocytes, with disassembly at minus ends and assembly at plus ends [10,31]. Switching between mechanism (b) and mechanism (c) has been directly observed in *Xenopus* egg extract spindles [32].

5. Kinetochore-Attached Microtubules Can 'Flux' Continuously toward the Poles

Microtubule-marking techniques have also revealed that kinetochore-attached microtubules in many spindles flow steadily toward the poles (Figure 4b,c). This poleward microtubule 'flux' is coupled to minus end disassembly at or near the poles [23,31–37]. Anaphase A in these cells is therefore a superposition of a kinetochore's movement relative to the microtubules and the microtubules' flux relative to the poles. The contribution of flux to poleward kinetochore movement varies widely depending on cell type (Table 1). In mitotic human cells, for example, flux accounts for about a third of anaphase A chromosome-to-pole movement, the remaining two-thirds of which is due to plus end disassembly [30]. In budding or fission yeast, there is apparently no flux, so anaphase A in these cells is probably explained entirely by plus end disassembly [26–29]. In contrast, flux appears to be solely responsible for anaphase A in plant (tobacco) cells [37] and in meiotic crane-fly spermatocytes [10,31]. In the crane-fly spermatocytes, kinetochore-attached microtubule plus ends assemble, rather than disassembling during anaphase A. The bottom line is that microtubule fibers linking kinetochores to poles can disassemble from either end, or from both ends. The questions about how load-bearing attachments are maintained and how the speeds of movement are coordinated with rates of filament disassembly apply to both ends of the microtubules.

Table 1. Speeds of chromosome-to-pole anaphase A motion, and microtubule-to-pole flux motion, measured in various spindle/cell types.

Spindle/Cell Type	Chromosome-to-Pole Speed (μm/min)	Speed Measured in Anaphase A?	Microtubule-to-Pole Flux Speed (μm/min)	Technique for Flux Measurement	Flux Measured in Anaphase A?	Experimental Condition	Fraction of Anaphase A Speed Due to Flux (%)	Reference
Sand dollar embryos	1	yes	1.8	photobleaching	yes	control	180	[38]
Newt lung cells	1.7 0.54	yes yes	0.44 0.18	photoactivation photoactivation	yes yes	early anaphase late anaphase	26 33	[33]
Newt lung cells	0.2	yes	0.2	photoactivation	yes	10 uM taxol, late anaphase	100	[39]
Pig kidney (LLC-PK) and rat kangaroo (PtK1) cells	1.2	yes	0.2	photoactivation	yes	early anaphase	17	[34]
Xenopus (meiotic) extract spindles	2 0.2 2	yes yes yes	2 0.2 2	photoactivation photoactivation photoactivation	yes yes no (metaphase)	control 1.5 mM AMPPNP 1 uM taxol	100 100 100	[40]
Xenopus (meiotic) extract spindles	2.8 0.7	yes yes	1.6 1.6	speckle speckle	yes yes	plus end depol. plus end polym.	57 229	[32]
Budding yeast	1.3	yes	-	-	-	CEN dots (14 kb LacO array)	-	[8]
Budding yeast	-	-	0	photobleaching	no (anaphase B)	ipMTs (not kMTs)	-	[26]
Budding yeast	0.3	yes	-	-	-	CEN dots (2–11 kb LacO arrays)	-	[41]
Budding yeast	0.3	yes	-	-	-	CEN dots (2 kb LacO array)	-	[42]
Fission yeast	-	-	0	photobleaching	no (anaphase B)	ipMTs (not kMTs)	-	[29]
Fission yeast	-	-	0	speckle	no (anaphase B)	ipMTs (not kMTs)	-	[28]
Fission yeast	-	-	0	photobleaching	no (anaphase B)	ipMTs (not kMTs)	-	[27]
Drosophila embryos	3.6	yes	3.2	speckle	yes	control, 18 °C	89	[35]
Drosophila embryos	6.4	yes	1.9	speckle	yes	control	30	[43]
Drosophila embryos	5.6 3.4 3.2	yes yes yes	2.2 3.4 0	speckle speckle speckle	yes yes no (metaphase)	control anti-KLP59C anti-KLP10A	39 100 0	[36]
Drosophila (S2) cells	1.2 0.6 0.8 0.7	yes yes yes yes	0.6 0.5 0.2 0.1	photobleaching photobleaching photobleaching photobleaching	yes yes yes yes	control katanin RNAi spastin RNAi fidgetin RNAi	50 83 25 14	[44]
Drosophila (S2) cells	1.7 0.7 1.7 0.8	yes yes yes yes	0.9 0.5 0.9 0.3	photobleaching photobleaching photobleaching photobleaching	yes yes yes yes	control KLP59D RNAi KLP59C RNAi KLP10A RNAi	53 71 53 38	[45]

Table 1. *Cont.*

Spindle/Cell Type	Chromosome-to-Pole Speed (µm/min)	Speed Measured in Anaphase A?	Microtubule-to-Pole Flux Speed (µm/min)	Technique for Flux Measurement	Flux Measured in Anaphase A?	Experimental Condition	Fraction of Anaphase A Speed Due to Flux (%)	Reference
Drosophila (S2) cells	0.8 0.7	yes yes	0.4 0.2	speckle speckle	yes yes	control CLASP & KLP10A RNAi	50 28	[46]
Drosophila spermatocytes (meiosis)	- 1.7 2.7	- yes yes	0.6 0.6 1	photobleaching photobleaching photobleaching	no (metaphase) yes yes	metaphase disjoining separated	- 35 37	[47]
Crane-fly spermatocytes (meiosis)	0.5	yes	0.9	speckle	yes	autosomal half-bivalents	180	[31]
Crane-fly spermatocytes (meiosis)	1.3	yes	0.8	speckle	yes	sex univalent, bipolar link cut	62	[48]
Crane-fly spermatocytes (meiosis)	0.7	no (metaphase)	0.7	speckle	no (metaphase)	autosomal, cut K-fragment	100	[49]
Human (U20S) cells	1.5 1.2	yes yes	0.5 0	photoactivation photoactivation	no (metaphase) no (metaphase)	control MCAK & Kif2a RNAi	33 0	[30]
Human (U20S) cells	- - -	- - -	0.5 0.3 0.3	photoactivation photoactivation photoactivation	no (metaphase) no (metaphase) no (metaphase)	control CLASPs RNAi Cenp-E RNAi	- - -	[50]
Human (U20S) cells	- -	- -	0.6 0.3	photoconversion photoconversion	no (metaphase) no (metaphase)	control Kif4A RNAi	- -	[51]
Human (U20S) cells	0.44 0.3	yes yes	0.9 0.6	photoactivation photoactivation	no (metaphase) no (metaphase)	control fidgetin siRNA	- -	[52]
Human (HeLa) cells	- -	- -	0.4 0.2	photoactivation photoactivation	no (metaphase) no (metaphase)	control ectopic MCAK at CEN	- -	[53]
Human (HeLa) cells	1.7 0.9 2.8	yes yes yes	- - -	- - -	- - -	control Kif18A overexpress. Kif18A siRNA	- - -	[15]
Tobacco (BY-2) cells	2.1	yes	2	photobleaching	yes	control	95	[37]

6. Anaphase in Some Cell Types Does Not Conform to the Canonical View

A modern student of mitosis reading the classical literature cannot help but notice how many more types of cells were being examined. The advent of genetic and molecular approaches enabled a terrific array of tools that could not previously have been imagined. But these state-of-the-art tools have been aimed at a much more limited set of model cell types. And even within this limited set, there are examples that do not conform to the canonical view. Anaphase chromosome separation in the acentrosomal meiosis I spindles of *C. elegans* oocytes is apparently independent of kinetochores [54]. Instead, the chromosomes seem to be pushed from behind by microtubules growing and/or sliding out from the equator. The univalent X Y sex chromosomes in meiosis I crane-fly spindles move toward one pole while retaining microtubule fiber attachments to both poles [48]. The fiber on the trailing side elongates, while the leading fiber shortens. Probably more cases that do not fit the 'normal' picture will emerge as more transcriptomes and genomes are sequenced, and as new genome-editing technologies, such as CRISPR [55], facilitate live imaging of fluorescent-marked spindles in less-studied cell types

7. Kinetochores Can Either Be Actively Pulling Poleward or Passively Slipping Anti-Poleward

For a true, mechanistic understanding of anaphase, it is not enough simply to describe the motions of the kinetochores, the microtubules, and the poles relative to one another. We need to understand where and how the motive forces are generated. Biophysically, 'force generation' (or 'force production') refers to the *active* processes by which chemical energy, usually in the form of nucleotide triphosphates, is converted into mechanical work, defined as force acting through a distance. The forces that draw kinetochores toward spindle poles must be generated somewhere within the kinetochores themselves, within the poles, or within the material connecting them.

The coupling of kinetochore movement to microtubule plus end disassembly strongly suggests that the kinetochore-microtubule interface is a site where force is actively generated. Compared to the early ablation studies that used UV-lamps [18], newer laser-equipped microscopes have enabled faster and more finely targeted ablations, making it possible in certain large cells (e.g., newt lung or PtK cells) to micro-surgically sever the centromeric chromatin connecting two sister kinetochores [56], or to selectively destroy one sister of a pair [57]. If a kinetochore moving poleward during metaphase is micro-surgically freed from its sister, it continues moving poleward (Figure 5a). However, if a kinetochore moving anti-poleward is freed, then it abruptly stops (Figure 5a,b), suggesting that its anti-poleward motion prior to the severing operation was a passive response to externally generated pulling forces (e.g., to forces generated by its poleward-moving sister) [57]. These observations, together with the highly coordinated oscillations of sisters in unperturbed cells [12], suggest that the force-producing machinery at a kinetochore can adopt two distinct states, an active state in which it generates pole-directed pulling force, and a 'neutral' state in which it remains stationary or passively slips anti-poleward in response to external forces. Such two-state behavior, with active minus end-directed pulling and passive plus end-directed slippage, is also observed when purified kinetochores are attached in vitro to dynamic microtubule tips ([58]; discussed further below). The behavior has implications for how a kinetochore's force-generating machinery might operate, both before and after the metaphase-anaphase transition.

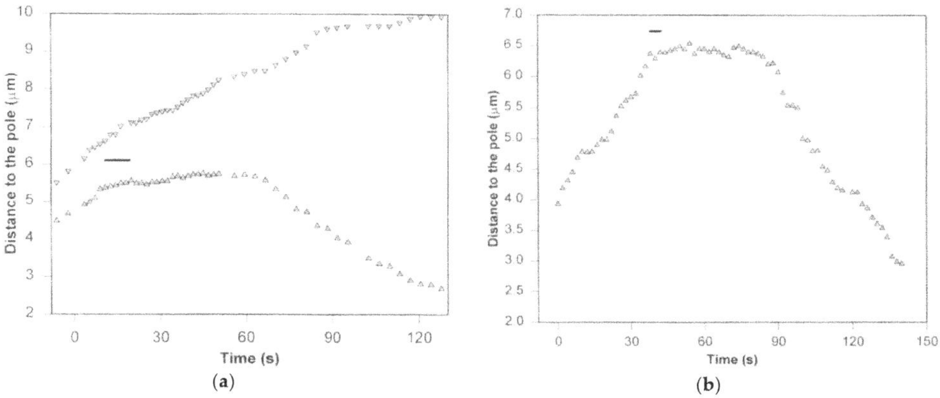

Figure 5. Kinetochores can adopt two distinct states, an active state that generates pole-directed pulling force, and a 'neutral' state that remains stationary or passively slips anti-poleward in response to external forces. (**a**) Motions of sister kinetochore regions in a metaphase PtK1 cell before, during (*horizontal bar*) and after micro-surgically separating the sisters. (**b**) Motion of a trailing kinetochore before, during (*horizontal bar*), and after selectively destroying its poleward moving sister kinetochore. In both cases the trailing kinetochore abruptly stops once it is micro-surgically freed from its sister. Then, after a ~20 s delay, it reverses its original directionality and begins to move poleward. These graphs are reprinted from [57], and are displayed under the terms of a Creative Commons License (Attribution-Noncommerical-Share Alike 3.0 Unported license, as described at http://creativecommons.org/licenses/by-nc-sa/3.0/).

8. Anaphase Spindle Generates More Force than Needed for Anaphase Chromosome Movement

It might seem natural to assume that the spindle forces normally generated during anaphase, when the chromosomes are undergoing their most obvious movements, are higher than during other phases of mitosis. As Mazia [1] (p. 142) noted, "human laziness leads us to associate movement with hard work". In anaphase, this assumption turns out to be false. However, the anaphase spindle is also capable of producing far more force than is normally necessary.

Classic microneedle experiments, performed almost four decades ago, still provide some of the best and most direct measurements of spindle forces in anaphase. Nicklas used extremely thin, calibrated glass needles to tug on individual chromosomes in meiotic grasshopper spermatocytes and to ask how much opposing force was required to completely halt their chromosome-to-pole motion. The stall force he measured was surprisingly high, 700 pN [59]. This value represents the apparent limit of force production by the anaphase spindle in these cells—i.e., the maximum poleward force that the spindle can exert on a chromosome, presumably through its kinetochore(s). Nicklas assumed this load was shared by a subset of 15 kinetochore-attached microtubules that extended all the way to spindle pole (out of a total of ~40 kinetochore-attached microtubules), leading to an often-cited estimate of 50 pN per microtubule [59]. This might be an overestimate, with the true value falling closer to 12 pN per microtubule, given the recent work suggesting that all kinetochore-attached microtubules, even those that do not extend all the way to a pole, are anchored within the spindle [60,61]. But in either case the forces during a normal, unperturbed anaphase are probably much, much lower still. Viscous drag calculations suggest that chromosome-to-pole movement is normally driven by forces of only 0.1 pN [62]. Elastic bending of chromosomes likewise suggests only 0.7 pN [63]. Thus, the anaphase spindle can apparently exert a maximum poleward force (700 pN) that exceeds the normal anaphase force by as much as 1000- or even 7000-fold.

9. Why Is the Anaphase Spindle 'Over-Engineered' to Produce Forces so Much Higher than Needed?

What could be the evolutionary advantage of such an exceedingly high force-generating capacity? High capacity for force production might be advantageous during anaphase for disentangling chromosomes that remain inappropriately intertwined, perhaps helping to promote the decatenation activity of topoisomerases. High force-generating capacity might also be important during earlier stages of mitosis, before anaphase. During prometaphase, force at kinetochores provides a regulatory cue that promotes the selective stabilization of properly bioriented chromosome-spindle attachments. (See [64–66] and the chapter in this volume by Grishchuk and Lampson [17].) Kinetochore force might also be important for silencing the 'wait' signals generated by the spindle assembly checkpoint, which control entry into anaphase (as discussed in the chapter in this volume by Joglekar [67]). Bioriented kinetochores congressing to the spindle equator in prometaphase spermatocytes support intermediate levels of force, around 50 pN [68], which is much higher than the feeble forces normally seen in anaphase, <1 pN, but still less than the maximal value of 700 pN. Thus, the spindle might have evolved to pull forcefully against kinetochores *prior* to anaphase, to ensure that when anaphase does occur, the chromosomes will segregate correctly. In other words, the spindle's capacity for producing very high forces during anaphase might be a byproduct of evolutionary pressure for high forces during earlier mitotic stages. Regardless of its evolutionary significance, the high force-generating capacity of the anaphase spindle has implications for the underlying mechanism of force production.

10. New Techniques Are Providing Force Estimates from a Wider Variety of Cell Types

Nicklas' microneedle measurements were truly ground-breaking and their relevance to current mitosis research persists even four decades later. However, it should be noted that their generality is uncertain. Grasshopper spermatocytes are especially amenable to chromosome micromanipulation, probably because they lack a robust cortical layer of cytoskeletal filaments and thus their outer plasma membrane can be severely indented by a microneedle without being punctured or torn. (The needles do not puncture the membrane during successful experiments—accidental punctures cause cytoplasmic leakage and rapid cell death.) New techniques are needed for measuring kinetochore forces in other types of cells that are not amenable to micromanipulation.

Fluorescence-based approaches have recently shown great promise. By tracking the positional fluctuations of fluorescent centromeric probes, kinetochore forces during metaphase in budding yeast have recently been estimated at 4 to 6 pN [69]. This estimate agrees well with Nicklas' prometaphase measurement of 50 pN, considering that the load on a grasshopper kinetochore is probably shared by numerous attached microtubules: Nicklas estimated 7 kinetochore-attached microtubules during prometaphase, each bearing 7 pN of load [68], whereas each kinetochore in budding yeast attaches just a single microtubule [21], bearing 4 to 6 pN. Calibrated fluorescence force-sensors inserted into the *Drosophila* kinetochore suggest somewhat higher loads during metaphase in this organism, 130 to 680 pN per kinetochore, or 12 to 62 pN per microtubule (assuming the load is shared by 11 microtubules) [70]. Thus, the forces sustained by kinetochore-microtubule junctions during normal prometaphase and metaphase might vary between 4 and ~60 pN, depending on the organism. How these pre-anaphase forces measured in yeast and *Drosophila* compare with the maximum force-generating capacity of their spindles is unknown, however, because the maximal force has only been measured in grasshopper spermatocytes.

Another potential approach for measuring kinetochore forces in living cells is to apply laser trapping. Calibrated laser traps have been used extensively for measuring forces produced by purified myosin, kinesin and dynein motors in vitro [71] and, more recently, to study isolated kinetochores and kinetochore subcomplexes coupled to microtubule tips in vitro (as discussed below). In a limited number of cases, laser traps have also been applied in living cells, to measure forces generated in vivo during the transport of small (and generally spherical) intracellular cargoes by kinesin and dynein motors in non-mitotic cells [72–75]. Because the standard methods for trap calibration cannot be applied in vivo, these studies have relied on external calibrations, performed after isolation of the trapped organelles (e.g., lipid droplets) from the cells [72,73,75], or they have used enhanced calibration methods that account for the viscoelastic behavior of cytoplasmic fluid [74,76]. Trap-induced photodamage, which is easily avoided in vitro by removal of dissolved oxygen [77], becomes a major concern whenever laser traps are applied in cells growing under aerobic conditions [78]. A recent study applying laser traps in meiotic spermatocytes from crane-fly and *Mesostoma* flatworms [79] suggests that the forces required to stall chromosome-to-pole movements in these cells might be ~100-fold lower than the 700 pN measured previously in grasshopper spermatocytes [59]. However, neither standard, nor enhanced trap calibration methods were used, and the chromosome movements were attenuated even after the laser trap was turned off, suggesting permanent photodamage rather than force-induced stalling.

11. Tip-Coupling: One of the Most Conserved Features of Mitosis and One of the Most Puzzling

The poleward movement of chromosomes coupled to shortening of microtubule plus ends is one of the most conserved features of mitosis. It is also one of the most puzzling. How is it possible for a kinetochore (or a spindle pole) to maintain a persistent and load-bearing grip on the end of a microtubule that is rapidly disassembling? Any proposed mechanism for anaphase A must explain this 'tip-coupling'. A general mechanism should also be capable of explaining the other observations discussed above, such as the possibility for transient reversals in kinetochore directionality, the switching between active poleward and passive anti-poleward states, and the levels of force at kinetochores.

12. Conventional Motors Are Found at Kinetochores but Might Not Be the Primary Basis for Tip-Coupling

Cytoplasmic dynein and kinesin-family motors were among the earliest molecules found to localize to centromeres [80–82] (closely following the seminal identification of CENP-A, -B, and -C [83]). Because ATP-powered motor enzymes are by themselves capable of moving along the sides of microtubule filaments, it is easy to imagine that they might represent the molecular basis for active force production at kinetochores. Minus end-directed motors anchored to a kinetochore could reach around the microtubule tip, moving along the sides of the filament and thereby dragging the chromosome poleward (Figure 6). Additional microtubule-modifying enzymes (microtubule depolymerases or severing enzymes) could explain how the motor-driven movement is coupled to plus end disassembly. Somehow the activities of these microtubule disassemblers would need to be coordinated with the motor enzymes.

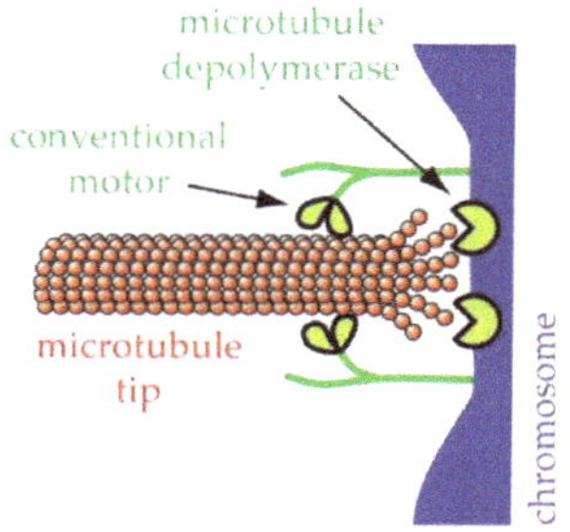

Figure 6. Model for kinetochore-microtubule tip-coupling based on conventional motor proteins and microtubule-regulators. Conventional ATP-powered, minus end-directed motor enzymes anchored at the kinetochore could reach around the tip of the microtubule, moving along the sides of the filament and thereby dragging the chromosome poleward (leftward in the diagram). The activities of additional microtubule depolymerases or severing enzymes, somehow coordinated with the conventional motor activity, could explain how poleward chromosome movement is coupled to plus end-disassembly.

There is good evidence that kinetochore-associated dynein contributes to anaphase A in certain cell types. Null mutations in the genes for zw10 or rod, components of the RZZ complex that links dynein to kinetochores, cause dramatic slowing of anaphase A chromosome-to-pole movement in *Drosophila* spermatocytes [84]. Acute inhibition of dynein by microinjection of excess p 50 'dynamitin' (a component of the dynein-activating complex, dynactin) or of anti-dynein antibodies similarly slows anaphase A chromosome-to-pole speeds by ~75% in *Drosophila* embryos [85]. However, microinjection-based inhibition of dynein in mammalian (PtK1) cells causes a much less-dramatic, ~33% slowing of anaphase A motion [86]. The chromosomes generally retain their attachments to dynamic microtubule plus ends [86], suggesting that kinetochore-associated dynein is dispensable for tip-coupling in these cells. Thus, while conventional motor proteins do play many vital roles during mitosis (especially for spindle assembly, prometaphase chromosome movements, and anaphase B, as discussed in the chapters in this volume by Kapoor [87], Goshima and Yamada [88], and Scholey et al. [3]), they do not seem to be the primary basis for tip-coupling. Dispensability of motor activity for tip-coupling in living cells is demonstrated most convincingly by studies of fission yeast, where poleward kinetochore movements coupled with microtubule disassembly can be directly observed even after all kinetochore-localized minus end-directed motors have been deleted [89]. Likewise, in budding yeast, disassembly-coupled kinetochore movements can continue in the absence of minus end-directed kinetochore motors [90]. More generally, deletion of various kinetochore-associated motors does not detach the kinetochores from the spindle [89,91–94]. These observations do not necessarily preclude a role for motors in tip-coupling, but they do argue against simple models in which tip-coupling is based primarily on a single type of conventional motor.

13. Kinetochores Also Contain Non-Motor Microtubule-Binding Elements

Our understanding of the biochemical composition and architecture of the kinetochore has grown immensely during the last decade (as discussed in the chapter in this volume by Musacchio and Desai [95]). The molecular details will not be repeated here, but the emerging view is that the kinetochore-microtubule interface includes an array of non-motor, microtubule binding proteins in addition to the conventional motors mentioned above. Foremost among these non-motor microtubule binders is the Ndc80 complex (Ndc80c), a fibrillar hetero-tetramer with one end that binds microtubules and another end that anchors stably into the core of the kinetochore [96–100]. Ndc80c localizes to the outer kinetochore layer, where microtubule tips are embedded, and its depletion causes widespread failure of kinetochore-microtubule attachment [101–103], suggesting a direct role in tip-coupling. Ndc80c is widely conserved. Its fibrillar structure contains hinge-points, enabling it to bend or fold [104–106]. Fluorescence measurements suggest that the relative abundance of Ndc80c (and other core subcomplexes) at individual kinetochores scales with the number of attached microtubules. Budding yeast kinetochores, which bind just one microtubule, are estimated to contain between 8 and 20 copies of Ndc80c [107,108]. Larger kinetochores that bind more microtubules have correspondingly more Ndc80c [109–111]. This scaling suggests modularity. The kinetochores of humans and other 'higher' eukaryotes might consist of large, parallel arrays of discrete microtubule-binding sites, each resembling a single budding yeast kinetochore [112].

Another microtubule-binding kinetochore element, specific to fungi, is the hetero-decameric Dam1 complex (Dam1c) [113–115]. Dam1c localizes to kinetochores in an Ndc80c-dependent manner and makes a major contribution to kinetochore-microtubule attachment in yeast [102,116]. Purified Dam1c spontaneously assembles into sixteen-membered, microtubule-encircling rings [117,118], which might function as sliding collars (as discussed below) [119,120]. The average number of Dam1 complexes per kinetochore is sufficient to form approximately one ring [107], or possibly two [108], per attached microtubule. Outside of fungi, the Ska complex has been proposed to provide a functionally similar activity [121,122], possibly via oligomerization, although it does not appear to form microtubule-encircling rings [122].

14. Toward an Integrated View of the Tip-Coupling Apparatus of the Kinetochore

The biochemical complexity of the kinetochore poses a major challenge for understanding how it functions. There are a variety of different microtubule-binding proteins likely to contribute, including the motor and non-motor proteins discussed above, and additional components as well. Unfortunately, our current understanding is too rudimentary to identify distinct roles for all of them. Current models for tip-coupling (and for other kinetochore functions as well, e.g., checkpoint signaling and error correction) emphasize the non-motor microtubule binders, especially Ndc80c and, in yeast, Dam1c. Kinetochore-anchored motor proteins are also very likely to be important. In principle, the kinetochore motors could participate in tip-coupling via their conventional ATP-powered walking along the sides of microtubules or, alternatively, they could participate in a manner independent of conventional walking motility [123–127]. That is, the kinetochore motors could function in tip-coupling essentially as fibrils that transiently bind and unbind from the microtubule, similarly to the non-motor microtubule binding fibril, Ndc80c. Another class of molecules likely to contribute are microtubule plus end-binders, such as those of the TOG (tumor overexpressed gene) family. TOG family proteins (Stu2 in budding yeast, XMAP215 in Xenopus, and chTOG in humans) localize to kinetochores [102,128–134] and contribute directly to tip-coupling in vitro [135,136]. The knockdown phenotypes for these plus end-binders, and for kinetochore motors, are often complex, suggesting roles in multiple different aspects of mitosis and making it difficult to assess specifically their roles in kinetochore tip-coupling in vivo.

An intriguing possibility is that the various microtubule-binders at kinetochores might interact with different structural features at the microtubule tip. For example, some might bind straight tubulins in the microtubule wall, while others might prefer curved protofilaments peeling out from the wall, and still others might even bind the longitudinal faces of tubulin dimers exposed uniquely at the extreme terminal subunits. More work is needed to test this idea. Especially useful would be better structural information about the relevant microtubule-binders, and more sophisticated biophysical methods for assessing the importance of specific microtubule contacts and specific tubulin conformations in kinetochore tip-coupling.

In the meantime, for the purpose of discussing potential biophysical mechanisms of tip-coupling, it seems sufficient at present to consider the kinetochore simply as a collection of flexible microtubule-binding fibrils, augmented in yeast (and possibly other organisms) by additional microtubule-binders that can potentially oligomerize into microtubule-encircling rings. This view is supported by the configuration of isolated yeast kinetochore particles seen in electron micrographs, which show 5 to 7 microtubule-binding fibrils connected to a central hub and sometimes associated with a microtubule-encircling ring [137]. It is also consistent with electron tomographic imaging of kinetochore-microtubule interfaces in vivo in multiple cell types [138,139].

15. Microtubules Could Be the Engines that Drive Poleward Chromosome Movement during Anaphase A

The tip-coupled movement of kinetochores implies force production at the kinetochore-microtubule interface. If conventional motor activity is dispensable, at least in some organisms, then how is energy transduced to drive this motility? Microtubules are likely to serve as the motors.

It is an old concept that anaphase A could be driven directly by the disassembly of spindle fibers. Inoue's observations using polarization microscopy showed not only that the spindle was composed of birefringent fibers, but also that poleward chromosome movement could be induced by artificial dissolution of the birefringent material, using cold-treatment for example [140]. Enthusiasm for a fiber-driven mechanism might have temporarily waned after the discovery of motor proteins at kinetochores [140]. However, it apparently regained traction when improvements in the biochemical handling of tubulin enabled in vitro reconstitution of movement driven by microtubule disassembly [123,141], without ATP-powered motor activity [126] (reviewed in [140]). Further support has come from the discoveries that non-motor microtubule binders within the kinetochores are vital for kinetochore-spindle attachment in vivo, and that they can reconstitute tip-coupling in vitro.

Microtubules are protein polymers composed of thousands of αβ-tubulins packed together in longitudinal rows, called 'protofilaments', that associate laterally to form a miniature tube [142]. In the presence of GTP, microtubules spontaneously self-assemble and they switch stochastically between periods of steady growth and rapid shortening, a behavior called 'dynamic instability' [13,14]. Dynamic instability is powered by GTP hydrolysis within αβ-tubulin. Growth occurs by addition of GTP-containing tubulins onto filament tips. Assembly triggers hydrolysis and phosphate release, so the body of a microtubule is composed primarily of GDP-tubulin, with 'caps' of GTP-tubulin at growing ends [143]. GDP-tubulin is intrinsically curved, but within the microtubule it is held straight—and therefore mechanically strained—by the bonds it forms with its lattice neighbors [144]. GTP-tubulin might be intrinsically straighter than GDP-tubulin [145], although recent work challenges this notion [146]. In any case, it is clear that some energy from GTP hydrolysis is retained within the GDP lattice [147,148], partly in the form of curvature-strain [143], and that this stored energy makes the microtubule unstable without protective end-caps. Severing the GTP-cap at a growing end triggers immediate disassembly [149]. During disassembly, the protofilaments first curl outward from the filament tip, releasing their curvature-strain, and then they break apart [144]. The energy released during tip disassembly can potentially be utilized to drive anaphase A chromosome-to-pole movement.

16. Purified Kinetochores and Sub-Complexes Are Excellent Tip-Couplers

Direct evidence that energy can indeed be harnessed from disassembling microtubules comes from in vitro motility assays using purified kinetochore sub-complexes or isolated kinetochore particles to reconstitute disassembly-driven movement. With time-lapse fluorescence microscopy, oligomeric assemblies of recombinant fluorescent-tagged Ndc80c [150] or Dam1c [120,151,152] can be seen to track with shortening microtubule tips. Attaching the complexes to microbeads allows their manipulation with a laser trap and shows that they can track even when opposing force is applied continuously (Figure 7). The earliest laser trap assays of this kind used tip-couplers made from recombinant Dam1c or Ndc80c alone, which tracked against one or two piconewtons [119,150]. Coupling performance improved with the incorporation of additional microtubule-binding kinetochore elements [153,154], with the use of native kinetochore particles isolated from yeast [58], and with the use of flexible tethers for linking sub-complexes to beads [155]. Further improvements seem likely, especially as continued advancements in kinetochore biochemistry enable reconstitutions of ever more complete and stable kinetochore assemblies [156–158]. However, the performance achieved in laser trap tip-coupling assays already provides a reasonably good match to physiological conditions. Native budding yeast kinetochore particles remain attached to dynamic microtubule tips for 50 min on average while continuously supporting 5 pN of tension [58,135]. These statistics compare favorably with the total duration of budding yeast mitosis, which is typically <1 h, and with the estimated levels of kinetochore force in this organism, 4 to 6 pN [69]. Opposing forces up to 29 pN are needed to halt the disassembly-driven movement of tip-couplers made of recombinant Dam1c linked to beads via long tethers [155]. This stall force compares favorably with the estimated maximum poleward force produced per kinetochore-attached microtubule during anaphase A, which is between 12 and 50 pN (as discussed above) [68].

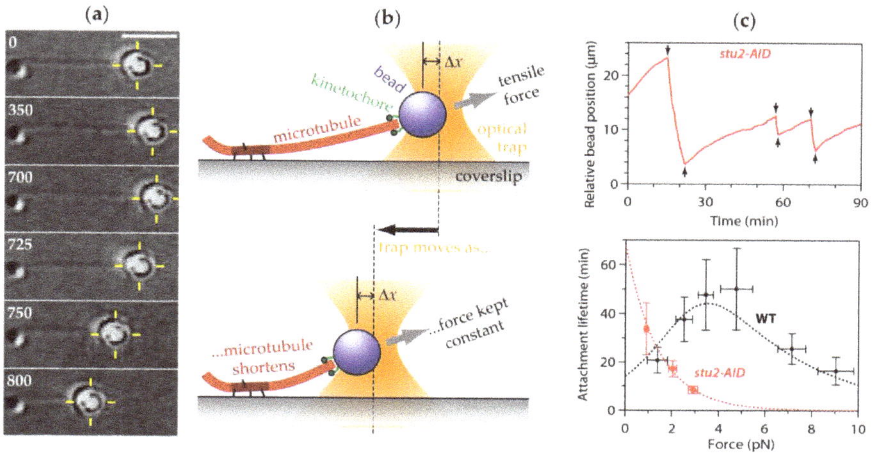

Figure 7. Laser trap assay for studying tip-coupling by purified kinetochore subcomplexes and native kinetochore particles. (**a**) Time-lapse images showing a bead decorated sparsely with native yeast kinetochore particles tracking with microtubule growth (0–700 s) and shortening (700–800 s). The laser trap (yellow crosshair) is moved automatically to keep a constant level of tension (here, ~1 pN) on the kinetochore as it moves with the microtubule tip. Scale bar, 4 μm. (**b**) Cartoon showing force clamp operation. The laser trap is servo-controlled to keep a fixed offset, Δx, between the trap and the bead, thereby maintaining a constant tensile force. (**c**) *Upper plot:* Record of position versus time for a native kinetochore isolated from yeast cells depleted of the TOG-family protein, Stu2. Arrows indicate switching of the microtubule tip from growth to shortening (↓, 'catastrophes') and from shortening back to growth (↑, 'rescues'). *Lower plot:* Mean attachment lifetime as a function of force for wild-type (WT, *black*) and Stu2-depleted (*stu2-AID*, *red*) kinetochore particles. Plots in (c) are adapted from [135], and are displayed with permission from Elsevier Publishing (http://www.sciencedirect.com/science/journal/00928674).

17. The Conformational Wave Model for Disassembly-Driven Movement

Two classes of models are proposed to explain disassembly-driven movement of kinetochores, conformational wave and biased diffusion (Figure 8). According to the conformational wave model, the kinetochore literally surfs on the wave of curling protofilaments that propagates down a microtubule as it disassembles. To drive movement, the protofilaments are proposed to pull directly on the kinetochore as they curl outward from a disassembling tip [141]. Evidence supporting this model is compelling but not definitive [159]. Oligomeric Dam1c rings seem to be ideal structures for harnessing protofilament curls [117,118,160,161], and Dam1c does indeed make a major contribution to the stability and strength of kinetochore-microtubule coupling in vitro [58,162], acting as a processivity factor to enhance Ndc80c-based coupling [154,163]. The contribution of Dam1c to tip-coupling is highest when it is flexibly tethered [155] and when free Dam1c is also present in solution [152,162], presumably because these conditions facilitate oligomerization of Dam1c into a microtubule-encircling ring. A partial Dam1 sub-complex that is specifically deficient in oligomerization forms tip attachments that are far less stable than those formed by the full, wild-type complex [162]. However, direct evidence that the enhancements in tip-coupling afforded by Dam1c oligomers depend on curling protofilaments is lacking. Complete microtubule-encircling rings are not strictly necessary for Dam1c-based tip-coupling [151,152]. In principle, Dam1c rings could function by biased diffusion (as discussed in [159,164]).

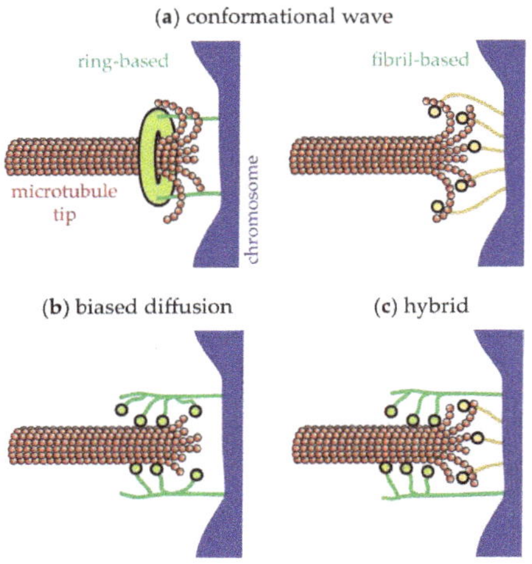

Figure 8. Models for tip-coupling without conventional motor activity. (**a**) Two versions of the conformational wave mechanism are shown, one (ring-based) in which elements of the kinetochore assemble into a microtubule encircling ring that is hooked by curling protofilaments, and another (fibril-based) where fibrillar kinetochore elements bind independently to the curling protofilaments. In either case, the curling action of the protofilaments exerts pulling force (directed leftward in the diagrams) on the chromosome. (**b**) In the biased diffusion mechanism, an array of kinetochore fibrils rapidly binds and unbinds the microtubule lattice at or near the tip. Thermal fluctuations of the chromosome that allow more fibrils to bind (leftward movements of the chromosome in the diagram) are favored by the energy of binding those elements. This biased thermal movement produces a thermodynamic pulling force. (**c**) A hybrid model is also shown, where force is produced by a combination of protofilament curling and biased thermal fluctuations. These diagrams are adapted from [159], and are displayed with permission from Elsevier Publishing (http://www.sciencedirect.com/science/journal/09628924).

Dam1c rings are not found outside fungi, but their absence does not necessarily rule out the conformational wave mechanism. Other molecules and structures could harness curling protofilaments. In humans and other eukaryotes, for example, the Ska complex might act as an attachment-stabilizer in a manner similar to Dam1c [121,122,165]. Ska complex does not appear to form oligomeric microtubule-encircling rings, but like Dam1c it can track with disassembling tips [166]. The Ska complex also dimerizes and might form lateral bridges between neighboring Ndc80 complexes [122]. Curling protofilaments might hook these lateral bridges. High-resolution electron tomograms show protofilaments curling outward from the tips of kinetochore-attached (and non-kinetochore) microtubules in mammalian (PtK1) spindles [138,167]. Sometimes fibrils can be discerned emanating from the kinetochores and connecting to the protofilament curls [138], suggesting the presence of a fibrillar protein with preferential affinity for curved protofilaments. Consistent with this possibility, the kinetochore protein Cenp-F contains an N-terminal microtubule-binding region that binds preferentially to ring- and curl-shaped tubulin oligomers (formed in the presence of dolastatin-10 and vinblastine, respectively) [168]. Beads decorated with N-terminal portions of Cenp-F can track with disassembling microtubule tips against forces of 3 pN [168], suggesting that its curl-binding activity could make a significant contribution to tip-coupling.

18. The Biased Diffusion Model for Disassembly-Driven Movement

Disassembly-driven kinetochore movement is also likely to depend partly on biased diffusion, a mechanism first proposed on purely theoretical grounds by Hill [169]. In this view the multiple microtubule-binding elements within a kinetochore form a diffusive attachment to the microtubule tip (Figure 8b). Thermal motions that bring more binding elements within reach of the tip are favored by the energy of binding those elements to the microtubule. Conversely, thermal motions away from the tip are disfavored because they reduce the number of binding elements that can reach the tip and thus they require some binding energy to be overcome. Hill showed theoretically that this bias is sufficient to allow persistent tracking with a disassembling microtubule tip, even against an external load.

Thermally driven diffusion along the microtubule lattice is a common property of many individual kinetochore proteins and subcomplexes. At the level of single molecules and small oligomers, Ndc80c [150], Dam1c [151], Ska complex [166], and Cenp-F [168] all bind and unbind quickly from microtubules and, while bound, diffuse rapidly over the lattice. When bound far from the microtubule tip and in the absence of external load their diffusive motion is random (the probability of movement in either direction is random) [150,151,166,168]. When they encounter a disassembling tip, a bias in their diffusion can be observed directly [150]. These behaviors fit strikingly well with the biased diffusion mechanism. Certain structural features of kinetochore subcomplexes also seem ideal for biased diffusion. Ndc80c [104], Dam1c [164], Ska complex [122], and Cenp-F [170] all appear to bind microtubules through flexible domains, which could allow some to bear load while others unbind and rebind in new locations, enabling a kinetochore to move or reorient on the microtubule without detaching. Diffusion along the microtubule lattice is negligibly slow for large assemblies of Dam1c [152] and for whole native kinetochore particles [58], but these observations do not rule out biased diffusion as a mechanism for tip-coupling by these assemblies. Large couplers that contain high numbers of microtubule-binders are not expected to diffuse detectably along the lattice, but they can nevertheless track robustly with a disassembling tip via pure biased diffusion [150,169]. Robust tip-tracking occurs in these cases, despite low mobility on the lattice, because the diffusional mobility increases as the tip begins to disassemble out from under the coupler. This in turn promotes lattice-directed movement and formation of new bonds, resulting in a steady state where the rate of new bond formation is balanced by the loss due to disassembly.

19. Movement Coupled to Tip Assembly

Reconstituted tip-couplers made from various combinations of kinetochore subcomplexes [119,154] and from native kinetochore particles [58,135,171,172] can also maintain persistent, tension-bearing attachments to *assembling* tips (e.g., see Figure 7). Their assembly-coupled movement in vitro is analogous to situations in vivo when kinetochores move anti-poleward in association with growing microtubule tips, such as during pre-anaphase chromosome oscillations, or during transient reversals of anaphase A chromosome-to-pole movement. The reconstituted couplers generally adopt a 'neutral' state, very much like that of kinetochores moving anti-poleward in vivo, requiring external tension to track with tip growth rather than being pushed autonomously by the growing tip. Affinity between the coupler and the microtubule creates a protein friction that resists movement along the filament [169]—an effect sometimes refered to as a 'slip clutch' [32]. Considering that curled protofilaments are much less prominent at assembling tips in vitro [144], and that the conformational wave mechanism is based on curled protofilaments, a purely conformational wave-based coupler would be expected to detach more quickly during assembly than during disassembly. But just the opposite is true: The reconstituted couplers usually detach far *less* quickly from assembling tips [58,135,171].

Based on electron tomographic studies of microtubule tips in cells, it has been suggested that protofilaments might curl out from *both* disassembling *and* assembling tips in vivo [138,139,173]. However, many of the kinetochore-attached plus ends examined in another electron tomographic study were apparently blunt, with straight protofilaments [167]. And in cells treated with nocodazole to promote tip disassembly, the same study found that kinetochore-attached microtubule ends were predominantly flared, with curling protofilaments [167], supporting the general view that curling protofilaments are restricted mainly to disassembling tips in vivo, as in vitro. Sheet-like extensions or blunt structures, not curls, have also been reported at assembling microtubule tips in mitotic and interphase cell extracts [174–176]. A purely conformational wave-based coupler should detach very quickly from these blunt microtubule ends. The biased diffusion mechanism has fewer structural constraints and could maintain a stable attachment independent of microtubule tip structure.

20. Mechanism of Poleward Flux Might Differ for Kinetochore-Attached Versus Non-Kinetochore Microtubules

Poleward microtubule flux contributes to anaphase A chromosome-to-pole motion in many organisms (Table 1). At a cellular level flux seems like a very close cousin to the movement of kinetochores relative to microtubule plus ends. Flux is coupled to disassembly of the pole-facing minus ends of spindle microtubules, just as kinetochore movement is coupled to plus end disassembly. Flux suggests force production at or near the depolymerizing minus ends, just as disassembly-coupled kinetochore movement suggests force production at plus ends. The speeds of both processes depend on some of the same types of microtubule regulatory molecules. Whether they share fundamentally similar mechanisms, however, is unclear.

The molecular and biophysical basis for poleward flux of non-kinetochore microtubules is reasonably well understood, but the same cannot be said for the flux of kinetochore-attached microtubules. Some non-kinetochore microtubules emanating from opposite spindle poles interdigitate within the central spindle to form antiparallel bundles—the so-called 'inter-polar microtubules' [3]. These bundles are held together by a collection of microtubule cross-linking proteins, including kinesin-5s, which are bipolar (tetrameric), processive, plus end-directed motors [3]. Individual purified kinesin-5 molecules can bind two antiparallel microtubules in vitro and simultaneously walk toward both plus ends, thereby driving outward protrusion of the minus ends [177]. Thus kinesin-5s appear to be perfectly suited for pushing inter-polar microtubules outward and driving their flux. But kinetochore-attached microtubules generally have parallel polarity [178], not antiparallel, and therefore their flux cannot be explained by a direct, antiparallel sliding action. Kinetochore-attached microtubules can associate laterally with non-kinetochore microtubules [20,60], and it has been suggested that perhaps the flux of kinetochore microtubules is driven indirectly, by the flux of their laterally associated neighbors (e.g., see [179]).

Alternatively, the mechanisms driving kinetochore-microtubule flux might differ from those driving non-kinetochore microtubule flux. Pharmacological inhibition of kinesin-5 dramatically slows flux in *Xenopus* extract spindles, in which a majority of microtubules are non-kinetochore-associated [180,181]. But in cultured mammalian (PtK1) cells, where a large proportion of microtubules are kinetochore-attached, kinesin-5 inhibition has only a minor effect on flux rates [179]. Furthermore, flux continues even when the spindles are monopolar, and therefore lacking antiparallel microtubules [179], indicating that neither kinesin-5 nor antiparallel microtubules are required for flux in these cells. Likewise, kinetochore-associated microtubule fibers that are mechanically detached and isolated from spindles in grasshopper spermatocytes flux in the apparent absence of antiparallel neighboring microtubules [182]. Thus, it seems that flux of kinetochore-attached microtubules can be driven by another mechanism, independent of the kinesin-5-dependent sliding of neighboring, antiparallel (inter-polar) microtubules.

21. Potential Biophysical Mechanisms for Kinetochore-Microtubule Flux

Flux generally depends on the activity of microtubule destabilizing enzymes that concentrate at spindle poles. Enzymes of the kinesin-13 family are ATP-powered depolymerases that catalyze the disassembly of microtubules by removal of tubulin subunits from their ends [183–185]. Kinesin-13s concentrate at poles in various spindle types, such as those in mitotic *Drosophila* cells [36], human cells [186,187], and frog cell extracts [188]. Depletion of the pole-localized *Drosophila* kinesin-13, KLP10A, specifically slows microtubule flux in this organism, and concomitantly reduces the speed of anaphase A chromosome-to-pole motion [36,45,46]. Similarly, the flux component of anaphase

A in mitotic human cells is slowed by co-depletion of a pole-localized and a centromere-associated kinesin-13, Kif2a and MCAK, respectively [30]. The AAA-family microtubule-severing enzymes, spastin and fidgetin, are also implicated in poleward microtubule flux in *Drosophila* [44,189] and human cells [52]. Their severing activity might be important for creating free microtubule minus ends (i.e., not capped by γ-tubulin rings) and thereby facilitating the catalysis of minus end disassembly by kinesin-13s. Collectively these observations indicate that microtubule destabilization activity at poles governs the rate of flux. But a governor is not necessarily a motor. The microtubule-destabilizers might or might not be directly involved in maintenance of load-bearing attachments between microtubules and spindle poles, or in the production of forces that drive flux. In some cells, microtubule depolymerizers also govern the speed of disassembly-coupled kinetochore movement [36,45,53], yet they are not usually considered to be the primary force-producers.

What then is the flux engine? One could envision a conformational wave- or biased diffusion-based tip-coupling that directly harnesses the energy released from minus end disassembly, analogous to the mechanisms discussed above for kinetochore motility. Whether spindle poles carry microtubule-binding elements with the properties necessary to support such tip-coupling is uncertain, but some evidence suggests so: The pole-localized *Drosophila* kinesin-13, KLP10A has been found to oligomerize into microtubule encircling rings [190,191], reminiscent of the Dam1c rings implicated in kinetochore tip-coupling. Other kinesin-13s, such as *Drosophila* KLP59C and human MCAK, which localize primarily near centromeres [36,192], can likewise form oligomeric rings around microtubules [190], and MCAK can function as a tip-coupler in vitro [193]. Together these observations suggest that kinesin-13s might function not only as depolymerizers but also as tip-couplers at spindle poles, and possibly at kinetochores as well.

Minus end-directed motors, particularly dynein, might also be involved in driving poleward microtubule flux. Dynein helps focus microtubules into poles in a variety of cell types. Pole focusing by motors is perhaps best understood in mitotic *Xenopus* egg extracts, where the minus end-directed movement of dynein oligomers can bring minus ends together to form polarized microtubule asters independently of centrosomal nucleation [194,195]. Dynein is also implicated in pole-focusing in *Drosophila* (S2 [196]) and mammalian cells (monkey kidney CV-1 [197,198]; rat-kangaroo PtK2 [60]; human RPE1 [61]). Pole-focusing by dynein probably requires oligomerization [199,200] via interaction with scaffolding proteins such as NuMA [195,198], and its importance for assembly and maintenance of bipolar spindles has been studied extensively.

Recent work implicates dynein in poleward movement specifically of kinetochore-attached microtubules. Bundles of kinetochore-attached microtubules that do not extend all the way to the spindle pole are sometimes seen in normal spindles [61,196] and can also be created artificially by laser micro-surgery [60,61]. When a microtubule fiber attached to one kinetochore of a bioriented pair is micro-surgically severed during metaphase, the cut fiber stub and its attached kinetochore initially recoil toward the sister kinetochore on the uncut side, as the chromatin linking the two sisters relaxes. This relaxation is expected due to the sudden loss of tension. But within a few tens of seconds the fiber stub is suddenly jerked poleward [60,61]. If a fiber is severed in early anaphase, the behavior is similar: There is no obvious initial recoil, presumably because sister chromatin cohesion is absent in anaphase, but the fiber stub and kinetochore are suddenly jerked poleward (Figure 9) [61], just as they are in metaphase. The poleward-facing ends of the fiber stubs lead these rapid poleward movements, apparently by associating laterally with nearby, uncut microtubules. Fluorescence imaging reveals rapid recruitment of NuMA and dynein to the newly created minus ends. Presumably this dynein drives poleward movement of the minus ends along neighboring pole-anchored microtubules (Figure 10). This activity was seen in both pre-anaphase and anaphase cells, and it shows that the spindle is capable of remarkable acts of self-healing. Whether a similar mechanism could drive the steady flux of kinetochore-attached microtubules during anaphase is uncertain. The idea seems attractive, although flux in *Drosophila* S2 cells has been shown to be independent of dynein [201]. Questions also remain about how depolymerase activity is engaged when the motors and minus ends reach the pole.

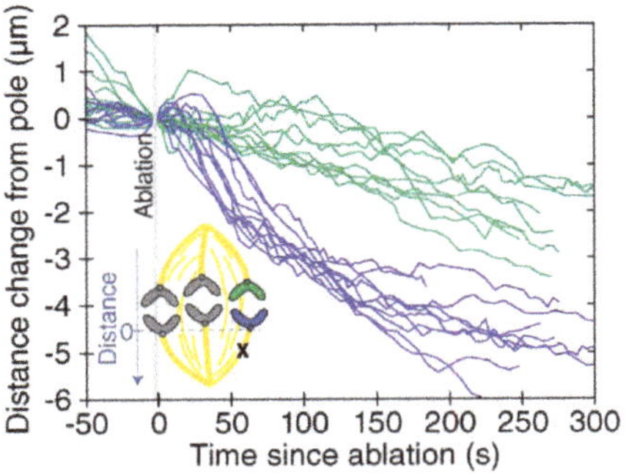

Figure 9. Change in distance from chromatids to poles before and after ablation of their kinetochore-associated microtubule fibers (k-fibers) during anaphase. Chromatids attached to ablated k-fibers (*blue traces*) are pulled toward poles faster than anaphase movement of their unmanipulated sisters (*green traces*) before resuming normal anaphase movement (at ~70 s). This graph is reprinted from [60], and is displayed under the terms of a Creative Commons License (Attribution-Noncommerical-Share Alike 3.0 Unported license, as described at http://creativecommons.org/licenses/by-nc-sa/3.0/).

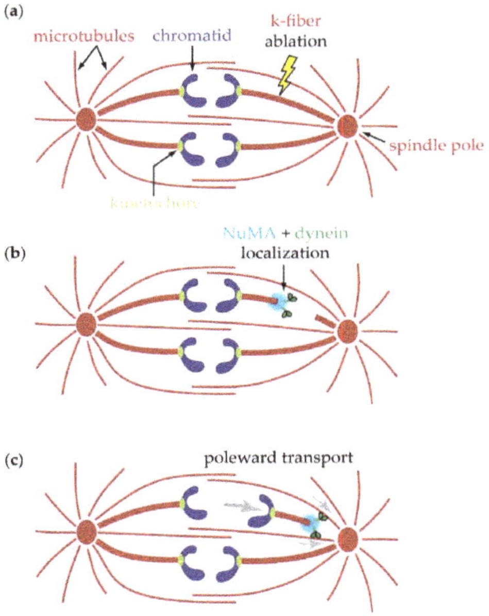

Figure 10. Spindle self-repair mechanism observed after micro-surgical ablation of kinetochore-associated microtubule fibers (k-fibers) in mammalian cells expressing fluorescent tubulin [60,61]. (**a**) Ablation of a k-fiber (*yellow lightning bolt*) during anaphase. (**b**) NuMA (*cyan*) and dynein/dynactin (*green*) rapidly localize to new microtubule minus ends on the k-fiber stub after ablation. (**c**) When the new minus end-localized dynein contacts neighboring microtubules, it walks processively along them, pulling the k-fiber stub as cargo and moving the attached chromosome. These diagrams are redrawn based on similar cartoons from [60], and are included here under the terms of a Creative Commons License (Attribution-Noncommerical-Share Alike 3.0 Unported license, as described at http://creativecommons.org/licenses/by-nc-sa/3.0/).

22. Loss of Tension by Itself Might Be Sufficient to Trigger Anaphase Chromosome-to-Pole Movement

Having surveyed possible mechanisms underlying chromosome-to-pole motion during anaphase A, it is interesting to return briefly, at the end of the chapter, to the beginning of anaphase. Anaphase begins abruptly. Cohesion between sister chromatids is proteolytically removed, essentially simultaneously from all sister pairs. Mechanical tension on all the kinetochores is suddenly lost. Is this sudden loss of tension, by itself, sufficient to trigger the poleward motion of kinetochores? Or is the inherent activity of the anaphase machinery modulated by regulatory cues at the metaphase-to-anaphase transition?

Ever since Östergren, a compelling hypothesis has been that the same mechanisms might account for both the alignment of chromosomes at metaphase and also their poleward movement at anaphase (e.g., see [140,202,203]). Micro-surgical studies support this view. When a kinetochore moving anti-poleward during metaphase is stopped by ablation of its sister (as described above), it stops only transiently, for ~20 s, and then begins to move poleward—i.e., with reversed, anaphase-like directionality [57]. This transition to poleward movement is apparently caused by the loss of tension when a chromatid is cut free from its sister. The anaphase-like poleward movement might be triggered in this case because micro-surgically severing the sisters closely mimics the normal trigger of anaphase, enzymatic removal of sister chromatid cohesion. Both operations cause a sudden loss of tension across the sisters.

In vitro reconstitutions of tip-coupling show directly that regulatory cues are not needed to trigger disassembly-driven kinetochore movement. Tension applied through Dam1c-based tip-couplers [204] or through native yeast kinetochore particles [58,135] promotes net growth of the attached microtubule. Tension speeds tip assembly, slows disassembly, inhibits switches from growth to shortening ('catastrophes'), and promotes the resumption of growth ('rescues') [58,135]. The effect of tension on catastrophe frequency is especially dramatic: At modest concentrations of free tubulin, the growth of a bare microtubule tip will typically persist for only a few minutes before a catastrophe occurs. Association of a relaxed kinetochore with the tip extends this uninterrupted growth time to ~8 min, but catastrophes are still relatively frequent. Applying a tension of 6 pN, however, can extend the uninterrupted growth time 13-fold, to over 100 min [58]. Thus, it is possible to experimentally induce a long period of assembly-coupled kinetochore movement by applying 6 pN of tension, and then to trigger disassembly-driven movement at will, simply by dropping the tension [205].

23. Phosphoregulatory Changes at the Metaphase-to-Anaphase Transition

While the simple loss of tension is sufficient to trigger an anaphase A-like switch in kinetochore directionality in vivo [57,140] and in vitro [58,135,205], it would be naïve to assume that the anaphase machinery is un-regulated during the true metaphase-to-anaphase transition in vivo. By now it is clear that multiple distinct mechanisms can underlie almost every aspect of mitosis. The same biochemical signaling cascade that brings about the sudden proteolytic destruction of sister cohesion also destroys cyclin B, thereby deactivating the cyclin-dependent kinase, CDK1, and causing a variety of global cellular changes associated with mitotic exit. Cyclin B and CDK1 are known to regulate microtubule dynamics (e.g., see [194,206]) and loss of cyclin B is proposed to stabilize inter-polar microtubules to promote anaphase B spindle elongation ([207]; as also discussed in the subsequent chapter on anaphase B [3]). If kinetochore-attached microtubules were similarly stabilized, the effect on anaphase A would be antagonistic, potentially slowing chromosome-to-pole movement by retarding disassembly at both plus and minus ends. However, evidence from budding yeast [206] and human tissue culture cells [208] indicates that the dephosphorylation associated with deactivation of CDK1 (or with activation of its antagonizing phosphatase, Cdc15) helps to promote, rather than antagonize anaphase A. In human cells, chemical inhibition of dephosphorylation converts the normally smooth chromosome-to-pole motion, with few reversals, into a much more oscillatory motion, with frequent reversals [208].

Another consequence of deactivating CDK1 is release of Aurora B kinase from centromeres (along with its co-members in the chromosomal passenger complex). Releasing Aurora B ensures that the sudden loss of kinetochore tension at anaphase onset does not activate the prometaphase error correction machinery, which would otherwise destabilize kinetochore-microtubule attachments. (Error correction is discussed in detail in the chapter by Grishchuk and Lampson [17].) This freeing of kinetochores from the influence of Aurora B should strengthen their attachments to spindle microtubules and, indeed, Nicklas noted in his early micromanipulation experiments that chromosomes became more difficult to detach as cells progressed from prometaphase into anaphase [209]. Freeing kinetochores from the influence of Aurora B might also affect the dynamics of kinetochore-attached microtubule plus ends: Aurora inhibitors stabilize kinetochore-attached microtubules in cells [210] and, conversely, phosphomimetic mutations at Aurora B target sites on Ndc80c and Dam1c destabilize kinetochore-attached plus ends in vitro [171,211]. Both observations implicate Aurora B in destabilization of kinetochore-attached plus ends. Thus, removal of Aurora B at anaphase onset should cause stabilization of the kinetochore-attached ends, which would be antagonistic toward anaphase A chromosome-to-pole movement. Perhaps the microtubule-stabilizing effects caused by loss of Aurora B are sufficiently counteracted by the destabilization due to loss of tension, or by other as-yet-unidentified regulatory events. Clearly more work is needed to understand how phosphoregulatory changes at anaphase onset regulate chromosome-to-pole motion.

24. Conclusions

Anaphase is the dramatic finale of mitosis when, after careful preparations are finished, the actual business of segregating duplicated chromosomes takes place in a beautifully orchestrated manner. Kinetochores are the main sites where forces are exerted on the chromosomes. The interfaces between kinetochores and microtubule plus ends are primary sites where forces are produced to drive anaphase A chromosome-to-pole movement. The microtubules themselves are likely to act as non-conventional motors, converting chemical energy from GTP hydrolysis into mechanical strain, storing this strain energy temporarily in their lattices, and then releasing it during disassembly. The released energy is harnessed in part by non-motor, microtubule-binding kinetochore elements, perhaps via surfing on waves of curling protofilaments. Meanwhile, in many cell types the kinetochore-attached microtubules are also transported steadily poleward, by mechanisms that are not yet well understood. This poleward flux supplements kinetochore tip-surfing. Chromosome-to-pole motion is likely triggered at the metaphase-to-anaphase transition in part by the simple loss of tension that occurs when cohesion between sister chromatids is suddenly lost, but additional phosphoregulatory influences are also important.

Acknowledgments: The author wishes to thank Luke Johnson for compiling the data and creating Table 1. The author is also grateful to Luke Johnson, Aida Llauró, Richard McIntosh, Juan Jesus Vicente, and two anonymous reviewers for their helpful comments and criticisms during the writing of this manuscript. The Asbury lab is currently supported by grants from the NIH (R01GM079373, P01GM105537) and the Packard Foundation (2006-30521).

Conflicts of Interest: The author declares no conflict of interest.

References

1. Mazia, D. Mitosis and the physiology of cell division. In *The Cell: Biochemistry, Physiology, Morphology, Vol III*; Brachet, J., Mirsky, A.E., Eds.; Academic Press: London, UK, 1961; pp. 77–412.
2. Strasburger, E. Die Controversen der indirecten Kerntheilung. *Arch. Mikrosk. Anat.* **1884**, *23*, 246–304. [CrossRef]
3. Scholey, J.M.; Civelekoglu-Scholey, G.; Brust-Mascher, I. Anaphase B. *Biology* **2016**, *5*, 51. [CrossRef] [PubMed]
4. Ris, H. The anaphase movement of chromosomes in the spermatocytes of the grasshopper. *Biol. Bull.* **1949**, *96*, 90–106. [CrossRef] [PubMed]

5. McIntosh, J.R.; Hays, T. A Brief History of Research on Mitotic Mechanisms. *Biology* **2016**, *5*, 55. [CrossRef] [PubMed]
6. Sharp, L.W. *Introduction to Cytology*; McGraw-Hill: New York, NY, USA; London, UK, 1934.
7. Rieder, C.L.; Davison, E.A.; Jensen, L.C.; Cassimeris, L.; Salmon, E.D. Oscillatory movements of monooriented chromosomes and their position relative to the spindle pole result from the ejection properties of the aster and half-spindle. *J. Cell Biol.* **1986**, *103*, 581–591. [CrossRef] [PubMed]
8. Straight, A.F.; Marshall, W.F.; Sedat, J.W.; Murray, A.W. Mitosis in living budding yeast: Anaphase A but no metaphase plate. *Science* **1997**, *277*, 574–578. [CrossRef] [PubMed]
9. Khodjakov, A.; Cole, R.W.; Bajer, A.S.; Rieder, C.L. The force for poleward chromosome motion in *Haemanthus* cells acts along the length of the chromosome during metaphase but only at the kinetochore during anaphase. *J. Cell Biol.* **1996**, *132*, 1093–1104. [CrossRef] [PubMed]
10. LaFountain, J.R., Jr.; Oldenbourg, R.; Cole, R.W.; Rieder, C.L. Microtubule flux mediates poleward motion of acentric chromosome fragments during meiosis in insect spermatocytes. *Mol. Biol. Cell* **2001**, *12*, 4054–4065. [CrossRef] [PubMed]
11. Rieder, C.L.; Salmon, E.D. Motile kinetochores and polar ejection forces dictate chromosome position on the vertebrate mitotic spindle. *J. Cell Biol.* **1994**, *124*, 223–233. [CrossRef] [PubMed]
12. Skibbens, R.V.; Skeen, V.P.; Salmon, E.D. Directional instability of kinetochore motility during chromosome congression and segregation in mitotic newt lung cells: A push-pull mechanism. *J. Cell Biol.* **1993**, *122*, 859–875. [CrossRef]
13. Mitchison, T.; Kirschner, M. Dynamic instability of microtubule growth. *Nature* **1984**, *312*, 237–242. [CrossRef] [PubMed]
14. Walker, R.A.; O'Brien, E.T.; Pryer, N.K.; Soboeiro, M.F.; Voter, W.A.; Erickson, H.P.; Salmon, E.D. Dynamic instability of individual microtubules analyzed by video light microscopy: Rate constants and transition frequencies. *J. Cell Biol.* **1988**, *107*, 1437–1448. [CrossRef] [PubMed]
15. Stumpff, J.; von Dassow, G.; Wagenbach, M.; Asbury, C.; Wordeman, L. The kinesin-8 motor Kif18A suppresses kinetochore movements to control mitotic chromosome alignment. *Dev. Cell* **2008**, *14*, 252–262. [CrossRef] [PubMed]
16. Cimini, D.; Moree, B.; Canman, J.C.; Salmon, E.D. Merotelic kinetochore orientation occurs frequently during early mitosis in mammalian tissue cells and error correction is achieved by two different mechanisms. *J. Cell Sci.* **2003**, *116*, 4213–4225. [CrossRef]
17. Grishchuk, E.L.; Lampson, M. Mechanisms to avoid and correct erroneous kinetochore-microtubule attachments. *Biology* **2017**, *6*, 1. [CrossRef]
18. Uretz, R.B.; Bloom, W.; Zirkle, R.E. Irradiation of parts of individual cells. II. Effects of an ultraviolet microbeam focused on parts of chromosomes. *Science* **1954**, *120*, 197–199. [CrossRef] [PubMed]
19. McNeill, P.A.; Berns, M.W. Chromosome behavior after laser microirradiation of a single kinetochore in mitotic PtK2 cells. *J. Cell Biol.* **1981**, *88*, 543–553. [CrossRef] [PubMed]
20. McDonald, K.L.; O'Toole, E.T.; Mastronarde, D.N.; McIntosh, J.R. Kinetochore microtubules in PTK cells. *J. Cell Biol.* **1992**, *118*, 369–383. [CrossRef]
21. Winey, M.; Mamay, C.L.; O'Toole, E.T.; Mastronarde, D.N.; Giddings, T.H., Jr.; McDonald, K.L.; McIntosh, J.R. Three-dimensional ultrastructural analysis of the Saccharomyces cerevisiae mitotic spindle. *J. Cell Biol.* **1995**, *129*, 1601–1615. [CrossRef] [PubMed]
22. Gorbsky, G.J.; Sammak, P.J.; Borisy, G.G. Chromosomes move poleward in anaphase along stationary microtubules that coordinately disassemble from their kinetochore ends. *J. Cell Biol.* **1987**, *104*, 9–18. [CrossRef] [PubMed]
23. Mitchison, T.; Evans, L.; Schulze, E.; Kirschner, M. Sites of microtubule assembly and disassembly in the mitotic spindle. *Cell* **1986**, *45*, 515–527. [CrossRef]
24. Shelden, E.; Wadsworth, P. Microinjection of biotin-tubulin into anaphase cells induces transient elongation of kinetochore microtubules and reversal of chromosome-to-pole motion. *J. Cell Biol.* **1992**, *116*, 1409–1420. [CrossRef] [PubMed]
25. Cimini, D.; Cameron, L.A.; Salmon, E.D. Anaphase spindle mechanics prevent mis-segregation of merotelically oriented chromosomes. *Curr. Biol.* **2004**, *14*, 2149–2155. [CrossRef] [PubMed]
26. Maddox, P.S.; Bloom, K.S.; Salmon, E.D. The polarity and dynamics of microtubule assembly in the budding yeast Saccharomyces cerevisiae. *Nat. Cell Biol.* **2000**, *2*, 36–41. [CrossRef] [PubMed]

27. Khodjakov, A.; La Terra, S.; Chang, F. Laser microsurgery in fission yeast; role of the mitotic spindle midzone in anaphase B. *Curr. Biol.* **2004**, *14*, 1330–1340. [CrossRef] [PubMed]
28. Sagolla, M.J.; Uzawa, S.; Cande, W.Z. Individual microtubule dynamics contribute to the function of mitotic and cytoplasmic arrays in fission yeast. *J. Cell Sci.* **2003**, *116*, 4891–4903. [CrossRef] [PubMed]
29. Mallavarapu, A.; Sawin, K.; Mitchison, T. A switch in microtubule dynamics at the onset of anaphase B in the mitotic spindle of Schizosaccharomyces pombe. *Curr. Biol.* **1999**, *9*, 1423–1426. [CrossRef]
30. Ganem, N.J.; Upton, K.; Compton, D.A. Efficient mitosis in human cells lacking poleward microtubule flux. *Curr. Biol.* **2005**, *15*, 1827–1832. [CrossRef] [PubMed]
31. LaFountain, J.R., Jr.; Cohan, C.S.; Siegel, A.J.; LaFountain, D.J. Direct visualization of microtubule flux during metaphase and anaphase in crane-fly spermatocytes. *Mol. Biol. Cell* **2004**, *15*, 5724–5732. [CrossRef] [PubMed]
32. Maddox, P.; Straight, A.; Coughlin, P.; Mitchison, T.J.; Salmon, E.D. Direct observation of microtubule dynamics at kinetochores in Xenopus extract spindles: Implications for spindle mechanics. *J. Cell Biol.* **2003**, *162*, 377–382. [CrossRef] [PubMed]
33. Mitchison, T.J.; Salmon, E.D. Poleward kinetochore fiber movement occurs during both metaphase and anaphase-A in newt lung cell mitosis. *J. Cell Biol.* **1992**, *119*, 569–582. [CrossRef] [PubMed]
34. Zhai, Y.; Kronebusch, P.J.; Borisy, G.G. Kinetochore microtubule dynamics and the metaphase-anaphase transition. *J. Cell Biol.* **1995**, *131*, 721–734. [CrossRef] [PubMed]
35. Maddox, P.; Desai, A.; Oegema, K.; Mitchison, T.J.; Salmon, E.D. Poleward microtubule flux is a major component of spindle dynamics and anaphase a in mitotic Drosophila embryos. *Curr. Biol.* **2002**, *12*, 1670–1674. [CrossRef]
36. Rogers, G.C.; Rogers, S.L.; Schwimmer, T.A.; Ems-McClung, S.C.; Walczak, C.E.; Vale, R.D.; Scholey, J.M.; Sharp, D.J. Two mitotic kinesins cooperate to drive sister chromatid separation during anaphase. *Nature* **2004**, *427*, 364–370. [CrossRef] [PubMed]
37. Dhonukshe, P.; Vischer, N.; Gadella, T.W., Jr. Contribution of microtubule growth polarity and flux to spindle assembly and functioning in plant cells. *J. Cell Sci.* **2006**, *119*, 3193–3205. [CrossRef] [PubMed]
38. Hamaguchi, Y.; Toriyama, M.; Sakai, H.; Hiramoto, Y. Redistribution of fluorescently labeled tubulin in the mitotic apparatus of sand dollar eggs and the effects of taxol. *Cell Struct. Funct.* **1987**, *12*, 43–52. [CrossRef] [PubMed]
39. Waters, J.C.; Mitchison, T.J.; Rieder, C.L.; Salmon, E.D. The kinetochore microtubule minus-end disassembly associated with poleward flux produces a force that can do work. *Mol. Biol. Cell* **1996**, *7*, 1547–1558. [CrossRef] [PubMed]
40. Desai, A.; Maddox, P.S.; Mitchison, T.J.; Salmon, E.D. Anaphase A chromosome movement and poleward spindle microtubule flux occur At similar rates in Xenopus extract spindles. *J. Cell Biol.* **1998**, *141*, 703–713. [CrossRef] [PubMed]
41. He, X.; Asthana, S.; Sorger, P.K. Transient sister chromatid separation and elastic deformation of chromosomes during mitosis in budding yeast. *Cell* **2000**, *101*, 763–775. [CrossRef]
42. Pearson, C.G.; Maddox, P.S.; Salmon, E.D.; Bloom, K. Budding yeast chromosome structure and dynamics during mitosis. *J. Cell Biol.* **2001**, *152*, 1255–1266. [CrossRef] [PubMed]
43. Brust-Mascher, I.; Scholey, J.M. Microtubule flux and sliding in mitotic spindles of Drosophila embryos. *Mol. Biol. Cell* **2002**, *13*, 3967–3975. [CrossRef] [PubMed]
44. Zhang, D.; Rogers, G.C.; Buster, D.W.; Sharp, D.J. Three microtubule severing enzymes contribute to the "Pacman-flux" machinery that moves chromosomes. *J. Cell Biol.* **2007**, *177*, 231–242. [CrossRef] [PubMed]
45. Rath, U.; Rogers, G.C.; Tan, D.; Gomez-Ferreria, M.A.; Buster, D.W.; Sosa, H.J.; Sharp, D.J. The Drosophila kinesin-13, KLP59D, impacts Pacman- and Flux-based chromosome movement. *Mol. Biol. Cell* **2009**, *20*, 4696–4705. [CrossRef] [PubMed]
46. Matos, I.; Pereira, A.J.; Lince-Faria, M.; Cameron, L.A.; Salmon, E.D.; Maiato, H. Synchronizing chromosome segregation by flux-dependent force equalization at kinetochores. *J. Cell Biol.* **2009**, *186*, 11–26. [CrossRef] [PubMed]
47. Savoian, M.S. Using Photobleaching to Measure Spindle Microtubule Dynamics in Primary Cultures of Dividing Drosophila Meiotic Spermatocytes. *J. Biomol. Tech.* **2015**, *26*, 66–73. [CrossRef] [PubMed]
48. LaFountain, J.R., Jr.; Cohan, C.S.; Oldenbourg, R. Pac-man motility of kinetochores unleashed by laser microsurgery. *Mol. Biol. Cell* **2012**, *23*, 3133–3142. [CrossRef]

49. LaFountain, J.R., Jr.; Cohan, C.S.; Oldenbourg, R. Functional states of kinetochores revealed by laser microsurgery and fluorescent speckle microscopy. *Mol. Biol. Cell* **2011**, *22*, 4801–4808. [CrossRef] [PubMed]
50. Maffini, S.; Maia, A.R.; Manning, A.L.; Maliga, Z.; Pereira, A.L.; Junqueira, M.; Shevchenko, A.; Hyman, A.; Yates, J.R., 3rd; Galjart, N.; et al. Motor-independent targeting of CLASPs to kinetochores by CENP-E promotes microtubule turnover and poleward flux. *Curr. Biol.* **2009**, *19*, 1566–1572. [CrossRef] [PubMed]
51. Wandke, C.; Barisic, M.; Sigl, R.; Rauch, V.; Wolf, F.; Amaro, A.C.; Tan, C.H.; Pereira, A.J.; Kutay, U.; Maiato, H.; et al. Human chromokinesins promote chromosome congression and spindle microtubule dynamics during mitosis. *J. Cell Biol.* **2012**, *198*, 847–863. [CrossRef] [PubMed]
52. Mukherjee, S.; Diaz Valencia, J.D.; Stewman, S.; Metz, J.; Monnier, S.; Rath, U.; Asenjo, A.B.; Charafeddine, R.A.; Sosa, H.J.; Ross, J.L.; et al. Human Fidgetin is a microtubule severing the enzyme and minus-end depolymerase that regulates mitosis. *Cell Cycle* **2012**, *11*, 2359–2366. [CrossRef] [PubMed]
53. Wordeman, L.; Wagenbach, M.; von Dassow, G. MCAK facilitates chromosome movement by promoting kinetochore microtubule turnover. *J. Cell Biol.* **2007**, *179*, 869–879. [CrossRef] [PubMed]
54. Dumont, J.; Oegema, K.; Desai, A. A kinetochore-independent mechanism drives anaphase chromosome separation during acentrosomal meiosis. *Nat. Cell Biol.* **2010**, *12*, 894–901. [CrossRef]
55. Doudna, J.A.; Charpentier, E. The new frontier of genome engineering with CRISPR-Cas9. *Science* **2014**, *346*, 1258096. [CrossRef] [PubMed]
56. Skibbens, R.V.; Rieder, C.L.; Salmon, E.D. Kinetochore motility after severing between sister centromeres using laser microsurgery: Evidence that kinetochore directional instability and position is regulated by tension. *J. Cell Sci.* **1995**, *108 Pt 7*, 2537–2548. [PubMed]
57. Khodjakov, A.; Rieder, C.L. Kinetochores moving away from their associated pole do not exert a significant pushing force on the chromosome. *J. Cell Biol.* **1996**, *135*, 315–327. [CrossRef] [PubMed]
58. Akiyoshi, B.; Sarangapani, K.K.; Powers, A.F.; Nelson, C.R.; Reichow, S.L.; Arellano-Santoyo, H.; Gonen, T.; Ranish, J.A.; Asbury, C.L.; Biggins, S. Tension directly stabilizes reconstituted kinetochore-microtubule attachments. *Nature* **2010**, *468*, 576–579. [CrossRef] [PubMed]
59. Nicklas, R.B. Measurements of the force produced by the mitotic spindle in anaphase. *J. Cell Biol.* **1983**, *97*, 542–548. [CrossRef] [PubMed]
60. Elting, M.W.; Hueschen, C.L.; Udy, D.B.; Dumont, S. Force on spindle microtubule minus ends moves chromosomes. *J. Cell Biol.* **2014**, *206*, 245–256. [CrossRef] [PubMed]
61. Sikirzhytski, V.; Magidson, V.; Steinman, J.B.; He, J.; Le Berre, M.; Tikhonenko, I.; Ault, J.G.; McEwen, B.F.; Chen, J.K.; Sui, H.; et al. Direct kinetochore-spindle pole connections are not required for chromosome segregation. *J. Cell Biol.* **2014**, *206*, 231–243. [CrossRef] [PubMed]
62. Nicklas, R.B. Chromosome Velocity during Mitosis as a Function of Chromosome Size and Position. *J. Cell Biol.* **1965**, *25*, 119–135. [CrossRef]
63. Marshall, W.F.; Marko, J.F.; Agard, D.A.; Sedat, J.W. Chromosome elasticity and mitotic polar ejection force measured in living Drosophila embryos by four-dimensional microscopy-based motion analysis. *Curr. Biol.* **2001**, *11*, 569–578. [CrossRef]
64. Sarangapani, K.K.; Asbury, C.L. Catch and release: How do kinetochores hook the right microtubules during mitosis? *Trends Genet. TIG* **2014**, *30*, 150–159. [CrossRef] [PubMed]
65. Nicklas, R.B. How cells get the right chromosomes. *Science* **1997**, *275*, 632–637. [CrossRef]
66. Nicklas, R.B.; Koch, C.A. Chromosome micromanipulation. 3. Spindle fiber tension and the reorientation of mal-oriented chromosomes. *J. Cell Biol.* **1969**, *43*, 40–50. [CrossRef]
67. Joglekar, A.P. A Cell Biological Perspective on Past, Present and Future Investigations of the Spindle Assembly Checkpoint. *Biology* **2016**, *5*, 44. [CrossRef] [PubMed]
68. Nicklas, R.B. The forces that move chromosomes in mitosis. *Annu. Rev. Biophys. Biophys. Chem.* **1988**, *17*, 431–449. [CrossRef] [PubMed]
69. Chacon, J.M.; Mukherjee, S.; Schuster, B.M.; Clarke, D.J.; Gardner, M.K. Pericentromere tension is self-regulated by spindle structure in metaphase. *J. Cell Biol.* **2014**, *205*, 313–324. [CrossRef] [PubMed]
70. Ye, A.A.; Cane, S.; Maresca, T.J. Chromosome biorientation produces hundreds of piconewtons at a metazoan kinetochore. *Nat. Commun.* **2016**, *7*, 13221. [CrossRef] [PubMed]
71. Greenleaf, W.J.; Woodside, M.T.; Block, S.M. High-resolution, single-molecule measurements of biomolecular motion. *Annu. Rev. Biophys. Biomol. Struct.* **2007**, *36*, 171–190. [CrossRef] [PubMed]

72. Leidel, C.; Longoria, R.A.; Gutierrez, F.M.; Shubeita, G.T. Measuring molecular motor forces in vivo: Implications for tug-of-war models of bidirectional transport. *Biophys. J.* **2012**, *103*, 492–500. [CrossRef] [PubMed]
73. Sims, P.A.; Xie, X.S. Probing dynein and kinesin stepping with mechanical manipulation in a living cell. *Chemphyschem* **2009**, *10*, 1511–1516. [CrossRef]
74. Hendricks, A.G.; Holzbaur, E.L.; Goldman, Y.E. Force measurements on cargoes in living cells reveal collective dynamics of microtubule motors. *Proc. Natl. Acad. Sci. USA* **2012**, *109*, 18447–18452. [CrossRef] [PubMed]
75. Shubeita, G.T.; Tran, S.L.; Xu, J.; Vershinin, M.; Cermelli, S.; Cotton, S.L.; Welte, M.A.; Gross, S.P. Consequences of motor copy number on the intracellular transport of kinesin-1-driven lipid droplets. *Cell* **2008**, *135*, 1098–1107. [CrossRef] [PubMed]
76. Jun, Y.; Tripathy, S.K.; Narayanareddy, B.R.; Mattson-Hoss, M.K.; Gross, S.P. Calibration of optical tweezers for in vivo force measurements: How do different approaches compare? *Biophys. J.* **2014**, *107*, 1474–1484. [CrossRef] [PubMed]
77. Neuman, K.C.; Chadd, E.H.; Liou, G.F.; Bergman, K.; Block, S.M. Characterization of photodamage to Escherichia coli in optical traps. *Biophys. J.* **1999**, *77*, 2856–2863. [CrossRef]
78. Gross, S.P. Application of optical traps in vivo. *Methods Enzymol.* **2003**, *361*, 162–174. [PubMed]
79. Ferraro-Gideon, J.; Sheykhani, R.; Zhu, Q.; Duquette, M.L.; Berns, M.W.; Forer, A. Measurements of forces produced by the mitotic spindle using optical tweezers. *Mol. Biol. Cell* **2013**, *24*, 1375–1386. [CrossRef] [PubMed]
80. Steuer, E.R.; Wordeman, L.; Schroer, T.A.; Sheetz, M.P. Localization of cytoplasmic dynein to mitotic spindles and kinetochores. *Nature* **1990**, *345*, 266–268. [CrossRef] [PubMed]
81. Yen, T.J.; Li, G.; Schaar, B.T.; Szilak, I.; Cleveland, D.W. CENP-E is a putative kinetochore motor that accumulates just before mitosis. *Nature* **1992**, *359*, 536–539. [CrossRef]
82. Pfarr, C.M.; Coue, M.; Grissom, P.M.; Hays, T.S.; Porter, M.E.; McIntosh, J.R. Cytoplasmic dynein is localized to kinetochores during mitosis. *Nature* **1990**, *345*, 263–265. [CrossRef] [PubMed]
83. Earnshaw, W.C.; Migeon, B.R. Three related centromere proteins are absent from the inactive centromere of a stable isodicentric chromosome. *Chromosoma* **1985**, *92*, 290–296. [CrossRef] [PubMed]
84. Savoian, M.S.; Goldberg, M.L.; Rieder, C.L. The rate of poleward chromosome motion is attenuated in Drosophila zw10 and rod mutants. *Nat. Cell Biol.* **2000**, *2*, 948–952. [PubMed]
85. Sharp, D.J.; Rogers, G.C.; Scholey, J.M. Cytoplasmic dynein is required for poleward chromosome movement during mitosis in Drosophila embryos. *Nat. Cell Biol.* **2000**, *2*, 922–930. [PubMed]
86. Howell, B.J.; McEwen, B.F.; Canman, J.C.; Hoffman, D.B.; Farrar, E.M.; Rieder, C.L.; Salmon, E.D. Cytoplasmic dynein/dynactin drives kinetochore protein transport to the spindle poles and has a role in mitotic spindle checkpoint inactivation. *J. Cell Biol.* **2001**, *155*, 1159–1172. [CrossRef] [PubMed]
87. Kapoor, T.M. Spindle Assembly. *Biology* **2017**, *6*, 8. [CrossRef] [PubMed]
88. Goshima, G.; Yamada, M. Mitotic spindle assembly in land plants: Molecules and mechanisms. *Biology* **2017**, *6*, 6. [CrossRef]
89. Grishchuk, E.L.; McIntosh, J.R. Microtubule depolymerization can drive poleward chromosome motion in fission yeast. *Embo J.* **2006**, *25*, 4888–4896. [CrossRef] [PubMed]
90. Tanaka, K.; Kitamura, E.; Kitamura, Y.; Tanaka, T.U. Molecular mechanisms of microtubule-dependent kinetochore transport toward spindle poles. *J. Cell Biol.* **2007**, *178*, 269–281. [CrossRef] [PubMed]
91. Tytell, J.D.; Sorger, P.K. Analysis of kinesin motor function at budding yeast kinetochores. *J. Cell Biol.* **2006**, *172*, 861–874. [CrossRef] [PubMed]
92. Weaver, B.A.; Bonday, Z.Q.; Putkey, F.R.; Kops, G.J.; Silk, A.D.; Cleveland, D.W. Centromere-associated protein-E is essential for the mammalian mitotic checkpoint to prevent aneuploidy due to single chromosome loss. *J. Cell Biol.* **2003**, *162*, 551–563. [CrossRef] [PubMed]
93. Kapoor, T.M.; Lampson, M.A.; Hergert, P.; Cameron, L.; Cimini, D.; Salmon, E.D.; McEwen, B.F.; Khodjakov, A. Chromosomes can congress to the metaphase plate before biorientation. *Science* **2006**, *311*, 388–391. [CrossRef] [PubMed]
94. Yang, Z.; Tulu, U.S.; Wadsworth, P.; Rieder, C.L. Kinetochore dynein is required for chromosome motion and congression independent of the spindle checkpoint. *Curr. Biol.* **2007**, *17*, 973–980. [CrossRef] [PubMed]

95. Musacchio, A.; Desai, A. Kinetochore assembly, structure, and function. *Biology* **2017**, *6*, 5. [CrossRef] [PubMed]
96. Wei, R.R.; Al-Bassam, J.; Harrison, S.C. The Ndc80/HEC1 complex is a contact point for kinetochore-microtubule attachment. *Nat. Struct. Mol. Biol.* **2007**, *14*, 54–59. [CrossRef] [PubMed]
97. Wei, R.R.; Schnell, J.R.; Larsen, N.A.; Sorger, P.K.; Chou, J.J.; Harrison, S.C. Structure of a central component of the yeast kinetochore: The Spc24p/Spc25p globular domain. *Structure* **2006**, *14*, 1003–1009. [CrossRef] [PubMed]
98. Wei, R.R.; Sorger, P.K.; Harrison, S.C. Molecular organization of the Ndc80 complex, an essential kinetochore component. *Proc. Natl. Acad. Sci. USA* **2005**, *102*, 5363–5367. [CrossRef]
99. Ciferri, C.; Pasqualato, S.; Screpanti, E.; Varetti, G.; Santaguida, S.; Dos Reis, G.; Maiolica, A.; Polka, J.; De Luca, J.G.; De Wulf, P.; et al. Implications for kinetochore-microtubule attachment from the structure of an engineered Ndc80 complex. *Cell* **2008**, *133*, 427–439. [CrossRef] [PubMed]
100. Alushin, G.M.; Ramey, V.H.; Pasqualato, S.; Ball, D.A.; Grigorieff, N.; Musacchio, A.; Nogales, E. The Ndc80 kinetochore complex forms oligomeric arrays along microtubules. *Nature* **2010**, *467*, 805–810. [CrossRef] [PubMed]
101. DeLuca, J.G.; Dong, Y.; Hergert, P.; Strauss, J.; Hickey, J.M.; Salmon, E.D.; McEwen, B.F. Hec1 and nuf2 are core components of the kinetochore outer plate essential for organizing microtubule attachment sites. *Mol. Biol. Cell* **2005**, *16*, 519–531. [CrossRef] [PubMed]
102. He, X.; Rines, D.R.; Espelin, C.W.; Sorger, P.K. Molecular analysis of kinetochore-microtubule attachment in budding yeast. *Cell* **2001**, *106*, 195–206. [CrossRef]
103. McCleland, M.L.; Gardner, R.D.; Kallio, M.J.; Daum, J.R.; Gorbsky, G.J.; Burke, D.J.; Stukenberg, P.T. The highly conserved Ndc80 complex is required for kinetochore assembly, chromosome congression, and spindle checkpoint activity. *Genes Dev.* **2003**, *17*, 101–114. [CrossRef] [PubMed]
104. Wang, H.W.; Long, S.; Ciferri, C.; Westermann, S.; Drubin, D.; Barnes, G.; Nogales, E. Architecture and flexibility of the yeast Ndc80 kinetochore complex. *J. Mol. Biol.* **2008**, *383*, 894–903. [CrossRef] [PubMed]
105. Joglekar, A.P.; Bloom, K.; Salmon, E.D. In vivo protein architecture of the eukaryotic kinetochore with nanometer scale accuracy. *Curr. Biol.* **2009**, *19*, 694–699. [CrossRef] [PubMed]
106. Tien, J.F.; Umbreit, N.T.; Zelter, A.; Riffle, M.; Hoopmann, M.R.; Johnson, R.S.; Fonslow, B.R.; Yates, J.R., 3rd; MacCoss, M.J.; Moritz, R.L.; et al. Kinetochore Biorientation in Saccharomyces cerevisiae Requires a Tightly Folded Conformation of the Ndc80 Complex. *Genetics* **2014**, *198*, 1483–1493. [CrossRef] [PubMed]
107. Joglekar, A.P.; Bouck, D.C.; Molk, J.N.; Bloom, K.S.; Salmon, E.D. Molecular architecture of a kinetochore-microtubule attachment site. *Nat. Cell Biol.* **2006**, *8*, 581–585. [CrossRef] [PubMed]
108. Lawrimore, J.; Bloom, K.S.; Salmon, E.D. Point centromeres contain more than a single centromere-specific Cse4 (CENP-A) nucleosome. *J. Cell Biol.* **2011**, *195*, 573–582. [CrossRef] [PubMed]
109. Emanuele, M.J.; McCleland, M.L.; Satinover, D.L.; Stukenberg, P.T. Measuring the stoichiometry and physical interactions between components elucidates the architecture of the vertebrate kinetochore. *Mol. Biol. Cell* **2005**, *16*, 4882–4892. [CrossRef] [PubMed]
110. Joglekar, A.P.; Bouck, D.; Finley, K.; Liu, X.; Wan, Y.; Berman, J.; He, X.; Salmon, E.D.; Bloom, K.S. Molecular architecture of the kinetochore-microtubule attachment site is conserved between point and regional centromeres. *J. Cell Biol.* **2008**, *181*, 587–594. [CrossRef] [PubMed]
111. Suzuki, A.; Badger, B.L.; Salmon, E.D. A quantitative description of Ndc80 complex linkage to human kinetochores. *Nat. Commun.* **2015**, *6*, 8161. [CrossRef] [PubMed]
112. Zinkowski, R.P.; Meyne, J.; Brinkley, B.R. The centromere-kinetochore complex: A repeat subunit model. *J. Cell Biol.* **1991**, *113*, 1091–1110. [CrossRef] [PubMed]
113. Cheeseman, I.M.; Brew, C.; Wolyniak, M.; Desai, A.; Anderson, S.; Muster, N.; Yates, J.R.; Huffaker, T.C.; Drubin, D.G.; Barnes, G. Implication of a novel multiprotein Dam1p complex in outer kinetochore function. *J. Cell Biol.* **2001**, *155*, 1137–1145. [CrossRef] [PubMed]
114. Cheeseman, I.M.; Enquist-Newman, M.; Muller-Reichert, T.; Drubin, D.G.; Barnes, G. Mitotic spindle integrity and kinetochore function linked by the Duo1p/Dam1p complex. *J. Cell Biol.* **2001**, *152*, 197–212. [CrossRef] [PubMed]
115. Hofmann, C.; Cheeseman, I.M.; Goode, B.L.; McDonald, K.L.; Barnes, G.; Drubin, D.G. Saccharomyces cerevisiae Duo1p and Dam1p, novel proteins involved in mitotic spindle function. *J. Cell Biol.* **1998**, *143*, 1029–1040. [CrossRef] [PubMed]

116. Janke, C.; Ortiz, J.; Tanaka, T.U.; Lechner, J.; Schiebel, E. Four new subunits of the Dam1-Duo1 complex reveal novel functions in sister kinetochore biorientation. *Embo J.* **2002**, *21*, 181–193. [CrossRef] [PubMed]
117. Miranda, J.J.; De Wulf, P.; Sorger, P.K.; Harrison, S.C. The yeast DASH complex forms closed rings on microtubules. *Nat. Struct. Mol. Biol.* **2005**, *12*, 138–143. [CrossRef] [PubMed]
118. Westermann, S.; Avila-Sakar, A.; Wang, H.W.; Niederstrasser, H.; Wong, J.; Drubin, D.G.; Nogales, E.; Barnes, G. Formation of a dynamic kinetochore- microtubule interface through assembly of the Dam1 ring complex. *Mol. Cell* **2005**, *17*, 277–290. [CrossRef] [PubMed]
119. Asbury, C.L.; Gestaut, D.R.; Powers, A.F.; Franck, A.D.; Davis, T.N. The Dam1 kinetochore complex harnesses microtubule dynamics to produce force and movement. *Proc. Natl. Acad. Sci. USA* **2006**, *103*, 9873–9878. [CrossRef] [PubMed]
120. Westermann, S.; Wang, H.W.; Avila-Sakar, A.; Drubin, D.G.; Nogales, E.; Barnes, G. The Dam1 kinetochore ring complex moves processively on depolymerizing microtubule ends. *Nature* **2006**, *440*, 565–569. [CrossRef] [PubMed]
121. Welburn, J.P.; Grishchuk, E.L.; Backer, C.B.; Wilson-Kubalek, E.M.; Yates, J.R., 3rd; Cheeseman, I.M. The human kinetochore Ska1 complex facilitates microtubule depolymerization-coupled motility. *Dev. Cell* **2009**, *16*, 374–385. [CrossRef] [PubMed]
122. Jeyaprakash, A.A.; Santamaria, A.; Jayachandran, U.; Chan, Y.W.; Benda, C.; Nigg, E.A.; Conti, E. Structural and functional organization of the Ska complex, a key component of the kinetochore-microtubule interface. *Mol. Cell* **2012**, *46*, 274–286. [CrossRef] [PubMed]
123. Coue, M.; Lombillo, V.A.; McIntosh, J.R. Microtubule depolymerization promotes particle and chromosome movement in vitro. *J. Cell Biol.* **1991**, *112*, 1165–1175. [CrossRef] [PubMed]
124. Lombillo, V.A.; Coue, M.; McIntosh, J.R. In vitro motility assays using microtubules tethered to Tetrahymena pellicles. *Methods Cell Biol.* **1993**, *39*, 149–165. [PubMed]
125. Lombillo, V.A.; Nislow, C.; Yen, T.J.; Gelfand, V.I.; McIntosh, J.R. Antibodies to the kinesin motor domain and CENP-E inhibit microtubule depolymerization-dependent motion of chromosomes in vitro. *J. Cell Biol.* **1995**, *128*, 107–115. [CrossRef] [PubMed]
126. Lombillo, V.A.; Stewart, R.J.; McIntosh, J.R. Minus-end-directed motion of kinesin-coated microspheres driven by microtubule depolymerization. *Nature* **1995**, *373*, 161–164. [CrossRef] [PubMed]
127. Gudimchuk, N.; Vitre, B.; Kim, Y.; Kiyatkin, A.; Cleveland, D.W.; Ataullakhanov, F.I.; Grishchuk, E.L. Kinetochore kinesin CENP-E is a processive bi-directional tracker of dynamic microtubule tips. *Nat. Cell Biol.* **2013**, *15*, 1079–1088. [CrossRef] [PubMed]
128. Gard, D.L.; Kirschner, M.W. A microtubule-associated protein from Xenopus eggs that specifically promotes assembly at the plus-end. *J. Cell Biol.* **1987**, *105*, 2203–2215. [CrossRef] [PubMed]
129. Hsu, K.S.; Toda, T. Ndc80 internal loop interacts with Dis1/TOG to ensure proper kinetochore-spindle attachment in fission yeast. *Curr. Biol.* **2011**, *21*, 214–220. [CrossRef] [PubMed]
130. Kalantzaki, M.; Kitamura, E.; Zhang, T.; Mino, A.; Novak, B.; Tanaka, T.U. Kinetochore-microtubule error correction is driven by differentially regulated interaction modes. *Nat. Cell Biol.* **2015**, *17*, 421–433. [CrossRef] [PubMed]
131. Ohkura, H.; Adachi, Y.; Kinoshita, N.; Niwa, O.; Toda, T.; Yanagida, M. Cold-sensitive and caffeine-supersensitive mutants of the Schizosaccharomyces pombe dis genes implicated in sister chromatid separation during mitosis. *EMBO J.* **1988**, *7*, 1465–1473. [PubMed]
132. Tanaka, K.; Mukae, N.; Dewar, H.; van Breugel, M.; James, E.K.; Prescott, A.R.; Antony, C.; Tanaka, T.U. Molecular mechanisms of kinetochore capture by spindle microtubules. *Nature* **2005**, *434*, 987–994. [CrossRef] [PubMed]
133. Tang, N.H.; Takada, H.; Hsu, K.S.; Toda, T. The internal loop of fission yeast Ndc80 binds Alp7/TACC-Alp14/TOG and ensures proper chromosome attachment. *Mol. Biol. Cell* **2013**, *24*, 1122–1133. [CrossRef] [PubMed]
134. Wang, P.J.; Huffaker, T.C. Stu2p: A microtubule-binding protein that is an essential component of the yeast spindle pole body. *J. Cell Biol.* **1997**, *139*, 1271–1280. [CrossRef] [PubMed]
135. Miller, M.P.; Asbury, C.L.; Biggins, S. A TOG Protein Confers Tension Sensitivity to Kinetochore-Microtubule Attachments. *Cell* **2016**, *165*, 1428–1439. [CrossRef] [PubMed]

136. Trushko, A.; Schaffer, E.; Howard, J. The growth speed of microtubules with XMAP215-coated beads coupled to their ends is increased by tensile force. *Proc. Natl. Acad. Sci. USA* **2013**, *110*, 14670–14675. [CrossRef] [PubMed]
137. Gonen, S.; Akiyoshi, B.; Iadanza, M.G.; Shi, D.; Duggan, N.; Biggins, S.; Gonen, T. The structure of purified kinetochores reveals multiple microtubule-attachment sites. *Nat. Struct. Mol. Biol.* **2012**, *19*, 925–929. [CrossRef] [PubMed]
138. McIntosh, J.R.; Grishchuk, E.L.; Morphew, M.K.; Efremov, A.K.; Zhudenkov, K.; Volkov, V.A.; Cheeseman, I.M.; Desai, A.; Mastronarde, D.N.; Ataullakhanov, F.I. Fibrils connect microtubule tips with kinetochores: A mechanism to couple tubulin dynamics to chromosome motion. *Cell* **2008**, *135*, 322–333. [CrossRef] [PubMed]
139. McIntosh, J.R.; O'Toole, E.; Zhudenkov, K.; Morphew, M.; Schwartz, C.; Ataullakhanov, F.I.; Grishchuk, E.L. Conserved and divergent features of kinetochores and spindle microtubule ends from five species. *J. Cell Biol.* **2013**, *200*, 459–474. [CrossRef] [PubMed]
140. Inoue, S.; Salmon, E.D. Force generation by microtubule assembly/disassembly in mitosis and related movements. *Mol. Biol. Cell* **1995**, *6*, 1619–1640. [CrossRef] [PubMed]
141. Koshland, D.E.; Mitchison, T.J.; Kirschner, M.W. Polewards chromosome movement driven by microtubule depolymerization in vitro. *Nature* **1988**, *331*, 499–504. [CrossRef] [PubMed]
142. Amos, L.; Klug, A. Arrangement of subunits in flagellar microtubules. *J. Cell Sci.* **1974**, *14*, 523–549. [PubMed]
143. Desai, A.; Mitchison, T.J. Microtubule polymerization dynamics. *Annu. Rev. Cell Dev. Biol.* **1997**, *13*, 83–117. [CrossRef] [PubMed]
144. Mandelkow, E.M.; Mandelkow, E.; Milligan, R.A. Microtubule dynamics and microtubule caps: A time-resolved cryo-electron microscopy study. *J. Cell Biol.* **1991**, *114*, 977–991. [CrossRef] [PubMed]
145. Muller-Reichert, T.; Chretien, D.; Severin, F.; Hyman, A.A. Structural changes at microtubule ends accompanying GTP hydrolysis: Information from a slowly hydrolyzable analogue of GTP, guanylyl (alpha,beta)methylenediphosphonate. *Proc. Natl. Acad. Sci. USA* **1998**, *95*, 3661–3666. [CrossRef] [PubMed]
146. Rice, L.M.; Montabana, E.A.; Agard, D.A. The lattice as allosteric effector: Structural studies of alphabeta- and gamma-tubulin clarify the role of GTP in microtubule assembly. *Proc. Natl. Acad. Sci. USA* **2008**, *105*, 5378–5383. [CrossRef] [PubMed]
147. Caplow, M.; Ruhlen, R.L.; Shanks, J. The free energy for hydrolysis of a microtubule-bound nucleotide triphosphate is near zero: All of the free energy for hydrolysis is stored in the microtubule lattice. *J. Cell Biol.* **1994**, *127*, 779–788. [CrossRef] [PubMed]
148. Caplow, M.; Shanks, J. Evidence that a single monolayer tubulin-GTP cap is both necessary and sufficient to stabilize microtubules. *Mol. Biol. Cell* **1996**, *7*, 663–675. [CrossRef] [PubMed]
149. Walker, R.A.; Inoue, S.; Salmon, E.D. Asymmetric behavior of severed microtubule ends after ultraviolet-microbeam irradiation of individual microtubules in vitro. *J. Cell Biol.* **1989**, *108*, 931–937. [CrossRef] [PubMed]
150. Powers, A.F.; Franck, A.D.; Gestaut, D.R.; Cooper, J.; Gracyzk, B.; Wei, R.R.; Wordeman, L.; Davis, T.N.; Asbury, C.L. The Ndc80 kinetochore complex forms load-bearing attachments to dynamic microtubule tips via biased diffusion. *Cell* **2009**, *136*, 865–875. [CrossRef] [PubMed]
151. Gestaut, D.R.; Graczyk, B.; Cooper, J.; Widlund, P.O.; Zelter, A.; Wordeman, L.; Asbury, C.L.; Davis, T.N. Phosphoregulation and depolymerization-driven movement of the Dam1 complex do not require ring formation. *Nat. Cell Biol.* **2008**, *10*, 407–414. [CrossRef] [PubMed]
152. Grishchuk, E.L.; Spiridonov, I.S.; Volkov, V.A.; Efremov, A.; Westermann, S.; Drubin, D.; Barnes, G.; Ataullakhanov, F.I.; McIntosh, J.R. Different assemblies of the DAM1 complex follow shortening microtubules by distinct mechanisms. *Proc. Natl. Acad. Sci. USA* **2008**, *105*, 6918–6923. [CrossRef] [PubMed]
153. Kudalkar, E.M.; Scarborough, E.A.; Umbreit, N.T.; Zelter, A.; Gestaut, D.R.; Riffle, M.; Johnson, R.S.; MacCoss, M.J.; Asbury, C.L.; Davis, T.N. Regulation of outer kinetochore Ndc80 complex-based microtubule attachments by the central kinetochore Mis12/MIND complex. *Proc. Natl. Acad. Sci. USA* **2015**, *112*, E5583–E5589. [CrossRef] [PubMed]
154. Tien, J.F.; Umbreit, N.T.; Gestaut, D.R.; Franck, A.D.; Cooper, J.; Wordeman, L.; Gonen, T.; Asbury, C.L.; Davis, T.N. Cooperation of the Dam1 and Ndc80 kinetochore complexes enhances microtubule coupling and is regulated by aurora B. *J. Cell Biol.* **2010**, *189*, 713–723. [CrossRef] [PubMed]

155. Volkov, V.A.; Zaytsev, A.V.; Gudimchuk, N.; Grissom, P.M.; Gintsburg, A.L.; Ataullakhanov, F.I.; McIntosh, J.R.; Grishchuk, E.L. Long tethers provide high-force coupling of the Dam1 ring to shortening microtubules. *Proc. Natl. Acad. Sci. USA* **2013**, *110*, 7708–7713. [CrossRef] [PubMed]
156. Weir, J.R.; Faesen, A.C.; Klare, K.; Petrovic, A.; Basilico, F.; Fischbock, J.; Pentakota, S.; Keller, J.; Pesenti, M.E.; Pan, D.; et al. Insights from biochemical reconstitution into the architecture of human kinetochores. *Nature* **2016**, *537*, 249–253. [CrossRef] [PubMed]
157. Petrovic, A.; Keller, J.; Liu, Y.; Overlack, K.; John, J.; Dimitrova, Y.N.; Jenni, S.; van Gerwen, S.; Stege, P.; Wohlgemuth, S.; et al. Structure of the MIS12 Complex and Molecular Basis of Its Interaction with CENP-C at Human Kinetochores. *Cell* **2016**, *167*, 1028–1040. [CrossRef] [PubMed]
158. Dimitrova, Y.N.; Jenni, S.; Valverde, R.; Khin, Y.; Harrison, S.C. Structure of the MIND Complex Defines a Regulatory Focus for Yeast Kinetochore Assembly. *Cell* **2016**, *167*, 1014–1027. [CrossRef] [PubMed]
159. Asbury, C.L.; Tien, J.F.; Davis, T.N. Kinetochores' gripping feat: Conformational wave or biased diffusion? *Trends Cell Biol.* **2011**, *21*, 38–46. [CrossRef] [PubMed]
160. Efremov, A.; Grishchuk, E.L.; McIntosh, J.R.; Ataullakhanov, F.I. In search of an optimal ring to couple microtubule depolymerization to processive chromosome motions. *Proc. Natl. Acad. Sci. USA* **2007**, *104*, 19017–19022. [CrossRef] [PubMed]
161. Salmon, E.D. Microtubules: A ring for the depolymerization motor. *Curr. Biol.* **2005**, *15*, R299–R302. [CrossRef] [PubMed]
162. Umbreit, N.T.; Miller, M.P.; Tien, J.F.; Cattin Ortola, J.; Gui, L.; Lee, K.K.; Biggins, S.; Asbury, C.L.; Davis, T.N. Kinetochores require oligomerization of Dam1 complex to maintain microtubule attachments against tension and promote biorientation. *Nat. Commun.* **2014**, *5*, 4951. [CrossRef] [PubMed]
163. Lampert, F.; Hornung, P.; Westermann, S. The Dam1 complex confers microtubule plus end-tracking activity to the Ndc80 kinetochore complex. *J. Cell Biol.* **2010**, *189*, 641–649. [CrossRef] [PubMed]
164. Miranda, J.J.; King, D.S.; Harrison, S.C. Protein arms in the kinetochore-microtubule interface of the yeast DASH complex. *Mol. Biol. Cell* **2007**, *18*, 2503–2510. [CrossRef] [PubMed]
165. Guimaraes, G.J.; Deluca, J.G. Connecting with Ska, a key complex at the kinetochore-microtubule interface. *EMBO J.* **2009**, *28*, 1375–1377. [CrossRef] [PubMed]
166. Schmidt, J.C.; Arthanari, H.; Boeszoermenyi, A.; Dashkevich, N.M.; Wilson-Kubalek, E.M.; Monnier, N.; Markus, M.; Oberer, M.; Milligan, R.A.; Bathe, M.; et al. The kinetochore-bound Ska1 complex tracks depolymerizing microtubules and binds to curved protofilaments. *Dev. Cell* **2012**, *23*, 968–980. [CrossRef] [PubMed]
167. VandenBeldt, K.J.; Barnard, R.M.; Hergert, P.J.; Meng, X.; Maiato, H.; McEwen, B.F. Kinetochores use a novel mechanism for coordinating the dynamics of individual microtubules. *Curr. Biol.* **2006**, *16*, 1217–1223. [CrossRef] [PubMed]
168. Volkov, V.A.; Grissom, P.M.; Arzhanik, V.K.; Zaytsev, A.V.; Renganathan, K.; McClure-Begley, T.; Old, W.M.; Ahn, N.; McIntosh, J.R. Centromere protein F includes two sites that couple efficiently to depolymerizing microtubules. *J. Cell Biol.* **2015**, *209*, 813–828. [CrossRef] [PubMed]
169. Hill, T.L. Theoretical problems related to the attachment of microtubules to kinetochores. *Proc. Natl. Acad. Sci. USA* **1985**, *82*, 4404–4408. [CrossRef] [PubMed]
170. Musinipally, V.; Howes, S.; Alushin, G.M.; Nogales, E. The microtubule binding properties of CENP-E's C-terminus and CENP-F. *J. Mol. Biol.* **2013**, *425*, 4427–4441. [CrossRef] [PubMed]
171. Sarangapani, K.K.; Akiyoshi, B.; Duggan, N.M.; Biggins, S.; Asbury, C.L. Phosphoregulation promotes release of kinetochores from dynamic microtubules via multiple mechanisms. *Proc. Natl. Acad. Sci. USA* **2013**, *110*, 7282–7287. [CrossRef] [PubMed]
172. Sarangapani, K.K.; Duro, E.; Deng, Y.; Alves Fde, L.; Ye, Q.; Opoku, K.N.; Ceto, S.; Rappsilber, J.; Corbett, K.D.; Biggins, S.; et al. Sister kinetochores are mechanically fused during meiosis I in yeast. *Science* **2014**, *346*, 248–251. [CrossRef] [PubMed]
173. Hoog, J.L.; Huisman, S.M.; Sebo-Lemke, Z.; Sandblad, L.; McIntosh, J.R.; Antony, C.; Brunner, D. Electron tomography reveals a flared morphology on growing microtubule ends. *J. Cell Sci.* **2011**, *124*, 693–698. [CrossRef] [PubMed]
174. Chretien, D.; Fuller, S.D.; Karsenti, E. Structure of growing microtubule ends: Two-dimensional sheets close into tubes at variable rates. *J. Cell Biol.* **1995**, *129*, 1311–1328. [CrossRef] [PubMed]

175. Arnal, I.; Karsenti, E.; Hyman, A.A. Structural transitions at microtubule ends correlate with their dynamic properties in Xenopus egg extracts. *J. Cell Biol.* **2000**, *149*, 767–774. [CrossRef] [PubMed]
176. Guesdon, A.; Bazile, F.; Buey, R.M.; Mohan, R.; Monier, S.; Garcia, R.R.; Angevin, M.; Heichette, C.; Wieneke, R.; Tampe, R.; et al. EB1 interacts with outwardly curved and straight regions of the microtubule lattice. *Nat. Cell Biol.* **2016**, *18*, 1102–1108. [CrossRef] [PubMed]
177. Kapitein, L.C.; Peterman, E.J.; Kwok, B.H.; Kim, J.H.; Kapoor, T.M.; Schmidt, C.F. The bipolar mitotic kinesin Eg5 moves on both microtubules that it crosslinks. *Nature* **2005**, *435*, 114–118. [CrossRef] [PubMed]
178. McIntosh, J.R.; Euteneuer, U. Tubulin hooks as probes for microtubule polarity: An analysis of the method and an evaluation of data on microtubule polarity in the mitotic spindle. *J. Cell Biol.* **1984**, *98*, 525–533. [CrossRef] [PubMed]
179. Cameron, L.A.; Yang, G.; Cimini, D.; Canman, J.C.; Kisurina-Evgenieva, O.; Khodjakov, A.; Danuser, G.; Salmon, E.D. Kinesin 5-independent poleward flux of kinetochore microtubules in PtK1 cells. *J. Cell Biol.* **2006**, *173*, 173–179. [CrossRef] [PubMed]
180. Miyamoto, D.T.; Perlman, Z.E.; Burbank, K.S.; Groen, A.C.; Mitchison, T.J. The kinesin Eg5 drives poleward microtubule flux in Xenopus laevis egg extract spindles. *J. Cell Biol.* **2004**, *167*, 813–818. [CrossRef] [PubMed]
181. Shirasu-Hiza, M.; Perlman, Z.E.; Wittmann, T.; Karsenti, E.; Mitchison, T.J. Eg5 causes elongation of meiotic spindles when flux-associated microtubule depolymerization is blocked. *Curr. Biol.* **2004**, *14*, 1941–1945. [CrossRef] [PubMed]
182. Chen, W.; Zhang, D. Kinetochore fibre dynamics outside the context of the spindle during anaphase. *Nat. Cell Biol.* **2004**, *6*, 227–231. [CrossRef] [PubMed]
183. Desai, A.; Verma, S.; Mitchison, T.J.; Walczak, C.E. Kin I kinesins are microtubule-destabilizing enzymes. *Cell* **1999**, *96*, 69–78. [CrossRef]
184. Cooper, J.R.; Wagenbach, M.; Asbury, C.L.; Wordeman, L. Catalysis of the microtubule on-rate is the major parameter regulating the depolymerase activity of MCAK. *Nat. Struct. Mol. Biol.* **2010**, *17*, 77–82. [CrossRef] [PubMed]
185. Hunter, A.W.; Caplow, M.; Coy, D.L.; Hancock, W.O.; Diez, S.; Wordeman, L.; Howard, J. The kinesin-related protein MCAK is a microtubule depolymerase that forms an ATP-hydrolyzing complex at microtubule ends. *Mol. Cell* **2003**, *11*, 445–457. [CrossRef]
186. Ganem, N.J.; Compton, D.A. The KinI kinesin Kif2a is required for bipolar spindle assembly through a functional relationship with MCAK. *J. Cell Biol.* **2004**, *166*, 473–478. [CrossRef] [PubMed]
187. Manning, A.L.; Ganem, N.J.; Bakhoum, S.F.; Wagenbach, M.; Wordeman, L.; Compton, D.A. The kinesin-13 proteins Kif2a, Kif2b, and Kif2c/MCAK have distinct roles during mitosis in human cells. *Mol. Biol. Cell* **2007**, *18*, 2970–2979. [CrossRef] [PubMed]
188. Wilbur, J.D.; Heald, R. Mitotic spindle scaling during Xenopus development by kif2a and importin alpha. *eLife* **2013**, *2*, e00290. [CrossRef] [PubMed]
189. Sharp, D.J.; Ross, J.L. Microtubule-severing enzymes at the cutting edge. *J. Cell Sci.* **2012**, *125*, 2561–2569. [CrossRef] [PubMed]
190. Tan, D.; Asenjo, A.B.; Mennella, V.; Sharp, D.J.; Sosa, H. Kinesin-13s form rings around microtubules. *J. Cell Biol.* **2006**, *175*, 25–31. [CrossRef] [PubMed]
191. Zhang, D.; Asenjo, A.B.; Greenbaum, M.; Xie, L.; Sharp, D.J.; Sosa, H. A second tubulin binding site on the kinesin-13 motor head domain is important during mitosis. *PLoS ONE* **2013**, *8*, e73075. [CrossRef] [PubMed]
192. Wordeman, L.; Mitchison, T.J. Identification and partial characterization of mitotic centromere-associated kinesin, a kinesin-related protein that associates with centromeres during mitosis. *J. Cell Biol.* **1995**, *128*, 95–104. [CrossRef] [PubMed]
193. Oguchi, Y.; Uchimura, S.; Ohki, T.; Mikhailenko, S.V.; Ishiwata, S. The bidirectional depolymerizer MCAK generates force by disassembling both microtubule ends. *Nat. Cell Biol.* **2011**, *13*, 846–852. [CrossRef] [PubMed]
194. Verde, F.; Dogterom, M.; Stelzer, E.; Karsenti, E.; Leibler, S. Control of microtubule dynamics and length by cyclin A- and cyclin B-dependent kinases in Xenopus egg extracts. *J. Cell Biol.* **1992**, *118*, 1097–1108. [CrossRef] [PubMed]
195. Merdes, A.; Ramyar, K.; Vechio, J.D.; Cleveland, D.W. A complex of NuMA and cytoplasmic dynein is essential for mitotic spindle assembly. *Cell* **1996**, *87*, 447–458. [CrossRef]

196. Maiato, H.; Rieder, C.L.; Khodjakov, A. Kinetochore-driven formation of kinetochore fibers contributes to spindle assembly during animal mitosis. *J. Cell Biol.* **2004**, *167*, 831–840. [CrossRef] [PubMed]
197. Khodjakov, A.; Cole, R.W.; Oakley, B.R.; Rieder, C.L. Centrosome-independent mitotic spindle formation in vertebrates. *Curr. Biol.* **2000**, *10*, 59–67. [CrossRef]
198. Gaglio, T.; Dionne, M.A.; Compton, D.A. Mitotic spindle poles are organized by structural and motor proteins in addition to centrosomes. *J. Cell Biol.* **1997**, *138*, 1055–1066. [CrossRef] [PubMed]
199. Surrey, T.; Nedelec, F.; Leibler, S.; Karsenti, E. Physical properties determining self-organization of motors and microtubules. *Science* **2001**, *292*, 1167–1171. [CrossRef] [PubMed]
200. Nedelec, F.; Surrey, T.; Karsenti, E. Self-organisation and forces in the microtubule cytoskeleton. *Curr. Opin. Cell Biol.* **2003**, *15*, 118–124. [CrossRef]
201. Maiato, H.; Khodjakov, A.; Rieder, C.L. Drosophila CLASP is required for the incorporation of microtubule subunits into fluxing kinetochore fibres. *Nat. Cell Biol.* **2005**, *7*, 42–47. [CrossRef] [PubMed]
202. Östergren, G. Luzula and the mechanism of chromosome movements. *Hereditas* **1949**, *35*, 445–468. [CrossRef]
203. Ostergren, G.; Mole-Bajer, J.; Bajer, A. An interpretation of transport phenomena at mitosis. *Ann. N. Y. Acad. Sci.* **1960**, *90*, 381–408. [CrossRef] [PubMed]
204. Franck, A.D.; Powers, A.F.; Gestaut, D.R.; Gonen, T.; Davis, T.N.; Asbury, C.L. Tension applied through the Dam1 complex promotes microtubule elongation providing a direct mechanism for length control in mitosis. *Nat. Cell Biol.* **2007**, *9*, 832–837. [CrossRef] [PubMed]
205. Llauró, A.; Asbury, C.L.; University of Washington, Seattle, WA, USA. unpublished observations. 2016.
206. Higuchi, T.; Uhlmann, F. Stabilization of microtubule dynamics at anaphase onset promotes chromosome segregation. *Nature* **2005**, *433*, 171–176. [CrossRef] [PubMed]
207. Brust-Mascher, I.; Scholey, J.M. Mitotic motors and chromosome segregation: The mechanism of anaphase B. *Biochem. Soc. Trans.* **2011**, *39*, 1149–1153. [CrossRef] [PubMed]
208. Su, K.C.; Barry, Z.; Schweizer, N.; Maiato, H.; Bathe, M.; Cheeseman, I.M. A Regulatory Switch Alters Chromosome Motions at the Metaphase-to-Anaphase Transition. *Cell Rep.* **2016**, *17*, 1728–1738. [CrossRef] [PubMed]
209. Nicklas, R.B.; Staehly, C.A. Chromosome micromanipulation. I. The mechanics of chromosome attachment to the spindle. *Chromosoma* **1967**, *21*, 1–16. [CrossRef] [PubMed]
210. Lampson, M.A.; Renduchitala, K.; Khodjakov, A.; Kapoor, T.M. Correcting improper chromosome-spindle attachments during cell division. *Nat. Cell Biol.* **2004**, *6*, 232–237. [CrossRef] [PubMed]
211. Umbreit, N.T.; Gestaut, D.R.; Tien, J.F.; Vollmar, B.S.; Gonen, T.; Asbury, C.L.; Davis, T.N. The Ndc80 kinetochore complex directly modulates microtubule dynamics. *Proc. Natl. Acad. Sci. USA* **2012**, *109*, 16113–16118. [CrossRef] [PubMed]

biology

MDPI

Review

Anaphase B

Jonathan M. Scholey [1,*], Gul Civelekoglu-Scholey [1] and Ingrid Brust-Mascher [2,*]

1 Department of Molecular and Cell Biology, University of California, Davis, CA 95616, USA; egcivelekogluscholey@ucdavis.edu
2 Department of Anatomy, Physiology and Cell Biology, School of Veterinary Medicine, University of California, Davis, CA 95616, USA
* Correspondence: jmscholey@ucdavis.edu (J.M.S.); ibrustmascher@ucdavis.edu (I.B.-M.); Tel.: +1-530-752-8854 (I.B.-M.)

Academic Editor: Richard McIntosh
Received: 28 September 2016; Accepted: 1 December 2016; Published: 8 December 2016

Abstract: Anaphase B spindle elongation is characterized by the sliding apart of overlapping antiparallel interpolar (ip) microtubules (MTs) as the two opposite spindle poles separate, pulling along disjoined sister chromatids, thereby contributing to chromosome segregation and the propagation of all cellular life. The major biochemical "modules" that cooperate to mediate pole–pole separation include: (i) midzone pushing or (ii) braking by MT crosslinkers, such as kinesin-5 motors, which facilitate or restrict the outward sliding of antiparallel interpolar MTs (ipMTs); (iii) cortical pulling by disassembling astral MTs (aMTs) and/or dynein motors that pull aMTs outwards; (iv) ipMT plus end dynamics, notably net polymerization; and (v) ipMT minus end depolymerization manifest as poleward flux. The differential combination of these modules in different cell types produces diversity in the anaphase B mechanism. Combinations of antagonist modules can create a force balance that maintains the dynamic pre-anaphase B spindle at constant length. Tipping such a force balance at anaphase B onset can initiate and control the rate of spindle elongation. The activities of the basic motor filament components of the anaphase B machinery are controlled by a network of non-motor MT-associated proteins (MAPs), for example the key MT cross-linker, Ase1p/PRC1, and various cell-cycle kinases, phosphatases, and proteases. This review focuses on the molecular mechanisms of anaphase B spindle elongation in eukaryotic cells and briefly mentions bacterial DNA segregation systems that operate by spindle elongation.

Keywords: anaphase B; mitotic motors; spindle elongation; poleward flux

1. Introduction and Historical Perspective

During anaphase, chromosomes are physically separated on the pre-assembled mitotic spindle machinery by a cell type-specific combination of (i) chromosome-to-pole motility (anaphase A) coupled to the pacman- and/or poleward flux-based depolymerization of kinetochore MTs (kMTs) and (ii) spindle elongation (anaphase B) mediated by cortical force generators and/or midzonal MT-MT sliding motors that respectively pull or push apart the spindle poles (Figures 1 and 2) [1–7]. Anaphase B spindle elongation appears to be broadly deployed among eukaryotes and in some systems, e.g., *S. cerevisiae* cells and early *C. elegans* embryos, it is the major mechanism of chromosome segregation [8,9]. Moreover, in some bacterial cells, mechanisms strikingly similar to eukaryotic anaphase B spindle elongation segregate DNA [10]. Underscoring the significance of the process, anaphase B spindle elongation contributes to the correction of mitotic chromosome attachment errors [11–13] and defects in the anaphase B component of chromosome segregation may contribute to human disease—for example a prolonged anaphase B in lymphocytes appears to correlate with an increased risk of cancer [14]. The focus of the current review is on understanding the basic molecular mechanisms of anaphase B spindle elongation. Reviews of aspects of this topic have been published previously e.g., [15–17].

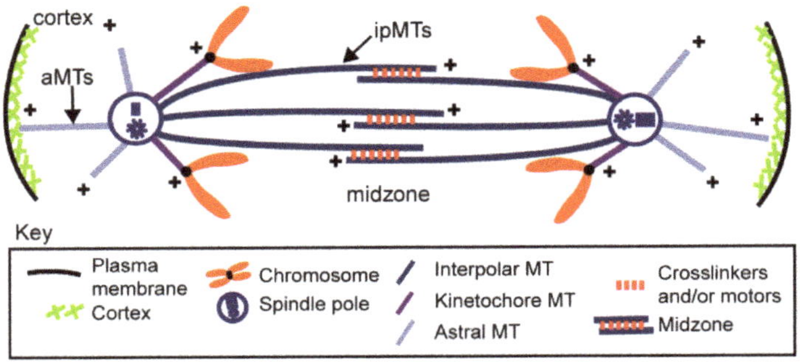

Figure 1. Basic structure of the anaphase B spindle. The major components driving anaphase B spindle elongation are shown, namely ipMTs and the spindle midzone as well as aMTs and the cell cortex, and the structural polarity of spindle MTs is indicated by marking their plus ends. For simplicity, branched augmin-nucleated and chromatin-nucleated MTs that do not reach the poles, as well as pole-nucleated MTs that do not reach kinetochores or the midzone, are not included. Also, anastral spindles lacking centrosomes at the poles are not represented here.

Anaphase B was clearly distinguished from anaphase A in the 1940s by Ris, who showed that spindle elongation in insect cells was more sensitive to inhibition by chloral hydrate than was chromosome-to-pole motion, providing evidence that the two components of chromosome segregation are driven by distinct molecular mechanisms [18,19]. However, anaphase B spindle elongation had apparently been described much earlier, for example by Druner, who proposed a midzonal pushing mechanism in 1894 (see [20] p. 22), and Boveri, who proposed a cortical pulling mechanism in 1888 (see [21] p. 41). Subsequent light microscopy studies have documented the kinetics of anaphase B spindle elongation in a variety of eukaryotic cell types (e.g., see Figure 2 in [3]).

An important advance was the proposal and subsequent testing of a "sliding filament" hypothesis for mitosis [22], in which it was postulated that mitotic motors slide apart adjacent MTs to drive many of the movements of the mitotic spindle that contribute to chromosome movements, in a manner analogous to class-II myosin filaments, which drive the sliding filament mechanism of muscle contraction [23]. Testing the sliding filament model promoted detailed electron microscopy of the organization of mitotic spindle MTs [24–27] (Figure 3) and a biochemical search for the motors that mediate MT-MT sliding [28–31].

Electron microscopic analysis of the three-dimensional ultrastructure of the mitotic spindle by McIntosh and colleagues showed that the sliding filament model could not explain all aspects of mitosis e.g., chromosome-to-pole movement during anaphase A, but such a mechanism could drive pole–pole separation during anaphase B spindle elongation [26,32]. This hypothesis was further supported by light microscopic observations of elongating spindles marked by photo-bleaching in living cells [33] and the reactivation of anaphase B spindle elongation in isolated diatom mitotic spindles [34,35]. The inhibition of isolated diatom spindle elongation by a pan-kinesin peptide antibody suggested that a kinesin motor drives anaphase B [36], a hypothesis supported by the characterization of purified kinesin-5 motors with the potential to act like miniature myosin filaments that could cross-link and slide apart antiparallel ipMTs in the spindle midzone [28,37]. While a significant body of evidence supports such a midzone pushing model, other work has shown that the pulling apart of the spindle poles by motors located on the cell cortex can provide an alternative or complementary mechanism to accomplish midzonal MT-MT sliding and anaphase B spindle elongation [38–41].

Here we review the contribution of these complementary midzone pushing and cortical pulling mechanisms to the sliding filament mechanism of anaphase B spindle elongation in eukaryotic cells (Figure 2). We also briefly mention the bipolar spindle comprising antiparallel bundles of actin-like filaments that elongates by filament polymerization to push apart clusters of R1 plasmids in *E. coli* bacteria, using a mechanism analogous to eukaryotic anaphase B [10].

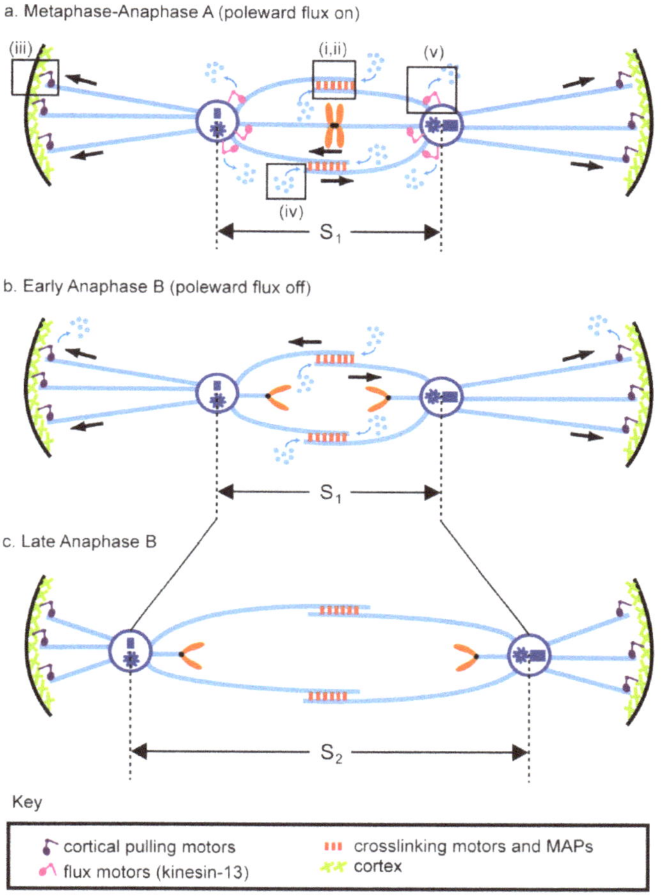

Figure 2. Anaphase B in an idealized and simplified mitotic spindle. The spindle is depicted (**a**) during metaphase-anaphase A (aka pre-anaphase B), when poleward flux is "on" maintaining the spindle at a constant length, S_1; (**b**) at the start of anaphase B, when flux is turned off so that the spindle can begin to elongate; and (**c**) at late anaphase, when the spindle has completed its elongation to length S_2. The major biochemical modules are shown, namely midzone (i) pushing or (ii) braking by MT crosslinkers, particularly kinesin-5 motors and Ase1p MAPs; (iii) cortical pulling by depolymerizing proteins and/or dynein motors attached to the cortex that respectively disassemble or translocate along aMTs to pull them and the attached poles outward; (iv) ipMT plus end dynamics, notably net polymerization; and (v) ipMT minus end depolymerization manifest as poleward flux. In most cells anaphase B starts after anaphase A (as depicted here), but there are exceptions, e.g., in mouse eggs anaphase B precedes anaphase A [42]. Unless otherwise indicated, in this and all other figures, arrows depict direction of movement of ipMTs and aMTs.

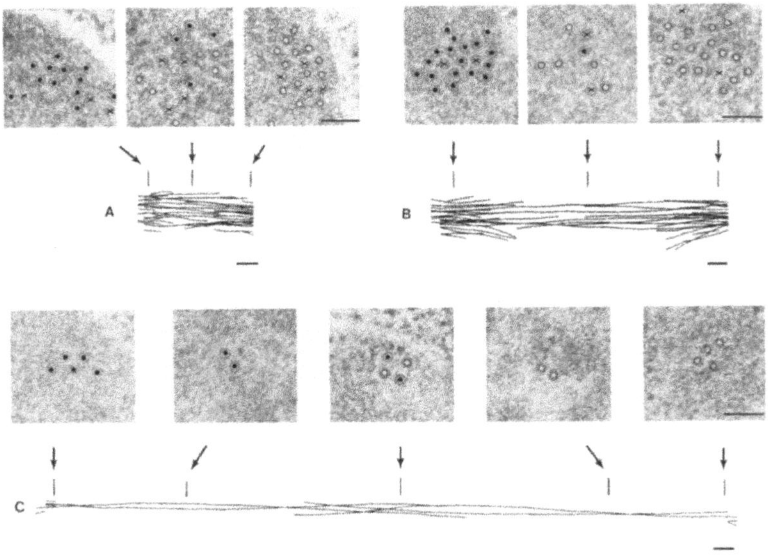

Figure 3. Electron microscopic analysis of anaphase B spindle elongation in budding yeast mitotic spindles showing the structural reorganization of ipMT bundles. 3D reconstructions of (**A**) short spindle; (**B**) early elongating; and (**C**) late elongating spindle. Sample cross sections taken at points indicated by arrows are shown for each reconstruction. In (**B**) kMTs have mostly depolymerized; in (**C**) they have completely depolymerized. Scale bar: 0.1 μm (Originally published as Figure 4 in reference [32], used with permission).

2. Dynamics of Anaphase B in Living Cells

Aspects of mitosis including anaphase B spindle dynamics have been studied using light microscopy for over a century in several systems [43], revealing that, during anaphase B, spindles typically elongate over distances of 1–10 μm at rates of 0.01–0.1 μm/s (Table 1). In favorable cases, anaphase B can be visualized without staining, revealing, for example, that isolated diatom spindles elongate at a rate (0.015 μm/s) [44] approaching that observed in vivo (0.04 μm/s) [45]. Nowadays, however, it is more common to use fluorescence microscopy, e.g., time lapse confocal microscopy of cells containing fluorescent proteins to probe spindle dynamics, e.g., spindle length as a function of time. In *Drosophila* syncytial embryos, for example, about 1000 spindles proceed through mitosis simultaneously in a very well-defined pattern, with anaphase B spindles elongating at a highly reproducible linear rate of ≈0.1 μm/s [46]. Anaphase B spindles in most organisms studied so far elongate at a single linear rate, although some spindles elongate in a biphasic manner at two distinct rates (Table 1). For example, in *Ustilago Maydis*, an initial slow elongation rate is followed by a second faster phase [39], while in *S. cerevisiae* an initial fast phase of spindle elongation is followed by a second slower rate [47].

Interestingly, changes in MT dynamics have been observed at the onset of anaphase B in several systems using FRAP (Fluorescence Recovery after Photobleaching), photoactivation, and FSM (fluorescence speckle microscopy) experiments that monitor fluorescent tubulin behavior. These techniques reveal that spindle MTs display rapid turnover reflecting two types of polymer dynamics coupled to GTP hydrolysis, namely; (i) dynamic instability, characterized by four parameters, the growth rate, shrinkage rate, catastrophe frequency, and rescue frequency; and (ii) poleward flux in which tubulin subunits polymerize at the MT plus ends facing the spindle equator and depolymerize at the minus ends around the poles as the MT polymer lattice slides polewards [48–55]. However, in many systems at the onset of anaphase, spindle MTs display changes in the kinetics of fluorescence recovery in FRAP experiments, reflecting changes in MT dynamics [33,56–59]. For example, in the fission yeast, *S. pombe*, there is no detectable recovery of fluorescence after photobleaching during anaphase B spindle elongation suggesting a dramatic decrease in MT dynamics [58]. In *Drosophila* embryos, poleward flux within ipMTs stops at the onset of anaphase B [60,61] and ipMT plus ends redistribute to the equator, making the ipMTs more stable [62].

Table 1. Rates and extent of spindle length changes during anaphase B.

Organism	Rate of Spindle Elongation	Extent of Spindle Elongation from Metaphase to Telophase	Reference(s)
Diatom	Live: 0.038 ± 0.005 μm/s Isolated: 0.015 ± 0.002 μm/s *	~2 μm 1.9 ± 0.17 μm	[45] [44]
Ustilago maydis	Slow: 0.02 ± 0.003 μm/s Fast: 0.09 ± 0.003 μm/s	~0.5 μm ~4.5 μm	[39]
Schizosaccharomyces pombe	~0.013 μm/s	7–10 μm	[58]
Saccharomyces cerevisiae	Fast: 0.018 ± 0.005 μm/s Slow: ~0.006 μm/s	~4 μm 2–6 μm	[47]
Drosophila syncytial embryo	0.08 ± 0.015 μm/s (cycle 12)	5 μm	[61]
S2 cell	0.017 μm/s	5 μm	[63]
C. elegans	0.107 ± 0.008 μm/s	8.33 ± 0.29 μm	[64]
LLC-Pk1 epithelial cells	0.049 ± 0.017 μm/s	8.73 ± 2.4 μm	[65]

* varies depending on assay conditions.

Another method for studying spindle MT behavior utilizes laser ablation experiments to sever one or more MT bundles within the spindle, monitor how the spindle responds, and infer where forces are generated [40,66–68]. For example, in *C. elegans* embryos, laser ablation of the spindle midzone causes the poles to move rapidly toward the cell cortex, revealing that the midzone is dispensable for anaphase B spindle elongation and instead acts as a brake [40] due to the braking action of bipolar kinesin-5 motors [64], or the combined action of the MT bundling protein Ase1p/PRC1 and kinesin-6 [69,70]. The role of this braking action is unclear but it may somehow contribute to the fidelity of spindle elongation. In contrast, in the fission yeast spindle, laser dissection showed that midzone pushing is necessary and sufficient for anaphase B spindle elongation [68,71].

3. Energetics of Anaphase B

How much force and energy are needed to drive anaphase B spindle elongation at the rates typically observed? It is estimated that very little force, much less than a piconewton (pN), is required to move spindle poles and chromosomes at the speeds observed against cytoplasmic viscous drag [72]. However, it is hard to make precise estimates because the viscosity of cytoplasm is difficult to measure due to its anisotropy and heterogeneity, e.g., [73]. Such a low force value suggests that the free energy released by the hydrolysis of far fewer than 100 ATP fuel molecules could support spindle pole movement over a distance of 10 μm (ignoring imperfections in mechanochemical coupling efficiency and ATP hydrolysis by coupled cell cycle regulatory kinase–phosphatase reactions), which is very small compared to the 10^7 ATP s^{-1} expended by a "typical" cell (see [74] for rudimentary calculations). On the other hand, famous experiments using calibrated microneedles revealed that insect spindles are capable of exerting far greater forces than this on anaphase chromosomes, approaching a nanonewton (0.7 nN) stall force [75]. The relative contribution of anaphase A and B to this force is unclear, although similar experiments done in echinoderm eggs suggest that the stall force required to specifically inhibit anaphase B spindle elongation is of a comparable magnitude [76]. Indeed, some researchers favor the idea that the force for anaphase A chromosome-to-pole motility could be generated by the same ipMT sliding filament mechanism that elongates the anaphase B spindle via passive crosslinks between the moving ipMTs and adjacent kMTs as discussed by [1]. Furthermore, the inhibition of bipolar kinesin-5 motors sometimes leads to defects in anaphase A as well as anaphase B, suggesting that they could participate in such a crosslinking mechanism [77]. Despite the anaphase spindle's capability for generating such a high stall force, it has a much lower specific power output than e.g., muscle or motile cilia, plausibly reflecting its adaptation to precision rather than power [78].

Some of the most direct experiments on ATP expenditure by the spindle during anaphase B have utilized in vitro cell models. For example, in permeabilized vertebrate cultured cells where midzone pushing and cortical pulling may cooperate, anaphase B spindle elongation, unlike anaphase A, requires ATP hydrolysis (half-maximal rate at ≈100–200 μM MgATP), whereas other nucleotides such as GTP cannot substitute for ATP [79]. Isolated diatom central spindles supplied with ATP fuel will elongate at a constant linear rate that is independent of tubulin polymerization which influences the extent but not the speed of elongation [44]. At least a thousand-fold less force (≈1 fN) is needed to elongate these spindles against viscous drag at the rates observed in vitro compared to in vivo because pole motility is opposed only by water (an isotropic liquid), whose viscosity is much lower than that of cytoplasm [44]. Consequently, the free energy of hydrolysis of a single ATP molecule can provide more than enough energy, with the fuel very likely being used by some type of kinesin motor [36]. Given the low ATP turnover involved, the suggestion that a striated muscle-type ATP-regenerating creatine kinase/phosphocreatine/ADP system plays a significant direct role in anaphase B spindle elongation may merit re-evaluation [80,81]. The reactivation of anaphase B in echinoderm eggs is different in that it requires GTP but not ATP, and the presence of assembly-competent tubulin affects both the rate and extent of elongation, suggesting a dominant role for MT polymerization [82]. Therefore potential force generators for anaphase B in different systems include not only MT-based motors like kinesins and dyneins, but also dynamic cytoskeletal filaments that can polymerize or depolymerize to exert pushing and pulling forces, respectively, and midzonal MT-crosslinking MAPs, which have recently been proposed to exert entropic expansion forces [83–85].

4. Structural Studies of the Anaphase B Spindle

Classic work using painstaking electron microscopy (EM) has elucidated the three-dimensional organization and polarity patterns of MTs within the mitotic spindles of several cell types, providing an important foundation for understanding the mechanism of anaphase B spindle elongation, as well as other aspects of mitosis (Figure 3). For example, early serial section electron microscopic analysis of human cultured cells during anaphase and telophase was consistent with the hypothesis that ipMT bundles consist of two sets of MTs that emanate from opposite poleward regions and overlap at the spindle midzone [27]. In some cases the polarity patterns of MTs within such ipMT bundles have been directly determined using the method of hook decoration [86]; for example, in endosperm cells of the plant *Haemanthus* that were fixed during anaphase, the two sets of opposite polarity ipMTs were shown to be oriented with their minus and plus ends facing the spindle poles and the midzone, respectively [24]. Although the number of ipMTs varies from less than 10 in budding yeast to hundreds in diatoms and cultured cells, the same overall structural organization is thought to apply to ipMT bundles within most spindles (Figure 1). This has been confirmed for budding yeast [32], fission yeast [87,88], and diatom [26], for example, where the minus ends of the overlapping ipMTs appear to physically interact with the spindle poles. The outward sliding of these ipMTs could therefore directly exert compressive forces on the poles to push them apart, leading to anaphase B spindle elongation. This structural organization further suggests that a plus end-directed antiparallel ipMT-crosslinking motor located in the midzone, with a functional organization equivalent to a bipolar myosin-II filament [89], could perform such a function e.g., see Figure 4b in [28]. The hypothesis that motors associated with interdigitating antiparallel ipMTs at the anaphase B spindle midzone could push apart the spindle poles in diatoms is supported by studies of the ATP-dependent reactivation of the elongation of isolated central spindle preparations [35]. Like the anaphase spindles seen in intact diatom cells, by EM these isolates display a robust midzone of ≈600 overlapping ipMTs that slide apart following ATP addition to drive pole–pole separation [34].

A somewhat different picture emerged from the detailed EM analysis of ipMT bundles during anaphase in PtK1 cells [90]. These bundles are also organized into two sets of opposite polarity with their plus ends overlapping at the midzone, but the minus ends of most of these ipMTs do not actually reach the poles, suggesting that they cannot directly push on the poles to drive spindle elongation. While it is possible that these ipMTs act indirectly, e.g., via an interaction between their minus ends and kinetochore MTs whose minus ends do contact the poles, it is perhaps more plausible to think that, in this system, the poles are pulled apart by a cortical pulling mechanism [90]. In this scenario it is possible that bipolar MT crosslinking motors within the spindle midzone could serve as brakes that restrict the rate of ipMT-MT sliding [64,65], thereby enhancing the fidelity and directionality of pole–pole separation [1].

The use of EM to directly visualize the mitotic motors that are predicted to crosslink and slide apart, or constrain the sliding apart, of antiparallel ipMTs in the spindle midzone has proven difficult and has yielded less definitive information. MT-MT cross-bridges have been seen in EM images of sectioned mitotic spindles but they are sometimes rather ill-defined, vary in average length from about 20 nm to about 60–65 nm, usually do not display an obvious regular axial spacing, and sometimes appear to form a "matrix" [25,26,32,37,87,91]. It is plausible to think that these cross-bridges could comprise non-motor MAPs such as Ase1p (aka PRC1), or mitotic motors such as the bipolar kinesin-5 (discussed in the next section). Consistent with the latter idea, for example, light microscopy of both living and fixed *Drosophila* embryo anaphase spindles suggests that kinesin-5 localizes along the entire length of ipMT bundles where it co-localizes via competitive binding with Ase1p at the spindle midzone (Figure 4) [37,92,93]. In this system, serial section EM is consistent with the basic conclusion of the earlier pioneer work [27] in suggesting that each of the anaphase B spindle's nine ipMT bundles contains about 30–40 MTs per half-spindle that are parallel near the poles and overlap in an antiparallel orientation for ≈2–3 μm at the midzone [37]. By immuno-EM, Au-coupled anti-kinesin-5 clearly decorated MTs all along these bundles and was sometimes seen to be associated with 60–65 nm long cross-bridges between adjacent MTs [37], a length similar to the 57–61 nm length of the purified *Drosophila* kinesin-5 rod [28,94]. While these results are consistent with the idea that kinesin-5 could form at least some of the cross-bridges seen in EM sections of ipMT bundles, they do not prove it and more definitive and comprehensive information concerning the identity and architecture of these structures would be useful, especially given the complex molecular composition of the spindle midzone [95,96]. This important, challenging problem merits further work.

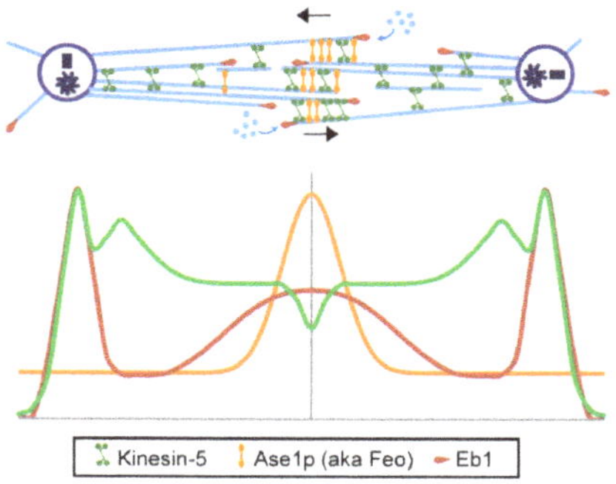

Figure 4. Anaphase B spindle in a *Drosophila* embryo. The upper drawing of the anaphase B spindle has a pole–pole axis corresponding with that of the drawing below showing the relative distribution of kinesin-5 motors, Ase1p crosslinkers, and the plus end binding protein Eb1 along the spindle (adapted from [93]).

5. Conserved Biochemical Modules Involved in Anaphase B

Current evidence suggests that a handful of conserved biochemical "modules" are deployed to different extents in a combinatorial fashion in distinct cell types to accomplish the elongation of the anaphase B spindle, including outward sliding of antiparallel ipMTs to push apart spindle poles, restriction of ipMT sliding by midzonal crosslinkers, growth of ipMT plus ends by MT polymerization, cortical forces that pull the spindle poles outwards and the use of poleward MT flux as a regulatory switch (Figure 5).

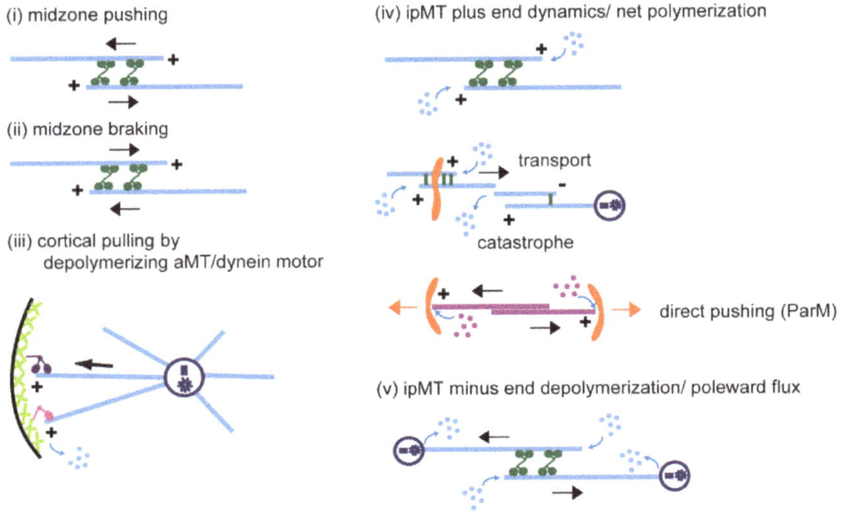

Figure 5. Anaphase B modules. The five biochemical modules depicted in Figure 2 (modules **i–v**) that are deployed to various extents in different systems are shown in more detail.

5.1. (Module i) Midzone Pushing: Pole–Pole Separation by Outward Sliding of Antiparallel ipMTs

Much of the work described in section 10.4 supports the idea that plus-end-directed bipolar mitotic motors could act at the spindle midzone to slide apart overlapping ipMTs and generate pushing forces to separate the anaphase B spindle poles. This model is especially appealing in the case of the diatom central spindle [26,35], where it was further supported by observations that the laser microbeam-induced destruction of ipMTs at the presumptive site of force generation in the spindle midzone, but not around the poles, inhibited spindle elongation [67]. Similar results supporting an ipMT pushing mechanism have been obtained using laser microsurgery of elongating fission yeast spindles [68,71]. In living, cultured PtK1 cells, the dynamics of anaphase B spindles containing fluorescent tubulin and marked by photobleaching was studied using light microscopy, leading to proposals that the sliding apart of ipMTs by a force generated at the zone of interdigitation at the midzone could contribute to spindle elongation [33] (although subsequent EM of these spindles yielded the caveats noted above; [90]). Circumstantial evidence in support of a bipolar kinesin-5-mediated midzonal ipMT-MT sliding model was obtained in *Drosophila* embryos where the spindle poles separate during anaphase B at a linear rate of ~0.1 μm/s, which, as expected, is almost exactly twice the rate at which tubulin speckles flux towards the opposite poles along ipMTs (0.05 μm/s) during pre-anaphase B and twice the speed at which purified fly embryo kinesin-5 moves MTs in motility assays (0.05 μm/s) [31,77]. We discuss the properties of the presumptive motors in more detail below.

5.2. (Module ii) Midzone Braking

MT-MT crosslinking MAPS and motors on the spindle midzone can also serve as brakes that restrict the rate and extent of pole–pole separation driven by antagonistic force generators e.g., cortical pulling motors [64,65,96–99].

5.3. (Module iii) Cortical Pulling Apart of the Anaphase B Spindle Poles

One interpretation of the EM work cited above, showing that the minus ends of ipMT bundles in PtK1 cells do not appear to reach the spindle poles [90], is that a mechanism other than ipMT mediated pushing forces could operate to separate the spindle poles in some cells, for example a force generator that acts at the cell cortex to exert pulling forces on the asters to pull apart the associated spindle poles [38]. Indeed, a significant body of evidence supports the existence of such external pulling forces that pull on the spindle poles to control pole–pole spacing and even to position the entire spindle [40,100,101]. Candidates for such force generators include cortically-anchored dynein or astral MT depolymerization, with kinesin-5 acting either as a supplementary pole-separating force generator or as a counteracting brake [39,64,102]. We assume here that direct contacts between aMTs and the cell cortex are required to exert pulling forces on the spindle poles [103] but it should be noted that cytoplasmic force generators may also somehow be able to pull asters outward in the absence of cortices e.g., [104].

5.4. (Module iv) ipMT Plus End Dynamics and Net Polymerization

MT polymer dynamics, characterized by dynamic instability and poleward MT flux, obviously play critical roles throughout mitosis [48,49]. For example, during anaphase B in some systems, overlapping ipMTs at the spindle midzone grow by polymerization of their plus ends as they slide apart. This was a conclusion of the clever photo-bleaching experiments done on elongating anaphase B spindles in PtK1 cells by Saxton and McIntosh [33], in which tubulin subunits were observed to add on to the plus ends of the outwardly sliding ipMTs. Further support for this idea was obtained using isolated diatom spindles that, when supplied with ATP fuel in the absence of free tubulin subunits, will elongate to an extent that is limited by the size of the original overlap zone [34,35]. However, when tubulin subunits are added to these preparations, the subunits polymerize onto the plus ends of pre-existing ipMTs causing the length of the initial overlap zone and the extent of subsequent ATP-induced spindle elongation to increase by a corresponding amount [105,106]. Thus the polymerization–depolymerization of ipMT plus end at the spindle midzone appears to play an important role in determining the extent of sliding apart of antiparallel ipMTs at the midzone and, in turn, the extent of anaphase B spindle elongation. In *Drosophila* embryo spindles, a spatial gradient of MT catastrophe frequencies (decreasing towards the equator) is established at the onset of anaphase B, causing ipMTs to polymerize at their plus ends and grow at the equator to invade the midzone [62]. A complementary mechanism occurs during early anaphase in cultured human cells, where augmin/γ-TuRC nucleates the branching polymerization of ipMT plus ends on pre-existing spindle MTs [107]. In both cases the growing ipMTs can then be crosslinked by MAPs and motors around the equator to produce a more robust anaphase spindle midzone, but the significance of this for the dynamics of spindle elongation remains to be determined.

A variant of the coupling of ipMT plus end polymerization to ipMT sliding occurs in some prokaryotes where polymerization generates compressive forces that directly push apart the poles. Bacterial cells are generally thought to lack filament-sliding motors analogous to kinesins, dyneins, and myosins, and in the R1 plasmid segregation system discussed in Section 6.3, for example, the bidirectional polymerization of antiparallel cytoskeletal filament bundles can directly generate forces that drive spindle elongation [10]. Interestingly, recent work suggests that an analogous pushing mechanism based on midzonal ipMT plus end polymerization may contribute, in a redundant fashion, to spindle elongation and chromosome segregation in *C. elegans* embryos [108,109].

Another biochemical mechanism based on ipMT plus ends dynamics, named "slide-and-cluster" has been proposed to control spindle length in *Xenopus* extract spindles [110]. Here, antiparallel MTs nucleated on chromatin that grow by plus-end polymerization, are slid apart, minus-end leading, by kinesin-5 motors, with further poleward transport of the MTs being facilitated by minus-end-directed motors that move them along pre-existing MT tracks. The dynamics of the sliding MT plus ends determines their lifetime since they can disappear via catastrophic depolymerization so that spindle length depends on both the lifetime and the rate of poleward MT transport of these spindle MTs. Whether this module contributes to anaphase B, something that is difficult to test in extracts, is discussed later (see theoretical models).

5.5. (Module v) ipMT Minus End Depolymerization: Poleward Flux as a Regulatory Switch for Anaphase B

It has long been recognized that the flux of tubulin towards the spindle poles is a striking feature of many mitotic spindles [50,53]. While most attention has focused on its role in anaphase A chromosome-to-pole motility, it may also play a critical regulatory role in chromosome segregation by turning on and off anaphase B spindle elongation, at least in some systems. In *Drosophila* embryo mitotic spindles, for example, there is evidence that the suppression of poleward tubulin flux within ipMT bundles due to the inhibition of ipMT minus end depolymerization can initiate and control the rate of anaphase B spindle elongation [60,77], but whether this is a system-specific or broadly utilized mechanism remains to be established.

5.6. Combination of Modules and the Force Balance Concept

Spindles in different cell-types utilize different combinations of antagonist or complementary modules to produce or modulate the force that drives anaphase B spindle elongation. For example, the net polymerization of ipMT plus end polymerization (module iv) can complement midzone pushing (module i) to enhance the extent of spindle elongation [105] whereas midzone braking (module ii) can antagonize cortical pulling (module iii) to slow down the rate of spindle elongation [64]. Importantly, anaphase B can be controlled by antagonistic modules that create a force balance of the type initially proposed by Ostergren to control metaphase chromosome position [111,112], as reviewed in Chapter 4. A good example of this was mentioned in the previous paragraph. The combination of midzone pushing (module i) and ipMT depolymerization (module v) during pre-anaphase B produces a force balance in which ipMTs undergo poleward flux and the spindle is maintained at a constant steady state length. When this force balance is tipped at anaphase B onset by the inhibition of ipMT minus end depolymerization at spindle poles, the opposing forces become unbalanced, allowing midzone pushing to exert net outward force on the spindle poles to elongate the anaphase B spindle [60,61,113].

6. Properties and Functions of the Molecular Nuts and Bolts of the Anaphase B Machinery

The anaphase B spindle is thought to comprise fairly "typical", structurally polar MTs that are assembled from $\alpha\beta$-tubulin dimers in a head-to-tail fashion with the β subunits facing the MT plus ends and that display both dynamic instability and poleward flux [48–50,114]. A full understanding of the mechanism of anaphase B requires an elucidation of the functions and mechanism of action of all the molecules that interact with these MTs, but in many cases this is difficult because the functional perturbation of key molecules can also interfere with earlier phases of mitosis, thereby obscuring later roles in anaphase B. This is compounded by the well-known existence of functional redundancy between different mitotic mechanisms [115], combined with the fact that many key molecules localize to multiple sites in the spindle e.g., to the kinetochore, midzone, and cortex, making their site of action in anaphase B difficult to discern.

6.1. Molecules of the Central Spindle

6.1.1. Antiparallel ipMT-Crosslinking MAPs of the Ase1p Family

The anaphase B-specific functions of homodimeric MT-MT crosslinking MAPs of the Ase1p family (i.e., anaphase spindle elongating protein 1; aka PRC1, MAP65, Feo) are unusually obvious because these MAPs only localize to the spindle midzone following metaphase and their function is required for anaphase B spindle elongation [116,117]. Three properties of Ase1p MAPs underlie their critical role in organizing the midzone and facilitating anaphase B spindle elongation. First, Ase1p dimers preferentially crosslink ipMTs into antiparallel orientations where they oligomerize to form a "matrix" between pairs of bundled MTs [84,117,118]. This matrix may correspond to the "osmiophilic matrix" seen in EMs of stained diatom spindles [26] (reviewed in [96]). This bundling of MTs by Ase1p may be facilitated by its structure—it has been proposed that Ase1p is a flexible molecule in solution, which adopts a more rigid conformation only when bundling antiparallel MTs [119]. Second, the resulting Ase1p complex can serve as a key "regulatory hub" for controlling the cell-cycle-dependent localization of various motors and other proteins to the midzone in a system-specific fashion (reviewed in [17]). For example, at anaphase onset in budding yeast, the cell-cycle-regulated de-phosphorylation of Ase1p allows it to recruit kinesin-5 sliding motors to the midzone to drive spindle elongation [120] whereas in *Drosophila* embryos the Ase1p family member, Feo, partially restricts the association of kinesin-5 sliding motors to the anaphase B spindle midzone (Figure 4) [93]. In the latter system, the dissociation of Ase1p from the midzone permits more kinesin-5 to bind in its place, but anaphase B spindle elongation is then impaired, suggesting that the Ase1p-mediated spindle midzone organization is required to facilitate the kinesin-5-mediated ipMT sliding filament mechanism that underlies anaphase B [93]. In light of these results, it is tempting to speculate that the kinesin-anchoring midzone matrix associated with anaphase B in Diatoms may comprise Ase1p [121]. Finally, striking new work suggests that diffusible Ase1p crosslinkers can also directly generate forces for ipMT-MT sliding via an ATP hydrolysis-independent entropic expansion mechanism that could, for example, control the length of the antiparallel overlaps at the midzone during anaphase B spindle elongation [85,122].

In the context of midzone organization by an Ase1p "matrix," it is worth noting that the role of a spindle matrix distinct from MT, MAPs and MT-based motors but capable of augmenting the activities of these well-characterized spindle components during mitosis continues to draw attention e.g., [123]. However, apart from the aforementioned work on Ase1p, we are not aware of any evidence that such an entity operates during anaphase B spindle elongation. For example, in *Drosophila* embryos, a lamin B spindle envelope that has been proposed to form a matrix that augments the activities of mitotic motors during earlier phases of mitosis is disassembled prior to anaphase B onset [124].

6.1.2. MT Crosslinking and Sliding Motors; Kinesin-5 Plus Kinesins-4, -6, -8, and -12

Once ipMTs have been organized into an ordered array of antiparallel bundles at the midzone by Ase1p crosslinkers, they can be slid apart by various combinations of ipMT-MT crosslinking and sliding motors, most notably kinesin-5. Purified kinesin-5 motors display a bipolar, homotetrameric ultrastructure consisting of pairs of motor domains at opposite ends of a central 60 nm long rod [28,31,94]. A novel four-helix bundle called the "BASS" (or Bipolar ASSembly) domain, comprising a pair of intertwined antiparallel coiled-coil dimers stabilized by patches of hydrophobic and charged residues, directs the assembly of four motor subunits into these bipolar tetrameric minifilaments [94,125], whose four motor domains can move slowly and moderately processively towards the plus ends of MTs against substantial opposing forces [126–128]. This unique homotetrameric architecture is essential for kinesin-5 function during mitosis [129] plausibly because it allows kinesin-5 motors to dynamically interact with spindle MTs via a reaction-diffusion mechanism [92], preferentially binding MTs in the antiparallel orientation [130], and driving or constraining their sliding apart throughout mitosis by a sliding filament mechanism [30,98]. Kinesin-5 was discovered based on its essential role in early bipolar mitotic spindle assembly or maintenance [126,131] but subsequent work in several systems also supports a role in driving outward ipMT sliding and spindle elongation during anaphase B [77,132–135]. In some cases, however, kinesin-5 appears to serve as a brake that restricts spindle elongation [64,65,97,99]. Cutting-edge optical trap motility assays provide mechanistic insights into how bipolar kinesin-5 motors can switch between the generation of both outward sliding and inward braking forces on crosslinked MTs [98]. It should also be noted that budding and fission yeast kinesin-5 motors are capable of reversing their polarity of MT-based motility [136–138], but whether this is significant for the mechanism of anaphase B is, to our knowledge, unknown.

Members of the kinesin-4, -6, -8, and -12 families of plus-end-directed motors could also contribute to ipMT-MT crosslinking and sliding during anaphase B. For example, the kinesin-4 KLP3A organizes midzonal ipMT bundles in fly spindles and somehow couples the downregulation of poleward flux to the onset of spindle elongation at anaphase [60,139]. Kinesins-6 and -8 have been reported to mediate antiparallel MT-MT sliding based on in vitro assays [140,141] and they could therefore augment or replace the function of kinesin-5 in driving anaphase B spindle elongation. Kinesin-8 is best known as a MT-translocating, length-dependent MT depolymerase [142], which influences spindle assembly and length control throughout mitosis. In budding yeast, for example, a complex interplay between its MT-depolymerizing and MT-MT sliding activity appears to contribute to both spindle elongation and disassembly during anaphase B [141,143]. Kinesin-6 dimers, on the other hand, often co-assemble with two subunits of a G-protein cofactor, containing a GTPase-activating domain for Rho-family GTPases (RhoGAP domain), which, in contrast to the prevailing view [144] was recently reported to be dispensable for MT-MT bundling but is required for MT motor activity [145]. The resulting heterotetrameric complex is usually thought to organize the anaphase spindle midzone to control normal cleavage furrow assembly and cytokinesis [95,146]. In fission yeast, however, kinesin-6 is proposed to form homotetramers, based on chemical crosslinking, that bind Ase1p in a phosphorylation-dependent manner to form a complex that interacts with and slides apart antiparallel ipMTs at the midzone to drive anaphase B spindle elongation [147]. Finally vertebrate kinesin-12 can crosslink and slide adjacent MTs in vitro [148] and is a candidate motor driving spindle elongation during *C. elegans* meiosis [149] but it preferentially crosslinks MTs into the parallel rather than antiparallel orientation and, although it substitutes for kinesin-5 function during spindle assembly, to our knowledge, a role in anaphase B has not been reported.

It is perhaps worth emphasizing that many MT crosslinking kinesin motors that organize and control the length of the spindle midzone also play key roles in organizing the cleavage furrow for cytokinesis (although the specific roles of some of them in anaphase B spindle elongation e.g., kinesin-4 recruited to the midzone by Ase1p/PRC1 [150–153] requires clarification). The role of midzonal kinesins in cytokinesis, which is not a focus of the current manuscript, has been well covered in a recent review [146].

6.1.3. Molecules Controlling MT Plus end Dynamics

A complex network of MT plus end tracking proteins (+TIPs) probably plays important roles in controlling ipMT plus end dynamics at the spindle midzone during anaphase B, but this is a topic that requires further work. +TIPs that enhance MT plus end polymerization include autonomous MT binding proteins such as the master regulator, EB (end binding) protein, and Tog-domain XMAP215 proteins, along with the "hitchhikers" that they recruit to MT plus ends such as CLASP, another Tog protein, whereas proteins such as kinesin-13 and kinesin-8 antagonize these proteins by depolymerizing MT plus ends [114]. Impressive biochemical reconstitution experiments are being used to study these complexes [154] and evidence supporting their possible role in anaphase B emerges from observations that in *C. elegans* zygotes and mammalian cells, for example, CLASPs localize to the anaphase spindle midzone where they may promote ipMT polymerization during spindle elongation [155,156]. Moreover, based on the effects of CLASPs on kMT dynamics in *Drosophila* cells, it is plausible to think that +TIPs could promote ipMT plus end polymerization to contribute to poleward flux during pre-anaphase B when the spindle maintains a constant steady state length, as well as to anaphase B spindle elongation [157], but this requires testing. Also requiring further functional analysis are the results of in vitro assays which suggest that bipolar kinesin-5 motors not only slide apart ipMTs at the midzone but may also contribute to +TIP activity by stimulating the polymerization of ipMT plus ends at the midzone [158]. In *Drosophila* embryos, the transition from poleward flux to spindle elongation is accompanied by the rapid formation of a spatial gradient of MT plus end catastrophe events, decreasing in the anti-poleward direction, which causes ipMT plus ends to grow towards the equator and augment the midzone where outward ipMT-MT sliding forces are generated [62]. In human cells the augmin/γ-TuRC complex nucleates the branching polymerization of MTs to promote robust central spindle assembly [107].

Another possible mechanism for assembling MT plus ends at the midzone may merit investigation. The anterograde heterotrimeric kinesin-2 motor [159] is understood to deliver tubulin subunits to the MT plus ends of growing ciliary axonemes [160], but it has also been localized to the midzone of anaphase sea urchin embryos, where its role is unknown [161]. It is tempting to speculate that kinesin-2 may also deliver tubulins for assembly at the plus ends of overlapping ipMTs in the anaphase spindle midzone, but this idea has not, to our knowledge, been tested so far.

6.1.4. Molecules Controlling ipMT Minus End Dynamics

A topic that merits further work is how ipMT minus end dynamics contribute to anaphase B, especially at the poles, where compressive forces are exerted [114]. One well-established minus-end regulator, γ-TuRC is understood to nucleate MT polymerization at centrosomes, but it can also bind MT minus ends throughout the spindle whereupon a slide-and-cluster mechanism (which could contribute to anaphase B spindle elongation—see theoretical models) transports the MTs to the poles [162]. There is better evidence that minus-end-targeting −TIPs such as CAMSAPs and Patronin [114] play important roles in anaphase B, at least in some systems. These proteins stabilize MT minus ends against the depolymerizing activity of kinesin-13 [163–165] and in *Drosophila* embryos, the Patronin-mediated inhibition of kinesin-13-dependent ipMT minus end depolymerization at the poles occurs in response to cyclin B degradation [113,166]. This induces a switch from poleward flux to anaphase B spindle elongation by allowing outwardly sliding ipMTs to push apart the spindle poles, but whether this switch is used elsewhere is unknown.

6.1.5. Chromosomal Proteins Required for Anaphase B Spindle Elongation

An important and sizeable set of proteins known as chromosomal passenger proteins translocate from the chromosomes to the spindle midzone at anaphase onset and are understood to perform key roles in coordinating progress through anaphase B, for example by stabilizing the spindle midzone and recruiting proteins required for telophase, cytokinesis and mitotic exit [167]. In *C. elegans* embryos, it was recently reported that a subset of kinetochore proteins is required for spindle midzone assembly and normal anaphase B spindle elongation [156]. Sorting out the precise relationships between these chromosomal proteins and anaphase B is a cutting-edge problem in mitosis research.

6.2. Molecules of the Cortical Pulling Machinery

6.2.1. Attachment of MT Plus Ends to the Cortex

The cortical pulling mechanism (Figures 2 and 5 (iii)) requires that astral MT plus ends interact with the cell cortex, and this is again mediated by the +TIP network, which includes MT-depolymerases e.g., kinesin-13, MT polymerases, e.g., CLASPs, and motors, most notably dynein-dynactin, some of which interact directly with the lipid bilayer but usually bind via membrane-associated adaptors or via the cortical actin cytoskeleton [114].

6.2.2. Cortical Force Generators

The spindle poles can potentially be pulled apart by; (a) a polymer ratchet mechanism in which the plus ends of astral MTs linked to the cell cortex depolymerize to pull the poles outward [49,102]; and (b) a motor-dependent mechanism in which dynein anchored at the cell cortex walks towards the minus ends of astral MTs to pull the poles outward [168,169]. For example, observations and simulations of astral MT depolymerization following contact of their plus ends with the *C. elegans* embryo cortex suggest that cortical adaptors may couple aMT disassembly to the generation of forces for pulling apart the poles [103]. Moreover, single-molecule TIRF microscopy in fission yeast cells has allowed the direct visualization of dynein dynamics as it diffuses in the cytoplasm, attaches to an aMT, undergoes 1-D diffusion along the aMT to reach the plus end, whereupon it "off-loads" and binds to the cell cortex [170]. Once anchored at the cell cortex, dynein may pull the spindle poles outward in two ways, first by using ATP hydrolysis to walk towards the minus ends of the aMT [168] and second, based on clever in vitro experiments, by inducing the catastrophic depolymerization of captured aMTs to generate pulling forces on the spindle poles [171]. The resulting pulling forces may then participate in many aspects of mitosis and mitotic spindle positioning e.g., [172,173] including anaphase B spindle elongation [38,39].

6.3. Molecules Involved in Prokaryotic Anaphase B

Understanding of prokaryotic DNA segregation by the bacterial cytoskeletal machinery has advanced enormously in the past two decades. It is now thought that, in different bacteria, at least three types of dynamic polymer can mediate this process. These polymers are assembled from the actin-like protein ParM, the ATPase ParA, or the tubulin-like protein, TubZ [174]. Of these, ParM filaments appear to segregate R1 plasmids in *E. Coli* by an antiparallel array of filaments that resembles the overlapping MTs of the spindle midzone that participate in eukaryotic anaphase B [10]. Briefly, bundles of ParM filaments use ATP hydrolysis to undergo dynamic instability, allowing them to search and capture the centromeres (*parC*) of "sister" plasmids harboring bound ParR adaptor protein, and to push the captured plasmids to opposite ends of the bacterial cell before depolymerizing again. Although the left-handed parallel two-stranded ParM filament is structurally polar, the two ends elongate at equivalent rates, and adjacent filaments display at least a 5:1 preference for pairing in the antiparallel orientation in vitro [175]. A striking reconstitution of the R1 plasmid segregation system has been accomplished, in which ParM filaments polymerize bidirectionally to push apart ParR/*parC* coated beads [176].

7. Cell Cycle Control of Anaphase B

The cell cycle control of anaphase B spindle elongation has been reviewed recently by [17] and subsequent work suggests that this is likely to be complex [177]. Briefly, it is understood that anaphase onset occurs following the dephosphorylation of cyclin dependent kinase (cdk1) substrates leading to inactivation of the spindle assembly checkpoint (the SAC, as reviewed in chapter 5, which ensures proper metaphase chromosome alignment) and the loss of cohesion between sister chromatids that can then separate and move poleward [178,179]. In animal cells, the sequential ubiquitin-dependent proteolytic degradation of distinct mitotic cyclins is required for progression through mitosis, with cyclin A degradation allowing progression through metaphase and cyclin B degradation, which occurs immediately after the SAC is satisfied, being necessary for anaphase B [179]. A third cyclin B3 has been found in *Drosophila* whose destruction follows that of cyclin B to promote anaphase B spindle elongation [180,181].

In many organisms, anaphase A and B start simultaneously after chromatids disjoin, i.e., kMT shortening moves chromosomes poleward and the spindle elongates at the same time. In *Drosophila* embryos, however, there is a distinct transition between anaphase A and anaphase B which is regulated by cyclin B degradation [62]. In this system the patronin-dependent downregulation of poleward flux is proposed to function as a regulator of the onset and the rate of anaphase B [60–62,113]. As emphasized by [17], the MT bundler Ase1p serves as a regulatory hub for anaphase B in many model organisms. Ase1p accumulates at the anaphase spindle midzone in a phosphorylation- and motor-dependent manner, where it not only controls midzone organization via MT-MT crosslinking but also recruits other key anaphase B proteins. For example, in yeast and *Drosophila* embryos, Ase1p mediates the recruitment of several ipMT sliding motors and this activity is required for proper spindle midzone organization and elongation during anaphase B [93,120,147].

Thus, the regulation of anaphase B spindle elongation is complex. It occurs at several levels and involves ubiquitin-dependent proteolytic degradation, cdk and phosphatase-dependent phosphorylation-dephosphorylation cycles, the key midzone regulatory MAP, Ase1p, as well as the midzonal motors which localize key molecules to the midzone, and whose localization is, in turn, dependent on the regulatory molecules that they localize. In addition, at least in some systems, the up- and downregulation of poleward flux plays important regulatory roles associated with anaphase B. Moreover, exciting recent work indicates that the control of anaphase B involves the dynamic cooperation and antagonism between functionally interdependent kinases, phosphatases, MAPs and motors capable of producing spatially-controlled feedback loops that coordinate the dynamic turnover of phosphorylation sites to orchestrate spindle midzone assembly and elongation [177,182]. A similar dynamic complexity is likely to control the operation of the cortical pulling machinery involved in anaphase B as well.

8. Anaphase B in Model Systems

Here we briefly survey how the aforementioned modules and molecules are deployed and combined in a few model organisms to mediate anaphase B (Figure 6 and Table 1). It is likely that the study of "non-mainstream" organisms could further uncover different, more exotic mechanisms of anaphase B spindle elongation [183].

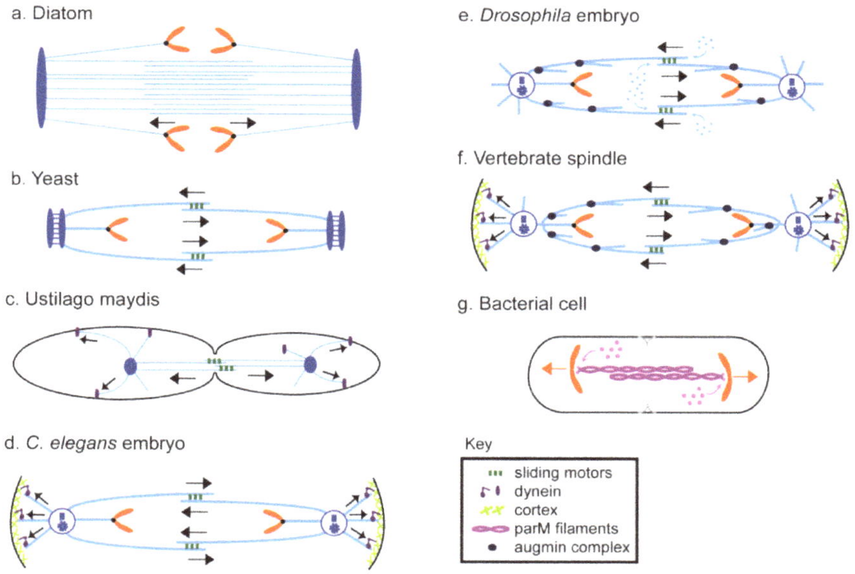

Figure 6. Anaphase B spindle design in different systems. The simplified drawings depict spindles from (**a**) diatoms; (**b**) budding yeast; (**c**) the rust fungus *Ustilago*; (**d**) early *C. elegans* embryos; (**e**) *Drosophila* syncytial embryos; (**f**) vertebrate cultured cells; and (**g**) bacterial cells. Not drawn to scale.

8.1. Diatoms

The diatom mitotic spindle (Figure 6a) has been a key model system for understanding the mechanism of anaphase B spindle elongation. It is unusual in having an almost paracrystalline array of hundreds of ipMTs that overlap to form the central spindle which links the spindle poles, and is spatially separated from the kMTs. The hypothesis that the central spindle uses an antiparallel ipMT sliding filament mechanism, coupled to ipMT plus end polymerization, to push apart the spindle poles has gained very strong support from seminal work involving detailed electron microscopy [26], live cell functional perturbation [67] combined with the in vitro reactivation of spindle elongation in isolated central spindles [35]; reviewed by [16]. The observation that pan-kinesin peptide antibodies inhibit spindle elongation in vitro suggests that one of the aforementioned MT-MT sliding kinesins uses ATP hydrolysis to drive ipMT-MT outward sliding in this system [36].

8.2. Fungi

8.2.1. Yeast

Anaphase B spindle elongation represents the major mechanism underlying chromosome segregation in *S. cerevisiae* and *S. pombe* with anaphase A contributing relatively little [58,132]. Budding yeast (Figure 6b) is unusual in possessing two members of the kinesin-5 family that cooperate to slide apart antiparallel ipMTs to push apart the spindle pole bodies, with one driving an early rapid phase of spindle elongation and the second driving a slower late phase [32,132]. Ase1p and kinesin-8 also play significant roles in this sliding filament mechanism [117,141], while cytoplasmic aMTs are dispensable for anaphase B [184] and mainly contribute to pulling the spindle/nucleus into the bud neck [185]. Similarly, a cortical pulling mechanism plays very little role in fission yeast anaphase B, which depends virtually exclusively on a midzonal ipMT-MT pushing mechanism [68,71,87], although in this case mediated predominantly by kinesin-6-driven MT-MT sliding, with forces generated by kinesin-5 playing at most a minor role [147]. In both types of yeast, poleward MT flux is thought not to occur and therefore does not contribute to the regulation of anaphase B.

8.2.2. Filamentous and Smut Fungi

Mitosis has been less extensively studied in other divisions of fungi, yet the studies that have been done provide some of the clearest available evidence so far for a cortical pulling mechanism for spindle elongation. For example, laser microbeam studies of living cells of filamentous fungi suggest that outward pulling forces acting on aMTs drive pole–pole separation during anaphase B, with the central spindle serving to constrain the rate of spindle elongation [38]. Subsequently, laser microsurgery and genetic analysis done on the smut fungus, *Ustilago maydis* (Figure 6c), supports the hypothesis that such outward pulling forces are generated by cortical dynein, which drives fast spindle elongation, with kinesin-5 on the midzone driving an initial slow phase of anaphase B [39].

8.3. Plants

The mechanism of mitosis in plant cells has long been a topic of great interest [186]. Many plant spindles lack centrosomes (or other discrete MT organizing centers) and aMTs at their poles but, based on EM studies of Haemanthus, they are thought to have "conventional" ipMT bundles constructed from two sets of opposite polarity ipMTs that interdigitate at the midzone and appear to slide apart during anaphase B [24]. Live cell imaging of mitosis in cultured tobacco cells suggests that the spindle elongates, possibly by an ipMT-MT pushing force from about 15 to 20 μm during anaphase B, contributing ≈40% of the final chromosome separation distance [187]. Plant genomes appear to contain large numbers of kinesin and Ase1p family members that could cooperate to drive such a midzonal sliding filament mechanism, e.g., the Arabidopsis genome encodes nine Ase1p MAPs and 61 kinesins, some of which localize to the anaphase spindle midzone, and much effort is aimed at determining their largely unknown functions [188,189]. For example, using fluorescent reporter tagging and live cell imaging in moss, it was found that 43 of 72 kinesins localize to the mitotic spindle and of these almost 30 localized to the antiparallel MT bundles of the anaphase spindle midzone and/or phragmoplast [190] Dissecting the precise functions of these motors, including any roles in anaphase B spindle elongation, represents a daunting yet exciting challenge for plant cell biologists.

8.4. *Animals*

8.4.1. *Caenorhabditis Elegans*

Cortical pulling forces acting on astral MTs represent the major mechanism for exerting outward-directed forces on the spindle poles in *C. elegans* embryos (Figure 6d). These forces, which have been characterized using elegant biophysical experiments [40,100], are thought to underlie both anaphase B spindle elongation and spindle positioning leading to developmentally important asymmetric cell divisions [191,192]. These pulling forces are very plausibly generated by a combination of aMT depolymerization and dynein motors that walk towards the minus ends of aMTs [103,171,191,193], with kinesin-5 at the spindle midzone serving as an antagonistic brake that constrains the rate of pole–pole separation [64]. Interestingly, recent work suggests that a direct interaction between the MT crosslinkers, kinesin-6, and Ase1p/PRC1, could augment the action of kinesin-5 by reinforcing the mechanical resilience of the central spindle to facilitate the dual functions of the midzone during anaphase B and cytokinesis [70]. It is also intriguing that, while cortical pulling represents the dominant mechanism driving spindle elongation in this system, centrosome ablation experiments have revealed the existence of a normally cryptic, redundant mechanism in which ipMT plus end polymerization at the spindle midzone generates an outward force that drives spindle elongation and chromosome segregation during anaphase [108].

Underscoring the diversity of anaphase B mechanisms operating, even within different cells of the same organism, a novel spindle elongation-dependent chromosome segregation mechanism has been found in female meiotic spindles. First, during metaphase, the spindle poles move inward and attach to paired chromosomes; then, during anaphase, the poles are separated carrying the attached homologs with them [109,149]. Since these spindles are anastral, cortical pulling is unlikely, and the authors propose that a motor-driven midzone pushing mechanism drives pole–pole separation, independent of both dynein and kinesin-5 function, but requiring ipMT polymerization. Candidate force generators include one of the aforementioned MT-MT sliding kinesins, a Par-M-type polymerization mechanism and/or an Ase1p-dependent entropic expansion mechanism.

8.4.2. *Drosophila*

Anaphase B spindle elongation in *Drosophila* embryos (Figures 4 and 6e) during cycles 10–13 is thought to depend on a persistent kinesin-5-generated interpolar (ip) microtubule (MT) sliding filament mechanism that "engages" to push apart the spindle poles when poleward flux is turned off [15,60,61]. Based on serial section EM, the spindle poles are linked by "conventional" ipMT bundles whose MTs are crosslinked into a mechanical continuum, possibly by augmin [37,60,194]. Thus, pre-anaphase B spindles are characterized by a force balance in which the outward, kinesin-5-driven sliding of these ipMT bundles is balanced by the kinesin-13-catalyzed depolymerization of their minus ends when they reach the poles, producing poleward flux and maintaining the spindle at a steady length [77,166]. Following cyclin B degradation to initiate anaphase B [180], however, the MT minus-end capping protein, Patronin [163] counteracts kinesin-13 activity at spindle poles to turn off ipMT minus end depolymerization so that poleward flux ceases and the outwardly sliding ipMTs can then elongate the spindle [113]. At the same time, ipMTs display net growth and recruit MT-MT crosslinkers to build a more robust midzone where ipMT-MT sliding forces are generated [62]. Notable among these is Ase1p (aka Feo), which is required for normal spindle elongation as it controls the organization, stability, and motor composition of the midzone, thereby facilitating the kinesin-5-driven sliding filament mechanism underlying proper spindle elongation and chromosome segregation [93].

Here again diversity is found. For example, in cultured S2 cells, the basic mechanism involving an inverse correlation between poleward flux and spindle elongation is observed, but quantitative differences exist, for example flux is only partially turned off at anaphase B onset leading to more variance in rates of spindle elongation [63]. Moreover, mitotic spindles in extracts prepared from embryos during cycles 6–7 lack cortices but appear to utilize cytoplasmic astral pulling forces during anaphase spindle elongation [104]. The significance of these differences is unclear.

8.4.3. Vertebrates

The observation using EM that the minus ends of ipMT bundles in vertebrate-cultured cells do not reach the poles suggests that they do not directly push the poles apart [90], even though ipMTs slide apart and polymerize at the midzone as the anaphase B spindle elongates [33], plausibly assisted by the augmin-/γ-TuRC-dependent branching polymerization of ipMT plus ends [107] (Figure 6f). Various studies support the idea that forces exerted at the cortex, presumably by dynein and/or aMT depolymerization, pull the poles apart while these forces are resisted by antagonistic forces generated at the midzone [66,173]. Moreover, inhibitors that weaken or enhance MT binding by kinesin-5 lead to an increased or decreased rate of spindle elongation, respectively, suggesting that midzonal kinesin-5 acts as a brake that restricts anaphase B [65]. These studies were done mainly on mammalian cultured cells, and of course may not be applicable to all vertebrate cells, where it is well established that great diversity exists e.g., anaphase B is unusual in preceding anaphase A in mouse eggs [42]. It is also worth noting that *Xenopus* extract spindles, which have been so influential in studies of many aspects of the mechanism of mitosis [110,195,196], have contributed less to our understanding of anaphase B spindle elongation, which is not a robust feature of extract spindles, possibly because of the absence of cortices [197].

8.5. Prokaryotes

The R1 plasmid segregation system of *E. coli* represents a striking example of an anaphase B-like DNA segregation system operating in bacteria (Figure 6g). A large body of work suggests that an antiparallel bundle of ParM filaments, with their plus-ends facing outward and attached to ParR-coated centromeric DNA, can polymerize and exert force by a polymer ratchet mechanism to push pairs of sister plasmids to opposite ends of the cell [10,174–176]. Thus the bidirectional polymerization, rather than the sliding apart, of bundles of antiparallel cytoskeletal polymers drives spindle elongation.

9. Theoretical Models of Anaphase B

Because of the molecular complexity of the machinery that mediates anaphase B spindle elongation, it is appreciated that a full understanding of the molecular mechanism of anaphase B will require theoretical/quantitative modeling [198]. Theoretical models (Figure 7) that incorporate realistic properties of the spindle and its molecular components can yield solutions that illuminate e.g., the factors that govern the mechanical design of the elongating spindle, describe its dynamic evolution, in plots of spindle length versus time, and display spindle dynamics in a visually accessible manner using computer simulations. Model solutions can also illuminate features of anaphase B that cannot be revealed through intuition and, most useful of all, yield predictions that can be tested experimentally.

An early example of a theoretical model that has strongly influenced our thinking about the mechanism of anaphase B spindle elongation was the sliding filament model [22]. This qualitative model was initially proposed in order to explain the mechanism of mitosis, including both anaphase A and B, based on the sliding apart of adjacent, structurally polar spindle MTs driven by ATP-hydrolyzing mechanochemical cross-bridges (aka mitotic motors). The model was useful because it made precise and unique predictions about the relative polarity patterns of the spindle MTs, leading to the development of methods for carefully testing these predictions by EM (see above). The results essentially ruled out the sliding filament mechanism for chromosome-to-pole motion during anaphase A, which is now understood to depend upon some kind of "pacman-flux" mechanism associated with the shortening of kMTs, but they do strongly support an outward ipMT sliding filament mechanism for anaphase B spindle elongation.

A different type of theoretical approach illuminates the mechanical design principles of the "beam-like" fission yeast mitotic spindle undergoing anaphase B spindle elongation [88]. These spindles use an antiparallel ipMT-MT sliding filament mechanism generated at the midzone to elongate against opposing compressive forces that could potentially cause spindle buckling. Theoretical considerations of the physical principles operating in these spindles, combined with computer simulations and structural analysis via EM tomography, led the authors to propose that the compressive strength of the elongating spindles is optimized to support the drag forces that resist spindle elongation. This is accomplished by crosslinking the ipMTs into rigid, paracrystalline arrays that display square and hexagonal symmetry within and outside the central midzone, respectively, and by preserving the optimal number and net length of the ipMTs.

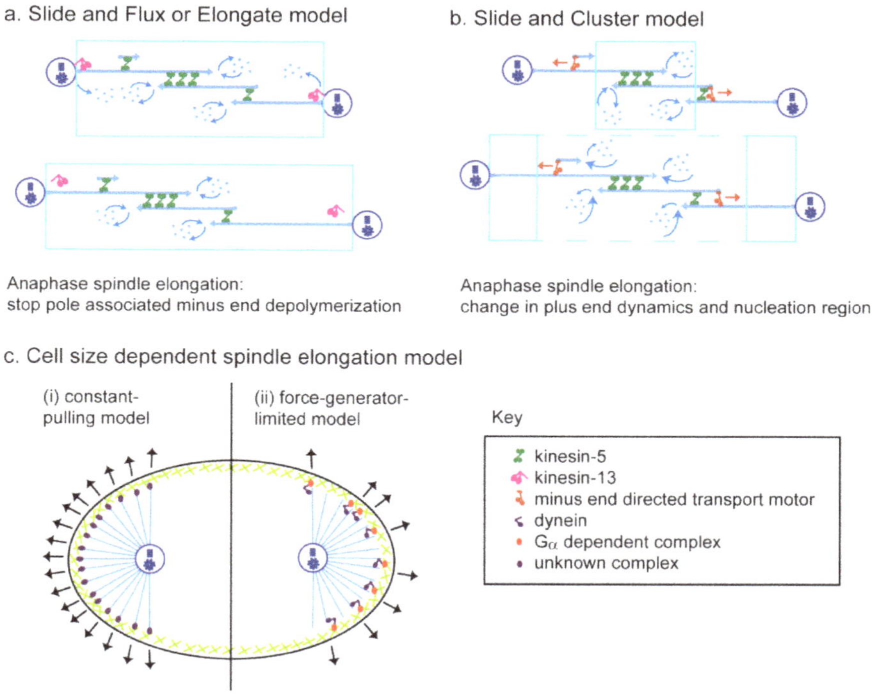

Figure 7. Theoretical models of anaphase B. Three of the models discussed in the text are drawn here, namely (**a**) the slide and flux-or-elongate (SAFE) model; (**b**) the slide-and-cluster (SAC) model; and (**c**) the cell-size dependent spindle elongation model. ((**a**) and (**b**) are based on [199] and (**c**) is based on [192].)

A "slide and flux-or-elongate" (SAFE) model based on spindle geometry and the biochemical properties of spindle components was used to model the dynamics of the transition between the pre-anaphase B (i.e., metaphase/anaphase A) spindle that maintains a constant steady state length and anaphase B spindle elongation in *Drosophila* embryos [60] (Figure 7a). Where possible, this force balance model was guided by experimental data on e.g., the geometry of embryo spindles including the three-dimensional organization and dynamic properties of MTs within ipMT bundles and the force velocity properties of mitotic motors. These properties were incorporated into a set of differential equations that describe the kinematics of the spindle poles, the kinematics of the overlapping ipMTs at the spindle midzone, and the ipMT-generated forces acting on the spindle poles. The solution of these model equations faithfully recapitulates spindle dynamics throughout the pre-anaphase B steady state, when the plus ends of the overlapping ipMTs at the midzone are slid apart and depolymerize at their minus ends, giving rise to poleward ipMT flux and anaphase B spindle elongation, which is initiated by the cessation of ipMT depolymerization at the poles, turning off flux and allowing the sliding ipMTs to push apart the poles. The model solution describes how bipolar kinesin-5 motors slide apart the dynamically unstable ipMTs at the midzone to produce the steady, linear rate of elongation that is observed in vivo and showed that the elongation rate is determined solely by the rate of ipMT sliding combined with the extent of suppression of ipMT depolymerization at the poles. It also made interesting, experimentally testable predictions, e.g., that the rate of elongation was robust to substantial changes in the number of sliding motors and ipMTs and to the dynamic instability properties of the ipMTs, but it suggested that the plus ends of these dynamic ipMTs must undergo net polymerization in order to sustain the robust, linear spindle elongation observed. Models superficially similar to this model have also been proposed to account for spindle length control in *Xenopus* extracts, where consistent anaphase B is not observed [200] and in cultured human cells, where an Aurora B gradient regulates the minus end depolymerization, length, and alignment of ipMTs that slide apart at their plus ends to determine the length of the central spindle that forms during late anaphase [201].

A very different force balance model, the "slide and cluster (SAC) model" [110] (Figure 7b), somewhat surprisingly, can also account for important features of anaphase B spindle elongation. In this model, which was initially developed to account for the control of metaphase meiotic spindle steady state length in *Xenopus* extracts, antiparallel MTs are nucleated around chromatin at the spindle midzone and are slid outward by kinesin-5. Around the equator, a minus-end-directed motor assists kinesin-5 by transporting the nucleated MTs, minus ends leading, along parallel MT tracks towards the spindle poles where it then opposes kinesin-5 and clusters the transported MTs to focus the poles. In this model, spindle length is determined by the rate of transport of the poleward sliding MTs combined with their lifetime, which in turn depends on the dynamic instability properties of their plus (not minus) ends. Thus, it was postulated, for example, that a change in plus end dynamics of these MTs, e.g., a decrease in their catastrophe frequency, could induce spindle elongation to a new steady state length, mimicking anaphase B and this idea was recently tested [199]. Despite the considerable differences in architecture that exist between *Xenopus* extract and *Drosophila* embryo spindles, quantitative computational modeling suggests that the SAC model can explain many, but not all, aspects of anaphase B spindle dynamics in *Drosophila* embryos almost as well as the SAFE model can. It was thus concluded that the SAFE model provides a more realistic description of the underlying molecular mechanism of anaphase B spindle elongation, at least in *Drosophila* embryos.

A quantitative model has been proposed to describe cell size-dependent anaphase B spindle elongation in *C. elegans* early embryos, invoking cortical force generators acting on astral MTs to pull apart the spindle poles [192] (Figure 7c). In this system, quantitative measurements revealed that the rate and extent of anaphase B spindle elongation, which govern the post-anaphase B spindle length, correlate with cell size. Two models for the cortical pulling mechanism controlling anaphase B spindle elongation were considered: (i) the "constant pulling model" for heterotrimeric G-protein (Gα)-independent spindle elongation in which astral MTs are pulled outward with a constant force; and (ii) the "force generator-limited model" for Gα-dependent spindle elongation, in which the density of force generators per unit area of cortex is constant and independent of cell size. In computer simulations, these two models could account for the observed dynamics of anaphase B spindle elongation that are seen in vivo. Specifically, simulations of the first model reproduced only the cell-size dependency of the extent of spindle elongation but not its speed, which remained constant with cell size, a situation that agrees with observations of Gα-disrupted cells. However, simulations using a combination of the first and second models reproduced observations in wild-type embryonic spindles, in which the extent and the speed of spindle elongation, as well as the resulting post-anaphase B spindle length, are all cell size-dependent.

10. Concluding Remarks

Anaphase B spindle elongation represents arguably one of the simplest sub-routines in the mechanism of mitosis, yet an elaborate and diverse machinery has evolved to accomplish pole–pole separation over distances of a few microns in the various systems that have so far been studied. Indeed, more variety than we currently appreciate may exist as more exotic mitotic mechanisms are uncovered in different cells and organisms [183], a view supported by the striking discovery that *C. elegans* meiotic spindles utilize a novel anaphase B mechanism to segregate chromosomes [149]. We have found it useful to describe the currently known mechanisms of anaphase B in terms of the differential deployment of a few conserved biochemical modules in different cell types, i.e., midzone pushing and braking via ipMT-MT crosslinking; cortical pulling, ipMT plus end dynamics/net polymerization, and minus end depolymerization/poleward flux. One aim of the research on this topic is to elucidate the molecular mechanism of these fundamental processes in atomic detail at high temporal resolution, where progress is being made, e.g., [125,128,175,202]. Progress is also being made in the reconstitution of some of these basic processes from purified components, which represents a powerful direction for investigating their underlying molecular mechanisms [154,176]. Although not a major focus of the current review, the regulation of anaphase B represents an important area for future studies. Current evidence suggests that this may be a very complex and dynamic process, making mathematical modeling an essential tool for understanding the mechanisms at work [177]. In addition, given the elaborate and diverse nature of the cytoskeleton-based mechanisms that mediate anaphase B and other aspects of mitosis and cell division among present-day eukaryotic and bacterial cells, a fascinating unknown is how these machineries and mechanisms originated and evolved from the purely physical mechanisms that were thought to operate in dividing ancestral protocells on the early Earth [203].

Acknowledgments: We thank Eileen O'Toole, Stan Cohn, and Haifeng Wang for helpful suggestions. Our work on mitosis in *Drosophila* was supported by NIH grant GM55507 to Jonathan M. Scholey.

Author Contributions: The three authors together planned the overall scope and organization of the text and figures. Jonathan M. Scholey wrote the initial version of the article incorporating drafts of sections 2 and 7 plus the figures which were done by Ingrid Brust-Mascher. All three authors read and edited the completed manuscript.

Conflicts of Interest: The authors declare no conflict of interest.

References

1. McIntosh, J.R.; Molodtsov, M.I.; Ataullakhanov, F.I. Biophysics of mitosis. *Q. Rev. Biophys.* **2012**, *45*, 147–207. [CrossRef] [PubMed]
2. Walczak, C.E.; Cai, S.; Khodjakov, A. Mechanisms of chromosome behaviour during mitosis. *Nat. Rev. Mol. Cell Biol.* **2010**, *11*, 91–102. [CrossRef] [PubMed]
3. Goshima, G.; Scholey, J.M. Control of mitotic spindle length. *Annu. Rev. Cell Dev. Biol.* **2010**, *26*, 21–57. [CrossRef] [PubMed]
4. Maiato, H.; Lince-Faria, M. The perpetual movements of anaphase. *Cell Mol. Life Sci.* **2010**, *67*, 2251–2269. [CrossRef] [PubMed]
5. Gadde, S.; Heald, R. Mechanisms and molecules of the mitotic spindle. *Curr. Biol.* **2004**, *14*, R797–R805. [CrossRef] [PubMed]
6. Cross, R.A.; McAinsh, A. Prime movers: The mechanochemistry of mitotic kinesins. *Nat. Rev. Mol. Cell Biol.* **2014**, *15*, 257–271. [CrossRef] [PubMed]
7. Petry, S. Mechanisms of mitotic spindle assembly. *Annu. Rev. Biochem.* **2016**, *85*, 659–683. [CrossRef] [PubMed]

8. Bouck, D.C.; Joglekar, A.P.; Bloom, K.S. Design features of a mitotic spindle: Balancing tension and compression at a single microtubule kinetochore interface in budding yeast. *Annu. Rev. Genet.* **2008**, *42*, 335–359. [CrossRef] [PubMed]
9. Maddox, P.S.; Oegema, K.; Desai, A.; Cheeseman, I.M. "Holo"er than thou: Chromosome segregation and kinetochore function in *C. elegans*. *Chromosome Res.* **2004**, *12*, 641–653. [CrossRef] [PubMed]
10. Gayathri, P.; Fujii, T.; Moller-Jensen, J.; van den Ent, F.; Namba, K.; Lowe, J. A bipolar spindle of antiparallel ParM filaments drives bacterial plasmid segregation. *Science* **2012**, *338*, 1334–1337. [CrossRef] [PubMed]
11. Cimini, D.; Cameron, L.A.; Salmon, E.D. Anaphase spindle mechanics prevent mis-segregation of merotelically oriented chromosomes. *Curr. Biol.* **2004**, *14*, 2149–2155. [CrossRef] [PubMed]
12. Courtheoux, T.; Gay, G.; Gachet, Y.; Tournier, S. Ase1/PRC1-dependent spindle elongation corrects merotely during anaphase in fission yeast. *J. Cell Biol.* **2009**, *187*, 399–412. [CrossRef] [PubMed]
13. Pidoux, A.L.; Uzawa, S.; Perry, P.E.; Cande, W.Z.; Allshire, R.C. Live analysis of lagging chromosomes during anaphase and their effect on spindle elongation rate in fission yeast. *J. Cell Sci.* **2000**, *113 Pt 23*, 4177–4191. [PubMed]
14. Ford, J.H. Protraction of anaphase B in lymphocyte mitosis with ageing: Possible contribution to age-related cancer risk. *Mutagenesis* **2013**, *28*, 307–314. [CrossRef] [PubMed]
15. Brust-Mascher, I.; Scholey, J.M. Mitotic motors and chromosome segregation: The mechanism of anaphase B. *Biochem. Soc. Trans.* **2011**, *39*, 1149–1153. [CrossRef] [PubMed]
16. Cande, W.Z.; Hogan, C.J. The mechanism of anaphase spindle elongation. *Bioessays* **1989**, *11*, 5–9. [CrossRef] [PubMed]
17. Roostalu, J.; Schiebel, E.; Khmelinskii, A. Cell cycle control of spindle elongation. *Cell Cycle* **2010**, *9*, 1084–1090. [CrossRef] [PubMed]
18. Ris, H. A quantitative study of anaphase movement in the aphid tamalia. *Biol. Bull.* **1943**, *85*, 164–178. [CrossRef]
19. Ris, H. The anaphase movement of chromosomes in the spermatocytes of the grasshopper. *Biol. Bull.* **1949**, *96*, 90–106. [CrossRef] [PubMed]
20. Rapaport, R. *Cytokinesis in Animal Cells*; Cambridge University Press: Cambridge, UK, 1996; p. 386.
21. Schrader, F. *Mitosis. The Movement of Chromosomes during Cell Division*, 3rd ed.; Columbia University Press: New York, NY, USA, 1949; p. 110.
22. McIntosh, J.R.; Hepler, P.K.; van Wie, D.G. Model for mitosis. *Nature* **1969**, *224*, 659–663. [CrossRef]
23. Huxley, H.; Hanson, J. Changes in the cross-striations of muscle during contraction and stretch and their structural interpretation. *Nature* **1954**, *173*, 973–976. [CrossRef] [PubMed]
24. Euteneuer, U.; Jackson, W.T.; McIntosh, J.R. Polarity of spindle microtubules in haemanthus endosperm. *J. Cell Biol.* **1982**, *94*, 644–653. [CrossRef] [PubMed]
25. Hepler, P.K.; McIntosh, J.R.; Cleland, S. Intermicrotubule bridges in mitotic spindle apparatus. *J. Cell Biol.* **1970**, *45*, 438–444. [CrossRef] [PubMed]
26. McDonald, K.; Pickett-Heaps, J.D.; McIntosh, J.R.; Tippit, D.H. On the mechanism of anaphase spindle elongation in diatoma vulgare. *J. Cell Biol.* **1977**, *74*, 377–388. [CrossRef] [PubMed]
27. McIntosh, J.R.; Landis, S.C. The distribution of spindle microtubules during mitosis in cultured human cells. *J. Cell Biol.* **1971**, *49*, 468–497. [CrossRef] [PubMed]
28. Kashina, A.S.; Baskin, R.J.; Cole, D.G.; Wedaman, K.P.; Saxton, W.M.; Scholey, J.M. A bipolar kinesin. *Nature* **1996**, *379*, 270–272. [CrossRef] [PubMed]
29. Scholey, J.M.; Porter, M.E.; Grissom, P.M.; McIntosh, J.R. Identification of kinesin in sea urchin eggs, and evidence for its localization in the mitotic spindle. *Nature* **1985**, *318*, 483–486. [CrossRef] [PubMed]
30. Kapitein, L.C.; Peterman, E.J.; Kwok, B.H.; Kim, J.H.; Kapoor, T.M.; Schmidt, C.F. The bipolar mitotic kinesin Eg5 moves on both microtubules that it crosslinks. *Nature* **2005**, *435*, 114–118. [CrossRef] [PubMed]
31. Cole, D.G.; Saxton, W.M.; Sheehan, K.B.; Scholey, J.M. A "slow" homotetrameric kinesin-related motor protein purified from *Drosophila* embryos. *J. Biol. Chem.* **1994**, *269*, 22913–22916. [PubMed]
32. Winey, M.; Mamay, C.L.; O'Toole, E.T.; Mastronarde, D.N.; Giddings, T.H., Jr.; McDonald, K.L.; McIntosh, J.R. Three-dimensional ultrastructural analysis of the *Saccharomyces cerevisiae* mitotic spindle. *J. Cell Biol.* **1995**, *129*, 1601–1615. [CrossRef] [PubMed]
33. Saxton, W.M.; McIntosh, J.R. Interzone microtubule behavior in late anaphase and telophase spindles. *J. Cell Biol.* **1987**, *105*, 875–886. [CrossRef] [PubMed]

34. Cande, W.Z.; McDonald, K. Physiological and ultrastructural analysis of elongating mitotic spindles reactivated in vitro. *J. Cell Biol.* **1986**, *103*, 593–604. [CrossRef] [PubMed]
35. Cande, W.Z.; McDonald, K.L. In vitro reactivation of anaphase spindle elongation using isolated diatom spindles. *Nature* **1985**, *316*, 168–170. [CrossRef] [PubMed]
36. Hogan, C.J.; Wein, H.; Wordeman, L.; Scholey, J.M.; Sawin, K.E.; Cande, W.Z. Inhibition of anaphase spindle elongation in vitro by a peptide antibody that recognizes kinesin motor domain. *Proc. Natl. Acad. Sci. USA* **1993**, *90*, 6611–6615. [CrossRef] [PubMed]
37. Sharp, D.J.; McDonald, K.L.; Brown, H.M.; Matthies, H.J.; Walczak, C.; Vale, R.D.; Mitchison, T.J.; Scholey, J.M. The bipolar kinesin, KLP61F, cross-links microtubules within interpolar microtubule bundles of *Drosophila* embryonic mitotic spindles. *J. Cell Biol.* **1999**, *144*, 125–138. [CrossRef] [PubMed]
38. Aist, J.R.; Bayles, C.J.; Tao, W.; Berns, M.W. Direct experimental evidence for the existence, structural basis and function of astral forces during anaphase B in vivo. *J. Cell Sci.* **1991**, *100 Pt 2*, 279–288. [PubMed]
39. Fink, G.; Schuchardt, I.; Colombelli, J.; Stelzer, E.; Steinberg, G. Dynein-mediated pulling forces drive rapid mitotic spindle elongation in *ustilago maydis*. *EMBO J.* **2006**, *25*, 4897–4908. [CrossRef] [PubMed]
40. Grill, S.W.; Gonczy, P.; Stelzer, E.H.; Hyman, A.A. Polarity controls forces governing asymmetric spindle positioning in the *Caenorhabditis elegans* embryo. *Nature* **2001**, *409*, 630–633. [CrossRef] [PubMed]
41. Waters, J.C.; Cole, R.W.; Rieder, C.L. The force-producing mechanism for centrosome separation during spindle formation in vertebrates is intrinsic to each aster. *J. Cell Biol.* **1993**, *122*, 361–372. [CrossRef] [PubMed]
42. FitzHarris, G. Anaphase B precedes anaphase A in the mouse egg. *Curr. Biol.* **2012**, *22*, 437–444. [CrossRef] [PubMed]
43. Yanagida, M. The role of model organisms in the history of mitosis research. *Cold Spring Harb. Perspect. Biol.* **2014**, *6*, a015768. [CrossRef] [PubMed]
44. Baskin, T.I.; Cande, W.Z. Kinetic analysis of mitotic spindle elongation in vitro. *J. Cell Sci.* **1990**, *97 Pt 1*, 79–89. [PubMed]
45. Cohn, S.A.; Pickett-Heaps, J.D. The effects of colchicine and dinitrophenol on the in vivo rates of anaphase A and B in the diatom *surirella*. *Eur. J. Cell Biol.* **1988**, *46*, 523–530. [PubMed]
46. Brust-Mascher, I.; Scholey, J.M. Mitotic spindle dynamics in *Drosophila*. *Int. Rev. Cytol.* **2007**, *259*, 139–172. [PubMed]
47. Yeh, E.; Skibbens, R.V.; Cheng, J.W.; Salmon, E.D.; Bloom, K. Spindle dynamics and cell cycle regulation of dynein in the budding yeast, *Saccharomyces cerevisiae*. *J. Cell Biol.* **1995**, *130*, 687–700. [CrossRef] [PubMed]
48. Desai, A.; Mitchison, T.J. Microtubule polymerization dynamics. *Annu. Rev. Cell Dev. Biol.* **1997**, *13*, 83–117. [CrossRef] [PubMed]
49. Inoue, S.; Salmon, E.D. Force generation by microtubule assembly/disassembly in mitosis and related movements. *Mol. Biol. Cell* **1995**, *6*, 1619–1640. [CrossRef] [PubMed]
50. Rogers, G.C.; Rogers, S.L.; Sharp, D.J. Spindle microtubules in flux. *J. Cell Sci.* **2005**, *118*, 1105–1116. [CrossRef] [PubMed]
51. Salmon, E.D.; Leslie, R.J.; Saxton, W.M.; Karow, M.L.; McIntosh, J.R. Spindle microtubule dynamics in sea urchin embryos: Analysis using a fluorescein-labeled tubulin and measurements of fluorescence redistribution after laser photobleaching. *J. Cell Biol.* **1984**, *99*, 2165–2174. [CrossRef] [PubMed]
52. Zhai, Y.; Kronebusch, P.J.; Simon, P.M.; Borisy, G.G. Microtubule dynamics at the g2/m transition: Abrupt breakdown of cytoplasmic microtubules at nuclear envelope breakdown and implications for spindle morphogenesis. *J. Cell Biol.* **1996**, *135*, 201–214. [CrossRef] [PubMed]
53. Mitchison, T.J. Polewards microtubule flux in the mitotic spindle: Evidence from photoactivation of fluorescence. *J. Cell Biol.* **1989**, *109*, 637–652. [CrossRef] [PubMed]
54. Mitchison, T.J.; Salmon, E.D. Poleward kinetochore fiber movement occurs during both metaphase and anaphase-A in newt lung cell mitosis. *J. Cell Biol.* **1992**, *119*, 569–582. [CrossRef] [PubMed]
55. Waterman-Storer, C.M.; Salmon, E.D. Fluorescent speckle microscopy of microtubules: How low can you go? *FASEB J.* **1999**, *13*, S225–S230. [PubMed]
56. Higuchi, T.; Uhlmann, F. Stabilization of microtubule dynamics at anaphase onset promotes chromosome segregation. *Nature* **2005**, *433*, 171–176. [CrossRef] [PubMed]
57. Maddox, P.S.; Bloom, K.S.; Salmon, E.D. The polarity and dynamics of microtubule assembly in the budding yeast *Saccharomyces cerevisiae*. *Nat. Cell Biol.* **2000**, *2*, 36–41. [CrossRef] [PubMed]

58. Mallavarapu, A.; Sawin, K.; Mitchison, T. A switch in microtubule dynamics at the onset of anaphase B in the mitotic spindle of *Schizosaccharomyces pombe*. *Curr. Biol.* **1999**, *9*, 1423–1426. [CrossRef]
59. Zhai, Y.; Kronebusch, P.J.; Borisy, G.G. Kinetochore microtubule dynamics and the metaphase-anaphase transition. *J. Cell Biol.* **1995**, *131*, 721–734. [CrossRef] [PubMed]
60. Brust-Mascher, I.; Civelekoglu-Scholey, G.; Kwon, M.; Mogilner, A.; Scholey, J.M. Model for anaphase B: Role of three mitotic motors in a switch from poleward flux to spindle elongation. *Proc. Natl. Acad. Sci. USA* **2004**, *101*, 15938–15943. [CrossRef] [PubMed]
61. Brust-Mascher, I.; Scholey, J.M. Microtubule flux and sliding in mitotic spindles of *Drosophila* embryos. *Mol. Biol. Cell* **2002**, *13*, 3967–3975. [CrossRef] [PubMed]
62. Cheerambathur, D.K.; Civelekoglu-Scholey, G.; Brust-Mascher, I.; Sommi, P.; Mogilner, A.; Scholey, J.M. Quantitative analysis of an anaphase B switch: Predicted role for a microtubule catastrophe gradient. *J. Cell Biol.* **2007**, *177*, 995–1004. [CrossRef] [PubMed]
63. De Lartigue, J.; Brust-Mascher, I.; Scholey, J.M. Anaphase B spindle dynamics in *Drosophila* s2 cells: Comparison with embryo spindles. *Cell Div.* **2011**. [CrossRef] [PubMed]
64. Saunders, A.M.; Powers, J.; Strome, S.; Saxton, W.M. Kinesin-5 acts as a brake in anaphase spindle elongation. *Curr. Biol.* **2007**, *17*, R453–R454. [CrossRef] [PubMed]
65. Collins, E.; Mann, B.J.; Wadsworth, P. Eg5 restricts anaphase B spindle elongation in mammalian cells. *Cytoskeleton (Hoboken)* **2014**, *71*, 136–144. [CrossRef] [PubMed]
66. Aist, J.R.; Liang, H.; Berns, M.W. Astral and spindle forces in PTK2 cells during anaphase B: A laser microbeam study. *J. Cell Sci.* **1993**, *104 Pt 4*, 1207–1216. [PubMed]
67. Leslie, R.J.; Pickett-Heaps, J.D. Ultraviolet microbeam irradiations of mitotic diatoms: Investigation of spindle elongation. *J. Cell Biol.* **1983**, *96*, 548–561. [CrossRef] [PubMed]
68. Khodjakov, A.; La Terra, S.; Chang, F. Laser microsurgery in fission yeast; role of the mitotic spindle midzone in anaphase B. *Curr. Biol.* **2004**, *14*, 1330–1340. [CrossRef] [PubMed]
69. Verbrugghe, K.J.; White, J.G. SPD-1 is required for the formation of the spindle midzone but is not essential for the completion of cytokinesis in *C. elegans* embryos. *Curr. Biol.* **2004**, *14*, 1755–1760. [CrossRef] [PubMed]
70. Lee, K.Y.; Esmaeili, B.; Zealley, B.; Mishima, M. Direct interaction between centralspindlin and PRC1 reinforces mechanical resilience of the central spindle. *Nat. Commun.* **2015**. [CrossRef] [PubMed]
71. Tolic-Norrelykke, I.M.; Sacconi, L.; Thon, G.; Pavone, F.S. Positioning and elongation of the fission yeast spindle by microtubule-based pushing. *Curr. Biol.* **2004**, *14*, 1181–1186. [CrossRef] [PubMed]
72. Nicklas, R.B. Chromosome velocity during mitosis as a function of chromosome size and position. *J. Cell Biol.* **1965**, *25*, 119–135. [CrossRef]
73. Hiramoto, Y. Mechanical properties of the protoplasm of the sea urchin egg. II. Fertilized egg. *Exp. Cell Res.* **1969**, *56*, 209–218. [CrossRef]
74. Scholey, J.M. Compare and contrast the reaction coordinate diagrams for chemical reactions and cytoskeletal force generators. *Mol. Biol. Cell* **2013**, *24*, 433–439. [CrossRef] [PubMed]
75. Nicklas, R.B. Measurements of the force produced by the mitotic spindle in anaphase. *J. Cell Biol.* **1983**, *97*, 542–548. [CrossRef] [PubMed]
76. Hiramoto, Y.; Nakano, Y. Micromanipulation studies of the mitotic apparatus in sand dollar eggs. *Cell Motil. Cytoskelet.* **1988**, *10*, 172–184. [CrossRef] [PubMed]
77. Brust-Mascher, I.; Sommi, P.; Cheerambathur, D.K.; Scholey, J.M. Kinesin-5-dependent poleward flux and spindle length control in *Drosophila* embryo mitosis. *Mol. Biol. Cell* **2009**, *20*, 1749–1762. [CrossRef] [PubMed]
78. Nicklas, R.B. A quantitative comparison of cellular motile systems. *Cell Motil.* **1984**, *4*, 1–5. [CrossRef] [PubMed]
79. Cande, W.Z. Nucleotide requirements for anaphase chromosome movements in permeabilized mitotic cells: Anaphase B but not anaphase A requires ATP. *Cell* **1982**, *28*, 15–22. [CrossRef]
80. Cande, W.Z. Creatine kinase role in anaphase chromosome movement. *Nature* **1983**, *304*, 557–558. [CrossRef] [PubMed]
81. Koons, S.J.; Eckert, B.S.; Zobel, C.R. Immunofluorescence and inhibitor studies on creatine kinase and mitosis. *Exp. Cell Res.* **1982**, *140*, 401–409. [CrossRef]
82. Palazzo, R.E.; Lutz, D.A.; Rebhun, L.I. Reactivation of isolated mitotic apparatus: Metaphase versus anaphase spindles. *Cell Motil. Cytoskelet.* **1991**, *18*, 304–318. [CrossRef] [PubMed]

83. Civelekoglu-Scholey, G.; Scholey, J.M. Mitotic force generators and chromosome segregation. *Cell Mol. Life Sci.* **2010**, *67*, 2231–2250. [CrossRef] [PubMed]
84. Janson, M.E.; Loughlin, R.; Loiodice, I.; Fu, C.; Brunner, D.; Nedelec, F.J.; Tran, P.T. Crosslinkers and motors organize dynamic microtubules to form stable bipolar arrays in fission yeast. *Cell* **2007**, *128*, 357–368. [CrossRef] [PubMed]
85. Lansky, Z.; Braun, M.; Ludecke, A.; Schlierf, M.; ten Wolde, P.R.; Janson, M.E.; Diez, S. Diffusible crosslinkers generate directed forces in microtubule networks. *Cell* **2015**, *160*, 1159–1168. [CrossRef] [PubMed]
86. Heidemann, S.R.; McIntosh, J.R. Visualization of the structural polarity of microtubules. *Nature* **1980**, *286*, 517–519. [CrossRef] [PubMed]
87. Ding, R.; McDonald, K.L.; McIntosh, J.R. Three-dimensional reconstruction and analysis of mitotic spindles from the yeast, *Schizosaccharomyces pombe*. *J. Cell Biol.* **1993**, *120*, 141–151. [CrossRef] [PubMed]
88. Ward, J.J.; Roque, H.; Antony, C.; Nedelec, F. Mechanical design principles of a mitotic spindle. *Elife* **2014**, *3*, e03398. [CrossRef] [PubMed]
89. Huxley, H.E. Electron microscope studies on the structure of natural and synthetic protein filaments from striated muscle. *J. Mol. Biol.* **1963**, *7*, 281–308. [CrossRef]
90. Mastronarde, D.N.; McDonald, K.L.; Ding, R.; McIntosh, J.R. Interpolar spindle microtubules in PTK cells. *J. Cell Biol.* **1993**, *123*, 1475–1489. [CrossRef] [PubMed]
91. Wilson, H.J. Arms and bridges on microtubules in the mitotic apparatus. *J. Cell Biol.* **1969**, *40*, 854–859. [CrossRef] [PubMed]
92. Cheerambathur, D.K.; Brust-Mascher, I.; Civelekoglu-Scholey, G.; Scholey, J.M. Dynamic partitioning of mitotic kinesin-5 cross-linkers between microtubule-bound and freely diffusing states. *J. Cell Biol.* **2008**, *182*, 429–436. [CrossRef] [PubMed]
93. Wang, H.; Brust-Mascher, I.; Scholey, J.M. The microtubule cross-linker Feo controls the midzone stability, motor composition, and elongation of the anaphase B spindle in *Drosophila* embryos. *Mol. Biol. Cell* **2015**, *26*, 1452–1462. [CrossRef] [PubMed]
94. Acar, S.; Carlson, D.B.; Budamagunta, M.S.; Yarov-Yarovoy, V.; Correia, J.J.; Ninonuevo, M.R.; Jia, W.; Tao, L.; Leary, J.A.; Voss, J.C.; et al. The bipolar assembly domain of the mitotic motor kinesin-5. *Nat. Commun.* **2013**. [CrossRef] [PubMed]
95. Glotzer, M. The 3ms of central spindle assembly: Microtubules, motors and maps. *Nat. Rev. Mol. Cell Biol.* **2009**, *10*, 9–20. [CrossRef] [PubMed]
96. Peterman, E.J.; Scholey, J.M. Mitotic microtubule crosslinkers: Insights from mechanistic studies. *Curr. Biol.* **2009**, *19*, R1089–R1094. [CrossRef] [PubMed]
97. Rozelle, D.K.; Hansen, S.D.; Kaplan, K.B. Chromosome passenger complexes control anaphase duration and spindle elongation via a kinesin-5 brake. *J. Cell Biol.* **2011**, *193*, 285–294. [CrossRef] [PubMed]
98. Shimamoto, Y.; Forth, S.; Kapoor, T.M. Measuring pushing and braking forces generated by ensembles of kinesin-5 crosslinking two microtubules. *Dev. Cell* **2015**, *34*, 669–681. [CrossRef] [PubMed]
99. Tikhonenko, I.; Nag, D.K.; Martin, N.; Koonce, M.P. Kinesin-5 is not essential for mitotic spindle elongation in *Dictyostelium*. *Cell Motil. Cytoskelet.* **2008**, *65*, 853–862. [CrossRef] [PubMed]
100. Grill, S.W.; Howard, J.; Schaffer, E.; Stelzer, E.H.; Hyman, A.A. The distribution of active force generators controls mitotic spindle position. *Science* **2003**, *301*, 518–521. [CrossRef] [PubMed]
101. Grill, S.W.; Hyman, A.A. Spindle positioning by cortical pulling forces. *Dev. Cell* **2005**, *8*, 461–465. [CrossRef] [PubMed]
102. Grishchuk, E.L.; Molodtsov, M.I.; Ataullakhanov, F.I.; McIntosh, J.R. Force production by disassembling microtubules. *Nature* **2005**, *438*, 384–388. [CrossRef] [PubMed]
103. Kozlowski, C.; Srayko, M.; Nedelec, F. Cortical microtubule contacts position the spindle in *C. elegans* embryos. *Cell* **2007**, *129*, 499–510. [CrossRef] [PubMed]
104. Telley, I.A.; Gaspar, I.; Ephrussi, A.; Surrey, T. Aster migration determines the length scale of nuclear separation in the *Drosophila* syncytial embryo. *J. Cell Biol.* **2012**, *197*, 887–895. [CrossRef] [PubMed]
105. Masuda, H.; Cande, W.Z. The role of tubulin polymerization during spindle elongation in vitro. *Cell* **1987**, *49*, 193–202. [CrossRef]
106. Masuda, H.; McDonald, K.L.; Cande, W.Z. The mechanism of anaphase spindle elongation: Uncoupling of tubulin incorporation and microtubule sliding during in vitro spindle reactivation. *J. Cell Biol.* **1988**, *107*, 623–633. [CrossRef] [PubMed]

107. Uehara, R.; Goshima, G. Functional central spindle assembly requires de novo microtubule generation in the interchromosomal region during anaphase. *J. Cell Biol.* **2010**, *191*, 259–267. [CrossRef] [PubMed]
108. Nahaboo, W.; Zouak, M.; Askjaer, P.; Delattre, M. Chromatids segregate without centrosomes during *Caenorhabditis elegans* mitosis in a ran- and clasp-dependent manner. *Mol. Biol. Cell* **2015**, *26*, 2020–2029. [CrossRef] [PubMed]
109. Dumont, J.; Oegema, K.; Desai, A. A kinetochore-independent mechanism drives anaphase chromosome separation during acentrosomal meiosis. *Nat. Cell Biol.* **2010**, *12*, 894–901. [CrossRef] [PubMed]
110. Burbank, K.S.; Mitchison, T.J.; Fisher, D.S. Slide-and-cluster models for spindle assembly. *Curr. Biol.* **2007**, *17*, 1373–1383. [CrossRef] [PubMed]
111. Hays, T.S.; Wise, D.; Salmon, E.D. Traction force on a kinetochore at metaphase acts as a linear function of kinetochore fiber length. *J. Cell Biol.* **1982**, *93*, 374–389. [CrossRef] [PubMed]
112. Ostergren, G. Considerations on some elementary features of mitosis. *Hereditas* **1950**, *36*, 1–19. [CrossRef]
113. Wang, H.; Brust-Mascher, I.; Civelekoglu-Scholey, G.; Scholey, J.M. Patronin mediates a switch from kinesin-13-dependent poleward flux to anaphase B spindle elongation. *J. Cell Biol.* **2013**, *203*, 35–46. [CrossRef] [PubMed]
114. Akhmanova, A.; Steinmetz, M.O. Control of microtubule organization and dynamics: Two ends in the limelight. *Nat. Rev. Mol. Cell Biol.* **2015**, *16*, 711–726. [CrossRef] [PubMed]
115. Goldstein, L.S. Functional redundancy in mitotic force generation. *J. Cell Biol.* **1993**, *120*, 1–3. [CrossRef] [PubMed]
116. Pellman, D.; Bagget, M.; Tu, Y.H.; Fink, G.R.; Tu, H. Two microtubule-associated proteins required for anaphase spindle movement in saccharomyces cerevisiae. *J. Cell Biol.* **1995**, *130*, 1373–1385. [CrossRef] [PubMed]
117. Schuyler, S.C.; Liu, J.Y.; Pellman, D. The molecular function of Ase1p: Evidence for a map-dependent midzone-specific spindle matrix. Microtubule-associated proteins. *J. Cell Biol.* **2003**, *160*, 517–528. [CrossRef] [PubMed]
118. Kapitein, L.C.; Janson, M.E.; van den Wildenberg, S.M.; Hoogenraad, C.C.; Schmidt, C.F.; Peterman, E.J. Microtubule-driven multimerization recruits Ase1p onto overlapping microtubules. *Curr. Biol.* **2008**, *18*, 1713–1717. [CrossRef] [PubMed]
119. Subramanian, R.; Ti, S.C.; Tan, L.; Darst, S.A.; Kapoor, T.M. Marking and measuring single microtubules by PRC1 and kinesin-4. *Cell* **2013**, *154*, 377–390. [CrossRef] [PubMed]
120. Khmelinskii, A.; Roostalu, J.; Roque, H.; Antony, C.; Schiebel, E. Phosphorylation-dependent protein interactions at the spindle midzone mediate cell cycle regulation of spindle elongation. *Dev. Cell* **2009**, *17*, 244–256. [CrossRef] [PubMed]
121. Wein, H.; Bass, H.W.; Cande, W.Z. Dsk1, a kinesin-related protein involved in anaphase spindle elongation, is a component of a mitotic spindle matrix. *Cell Motil. Cytoskelet.* **1998**, *41*, 214–224. [CrossRef]
122. Odde, D.J. Mitosis, diffusible crosslinkers, and the ideal gas law. *Cell* **2015**, *160*, 1041–1043. [CrossRef] [PubMed]
123. Jiang, H.; Wang, S.; Huang, Y.; He, X.; Cui, H.; Zhu, X.; Zheng, Y. Phase transition of spindle-associated protein regulate spindle apparatus assembly. *Cell* **2015**, *163*, 108–122. [CrossRef] [PubMed]
124. Civelekoglu-Scholey, G.; Tao, L.; Brust-Mascher, I.; Wollman, R.; Scholey, J.M. Prometaphase spindle maintenance by an antagonistic motor-dependent force balance made robust by a disassembling lamin-b envelope. *J. Cell Biol.* **2010**, *188*, 49–68. [CrossRef] [PubMed]
125. Scholey, J.E.; Nithianantham, S.; Scholey, J.M.; Al-Bassam, J. Structural basis for the assembly of the mitotic motor kinesin-5 into bipolar tetramers. *Elife* **2014**, *3*, e02217. [CrossRef] [PubMed]
126. Sawin, K.E.; Le Guellec, K.; Philippe, M.; Mitchison, T.J. Mitotic spindle organization by a plus-end-directed microtubule motor. *Nature* **1992**, *359*, 540–543. [CrossRef] [PubMed]
127. Valentine, M.T.; Gilbert, S.P. To step or not to step? How biochemistry and mechanics influence processivity in kinesin and Eg5. *Curr. Opin. Cell Biol.* **2007**, *19*, 75–81. [CrossRef] [PubMed]
128. Valentine, M.T.; Fordyce, P.M.; Krzysiak, T.C.; Gilbert, S.P.; Block, S.M. Individual dimers of the mitotic kinesin motor Eg5 step processively and support substantial loads in vitro. *Nat. Cell Biol.* **2006**, *8*, 470–476. [CrossRef] [PubMed]

129. Hildebrandt, E.R.; Gheber, L.; Kingsbury, T.; Hoyt, M.A. Homotetrameric form of cin8p, a *Saccharomyces cerevisiae* kinesin-5 motor, is essential for its in vivo function. *J. Biol. Chem.* **2006**, *281*, 26004–26013. [CrossRef] [PubMed]
130. Van den Wildenberg, S.M.; Tao, L.; Kapitein, L.C.; Schmidt, C.F.; Scholey, J.M.; Peterman, E.J. The homotetrameric kinesin-5 KLP61F preferentially crosslinks microtubules into antiparallel orientations. *Curr. Biol.* **2008**, *18*, 1860–1864. [CrossRef] [PubMed]
131. Enos, A.P.; Morris, N.R. Mutation of a gene that encodes a kinesin-like protein blocks nuclear division in *A. nidulans*. *Cell* **1990**, *60*, 1019–1027. [CrossRef]
132. Straight, A.F.; Sedat, J.W.; Murray, A.W. Time-lapse microscopy reveals unique roles for kinesins during anaphase in budding yeast. *J. Cell Biol.* **1998**, *143*, 687–694. [CrossRef] [PubMed]
133. Saunders, W.S.; Koshland, D.; Eshel, D.; Gibbons, I.R.; Hoyt, M.A. *Saccharomyces cerevisiae* kinesin- and dynein-related proteins required for anaphase chromosome segregation. *J. Cell Biol.* **1995**, *128*, 617–624. [CrossRef] [PubMed]
134. Avunie-Masala, R.; Movshovich, N.; Nissenkorn, Y.; Gerson-Gurwitz, A.; Fridman, V.; Koivomagi, M.; Loog, M.; Hoyt, M.A.; Zaritsky, A.; Gheber, L. Phospho-regulation of kinesin-5 during anaphase spindle elongation. *J. Cell Sci.* **2011**, *124*, 873–878. [CrossRef] [PubMed]
135. Sharp, D.J.; Brown, H.M.; Kwon, M.; Rogers, G.C.; Holland, G.; Scholey, J.M. Functional coordination of three mitotic motors in *Drosophila* embryos. *Mol. Biol. Cell* **2000**, *11*, 241–253. [CrossRef] [PubMed]
136. Roostalu, J.; Hentrich, C.; Bieling, P.; Telley, I.A.; Schiebel, E.; Surrey, T. Directional switching of the kinesin cin8 through motor coupling. *Science* **2011**, *332*, 94–99. [CrossRef] [PubMed]
137. Gerson-Gurwitz, A.; Thiede, C.; Movshovich, N.; Fridman, V.; Podolskaya, M.; Danieli, T.; Lakamper, S.; Klopfenstein, D.R.; Schmidt, C.F.; Gheber, L. Directionality of individual kinesin-5 cin8 motors is modulated by loop 8, ionic strength and microtubule geometry. *EMBO J.* **2011**, *30*, 4942–4954. [CrossRef] [PubMed]
138. Edamatsu, M. Bidirectional motility of the fission yeast kinesin-5, cut7. *Biochem. Biophys. Res. Commun.* **2014**, *446*, 231–234. [CrossRef] [PubMed]
139. Kwon, M.; Morales-Mulia, S.; Brust-Mascher, I.; Rogers, G.C.; Sharp, D.J.; Scholey, J.M. The chromokinesin, KLP3a, dives mitotic spindle pole separation during prometaphase and anaphase and facilitates chromatid motility. *Mol. Biol. Cell* **2004**, *15*, 219–233. [CrossRef] [PubMed]
140. Nislow, C.; Lombillo, V.A.; Kuriyama, R.; McIntosh, J.R. A plus-end-directed motor enzyme that moves antiparallel microtubules in vitro localizes to the interzone of mitotic spindles. *Nature* **1992**, *359*, 543–547. [CrossRef] [PubMed]
141. Su, X.; Arellano-Santoyo, H.; Portran, D.; Gaillard, J.; Vantard, M.; Thery, M.; Pellman, D. Microtubule-sliding activity of a kinesin-8 promotes spindle assembly and spindle-length control. *Nat. Cell Biol.* **2013**, *15*, 948–957. [CrossRef] [PubMed]
142. Varga, V.; Leduc, C.; Bormuth, V.; Diez, S.; Howard, J. Kinesin-8 motors act cooperatively to mediate length-dependent microtubule depolymerization. *Cell* **2009**, *138*, 1174–1183. [CrossRef] [PubMed]
143. Rizk, R.S.; Discipio, K.A.; Proudfoot, K.G.; Gupta, M.L., Jr. The kinesin-8 kip3 scales anaphase spindle length by suppression of midzone microtubule polymerization. *J. Cell Biol.* **2014**, *204*, 965–975. [CrossRef] [PubMed]
144. Mishima, M.; Kaitna, S.; Glotzer, M. Central spindle assembly and cytokinesis require a kinesin-like protein/RhoGAP complex with microtubule bundling activity. *Dev. Cell* **2002**, *2*, 41–54. [CrossRef]
145. Tao, L.; Fasulo, B.; Warecki, B.; Sullivan, W. Tum/RacGAP functions as a switch activating the Pav/kinesin-6 motor. *Nat. Commun.* **2016**. [CrossRef] [PubMed]
146. Mishima, M. Centralspindlin in rappaport's cleavage signaling. *Semin. Cell Dev. Biol.* **2016**, *53*, 45–56. [CrossRef] [PubMed]
147. Fu, C.; Ward, J.J.; Loiodice, I.; Velve-Casquillas, G.; Nedelec, F.J.; Tran, P.T. Phospho-regulated interaction between kinesin-6 KLP9p and microtubule bundler Ase1p promotes spindle elongation. *Dev. Cell* **2009**, *17*, 257–267. [CrossRef] [PubMed]
148. Drechsler, H.; McAinsh, A.D. Kinesin-12 motors cooperate to suppress microtubule catastrophes and drive the formation of parallel microtubule bundles. *Proc. Natl. Acad. Sci. USA* **2016**, *113*, E1635–E1644. [CrossRef] [PubMed]
149. McNally, K.P.; Panzica, M.T.; Kim, T.; Cortes, D.B.; McNally, F.J. A novel chromosome segregation mechanism during female meiosis. *Mol. Biol. Cell* **2016**, *27*, 2576–2589. [CrossRef] [PubMed]

150. Bieling, P.; Telley, I.A.; Surrey, T. A minimal midzone protein module controls formation and length of antiparallel microtubule overlaps. *Cell* **2010**, *142*, 420–432. [CrossRef] [PubMed]
151. Hu, C.K.; Coughlin, M.; Field, C.M.; Mitchison, T.J. Kif4 regulates midzone length during cytokinesis. *Curr. Biol.* **2011**, *21*, 815–824. [CrossRef] [PubMed]
152. Nunes Bastos, R.; Gandhi, S.R.; Baron, R.D.; Gruneberg, U.; Nigg, E.A.; Barr, F.A. Aurora b suppresses microtubule dynamics and limits central spindle size by locally activating Kif4a. *J. Cell Biol.* **2013**, *202*, 605–621. [CrossRef] [PubMed]
153. Zhu, C.; Zhao, J.; Bibikova, M.; Leverson, J.D.; Bossy-Wetzel, E.; Fan, J.B.; Abraham, R.T.; Jiang, W. Functional analysis of human microtubule-based motor proteins, the kinesins and dyneins, in mitosis/cytokinesis using rna interference. *Mol. Biol. Cell* **2005**, *16*, 3187–3199. [CrossRef] [PubMed]
154. Bieling, P.; Laan, L.; Schek, H.; Munteanu, E.L.; Sandblad, L.; Dogterom, M.; Brunner, D.; Surrey, T. Reconstitution of a microtubule plus-end tracking system in vitro. *Nature* **2007**, *450*, 1100–1105. [CrossRef] [PubMed]
155. Pereira, A.L.; Pereira, A.J.; Maia, A.R.; Drabek, K.; Sayas, C.L.; Hergert, P.J.; Lince-Faria, M.; Matos, I.; Duque, C.; Stepanova, T.; et al. Mammalian clasp1 and clasp2 cooperate to ensure mitotic fidelity by regulating spindle and kinetochore function. *Mol. Biol. Cell* **2006**, *17*, 4526–4542. [CrossRef] [PubMed]
156. Maton, G.; Edwards, F.; Lacroix, B.; Stefanutti, M.; Laband, K.; Lieury, T.; Kim, T.; Espeut, J.; Canman, J.C.; Dumont, J. Kinetochore components are required for central spindle assembly. *Nat. Cell Biol.* **2015**, *17*, 697–705. [CrossRef] [PubMed]
157. Maiato, H.; Khodjakov, A.; Rieder, C.L. *Drosophila* clasp is required for the incorporation of microtubule subunits into fluxing kinetochore fibres. *Nat. Cell Biol.* **2005**, *7*, 42–47. [CrossRef] [PubMed]
158. Chen, Y.; Hancock, W.O. Kinesin-5 is a microtubule polymerase. *Nat. Commun.* **2015**. [CrossRef] [PubMed]
159. Cole, D.G.; Chinn, S.W.; Wedaman, K.P.; Hall, K.; Vuong, T.; Scholey, J.M. Novel heterotrimeric kinesin-related protein purified from sea urchin eggs. *Nature* **1993**, *366*, 268–270. [CrossRef] [PubMed]
160. Craft, J.M.; Harris, J.A.; Hyman, S.; Kner, P.; Lechtreck, K.F. Tubulin transport by IFT is upregulated during ciliary growth by a cilium-autonomous mechanism. *J. Cell Biol.* **2015**, *208*, 223–237. [CrossRef] [PubMed]
161. Henson, J.H.; Cole, D.G.; Terasaki, M.; Rashid, D.; Scholey, J.M. Immunolocalization of the heterotrimeric kinesin-related protein KRP(85/95) in the mitotic apparatus of sea urchin embryos. *Dev. Biol.* **1995**, *171*, 182–194. [CrossRef] [PubMed]
162. Lecland, N.; Luders, J. The dynamics of microtubule minus ends in the human mitotic spindle. *Nat. Cell Biol.* **2014**, *16*, 770–778. [CrossRef] [PubMed]
163. Goodwin, S.S.; Vale, R.D. Patronin regulates the microtubule network by protecting microtubule minus ends. *Cell* **2010**, *143*, 263–274. [CrossRef] [PubMed]
164. Hendershott, M.C.; Vale, R.D. Regulation of microtubule minus-end dynamics by camsaps and patronin. *Proc. Natl. Acad. Sci. USA* **2014**, *111*, 5860–5865. [CrossRef] [PubMed]
165. Jiang, K.; Hua, S.; Mohan, R.; Grigoriev, I.; Yau, K.W.; Liu, Q.; Katrukha, E.A.; Altelaar, A.F.; Heck, A.J.; Hoogenraad, C.C.; et al. Microtubule minus-end stabilization by polymerization-driven camsap deposition. *Dev. Cell* **2014**, *28*, 295–309. [CrossRef] [PubMed]
166. Rogers, G.C.; Rogers, S.L.; Schwimmer, T.A.; Ems-McClung, S.C.; Walczak, C.E.; Vale, R.D.; Scholey, J.M.; Sharp, D.J. Two mitotic kinesins cooperate to drive sister chromatid separation during anaphase. *Nature* **2004**, *427*, 364–370. [CrossRef] [PubMed]
167. Carmena, M.; Wheelock, M.; Funabiki, H.; Earnshaw, W.C. The chromosomal passenger complex (CPC): From easy rider to the godfather of mitosis. *Nat. Rev. Mol. Cell Biol.* **2012**, *13*, 789–803. [CrossRef] [PubMed]
168. Pavin, N.; Tolic-Norrelykke, I.M. Dynein, microtubule and cargo: A menage a trois. *Biochem. Soc. Trans.* **2013**, *41*, 1731–1735. [CrossRef] [PubMed]
169. Schmidt, H.; Carter, A.P. Review: Structure and mechanism of the dynein motor ATPase. *Biopolymers* **2016**, *105*, 557–567. [CrossRef] [PubMed]
170. Ananthanarayanan, V.; Schattat, M.; Vogel, S.K.; Krull, A.; Pavin, N.; Tolic-Norrelykke, I.M. Dynein motion switches from diffusive to directed upon cortical anchoring. *Cell* **2013**, *153*, 1526–1536. [CrossRef] [PubMed]
171. Laan, L.; Pavin, N.; Husson, J.; Romet-Lemonne, G.; van Duijn, M.; Lopez, M.P.; Vale, R.D.; Julicher, F.; Reck-Peterson, S.L.; Dogterom, M. Cortical dynein controls microtubule dynamics to generate pulling forces that position microtubule asters. *Cell* **2012**, *148*, 502–514. [CrossRef] [PubMed]

172. Gonczy, P.; Pichler, S.; Kirkham, M.; Hyman, A.A. Cytoplasmic dynein is required for distinct aspects of mtoc positioning, including centrosome separation, in the one cell stage *Caenorhabditis elegans* embryo. *J. Cell Biol.* **1999**, *147*, 135–150. [CrossRef] [PubMed]
173. Vaisberg, E.A.; Koonce, M.P.; McIntosh, J.R. Cytoplasmic dynein plays a role in mammalian mitotic spindle formation. *J. Cell Biol.* **1993**, *123*, 849–858. [CrossRef] [PubMed]
174. Gerdes, K.; Howard, M.; Szardenings, F. Pushing and pulling in prokaryotic DNA segregation. *Cell* **2010**, *141*, 927–942. [CrossRef] [PubMed]
175. Bharat, T.A.; Murshudov, G.N.; Sachse, C.; Lowe, J. Structures of actin-like ParM filaments show architecture of plasmid-segregating spindles. *Nature* **2015**, *523*, 106–110. [CrossRef] [PubMed]
176. Garner, E.C.; Campbell, C.S.; Weibel, D.B.; Mullins, R.D. Reconstitution of DNA segregation driven by assembly of a prokaryotic actin homolog. *Science* **2007**, *315*, 1270–1274. [CrossRef] [PubMed]
177. Bastos, R.N.; Cundell, M.J.; Barr, F.A. Kif4a and pp2a-b56 form a spatially restricted feedback loop opposing aurora b at the anaphase central spindle. *J. Cell Biol.* **2014**, *207*, 683–693. [CrossRef] [PubMed]
178. Musacchio, A. Spindle assembly checkpoint: The third decade. *Philos. Trans. R. Soc. Lond. B Biol. Sci.* **2011**, *366*, 3595–3604. [CrossRef] [PubMed]
179. Sullivan, M.; Morgan, D.O. Finishing mitosis, one step at a time. *Nat. Rev. Mol. Cell Biol.* **2007**, *8*, 894–903. [CrossRef] [PubMed]
180. Parry, D.H.; O'Farrell, P.H. The schedule of destruction of three mitotic cyclins can dictate the timing of events during exit from mitosis. *Curr. Biol.* **2001**, *11*, 671–683. [CrossRef]
181. Yuan, K.; O'Farrell, P.H. Cyclin b3 is a mitotic cyclin that promotes the metaphase-anaphase transition. *Curr. Biol.* **2015**, *25*, 811–816. [CrossRef] [PubMed]
182. Afonso, O.; Matos, I.; Pereira, A.J.; Aguiar, P.; Lampson, M.A.; Maiato, H. Feedback control of chromosome separation by a midzone Aurora B gradient. *Science* **2014**, *345*, 332–336. [CrossRef] [PubMed]
183. Drechsler, H.; McAinsh, A.D. Exotic mitotic mechanisms. *Open Biol.* **2012**. [CrossRef] [PubMed]
184. Sullivan, D.S.; Huffaker, T.C. Astral microtubules are not required for anaphase B in *Saccharomyces cerevisiae*. *J. Cell Biol.* **1992**, *119*, 379–388. [CrossRef] [PubMed]
185. Yeh, E.; Yang, C.; Chin, E.; Maddox, P.; Salmon, E.D.; Lew, D.J.; Bloom, K. Dynamic positioning of mitotic spindles in yeast: Role of microtubule motors and cortical determinants. *Mol. Biol. Cell* **2000**, *11*, 3949–3961. [CrossRef] [PubMed]
186. Bannigan, A.; Lizotte-Waniewski, M.; Riley, M.; Baskin, T.I. Emerging molecular mechanisms that power and regulate the anastral mitotic spindle of flowering plants. *Cell. Motil. Cytoskelet.* **2008**, *65*, 1–11. [CrossRef] [PubMed]
187. Hayashi, T.; Sano, T.; Kutsuna, N.; Kumagai-Sano, F.; Hasezawa, S. Contribution of anaphase B to chromosome separation in higher plant cells estimated by image processing. *Plant Cell Physiol.* **2007**, *48*, 1509–1513. [CrossRef] [PubMed]
188. Lee, Y.R.; Qiu, W.; Liu, B. Kinesin motors in plants: From subcellular dynamics to motility regulation. *Curr. Opin. Plant Biol.* **2015**, *28*, 120–126. [CrossRef] [PubMed]
189. Struk, S.; Dhonukshe, P. Maps: Cellular navigators for microtubule array orientations in arabidopsis. *Plant Cell Rep.* **2014**, *33*, 1–21. [CrossRef] [PubMed]
190. Miki, T.; Naito, H.; Nishina, M.; Goshima, G. Endogenous localizome identifies 43 mitotic kinesins in a plant cell. *Proc. Natl. Acad. Sci. USA* **2014**, *111*, E1053–E1061. [CrossRef] [PubMed]
191. Cowan, C.R.; Hyman, A.A. Asymmetric cell division in *C. elegans*: Cortical polarity and spindle positioning. *Annu. Rev. Cell Dev. Biol.* **2004**, *20*, 427–453. [CrossRef] [PubMed]
192. Hara, Y.; Kimura, A. Cell-size-dependent spindle elongation in the *Caenorhabditis elegans* early embryo. *Curr. Biol.* **2009**, *19*, 1549–1554. [CrossRef] [PubMed]
193. Nguyen-Ngoc, T.; Afshar, K.; Gonczy, P. Coupling of cortical dynein and G alpha proteins mediates spindle positioning in *Caenorhabditis elegans*. *Nat. Cell Biol.* **2007**, *9*, 1294–1302. [CrossRef] [PubMed]
194. Hayward, D.; Metz, J.; Pellacani, C.; Wakefield, J.G. Synergy between multiple microtubule-generating pathways confers robustness to centrosome-driven mitotic spindle formation. *Dev. Cell* **2014**, *28*, 81–93. [CrossRef] [PubMed]
195. Brugues, J.; Nuzzo, V.; Mazur, E.; Needleman, D.J. Nucleation and transport organize microtubules in metaphase spindles. *Cell* **2012**, *149*, 554–564. [CrossRef] [PubMed]

196. Walczak, C.E.; Vernos, I.; Mitchison, T.J.; Karsenti, E.; Heald, R. A model for the proposed roles of different microtubule-based motor proteins in establishing spindle bipolarity. *Curr. Biol.* **1998**, *8*, 903–913. [CrossRef]
197. Murray, A.W.; Desai, A.B.; Salmon, E.D. Real time observation of anaphase in vitro. *Proc. Natl. Acad. Sci. USA* **1996**, *93*, 12327–12332. [CrossRef] [PubMed]
198. Civelekoglu-Scholey, G.; Cimini, D. Modelling chromosome dynamics in mitosis: A historical perspective on models of metaphase and anaphase in eukaryotic cells. *Interface Focus* **2014**. [CrossRef] [PubMed]
199. Brust-Mascher, I.; Civelekoglu-Scholey, G.; Scholey, J.M. Mechanism for anaphase B: Evaluation of "slide-and-cluster" versus "slide-and-flux-or-elongate" models. *Biophys. J.* **2015**, *108*, 2007–2018. [CrossRef] [PubMed]
200. Loughlin, R.; Heald, R.; Nedelec, F. A computational model predicts *Xenopus* meiotic spindle organization. *J. Cell Biol.* **2010**, *191*, 1239–1249. [CrossRef] [PubMed]
201. Uehara, R.; Tsukada, Y.; Kamasaki, T.; Poser, I.; Yoda, K.; Gerlich, D.W.; Goshima, G. Aurora B and Kif2a control microtubule length for assembly of a functional central spindle during anaphase. *J. Cell Biol.* **2013**, *202*, 623–636. [CrossRef] [PubMed]
202. Kellogg, E.H.; Howes, S.; Ti, S.C.; Ramirez-Aportela, E.; Kapoor, T.M.; Chacon, P.; Nogales, E. Near-atomic cryo-EM structure of PRC1 bound to the microtubule. *Proc. Natl. Acad. Sci. USA* **2016**, *113*, 9430–9439. [CrossRef] [PubMed]
203. Chen, I.A. Cell division: Breaking up is easy to do. *Curr. Biol.* **2009**, *19*, R327–R328. [CrossRef] [PubMed]

biology

MDPI

Review

The Consequences of Chromosome Segregation Errors in Mitosis and Meiosis

Tamara Potapova [1] and Gary J. Gorbsky [2,*]

[1] Stowers Institute for Medical Research, Kansas City, MO 64110, USA; tpo@stowers.org
[2] Cell Cycle and Cancer Biology Research Program, Oklahoma Medical Research Foundation, Oklahoma City, OK 73104, USA
* Correspondence: GJG@omrf.org; Tel.: +1-405-271-8168

Academic Editor: J. Richard McIntosh
Received: 10 November 2016; Accepted: 26 January 2017; Published: 8 February 2017

Abstract: Mistakes during cell division frequently generate changes in chromosome content, producing aneuploid or polyploid progeny cells. Polyploid cells may then undergo abnormal division to generate aneuploid cells. Chromosome segregation errors may also involve fragments of whole chromosomes. A major consequence of segregation defects is change in the relative dosage of products from genes located on the missegregated chromosomes. Abnormal expression of transcriptional regulators can also impact genes on the properly segregated chromosomes. The consequences of these perturbations in gene expression depend on the specific chromosomes affected and on the interplay of the aneuploid phenotype with the environment. Most often, these novel chromosome distributions are detrimental to the health and survival of the organism. However, in a changed environment, alterations in gene copy number may generate a more highly adapted phenotype. Chromosome segregation errors also have important implications in human health. They may promote drug resistance in pathogenic microorganisms. In cancer cells, they are a source for genetic and phenotypic variability that may select for populations with increased malignance and resistance to therapy. Lastly, chromosome segregation errors during gamete formation in meiosis are a primary cause of human birth defects and infertility. This review describes the consequences of mitotic and meiotic errors focusing on novel concepts and human health.

Keywords: aneuploidy; polyploidy; microtubule; chromosome instability; cancer; birth defects; fertility; drug resistance; centromere; kinetochore

1. Introduction

Other papers in the Special Issue "Mechanisms of Mitotic Chromosome Segregation" have explored the events of cell division and how defects might generate errors in the transmission of chromosomes to progeny cells. The defects are diverse in origin, including abnormalities in chromosome structure and function resulting in chromosomes that lag in anaphase or exhibit incomplete separation of sister chromatids. Spindle abnormalities such as multipolar spindles and defects in cytokinesis are additional sources of abnormal chromosome segregation. Finally, errors in cell cycle regulation, including delays during division and defects in cell cycle checkpoints also lead to missegregation. In this concluding chapter, we delve into the consequences of mitotic and meiotic errors. These can be benign or severe, depending on the degree and nature of the error, on the genetic background of the cell, and on the precise role of the cell in question. It is important to recognize that segregation abnormalities may not always generate aneuploidy. Even for a single chromosome that undergoes premature chromatid separation, random assortment will generate proper segregation to the two daughter cells 50% of the time. Outcomes of improper segregation are also influenced by stochastic variables that cause cells in seemingly identical situations to take different paths in response to identical errors [1]. Cell cycle checkpoints can sometimes identify impending errors and provide corrective countermeasures that lead to normal division. If checkpoints fail to correct the problem but division proceeds, then daughter cells are born with a genetic imbalance of one or more whole chromosomes, segments of chromosomes, or entire sets of chromosomes. In some instances, departure from conventional cell cycle patterns that lead to abnormal chromosome content are an aspect of normal development. This is often true for polyploidy, while aneuploidy is more often a result of errors in chromosome segregation. In most normal tissue cells, a surveillance system, highly dependent on the p53 tumor suppressor, is active in responding to the presence of abnormal chromosome content and can halt the cycling of the cell, cause cell death, or induce senescence (Figure 1) [2–5].

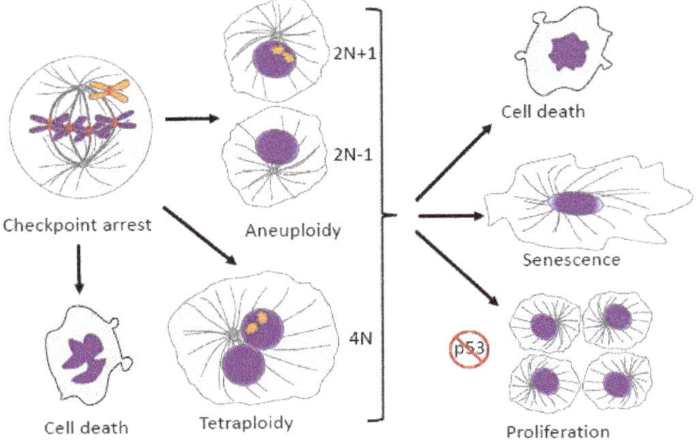

Figure 1. Mitotic defects have several potential outcomes. Failed alignment of chromosomes leads to mitotic arrest/delay enforced by the spindle checkpoint. If the failed alignment is not corrected, cells can follow several fates. They can undergo cell death directly from mitotic arrest. Cells may also suffer various kinds of abnormalities during mitotic exit, leading to the formation of aneuploid progeny. Alternatively, cells may exit mitosis without proper chromosome segregation and cytokinesis, resulting in a formation of a single tetraploid cell. Aneuploid or polyploid daughter cells may undergo cell death, cessation of proliferation and senescence, or continued proliferation. In most cases continued proliferation requires suppression or inactivation of the p53 tumor suppressor pathway.

Cells that lose or gain less than a whole set of chromosomes during cell division are termed aneuploid. Cells with a tendency to lose or gain chromosomes at a high rate are said to exhibit chromosome instability (see Table 1 for definitions). Certain genotypes may be inherently prone to continuous chromosomal instability, producing a diverse brood of aneuploid progeny. Alternatively, cells can be aneuploid but relatively stable in chromosome content. Cells may also undergo an increase in a whole set of chromosomes, a condition termed polyploidy. Polyploid cells frequently contain more than two centrosomes. In subsequent cell divisions, the centrosomes sometimes generate multipolar spindles where chromosomes are segregated to three or more daughter cells, resulting in aneuploid cells with variable numbers of chromosomes. The full consequences of chromosome segregation errors are vast in scope, since they affect many aspects of cell physiology, tissue homeostasis, and the adaptability of cells and organisms.

Chromosome segregation requires the coordination of two major pathways: chromosome movement and cell cycle regulation during M phase. A major contributor to this coordination is the mitotic spindle checkpoint. As detailed in another contribution in this series, defects in mitotic spindle assembly and chromosome alignment activate the spindle checkpoint, which delays cells in M phase. Optimally this delay allows the recovery of the normal spindle and balanced chromosome segregation. However, the delay can have multiple consequences. Mammalian cells arrested in M phase eventually exhibit markers indicative of DNA damage [6,7]. Cells in which the spindle checkpoint is purposefully activated by application of microtubule drugs often undergo apoptotic cell death, either directly in mitosis or after exiting M phase into G1 (Figure 1) [1,8]. One critical pathway in regulating cell death during M phase arrest is Cdk1-dependent phosphorylation and the subsequent degradation of Mcl-1, an anti-apoptotic member of the Bcl-2 family [9–12]. Certain aspects of apoptotic signaling are suppressed during M phase, but partial activation of these pathways may lead to cell death in the subsequent G1 phase [13]. In cells with normal p53 function, even a relatively short delay in M phase may lead cells to cease cycling after entering G1 [14]. Cells with chromosome segregation defects that escape apoptosis produce progeny with altered chromosome content. These cells may continue to cycle, particularly if p53 is inactivated. Chromosome segregation errors result in aneuploid or polyploid cells and are generally detrimental to both the cell and the organism. However, in some instances, changes in ploidy are programmed in normal development and physiology. At times, even accidental diversions from euploidy can generate beneficial evolutionary adaptations, particularly in single-cell organisms. In this review, we describe the consequences of aneuploidy and polyploidy due to segregation errors in mitosis and meiosis, focusing on recent novel ideas and on topics pertinent to human health.

Table 1. Definitions.

Ploidy is the number of sets of chromosomes in a cell or in an organism.
Haploid number refers to one set of chromosomes (1N), as in gametes or certain strains of budding yeast.
Diploid number refers to two sets of chromosomes (2N) that are homologous (one from each parent). Most animals are diploid.
Polyploid denotes a cell with more than two sets of chromosomes (triploid – 3N, tetraploid – 4N, pentaploid – 5N, etc.).
Euploid denotes the normal chromosome number in a species, usually an exact multiple of the haploid number (i.e.; human euploid genome contains 46 chromosomes – 2× the haploid number).
Chromosomal Instability is the tendency of a cell to gain or lose chromosomes or large segments of chromosomes. It is often abbreviated CIN.
Aneuploidy denotes the state of a cell having a chromosome number that deviates from a multiple of the haploid, i.e.; when there are extra or missing single chromosomes.
Whole chromosomal aneuploidy is having entire chromosomes gained or lost.
Segmental aneuploidy is having large regions of chromosomes deleted, duplicated or translocated from one chromosome to another. Cancer cells often exhibit both whole chromosome aneuploidy and segmental aneuploidy.
Trisomy refers to a diploid genome having gained an additional chromosome (2N + 1). Trisomy 21 indicates an extra chromosome 21 in a diploid genome.

2. Aneuploidy in Mitosis

2.1. *Effects of Aneuploidy on Gene Dosage*

In diploid organisms, apart from special instances such as the sex chromosomes of animals, genes are present in two copies that are both transcribed. Gain or loss of one copy changes the amount of gene product produced, a property called gene dosage. Unlike the doubling of a whole genome in polyploidy, where the increase of the gene dosage is equivalent for all chromosomes, loss or gain of an individual chromosome or chromosome fragment causes unbalanced changes in the cellular proteome. Studies in fungi and mammalian systems have shown that changes in mRNA and protein levels from genes on aneuploid chromosomes are roughly proportional to the changes in the chromosome copy number [15–19]. Aneuploidy of large, gene-rich chromosomes can cause changes in the expression of thousands of genes. Moreover, transcription factors encoded on aneuploid chromosomes will alter gene expression on other chromosomes [19]. Consequently, aneuploidy can cause a diverse spectrum of changes in the proteome of the cell, depending on the specific chromosomes lost or gained. Finally, aneuploidy itself may drive further chromosome instability, a topic treated in more detail below.

2.2. Effects of Aneuploidy on Cellular Fitness

The euploid karyotype is a product of natural selection for the best fitness for a species in an ecological niche. Aneuploidy will alter cellular physiology in many ways, depending on which chromosomes are extra or missing. Changes in the chromosome copy number alter the production of proteins (Figure 2). Imbalances in protein levels will occur particularly in protein complexes where the genes encoding individual components of the complex reside on different chromosomes. These imbalances will disrupt protein homeostasis, increase the workload for chaperones, and overload the protein degradation machinery, resulting in a form of toxicity referred to as proteotoxic stress [18,20]. Therefore, aneuploidy imposes a distinct fitness cost. In the ordinary environment, engineered aneuploid yeast strains and viable aneuploid mouse cells often grow slower than their euploid counterparts [21–23].

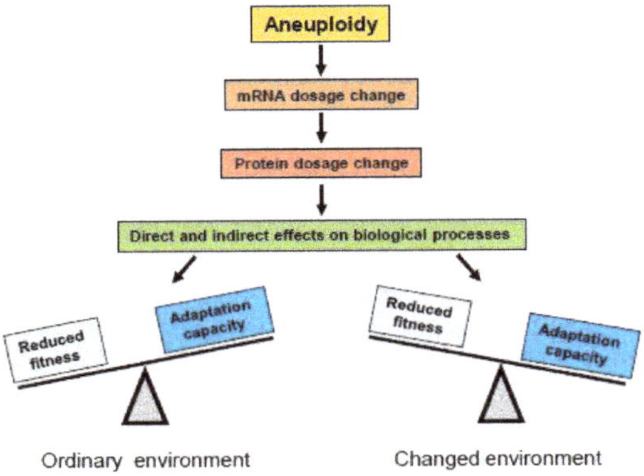

Figure 2. Aneuploidy produces changes in mRNA dosage which lead to changes in protein dosage for genes on the gained or lost chromosome(s). Changed protein levels can have direct effects on biological processes in which they are involved, or change the stoichiometry of protein complexes of which they are components, causing changes in their function. Changes in gene dosage of regulatory proteins like transcription factors may also exert indirect effects on biological processes by altering expression of their target genes on other chromosomes. In most cases, alterations in protein levels are disadvantageous or detrimental for organisms adapted to their ordinary environment (left), where the euploid karyotype provides best fitness. However, under conditions of environmental change, rapid alterations in expression of many genes may provide adaptive potential and be selected (right).

2.3. Aneuploidy in Fungi

Despite the overall fitness cost of carrying extra chromosomes, variation in the chromosome copy number can be found in many fungal species. The immediate shifts in the dosage of several genes conveyed by aneuploidy may confer rapid advantages in new environments, compared to the slower method via selection for adaptive mutations in individual genes. Fungi display remarkable genomic plasticity and can tolerate large-scale genomic changes. Variations in the chromosome copy number in *S. cerevisiae* have been detected frequently in association with domestication and adaptation to specific, often suboptimal, environments [24–28]. Aneuploidies that have deleterious phenotypes are quickly eliminated from populations by selection, leaving viable aneuploidies where the benefits of the presence of extra chromosome(s) outweigh the fitness cost (Figure 2). For instance, aneuploidy is common in laboratory strains of *S. cerevisiae* exposed to genetic transformation techniques, and in wild strains from diverse natural environments [29,30]. It was estimated that in the laboratory deletion collection of *S. cerevisiae* mutant strains, approximately 8% are aneuploid [31]. Aneuploidy appears to be more common in diploid versus haploid strains [32], consistent with the idea that smaller gene dosage changes are more tolerable. In industrial strains of *S. cerevisiae* cultured in fermenters or bioreactors, whole chromosome aneuploidies have repeatedly emerged in response to suboptimal conditions such as glucose or phosphate stress [26,32]. Therefore, the presence and frequency of aneuploidy appears to be strongly influenced by the environment. Karyotypic abnormalities are also frequently found in hospital isolates of pathogenic fungi *Candida albicans* and *Cryptococcus neoformans*, and aneuploidy in these pathogens has been associated with increased virulence and drug resistance [33–37].

2.4. Aneuploidy in Mammalian Cells

Aneuploidies commonly emerge in mammalian cell cultures, but this phenomenon has been traditionally viewed as an annoyance rather than a topic worthy of study. Most tissue culture cell lines are highly aneuploid, in part because many were originally derived from aneuploid tumors, but also because of fluctuations in ploidy with continued culture. Recently, the de novo generation of aneuploidy in cultured cells has emerged as an important concern for the therapeutic use of pluripotent human stem cells, where the correct karyotype is essential. Pluripotent stem cells acquire numerical and structural chromosomal aberrations during prolonged cell culture, and concurrently show initial signs of malignant transformation [38–40]. Human stem cell lines frequently gain a copy of chromosome 12. Trisomy 12 increases cell proliferation in culture and corresponds to global changes in the transcriptome, making the gene expression profile of aneuploid stem cells similar to germ cell tumors [38]. While gaining an extra copy of chromosome 12 allows trisomic stem cells to thrive in tissue culture conditions, it has obvious detrimental consequences for the therapeutic use of these cells. A recent study generated a panel of trisomic mouse embryonic stem cells lines, each carrying an extra copy of single chromosomes 6, 8, 11, 12, or 15. Most of these trisomic cell lines proliferated at a high rate in culture, but showed a reduced ability to differentiate and an increased potential to form teratomas, possibly due to deregulated gene expression [41].

Retinal Pigment Epithelium-1 (RPE-1) cells are a diploid line of human retinal pigment epithelium cells that are engineered to express telomerase, rendering them immortal [42]. They are widely used as a model of non-malignant human diploid cells. In continuous passage in culture, RPE-1 cells will spontaneously gain an extra copy of chromosome 12 [43,44], but for a different reason than pluripotent stem cells. One of the copies of chromosome 12 in this cell line carries a mutation in the RAS gene [45], which may provide these cells a growth advantage. A study of immortalized human colonic epithelial cells carrying an extra copy of chromosome 7 showed better growth in serum-free medium compared to the isogenic diploid cells [46]. The growth advantage of the trisomic cells was postulated to stem from overexpression of the epidermal growth factor receptor gene located on chromosome 7. Aneuploid cells derived from the colorectal cancer cell line DLD1 engineered to be trisomic for chromosomes 7 or 13 also demonstrated selective growth advantages under various suboptimal conditions, such as serum deprivation, hypoxia, or exposure to the chemotherapeutic agent 5-fluorouracil [47]. Thus, while aneuploidy may be deleterious under conditions where the euploid karyotype provides the best fitness, it can provide a selective advantage when conditions change, allowing adaptation to a novel environment.

2.5. Aneuploidy as a Driver for Genomic and Chromosomal Instability

Chromosomal instability refers to an increased propensity for chromosome segregation errors, resulting in aneuploidy and genomic imbalances [48,49]. Aneuploidy itself may foster chromosome instability by perturbing the stoichiometry of proteins involved in chromosome segregation [50,51]. Studies in aneuploid budding yeast showed that unbalanced changes in the copy number of chromosome VII and X perturbed the ratio of essential spindle checkpoint proteins Mad1 and Mad2, whose genes are located, respectively, on these two chromosomes. Maintaining the 1:1 stoichiometry of the Mad1:Mad2 ratio appears crucial for monitoring kinetochore attachments [52]. Changes in Mad1:Mad2 ratios compromised the spindle checkpoint and increased chromosomal instability [53]. Importantly, aneuploid strains that gained copies of chromosomes VII and X simultaneously were karyotypically more stable than strains where these chromosomes were gained individually. Changing stoichiometric ratios of components of protein complexes or individual proteins with essential structural or regulatory functions can have severe consequences. Dosage imbalance among components of protein complexes can induce defects in the maintenance of genomic fidelity, affecting mitosis, cytokinesis, DNA replication, DNA repair, etc. In line with this hypothesis, yeast strains aneuploid in different chromosomes show various degrees of karyotypic and genomic instability, likely due to products from genes encoded on the aneuploid chromosomes [53,54]. Studies in human cells also suggest that aneuploidy itself may increase chromosome instability by affecting chromosome segregation in mitosis or by inducing defects in replication [55,56].

2.6. Aneuploidy and Cancer

In the late 19th century, it was recognized that tumor cells often exhibit abnormal, asymmetric mitotic figures [57]. In the early 20th century, Boveri proposed that abnormal chromosome content was the source for tumor malignancy, the earliest molecular hypothesis for the origin of cancer [58]. It is estimated that approximately 86% of solid tumors and 72% of hematopoietic cancers exhibit aneuploidy [59]. The question of whether aneuploidy is a cause or consequence of cancer has generated considerable controversy [60,61]. In general, most cancers display various degrees of genomic instability, including point mutations, chromosomal rearrangements, and changes in whole chromosome ploidy [62]. High levels of genomic instability generally correlate with more aggressive tumors and poorer patient prognosis. The environment of cancers can be thought of as complex cellular ecosystems that are constantly evolving, responding to challenges such as depletion of oxygen and nutrients, immune assault, and medical attempts at therapy [63]. Cancer cells must adapt to challenges in their microenvironment, and aneuploidy serves as an enabling factor in tumor evolution. An example of the adaptive evolution of cancer cells via aneuploidy is the loss of the heterozygosity of tumor suppressor genes. Inactivating mutations occur in many tumor suppressor genes, such the Retinoblastoma gene (Rb), but a single wild-type copy maintains suppressor function. Loss of the chromosome containing the wild-type copy leads to the loss of tumor suppressor function in the cancer cells.

Mouse models have aided in testing the role of aneuploidy in the origin and progression of cancer. Either underexpression or overexpression of mitotic regulators fosters both aneuploidy and increased cancer predisposition. For example, mice engineered to underexpress or overexpress most components of the spindle checkpoint pathway exhibit aneuploidy and tissue-specific increases in cancer incidence [64–68]. Complete ablation of the *BUB1B* gene that encodes the BubR1 checkpoint protein is embryonically lethal, but hypomorphs show increased aneuploidy, increased susceptibility to carcinogen-induced tumors, and accelerated aging phenotypes [69–73]. Surprisingly, in contrast to the usual consequences of overexpression of spindle checkpoint proteins, overproduction of BubR1 protects against cancer and other aging phenotypes and extends lifespan [74,75]. In humans, a rare genetic disease called Mosaic Variegated Aneuploidy stems from mutations in the *BUB1B* gene, and afflicted individuals show a very high proportion of aneuploid tissue cells. These patients suffer from a variety of serious pathologies, including growth defects, microcephaly, and increased cancer incidence [76–78].

Mouse embryos, heterozygous for a deletion of the gene encoding the mitotic kinesin protein, Cenp-E, show a weakened spindle checkpoint, and their cells will often enter anaphase in the presence of one or a few unaligned chromosomes [79]. The animals develop normally but are more prone to developing certain types of spontaneous tumors, such as lymphomas in the spleen and pulmonary adenomas in the lung. However, they are partially protected from other cancers, such as liver tumors [79]. Thus, depending on the context, aneuploidy can promote or inhibit oncogenesis. Crossing Cenp-E heterozygotes with other mutants that further increase the rate of chromosome missegregation led to tumor suppression, suggesting that the amount of chromosome missegregation may be important, whereby low rates promote tumor growth and high rates suppress it [80].

The potential biphasic effect of chromosome missegregation, to promote tumorigenesis at low levels and inhibit tumorigenesis at high levels, may have significance for the use of anti-mitotic drugs in cancer therapy. Taxol, the common name for the drug paclitaxel, is one of the most widely prescribed anti-cancer drugs. It binds and hyperstabilizes microtubules both in the test tube and in cells [81,82]. In cell culture, at moderate concentrations, it arrests cells in mitosis by activation of the spindle checkpoint [83,84]. Thus, for many years, the common assumption was that mitotic arrest was the mechanism underlying Taxol's effectiveness in cancer therapy. However, the relatively low mitotic index in tumors in humans compared with Taxol's rapid ability to shrink some tumors led to proposals that Taxol's medical effectiveness might stem from targeting interphase tumor cells or the tumor environment [85,86]. A combined clinical and cell culture study led to the proposal that Taxol kills tumor cells in patients, not by mitotic arrest, but by increasing the propensity of tumor cells to undergo multipolar mitosis, leading to massive chromosome missegregation and tumor cell death [87]. Thus, while low levels of chromosome missegregation may be dangerous in promoting cancer, therapeutically driving missegregation to very high levels may conversely be an effective anti-cancer strategy.

Human cancers also exhibit genome instability due to dysfunction of chromosome telomeres, which may become too short after multiple rounds of replication (telomere attrition), or lose structural features such as telomere caps due to enzymatic defects [88]. Telomere defects generate broken ends on chromosomes, which may then fuse to similar broken ends on other chromosomes and generate dicentric chromosomes, in which a single chromosome will contain two widely separated centromeres. After this dicentric chromosome replicates and condenses in mitosis, each chromatid will harbor two centromeres that may orient toward opposite poles. During anaphase such a chromatid will form a chromosome bridge. This bridge can be severed during anaphase and form new broken ends. This mechanism can propagate breakage-fusion-bridge cycles that lead to complex chromosome rearrangements characteristic of tumor cells [89].

2.7. Cohesion Fatigue and Centromere Fission

Another potential source of aneuploidy in tumor cells is cohesion fatigue in cells that are delayed or arrested at metaphase. Normally metaphase is transient, lasting only a few minutes, and is followed by the onset of anaphase where the protease, Separase, severs Rad21, a component of the cohesin complex that holds sister chromatids together [90]. However, even when most chromosomes are aligned at the spindle midplane, anaphase onset may be delayed by the spindle checkpoint, activated by the failure of one or a few chromosomes to align in a timely manner [91]. Alternatively, defects in the expression of mitotic regulators, sometimes as a consequence of oncogenic changes, may induce delays at metaphase [68,92–96]. During the delay, chromosomes at the metaphase plate begin to separate asynchronously due to pulling forces of kinetochores on spindle microtubules, a process termed cohesion fatigue [92,97–99]. Cohesion fatigue is a general phenomenon, independent of the mechanism used to induce the metaphase delay [97–99]. The timing of chromatid separation varies among cell types and among the individual chromosomes within a single cell. Separation initiates at the kinetochores and then spreads distally along the chromosome arms [97].

In cancer cells, numerical and segmental aneuploidies are generally found together (Figure 3). Individually, many of the postulated origins of aneuploidy such as telomere erosion, DNA replication errors, DNA repair defects, or cytokinesis failure cannot in a simple way account for both numerical and structural chromosome abnormalities. As a result, many theories suggest independent origins for numerical and segmental aneuploidy. However, mitotic segregation defects can simultaneously generate both types of aneuploidy [100]. A common mitotic error is merotelic attachment whereby an individual kinetochore attaches to microtubules from both spindle poles. These attachment errors had been thought to be particularly dangerous because merotelic attachments are not well detected by the spindle checkpoint and allow progression to anaphase [101,102]. However, merotely that occurs during prometaphase may have a relatively minor effect on chromosome segregation because the merotelic kinetochore is more strongly attached to microtubules from the proper pole. Such merotelic chromatids are resolved properly in anaphase and do not result in chromosome missegregation [103,104]. In contrast, defects that occur after complete or partial chromatid separation in cohesion fatigue are likely to have severe consequences since the unattached kinetochore may be more prone to near equal attachment to both spindle poles. Kinetochores from partially or fully separated chromatids may attach to microtubules from both poles. Single chromatids derived from unreplicated DNA or from cohesion fatigue will sometimes congress to the spindle equator [101,105]. Alternatively, the two unpaired chromatids of a single chromosome may orient to the same pole, eventually resulting in numerical aneuploidy (Figure 4).

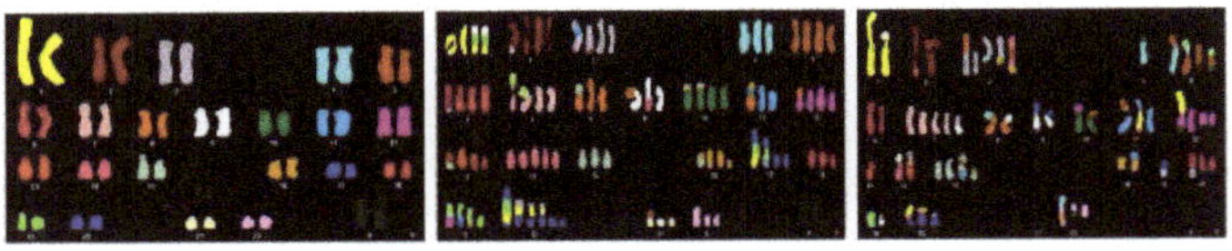

Figure 3. Cell lines derived from cancers exhibit numerical and segmental aneuploidy. Spectral karyotype comparison of normal human mammary epithelial cells (HMEC) and two breast cancer cell lines (MCF-7 and SUM149PT) that exhibit extensive numerical and segmental aneuploidy. Image reproduced from [106].

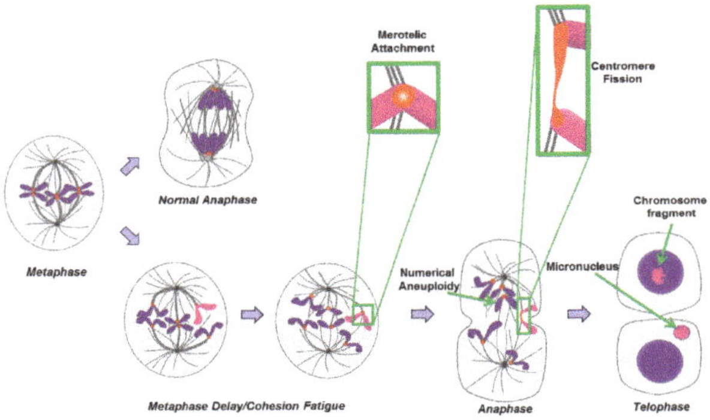

Figure 4. Numerical and segmental aneuploidy as an outcome of cohesion fatigue and centromere fission. A cell at metaphase will normally undergo balanced chromosome segregation in normal anaphase (upper path). If metaphase is delayed (lower path), chromatids may begin to undergo cohesion fatigue and separate. When sister chromatids separate they may both move to one of the two spindle poles leading to numerical aneuploidy following anaphase and mitotic exit. In other cases the kinetochore of an individual chromatid may undergo merotelic attachment to microtubules from both spindle poles (exemplified by sequential stages for the pink chromatid and detailed in the green boxes). Under this circumstance, spindle forces or cytokinesis may sever the chromatid resulting in chromosome fragments that can attach to other chromosomes resulting in segmental deletions, duplications, translocations and the formation of micronuclei.

The study of cultured cells from mutant mice and analysis of human cancers of recent origin have led to a proposal that segmental aneuploidy defects involve whole chromosome arms that arise through "centromere fission", breaks at or near centromeres [107,108]. Light and electron micrographs show that, under certain conditions, merotelically attached kinetochores undergo extreme stretching and possible severing [109]. Thus, merotelic attachment following partial or full chromatid separation in cohesion fatigue could generate both breaks at or near centromeres that give rise to duplications, deletions, and translocations as well as micronuclei (Figure 5). The potential consequences of micronuclei formation are described in detail below (Section 3). The fusion of broken arms containing centromeres to intact chromosomes may generate dicentric chromosomes that would then undergo rounds of breakage-fusion-bridge cycles to generate more complex and varied segmental chromosome defects. Cohesion fatigue in a large proportion of chromosomes, followed by subsequent cell division, is unlikely to generate viable daughter cells. However, partial or complete chromatid separation, occurring in just one or a few chromosomes, after shorter metaphase delays, may be a common initiating event for the numerical and segmental aneuploidies seen in cancer.

2.8. Modern Analysis and Implications of Cancer Aneuploidy

For many years, aneuploidy in tumors was studied using cytogenetic methods, which can accurately detect large karyotypic alterations but are less accurate in identifying small alterations. Comparative genomic hybridization and, more recently, next-generation sequencing techniques have enabled the detection of large and small copy number variations with higher resolution. Progress in sequencing technology has enabled large-scale cancer studies. The largest multi-institutional collaboration project The Cancer Genome Atlas (TCGA) has generated genomic data across many types of cancer and has shown that nearly all cancers have significant chromosome aberrations [110]. Chromosomal regions affected by copy number alterations in cancer vary widely in size. Among copy number variations, small (focal) amplifications or deletions are most frequent, followed by large-scale alterations: gain/loss of a chromosome arm or a whole chromosome. Interestingly, large-scale alterations typically show a low amplitude of amplification (i.e., gain or loss of one copy), but focal amplifications overall have a higher amplitude, indicating that small chromosomal segments can be duplicated multiple times [111].

Modern computational methods to analyze aneuploidy in multiple human tumor samples have generated a list of 70 genes highly correlated with chromosome instability, called the CIN70, and high expression of these genes correlates with adverse patient outcome [112,113]. Interestingly, a study of over 2000 breast cancer patients revealed a biphasic effect where the patients with the very highest CIN70 scores showed a better prognosis than those with intermediate scores [114]. On the other hand, for pathologists, frequent abnormal or atypical mitotic figures in human cancer tumors generally signify high malignancy. Aggressive cancers also exhibit abnormally shaped interphase nuclei and micronuclei that are post-mitotic footprints of severe mitotic problems [115]. In such cancers, the degree of karyotypic instability is likely to be very high.

Many copy number aberrations are unique, while others are recurrent. Recurrent copy number variations may indicate that alteration of this chromosomal segment functions as a driver in cancer progression. For instance, amplifications frequently encompass genomic regions that contain canonical oncogenes such as Cyclin D1, c-Myc, and ErbB-1. Deletions often include regions encoding canonical tumor suppressors such as *p16INK4A/p14ARF* and *PTEN*. Amplifications or deletions in these cases will result in corresponding changes in gene dosages of oncogenes or tumor suppressors, giving these karyotypes a selective proliferative advantage [116,117]. A recent computational analysis of the gene copy number in thousands of tumor-normal tissue pairs identified hundreds of novel potential oncogenes and tumor suppressor genes, many of which were correlated with whole chromosome and large segmental aneuploidies in the tumors [118]. However, verification of these guilt-by-association indications will require functional tests.

As a consequence of genomic instability, cells within a tumor can diverge and form distinct subpopulations, resulting in genomic heterogeneity within a tumor [119–121]. Clonal diversity is a prominent feature in many cancers, especially advanced ones, and is likely to play a role in tumor evolution and resistance to therapy [122]. Genomic studies of intra-tumor heterogeneity in a complex population of cells have been very challenging, but became possible with the development of single-cell sequencing methods that allowed dissecting complex chromosome rearrangements in individual cells [123]. Single-cell sequencing studies in breast cancer, investigating the evolutionary dynamics of copy number variations, found that complex aneuploidy appears early in tumor evolution and propagates by clonal expansion [124,125]. This finding contradicts the long-standing belief that complex aneuploid karyotypes develop gradually over time, but speaks to the idea that large changes in karyotype can sometimes be instantly advantageous in a specific tumor microenvironment, leading to cancer progression. Rapid and random changes in dosages of multiple genes on multiple chromosomes have a potential of giving the cancer cell the karyotype for better fitness.

2.9. *Aneuploidy and Drug Resistance*

Aneuploidy may promote the emergence of antibiotic-resistant infections and chemotherapy-resistant cancers. The budding yeast *S. cerevisiae* has traditionally been used as a model to study mechanisms of adaptations to various stresses. Yeast can rapidly adapt to unfavorable conditions by changing their karyotype. For instance, the gain of chromosome XV in budding yeast confers resistance to the antibiotic radicicol (inhibitor of chaperone protein HSP90), while the loss of chromosome XVI confers resistance to another antibiotic, tunicamycin [126,127]. Importantly, this resistance was related to the dosage of certain genes encoded on the extra chromosomes. In the case of radicicol, resistance was caused by the overexpression of two genes encoded on chromosome XV, *STI1* (co-chaperone of Hsp90) and *PDR5* (a pleiotropic drug efflux pump).

C. albicans and *C. neoformans*, two human fungal pathogens, frequently become aneuploid during infection and after antifungal treatments [128,129]. Drug resistance is a very common and serious problem with these pathogens. One of the first-line antibiotics for *Candida* fungal infections is fluconazole, a triazole antifungal medication that interferes with the sterol biosynthesis pathway, leading to defects in cell wall synthesis. Resistance to fluconazole has been frequently observed in clinical isolates. Sometimes this resistance can be attributed to mutations in the sterol biosynthesis pathway. However, almost half of the clinical isolates that are resistant to fluconazole carry extra copies of chromosome 5 [36]. In this case, fluconazole resistance can be narrowed down to two genes encoded on *C. albicans* chromosome 5: *ERG11* (lanosterol 14-alpha-demethylase, a component of the sterol biosynthesis pathway targeted by fluconazole), and *TAC1*, a regulator of drug efflux pumps [34].

In mammals, aneuploidy likely drives some chemotherapeutic drug resistance in cancers. Chromosomally unstable cancer cells tend to develop increased resistance to various chemotherapeutic drugs when compared to their stable counterparts [130–132]. The specific molecular mechanisms for drug-resistant phenotypes are not yet clearly tracked to specific genes. Prolonged exposure of human colorectal cancer cell lines to the anti-cancer drug Irinotecan led to the selection of cells containing an extra copy of chromosome 14 [133]. It was also reported that resistance to the anti-cancer drug 5-fluorouracil was increased in derivative colorectal cell lines experimentally engineered to be trisomic [47]. Findings in aneuploid rodent cell lines are consistent with the idea that multidrug resistance can occur through selection of novel aneuploidies [131,134]. Thus, as in the case of other environmental challenges, drug challenges create a novel niche to allow cells with altered chromosome constitutions selective survival and growth advantages [135].

3. Micronuclei

3.1. *Footprint of Mitotic Error*

Micronuclei, as an outcome of mitotic errors in higher eukaryotes, have long been observed, but only in recent years have researchers begun to pay close attention to their causes and consequences. A micronucleus is a tiny nucleus that forms from a lagging chromosome or a fragment of a chromosome that fails to incorporate into the main nucleus. When segregated sister chromatids de-condense and the nuclear envelope re-forms around them in telophase (the last stage of mitosis), spatially isolated chromosomes or chromosome fragments also de-condense, forming a small round nucleus enclosed by its own nuclear membrane (Figure 5). Micronuclei can be viewed as a footprint of chromosome missegregation that persists after mitotic exit and can be visualized in interphase cells. For pathologists, micronuclei often serve as a marker of chromosomal instability in aggressive cancers, and also as a tool to assess the genotoxicity of various chemicals [136].

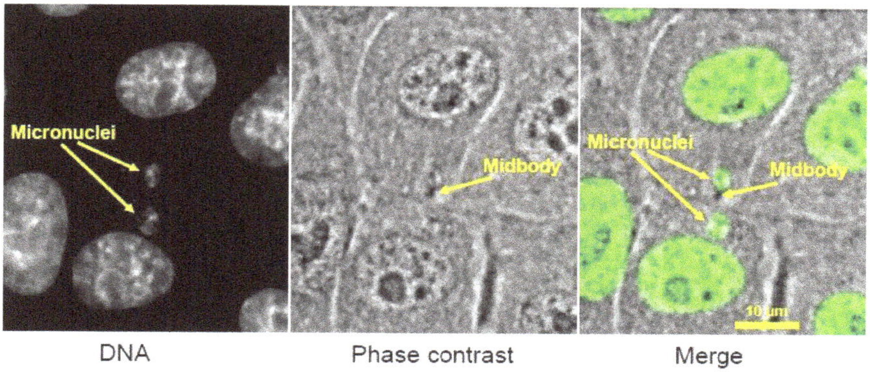

Figure 5. Micronuclei present in two daughter LLC-Pk cells, a cell line derived from porcine kidney. Micronuclei likely formed from lagging chromosomes that were partially trapped in the cytokinetic cleavage furrow. The midbody, the remnant of the cleavage furrow, bisects the region between the two micronuclei. (micrographs courtesy of Hem Sapkota.)

3.2. Causes and Consequences of Chromosome Entrapment in Micronuclei

It has been reported that cytokinesis can directly generate certain structural disruptions (chromosome breakage, nuclear envelope rupture) due to entrapment of chromatin from lagging chromosomes or chromosome bridges in the cleavage furrow [100,137]. Entrapment of chromatin in the cleavage furrow appears to be a common mechanism for generating micronuclei [138]. Further DNA damage in micronuclei occurs in the subsequent interphase. The nuclear envelope around micronuclei is abnormal. Its nuclear-cytoplasmic trafficking functions are defective, and the nuclear envelope may even undergo catastrophic collapse [139,140]. The deficient nuclear-cytoplasmic trafficking prevents proper communication of the micronucleus with the rest of the cell, including propagation of the DNA damage signaling. Thus, DNA damage in micronuclei is unable to elicit a robust cellular DNA damage checkpoint response [139,140]. Failure of the DNA damage checkpoint allows cells with micronuclei containing damaged DNA to reenter mitosis. The formation of micronuclei has been linked to chromothripsis, a form of genomic instability where an individual chromosome breaks into many fragments that religate at random [141,142]. Advances in single-cell sequencing methods have allowed direct demonstration that chromosomes trapped in micronuclei can undergo chromothripsis [44]. Defective DNA replication in micronuclei has been postulated to lead to chromosome fragmentation followed by random relegation [139,143]. A recent study used inducible inactivation of the Y chromosome centromere to control the timing of micronucleus formation, demonstrating that chromothripsis in this system occurs over multiple cell cycles [144]. A chromosome within a micronucleus first undergoes defective replication or fails to repair DNA damage, then becomes fragmented as the chromosome condenses in early mitosis. When incorporated back into the parent nucleus at mitotic exit, the fragments are re-ligated at random by the DNA repair mechanisms.

4. Aneuploidy in Meiosis

4.1. Causes of Aneuploidy in Meiosis

Meiosis is a specialized form of cell division. It consists of two chromosome segregation events that generate haploid gametes. In humans, aneuploidy in meiosis is a major cause of infertility, miscarriage, and congenital birth defects. It is estimated that 25% to 70% of human conceptions result in aneuploid embryos, most of which are spontaneously eliminated very early in development [145]. In the first division, termed meiosis I, the homologous chromosomes pair, undergo recombination, and segregate from each other (Figure 6). Homologous chromosomes are held together through the chiasmata formed by recombination. Defects in the assembly, maintenance, or positioning of the chiasmata on the chromosomes can result in failure of homologs to orient and move to opposite poles in meiosis I [146,147]. The second division, meiosis II, is similar to mitosis, where sister chromatids segregate. Premature separation of sister chromatids, depicted in Figure 6B, is an important contributor to meiotic aneuploidy.

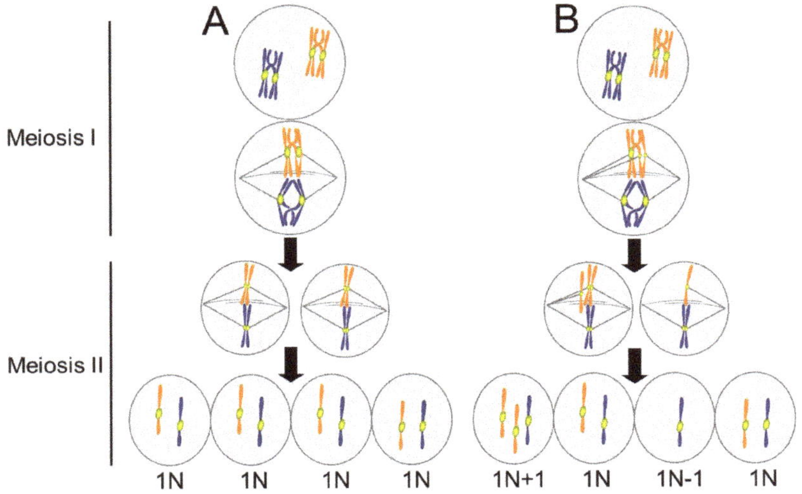

Figure 6. Meiotic errors lead to aneuploid gametes. (**A**) Normal meiosis consists of two chromosome segregation events without an intervening S phase. In Meiosis I, homologous chromosomes pair and undergo recombination, forming crossovers. In anaphase of Meiosis I, the homologous chromosomes segregate. In Meiosis II sister chromatids separate. The final product is four haploid (1N) cells; (**B**) Defects in meiosis result in aneuploidy. In the example shown, the chromatids of one chromosome separate prematurely and segregate to opposite poles resulting in an imbalance of chromatids in the two cells produced by Meiosis I. When these cells undergo Meiosis II, each produces one normal haploid (1N) gamete and one aneuploid (1N + 1 or 1N − 1) gamete. For simplicity, only two chromosome pairs are depicted.

Most human aneuploidy results from defects during oogenesis. In males, checkpoints to detect and eliminate aneuploid cells during spermatogenesis are robust [148]. Studies in mice show that while segregation defects increase with age in germ cell division in males, checkpoint pathways successfully eliminate the cells before they mature into sperm [149]. In females, checkpoints during oogenesis appear to be weaker. The spindle checkpoint in mouse oocytes does respond by delaying or arresting cell division in the presence of several misaligned chromosomes but appears to be incapable of detecting single chromosomes that fail to align properly [150–152]. The significant difference between male and female meiotic events appears as a consequence of differences in the cytoplasm-to-nuclear ratio, which is much higher in oocytes. Microsurgically bisecting the cytoplasm of oocytes decreases this ratio and increases the oocyte checkpoint response to unaligned chromosomes [153]. Thus, it seems that in situations where the cytoplasm-to-chromatin ratio is very high, the checkpoint signals are too diluted to be effective. In the case of *Xenopus* eggs, which can be over 1 mm in size, the spindle checkpoint signal from chromosomes in meiosis and in early embryos is too weak to effectively block cell cycle progression, even when the spindles are completely disrupted with microtubule drugs [154]. Concentrated extracts of *Xenopus* eggs can be made that recapitulate the cell cycle in the test tube [155]. These extracts similarly lack responsiveness to microtubule inhibitors unless they are supplemented with a high concentration of sperm nuclei, a source of chromosomes. The extracts then become responsive to microtubule drugs and arrest in M phase of the cell cycle [156]. The addition of the sperm nuclei decreases the cytoplasm-to-chromatin ratio, and the combined spindle checkpoint signaling from the concentrated chromosomes becomes competent to arrest the cell cycle in M phase.

4.2. The Maternal Age Effect

In humans, maternal age is a major risk factor for conception of aneuploid embryos. Studies from fertility clinics show that the proportion of aneuploid oocytes increases substantially in older women [157,158]. One suspected cause is age-related decreases in the already weak spindle checkpoint signaling in oocytes. Evidence showing decreased expression of checkpoint signaling proteins or checkpoint competence in older human and mouse oocytes is consistent with this hypothesis [69,159–161]. Even stronger evidence suggests that a significant contributor to the maternal age effect is compromised cohesion between sister chromatids. Cohesion is mediated by the Cohesin protein complex and is released by proteolytic cleavage of one component of the complex. During meiosis, cohesion between sister chromatids is released in two stages. In anaphase of meiosis I, Cohesin on the distal parts of the chromatids is cleaved to allow separation of the homologous chromosomes that have exchanged arms by recombination. The Cohesin near the kinetochores remains protected. At the onset of anaphase in meiosis II, the Cohesin holding sister kinetochores together is cleaved, allowing chromatids to move to opposite poles.

In meiosis I, sister kinetochores show wider separation in oocytes from older women, suggesting that the Cohesin between them is compromised [162]. Elegant imaging studies of oocytes from older mice show a tendency for the kinetochores of sister chromatids to separate prematurely during anaphase of meiosis I [163]. These separated chromatids then orient randomly in meiosis II (Figure 7). Why is sister chromatid cohesion compromised in oocytes from older mammals? According to current understanding, in mammals, the Cohesin complex can only be "established", which means made competent to hold sister chromatids together, immediately after replication in S phase [164]. In the ovaries of female mammals, S phase in oocytes occurs entirely during fetal life. Upon sexual maturity, those oocytes formed in the fetus then mature and are released over time during the female mammal's reproductive life. Studies in mice have confirmed the idea that the Cohesin complexes established during fetal life are necessary and sufficient for chromatid cohesion in the oocytes of mature females [165,166]. Extending to humans, this system would necessitate that the Cohesin proteins established on the oocyte chromosomes during S phase in the fetus remain intact for decades. Thus, it is reasonable to suspect that such incredibly long-lived proteins might suffer some degradation in function over the years. Indeed, evidence in mouse and human suggests that Cohesin and Cohesin regulators such as the Shugoshin 2 (Sgo2) protein in mammalian oocytes decay with age, allowing sister chromatids to separate prematurely and segregate randomly in meiosis II (Figure 7) [163,167–170]. However, a recent study of chromosome distributions in human oogenesis suggests that there may exist a mysterious rescue pathway whereby chromatids that prematurely separate in meiosis I are, in some unknown way, biased to segregate to opposite poles in meiosis II, thus generating an oocyte with a normal complement of chromosomes [171].

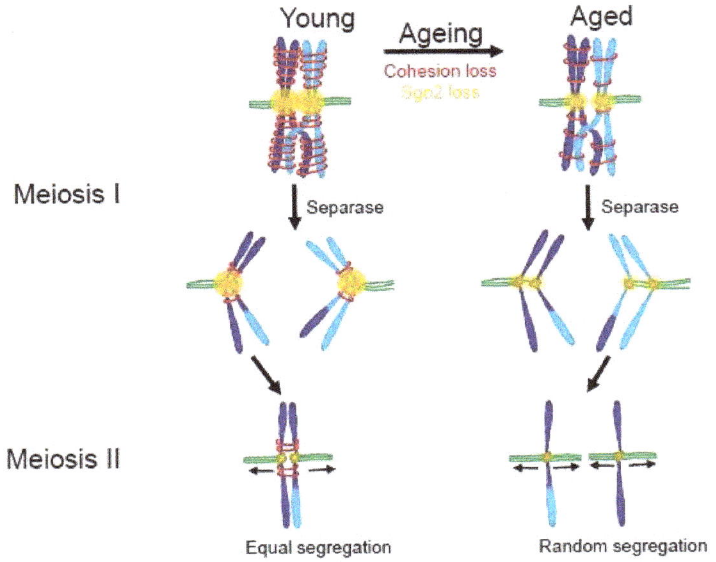

Figure 7. Loss of cohesin and the cohesin protector, Sgo2, in older mammals may lead to increased premature chromatid separation in meiosis. In young mammals (left) paired homologous chromosomes have high levels of cohesin and Sgo2 in Meiosis I (MI). At anaphase of Meiosis I, the protease, Separase, clips Cohesin on the distal chromosome arms allowing the homologous chromosomes to separate. Sgo2 protects Cohesin near the centromeres until Meiosis II (MII) ensuring that sister chromatids will orient to opposite poles. In aged mammals (right) there are diminished amounts of Cohesin and Sgo2. During Meiosis 1 Separase cleaves the majority of cohesin, including that near the centromeres. This allows sister chromatids to separate during anaphase of Meiosis I. In Meiosis II the individual chromatids may separate randomly to the spindle poles leading to a high incidence of aneuploidy. (Adapted from [163].)

4.3. Consequences of Aneuploidy in Meiosis

In mammals, whole chromosome aneuploidies on the level of the entire organism are highly detrimental for all chromosomes except sex chromosomes. Embryonic development appears to be extremely sensitive to the gene dosage imbalance caused by gains or losses of somatic chromosomes. In mice, gain or loss of any somatic chromosome is embryonic lethal. In humans, most somatic chromosome aneuploidies are also fatal. Most cases of meiotic aneuploidy result in spontaneous abortions in early pregnancy, causing more than half of miscarriages during the first trimester [172–174]. However, gains of certain small somatic chromosomes in humans can be viable. Viable somatic trisomies manifest in severe congenital diseases: Down syndrome (trisomy 21), Edwards syndrome (trisomy 18), and Patau syndrome (trisomy 13). Although the severity of phenotypes varies among patients, Down syndrome, caused by the gain of chromosome 21, the smallest somatic chromosome with the least number of genes, shows the highest viability. Today, the projected life expectancy of individuals with Down syndrome in developed countries is around 60 years [175], while in Edwards and Patau syndromes life expectancy is very short. Edwards and Patau syndromes are usually fatal before birth or within the first year of life, with only about 10% of patients surviving until 10 years of age.

Sex chromosome aneuploidies are viable with less severe phenotypes compared to somatic trisomies, likely because the gene dosage imbalances for genes encoded on sex chromosomes are reduced compared to genes encoded on somatic chromosomes. The absence of one sex chromosome

(45,XO) causes Turner syndrome, one of the most common chromosomal abnormalities in women [176]. Since gene expression from one of the copies of X chromosomes in females is almost completely inactivated in a process called X-inactivation [177], the dosage of most X chromosome genes in XO females is comparable to normal females [178]. Turner syndrome manifests mainly in infertility that can be accompanied by heart defects and learning disabilities. Because of X inactivation, women with triple X (XXX) syndrome also have a very mild phenotype [179].

Gains of X chromosomes in males causes Klinefelter's syndrome. Most frequently there is a gain of one extra copy of the X chromosome, resulting in the karyotype of 47,XXY. However, other karyotypes have been detected such as 48,XXXY, 48,XXYY and 49,XXXXY [180]. Because extra X chromosomes are inactivated in Klinefelter's males in the same way as in normal females, genes on these extra X chromosomes are mostly silenced and do not cause a severe gene dosage imbalance [181]. Individuals with Klinefelter's mainly suffer from infertility and low testosterone levels.

Y chromosome polysomies, particularly 47,XYY syndrome, also occur frequently and are frequently detected only by chance because the condition often lacks clinical manifestations. In this case, a phenotype is absent likely because the Y chromosome is very gene-poor, and genes located on the Y chromosome are not essential for viability. Therefore, increased dosage of the Y chromosome genes does not severely impair the development and physiology. Most 47,XYY males have normal sexual development and normal fertility.

4.4. Meiotic Aneuploidy and Cancer

Increased rates of chromosome instability lead to higher risks of certain malignancies in meiotic aneuploidy patients. For instance, Down syndrome patients have a high risk of acute myeloid leukemia, particularly the megakaryoblastic subtype [182], and trisomy 21 is frequently found in megakaryoblastic leukemia not associated with Down syndrome [183]. Human chromosome 21 contains genes encoding hematopoietic transcription factors ERG and ETS2, which are involved in megakaryopoiesis and have been shown to play a role in leukemia development. It is plausible that increased dosage of these two genes may play a role in predisposing Down syndrome patients to megakaryoblastic leukemia [184,185]. Edwards and Patau syndromes are usually fatal in the first year of life, but the few patients who survive with these trisomies for several years are predisposed to developing Wilms' tumor, a form of kidney cancer [186]. It is not clear whether genes located on the extra chromosomes play specific roles in the development of this cancer.

5. Polyploidy

5.1. Sources for Polyploidy

There are several ways for cells to become polyploid. Most are tied to incomplete cell division. Cytokinesis normally separates the cell into two after chromosome segregation. Failure of cytokinesis produces a single tetraploid binucleate cell. In the mammalian liver, embryonic cells are primarily mononucleate but, depending on species, the adult liver contains a high percentage of binucleate cells, at least some of which are thought to occur as a result of incomplete cytokinesis [187]. Cytokinesis failure can be caused by defects in many proteins involved in this process [188,189]. It can also be blocked experimentally by chemical inhibition of the acto-myosin system of the contractile ring, which normally constricts and separates daughter cells. For instance, cytochalasins block actin polymerization. The presence of a chromosome bridge between daughter cells may cause failure to complete the process termed abscission—the final step of cytokinesis [190,191].

Failure of the spindle checkpoint to maintain mitotic arrest in cells with disrupted mitotic spindles results in cells transiting from M phase to G1 without undergoing normal mitosis and cytokinesis. This abnormal cell cycle transition is sometimes termed "mitotic slippage". If cells exit mitosis with all chromosomes retained in one nucleus, the reformed interphase cell is tetraploid. Mitotic slippage occurs because of the gradual proteolysis of the main mitotic cyclin, Cyclin B1, which continues slowly even in checkpoint-arrested cells [192]. Cyclin B1 is the activator of the primary mitotic kinase, Cdk1. When the level of Cyclin B1 falls below a certain threshold, Cdk1 activity declines precipitously and cells exit M phase. Experimentally, mitotic slippage can be induced by treatment of cells with chemical inhibitors of Cdk1 or the spindle checkpoint kinase, Mps1 [193–195].

There are several instances where the generation of polyploid cells is a normal feature of differentiation. One form of this polyploidization is called endomitosis and occurs in megakaryocytes during their differentiation. Megakaryocytes are large cells with multilobed nuclei responsible for generating platelets, which mediate blood clotting. Cell cycles in megakaryocytes consists of rounds of DNA replication followed by brief entry and exit from a modified M phase. During this M phase, the nuclear envelope breaks down, chromosomes show partial condensation, and mitotic spindles form. However, the sister chromatids do not fully condense, nor do the cells undergo cytokinesis. Multiple rounds of these modified cell cycles result in massive polyploidy, a normal aspect of megakaryocyte differentiation [196,197]. Fully differentiated polyploid megakaryocytes become very large and eventually disintegrate into numerous platelets that enter circulation.

Endoreplication is another form of polyploidization where the genome replicates multiple times without intermittent entry into M phase [198]. These unusual cell cycles are called endocycles and the resulting polyploidy is called endopolyploidy. A well-known example of endoreplication is the polytene chromosomes formed in cells of the salivary glands of *Drosophila* larva. Repeated rounds of DNA synthesis without intermittent mitoses lead to the formation of multistranded chromosomes, each composed of hundreds of strands. This massive increase in DNA content increases protein production during larval development. These long, thick, multi-stranded polytene chromosomes have a characteristic appearance with easily distinguished "puffs" indicative of active transcription. Historically, polytene chromosomes may be the earliest observed banded chromosomes, paving the road for modern cytogenetics, the study of chromosome organization and function [199]. In mammals, endoreplication is an important aspect in the development of extra-embryonic tissues [200]. In trophoblast giant cells, endoreplication can reach very high ploidies, up to hundreds of folds in rodents [200]. The formation of these cells appears essential for normal embryo implantation and post-implantation functions of the placenta [201].

Another path to polyploidization, not tied to mitotic errors, is cell fusion. Cell fusion may be the source of at least some of the binucleate cells commonly present in the adult liver [202,203]. Several cancer-causing viruses, such as Hepatitis B and C, Epstein-Barr, and Human Papilloma Virus (HPV) are also fusogenic [204]. Unlike polyploidy which stems from mitotic errors or endoreplication, cells of different types can fuse and produce hybrid polyploid cells. Such hybrids may gain characteristics from both original cell types. For instance, the fusion of cancer cells with bone marrow–derived cells can produce malignant hybrid cancer cells with traits from bone marrow that may promote metastasis [205].

5.2. Polyploidy in Fungi

In many lower eukaryotes, the genome displays a high degree of plasticity in terms of both polyploidy and aneuploidy. Some fungi can proliferate as either haploid or diploid. Various types of fungi can also become polyploid through failed cell divisions, at least in laboratory settings. Subsequently, they propagate as polyploid or transition to being aneuploid by losing some of the extra chromosomes during subsequent cell divisions. Polyploidy, whole chromosome aneuploidy, and segmental aneuploidy are very common in domesticated populations of *S. cerevisiae* beer yeasts [24]. Laboratory strains of *S. cerevisiae* can be propagated at ploidies up to pentaploid. In certain conditions, polyploidization in yeast is adaptive. The encapsulated yeast *C. neoformans*, an opportunistic fungal pathogen, can cause pneumonia and meningitis in immunocompromised individuals. It can generate large polyploid cells called "titan cells". These cells are most frequently tetraploid or octoploid and are resistant to the host immune system and to the antifungal antibiotic fluconazole partly because they have a thicker cell wall and a sturdier, denser capsule [37,206]. Another opportunistically pathogenic yeast, *C. albicans*, routinely inhabits skin and mucous membranes. In immunodeficient individuals it can generate infections that, after exposure to fluconazole and other azole antifungals, can harbor tetraploid cells [207]. These tetraploid cells generate abnormal mitotic spindles in subsequent cell cycles, leading to aneuploidy in the progeny.

5.3. Polyploidy in Animals and Plants

Most species of the animal kingdom, with few exceptions, are diploid, and polyploidy of whole animals is unusual. In mammals, only one species is known to carry extra sets of chromosomes: the red viscacha rat, *Tympanoctomys barrerae* [208]. A few hundred cases of polyploid species are known among insects, reptiles, amphibians, crustaceans and fish [209]. Polyploidy can play a role in the evolution of animal species. One example is the commonly used lab species *Xenopus laevis*, the South African clawed toad. This frog appears to have evolved from the mating of two closely related species whose offspring retained both sets of chromosomes [210]. This condition is called allotetraploidy. Millions of years of evolution have generated many genetic changes, including deletions of duplicate genes, but a recent genome evolution study revealed that the species retains active alleles from both ancestors in over 56% of genes [211]. Chance genome doubling during mammalian embryogenesis typically leads to embryonic lethality. In humans, congenital triploidy and tetraploidy may account for up to 10% of spontaneous abortions [212]. Interestingly, a few case studies report live births of complete and mosaic tetraploid humans with severe developmental defects [213–215].

In stark contrast to the situation in animals, polyploidy is very common in plants, especially in angiosperms (flowering plants) [216]. The exact reasons for this tolerance are unknown, but it may be at least partially due to the absence of p53 in plants [217]. Polyploid plants are typically bigger than their diploid ancestors with correspondingly larger fruits. It is estimated that up to 30% of wild angiosperms are polyploid, and more than 50% of angiosperms that comprise agricultural and food crops can have ploidies ranging from triploid to octoploid and beyond [218]. Most agricultural crops are angiosperms. Polyploidy is very prominent in the grass family—the source of wheat, rice and corn. Other crops including potatoes, sugarcane, apples, strawberries, bananas, coffee are also polyploid [219]. This polyploidy is a product of human selection for plants with bigger, bulkier fruits, and is largely responsible for feeding modern humanity.

5.4. Polyploidy Can Lead to Aneuploidy

Polyploid animal cells are prone to chromosome missegregation during mitosis [220]. If polyploid cells contain supernumerary centrosomes, then cell divisions in these cells can be error-prone, because multipolar mitotic spindles that often form when there are more than two centrosomes cannot segregate sister chromatids equally (Figure 8). However, cells with extra centrosomes can and do generate bipolar mitotic spindles by clustering centrosomes to form two spindle poles [221]. Even with bipolar spindles, such cells still display an increased tendency for chromosome missegregation during anaphase [222].

An example of a normal tissue that progresses from polyploidy to aneuploidy is the adult mammalian liver where, in addition to the binucleate cells containing two diploid nuclei, binucleate and mononucleate cells polyploid and aneuploid cells are also present [223]. The percentage of binucleate cells declines after partial hepatectomy and regrowth, giving rise to both polyploid and aneuploid cells through abnormal divisions [223]. Notably, the existence of stable polyploid and aneuploid cells in normal mammalian liver without massive induction of cancer suggests that ploidy alterations on their own are not sufficient for oncogenesis, at least in hepatocytes.

As mentioned earlier, aneuploidy in fungi can be a consequence of polyploidy induced in response to anti-fungal agents. In the laboratory, tetraploid *S. cerevisiae* strains have a 200- to 1000-fold increase in chromosome loss rates relative to diploid cells, due, at least in part, to a higher incidence of syntelic kinetochore attachments, where both sister chromatids attach to the same spindle pole [224]. Polyploid titan cells in *C. neoformans* lung infections can rapidly produce diverse aneuploid progeny, which promotes adaptation to antibiotic therapy [37]. Similarly, in opportunistically pathogenic

C. *albicans*, tetraploid cells generated after exposure to antifungal agents can produce aneuploid progeny that acquire antibiotic resistance [207].

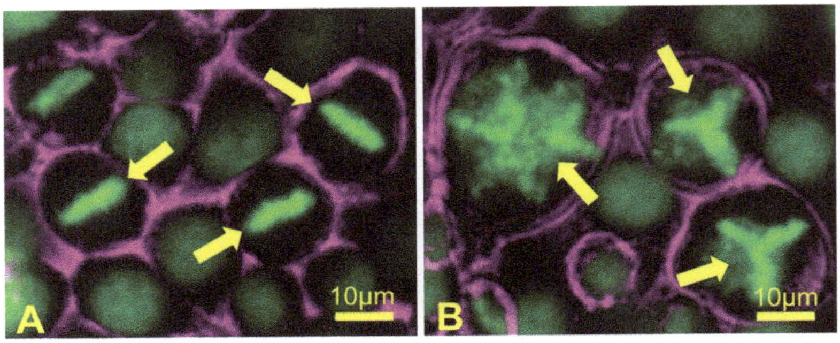

Figure 8. Failed cytokinesis can lead to multipolar spindle formation in the subsequent mitosis. (**A**) A field of control HeLa cells containing mitotic cells showing normal chromosome alignment at metaphase (arrows); (**B**) HeLa cells were treated for a short time with an actin polymerization inhibitor drug, which blocks cytokinesis and results in the formation of binucleate polyploid cells containing extra centrosomes. During the subsequent mitosis, polyploid cells form abnormal metaphase chromosome alignments (arrows) when the presence of extra centrosomes leads to assembly of multipolar mitotic spindles. When these cells then undergo anaphase and cytokinesis, chromosomes are segregated in a random, unequal manner, leading to the formation of daughter cells that are highly aneuploid. Chromosomes are depicted in green; cell surfaces are depicted in magenta.

5.5. Polyploidy in Cancer

Oncogenesis is a multi-step evolutionary progression that selects for traits that allow malignant cells to survive and proliferate. With the advances in whole genome sequencing technology, it has become possible to trace cancer genome changes during oncogenic progression. Tetraploidization appears to be a relatively common event in tumor evolution. Evidence for transient genome-doubling events has been detected during progression of malignancies, and tetraploidization has been associated with cancer aggressiveness and recurrence [225–228]. Tetraploidy may be an early event in tumorigenesis that fuels further genomic instability, leading to the selection of tumors with increased malignancy [229–232]. This idea has been tested in mice using a xenograft cancer model that began with non-malignant diploid or tetraploid xenograft cells. This approach demonstrated that tetraploid cells were tumorigenic when injected in immunodeficient mice and gave rise to malignant aneuploid cells, while isogenic diploid cells did not [233–235]. The aneuploid cells of tumors derived from tetraploid precursors demonstrated a wide diversity of karyotypes indicating a high degree of chromosomal instability.

Polyploid cells often contain extra centrosomes, and their cell divisions can be catastrophically error-prone because, as indicated previously, cells with extra centrosomes are prone to assemble multipolar mitotic spindles. Since multipolar spindles cannot segregate chromosomes equally, essentially all progeny become aneuploid. These aneuploid cells may also exhibit chromosome instability and acquire further numerical and segmental chromosomal aberrations [220,222,236–238]. However, it is also common that in cancer cells, multiple centrosomes can cluster to form bipolar spindles [239]. Strategies to inhibit centrosome clustering and thus purposefully drive spindle multipolarity have been proposed as potential cancer therapies [240,241]. Indeed, it has been proposed that the well-established anti-cancer drug Taxol may function in this manner [87]. On the other hand, in human tumors, the presence of abnormal mitotic figures such as multipolar spindles in biopsies is considered a feature of advanced malignancy. Understanding the positive and negative consequences of spindle multipolarity remains an important topic for future study.

Tetraploid cells in tissue culture can also reveal increased resistance to certain chemotherapeutic drugs compared to their parental diploid cells [242–246]. This effect is reminiscent of the elevated antibiotic resistance detected in polyploid fungi, although the mechanisms underlying this resistance in mammalian cells have not been discovered.

6. Ploidy Aberrations and P53

6.1. Ideas in Evolution and Cancer

In mammals, aneuploidy- and polyploidy-driven evolution of single cells is restrained by the tumor suppressor protein p53 [2,3,5,233,247,248]. p53 is a transcription factor that regulates the expression of various genes involved in stress responses, cell cycle arrest, and apoptosis. Yeast cells do not have the p53 gene, and homologues of mammalian p53 first appeared in protostomes (molluscs, annelids and arthropods) [249]. As mentioned previously, plants also lack p53 [217]. Animals with a large body size require many more cells and often exhibit longer lifespans than smaller animals. Thus, long-lived, large animals might be expected to have an increased susceptibility to cancer. However, no correlation between body size or lifespan and the occurrence of cancer can be found [250–252]. Interestingly, elephants possess 20 copies of the p53 gene and show a hyperactive p53-dependent DNA damage response, potentially contributing to cancer resistance in this large, long-lived animal [253].

More than half of human malignancies harbor mutations of the p53 gene [254], and together with alterations in other components of the p53 network, the p53 pathway is suppressed or inactivated in most human cancers [255]. Inactivation of the p53 pathway likely unleashes cancer evolution, enabling cancer cells with abnormal karyotypes to proliferate, limited only by their fitness in a given environment.

6.2. Concepts for Activation and Function

p53 is one of the most extensively studied proteins, yet it is still not clear what specific factor, or combination of factors, triggers its activation in cells with aberrant ploidies. Determining this has been challenging because transcriptional activation of p53 can be triggered by a wide variety of external and internal stresses [256–258], and the range of stresses that may occur in polyploid and aneuploid cells is also broad. The roles of p53 in sensing DNA damage as well as oxidative and proteotoxic stresses are well established. These stressors can accompany some cases of ploidy alterations. In addition, the stoichiometry of ribosomal proteins caused by changes in gene dosages can activate the p53 pathway by protecting p53 from ubiquitination by its key negative regulator, the ubiquitin ligase Mdm2, which targets it for degradation by the proteasome [259]. Moreover, polyploid and aneuploid cells frequently have an aberrant number of centrosomes, and recent studies show that the p53 may be activated by extra or missing centrosomes [260–262]. A recent study of chromosome missegregation in anaphase demonstrated that lagging or misaligned chromosomes stabilize p53 through retained phosphorylation of histone H3.3, suggesting that mitotic defects resulting in missegregated chromosomes can activate p53 directly [247].

When non-transformed cells in culture are induced to become polyploid by disruption of the actin or microtubule cytoskeleton, they usually block proliferation through the expression of p53 [2,233,263–268]. Direct imaging of cells in culture suggests that the p53-dependent arrest may be delayed for up to three cell cycles after the induction of polyploidy [269]. The pathways linking polyploid cell formation to p53 activation remain unclear. One mechanism may be activation of the Hippo tumor suppressor pathway, induced, at least in part, by extra centrosomes [262]. In tissue culture, proliferating lines of tetraploid cells can grow out from cultures experimentally induced to become tetraploid. These karyotypically stable tetraploid cells adapt to contain normal numbers of centrosomes and build bipolar mitotic spindles [43,262]. As yet, a full understanding of how aneuploid and polyploid cells circumvent p53-mediated arrest remains incomplete, but two recent studies indicated that overexpression of D-type Cyclins allows the continued proliferation of tetraploid cells despite the presence of wild-type p53 [43,270].

In a variety of cell lines, loss of function mutations of the Rb tumor suppressor caused significant chromosome segregation errors but only a modest increase in aneuploidy, unless p53 was also inactivated, whereupon aneuploidy was greatly enhanced [271]. In cells, heterozygous for an inactivating p53 mutation, loss of Rb function could increase the probability that a segregation error causes loss of the chromosome containing the wild-type p53 allele (loss of heterozygosity) and thus generate proliferative progeny permissive for further chromosome instability and increased aneuploidy. Of note, aneuploid cells also occur naturally in some tissues such as the adult liver [223] and the brain [272]. As discussed earlier, polyploidization is a part of a normal differentiation program in certain cell lineages. In tissue culture, human pluripotent stem cells and RPE-1 with normal p53 expression were found to gain extra copies of chromosome 12 and proliferate at a high rate [43]. It is not yet fully clear how, in various circumstances, acquisition of extra chromosomes in non-malignant p53-expressing cells in culture allows cells to evade detection or overcome activation of p53.

7. Conclusions and Perspectives for Human Health

The simple use of the terms mitotic or meiotic *errors* presupposes that such events are detrimental. In many, perhaps most cases, abrupt changes in chromosome content in humans will have unfavorable consequences, for example meiotic aneuploidies giving rise to abnormal embryos or cancer cells developing increased malignancy. However, studies in unicellular eukaryotes have demonstrated that aneuploidy and genomic instability can empower adaptive evolution. Here, there are distinct differences between single-cell eukaryotes and metazoans. In yeast, karyotypic diversity is limited only by the fitness cost and may allow exploitation of new environmental conditions. In multicellular organisms, proliferative competition of individual cells leads to cancer and compromises the fitness of the whole organism. However, there are clear instances where altered mitotic events have been subsumed into differentiation, providing evolutionary advantages in metazoans.

The study of the paths leading from segregation errors to adverse consequences has important potential for human health. For example, in individual cancers, what is the relationship among chromosome instability, aneuploidy, and malignancy? Which pathways—sister chromatid cohesion, cell cycle checkpoints, chromosome movement, and others—are affected? Can we take advantage of cancer's dependence on a specific auxiliary mitotic pathway to design therapies that are generally nontoxic to normal cells but lethal to the tumor? Can we design treatments that specifically target cells that are aneuploid? Can prevention of centrosome clustering be a viable cancer therapy? After many years of study, the complex role of p53 in tumor progression still holds secrets. How can these be revealed and exploited? What are the most important contributors of the maternal age effect? Is it possible to design interventions that might promote normal fertility and development? As we learn more about the mechanisms underlying mitosis and meiosis, we are sure to uncover more surprising insights into the complex interplay of the regulation of cell division with disease, health, and evolution.

Acknowledgments: We thank Hem Sapkota, Dean Dawson, Antonio D'Assoro, and Keith Jones for providing figures. Gary J. Gorbsky is supported by grant # R01GM111731 from the National Institute of General Medical Sciences, by the Oklahoma Center for Adult Stem Cell Research and by the McCasland Foundation. We thank Jennifer Gerton and the Stowers Institute for Medical Research for assistance in making this work possible.

Conflicts of Interest: The authors declare no conflicts of interest.

References

1. Gascoigne, K.E.; Taylor, S.S. Cancer cells display profound intra- and interline variation following prolonged exposure to antimitotic drugs. *Cancer Cell* **2008**, *14*, 111–122. [CrossRef] [PubMed]
2. Andreassen, P.R.; Lohez, O.D.; Lacroix, F.B.; Margolis, R.L. Tetraploid state induces p53-dependent arrest of nontransformed mammalian cells in G1. *Mol. Biol. Cell* **2001**, *12*, 1315–1328. [CrossRef] [PubMed]

3. Aylon, Y.; Oren, M. P53: Guardian of ploidy. *Mol. Oncol.* **2011**, *5*, 315–323. [CrossRef] [PubMed]
4. Duensing, A.; Duensing, S. Guilt by association? P53 and the development of aneuploidy in cancer. *Biochem. Biophys. Res. Commun.* **2005**, *331*, 694–700. [CrossRef] [PubMed]
5. Thompson, S.L.; Compton, D.A. Proliferation of aneuploid human cells is limited by a p53-dependent mechanism. *J. Cell Biol.* **2010**, *188*, 369–381. [CrossRef] [PubMed]
6. Dalton, W.B.; Nandan, M.O.; Moore, R.T.; Yang, V.W. Human cancer cells commonly acquire DNA damage during mitotic arrest. *Cancer Res.* **2007**, *67*, 11487–11492. [CrossRef] [PubMed]
7. Hayashi, M.T.; Cesare, A.J.; Fitzpatrick, J.A.; Lazzerini-Denchi, E.; Karlseder, J.A. Telomere-dependent DNA damage checkpoint induced by prolonged mitotic arrest. *Nat. Struct. Mol. Biol.* **2012**, *19*, 387–394. [CrossRef] [PubMed]
8. Bekier, M.E.; Fischbach, R.; Lee, J.; Taylor, W.R. Length of mitotic arrest induced by microtubule-stabilizing drugs determines cell death after mitotic exit. *Mol. Cancer. Ther.* **2009**, *8*, 1646–1654. [CrossRef] [PubMed]
9. Chu, R.; Terrano, D.T.; Chambers, T.C. Cdk1/cyclin B plays a key role in mitotic arrest-induced apoptosis by phosphorylation of Mcl-1, promoting its degradation and freeing Bak from sequestration. *Biochem. Pharmacol.* **2012**, *83*, 199–206. [CrossRef] [PubMed]
10. Harley, M.E.; Allan, L.A.; Sanderson, H.S.; Clarke, P.R. Phosphorylation of Mcl-1 by CDK1-cyclin B1 initiates its Cdc20-dependent destruction during mitotic arrest. *EMBO J.* **2010**, *29*, 2407–2420. [CrossRef] [PubMed]
11. Sakurikar, N.; Eichhorn, J.M.; Chambers, T.C. Cyclin-dependent kinase-1 (Cdk1)/cyclin B1 dictates cell fate after mitotic arrest via phosphoregulation of antiapoptotic Bcl-2 proteins. *J. Biol. Chem.* **2012**, *287*, 39193–39204. [CrossRef] [PubMed]
12. Sloss, O.; Topham, C.; Diez, M.; Taylor, S. Mcl-1 dynamics influence mitotic slippage and death in mitosis. *Oncotarget* **2016**, *7*, 5176–5192. [CrossRef]
13. Orth, J.D.; Loewer, A.; Lahav, G.; Mitchison, T.J. Prolonged mitotic arrest triggers partial activation of apoptosis, resulting in DNA damage and p53 induction. *Mol. Biol. Cell* **2012**, *23*, 567–576. [CrossRef] [PubMed]
14. Uetake, Y.; Sluder, G. Prolonged prometaphase blocks daughter cell proliferation despite normal completion of mitosis. *Curr. Biol.* **2010**, *20*, 1666–1671. [CrossRef] [PubMed]
15. Pavelka, N.; Rancati, G.; Zhu, J.; Bradford, W.D.; Saraf, A.; Florens, L.; Sanderson, B.W.; Hattem, G.L.; Li, R. Aneuploidy confers quantitative proteome changes and phenotypic variation in budding yeast. *Nature* **2010**, *468*, 321–325. [CrossRef] [PubMed]
16. Stingele, S.; Stoehr, G.; Peplowska, K.; Cox, J.; Mann, M.; Storchova, Z. Global analysis of genome, transcriptome and proteome reveals the response to aneuploidy in human cells. *Mol. Syst. Biol.* **2012**, *8*, 608–619. [CrossRef] [PubMed]
17. Upender, M.B.; Habermann, J.K.; McShane, L.M.; Korn, E.L.; Barrett, J.C.; Difilippantonio, M.J.; Ried, T. Chromosome transfer induced aneuploidy results in complex dysregulation of the cellular transcriptome in immortalized and cancer cells. *Cancer Res.* **2004**, *64*, 6941–6949. [CrossRef] [PubMed]
18. Torres, E.M.; Dephoure, N.; Panneerselvam, A.; Tucker, C.M.; Whittaker, C.A.; Gygi, S.P.; Dunham, M.J.; Amon, A. Identification of aneuploidy-tolerating mutations. *Cell* **2010**, *143*, 71–83. [CrossRef] [PubMed]
19. Rancati, G.; Pavelka, N.; Fleharty, B.; Noll, A.; Trimble, R.; Walton, K.; Perera, A.; Staehling-Hampton, K.; Seidel, C.W.; Li, R. Aneuploidy underlies rapid adaptive evolution of yeast cells deprived of a conserved cytokinesis motor. *Cell* **2008**, *135*, 879–893. [CrossRef] [PubMed]
20. Dephoure, N.; Hwang, S.; O'Sullivan, C.; Dodgson, S.E.; Gygi, S.P.; Amon, A.; Torres, E.M. Quantitative proteomic analysis reveals posttranslational responses to aneuploidy in yeast. *Elife* **2014**, *3*, e03023. [CrossRef] [PubMed]
21. Williams, B.R.; Prabhu, V.R.; Hunter, K.E.; Glazier, C.M.; Whittaker, C.A.; Housman, D.E.; Amon, A. Aneuploidy affects proliferation and spontaneous immortalization in mammalian cells. *Science* **2008**, *322*, 703–709. [CrossRef] [PubMed]
22. Siegel, J.J.; Amon, A. New insights into the troubles of aneuploidy. *Annu. Rev. Cell Dev. Biol.* **2012**, *28*, 189–214. [CrossRef] [PubMed]
23. Torres, E.M.; Sokolsky, T.; Tucker, C.M.; Chan, L.Y.; Boselli, M.; Dunham, M.J.; Amon, A. Effects of aneuploidy on cellular physiology and Cell Div.ision in haploid yeast. *Science* **2007**, *317*, 916–924. [CrossRef] [PubMed]

24. Gallone, B.; Steensels, J.; Prahl, T.; Soriaga, L.; Saels, V.; Herrera-Malaver, B.; Merlevede, A.; Roncoroni, M.; Voordeckers, K.; Miraglia, L.; et al. Domestication and divergence of saccharomyces cerevisiae beer yeasts. *Cell* **2016**, *166*, 1397–1410. [CrossRef] [PubMed]
25. Bergstrom, A.; Simpson, J.T.; Salinas, F.; Barre, B.; Parts, L.; Zia, A.; Nguyen Ba, A.N.; Moses, A.M.; Louis, E.J.; Mustonen, V.; et al. A high-definition view of functional genetic variation from natural yeast genomes. *Mol. Biol. Evol.* **2014**, *31*, 872–888. [CrossRef] [PubMed]
26. Dunham, M.J.; Badrane, H.; Ferea, T.; Adams, J.; Brown, P.O.; Rosenzweig, F.; Botstein, D. Characteristic genome rearrangements in experimental evolution of Saccharomyces cerevisiae. *Proc. Natl. Acad. Sci. USA* **2002**, *99*, 16144–16149. [CrossRef] [PubMed]
27. Dunn, B.; Richter, C.; Kvitek, D.J.; Pugh, T.; Sherlock, G. Analysis of the Saccharomyces cerevisiae pan-genome reveals a pool of copy number variants distributed in diverse yeast strains from differing industrial environments. *Genome Res.* **2012**, *22*, 908–924. [CrossRef] [PubMed]
28. Voordeckers, K.; Kominek, J.; Das, A.; Espinosa-Cantu, A.; De Maeyer, D.; Arslan, A.; Van Pee, M.; Van Der Zande, E.; Meert, W.; Yang, Y.; et al. Adaptation to high ethanol reveals complex evolutionary pathways. *PLoS Genet.* **2015**, *11*, e1005635. [CrossRef] [PubMed]
29. Kvitek, D.J.; Will, J.L.; and Gasch, A.P. Variations in stress sensitivity and genomic expression in diverse *S. cerevisiae* isolates. *PLoS Genet.* **2008**, *4*, e1000223.
30. Borneman, A.R.; Desany, B.A.; Riches, D.; Affourtit, J.P.; Forgan, A.H.; Pretorius, I.S.; Egholm, M.; Chambers, P.J. Whole-genome comparison reveals novel genetic elements that characterize the genome of industrial strains of Saccharomyces cerevisiae. *PLoS Genet.* **2011**, *7*, e1001287. [CrossRef] [PubMed]
31. Hughes, T.R.; Roberts, C.J.; Dai, H.; Jones, A.R.; Meyer, M.R.; Slade, D.; Burchard, J.; Dow, S.; Ward, T.R.; Kidd, M.J.; et al. Widespread aneuploidy revealed by DNA microarray expression profiling. *Nat. Genet.* **2000**, *25*, 333–337. [PubMed]
32. Gresham, D.; Desai, M.M.; Tucker, C.M.; Jenq, H.T.; Pai, D.A.; Ward, A.; DeSevo, C.G.; Botstein, D.; Dunham, M.J. The repertoire and dynamics of evolutionary adaptations to controlled nutrient-limited environments in yeast. *PLoS Genet.* **2008**, *4*, e1000303. [CrossRef] [PubMed]
33. Fries, B.C.; Casadevall, A. Serial isolates of Cryptococcus neoformans from patients with AIDS differ in virulence for mice. *J. Infect. Dis.* **1998**, *178*, 1761–1766. [CrossRef] [PubMed]
34. Selmecki, A.; Gerami-Nejad, M.; Paulson, C.; Forche, A.; Berman, J. An isochromosome confers drug resistance in vivo by amplification of two genes, ERG11 and TAC1. *Mol. Microbiol.* **2008**, *68*, 624–641. [CrossRef] [PubMed]
35. Selmecki, A.M.; Dulmage, K.; Cowen, L.E.; Anderson, J.B.; Berman, J. Acquisition of aneuploidy provides increased fitness during the evolution of antifungal drug resistance. *PLoS Genet.* **2009**, *5*, e1000705. [CrossRef] [PubMed]
36. Selmecki, A.; Forche, A.; Berman, J. Aneuploidy and isochromosome formation in drug-resistant Candida albicans. *Science* **2006**, *313*, 367–370. [CrossRef] [PubMed]
37. Gerstein, A.C.; Fu, M.S.; Mukaremera, L.; Li, Z.; Ormerod, K.L.; Fraser, J.A.; Berman, J.; Nielsen, K. Polyploid titan cells produce haploid and aneuploid progeny to promote stress adaptation. *mBio* **2015**, *6*, e01340-15. [CrossRef] [PubMed]
38. Ben-David, U.; Arad, G.; Weissbein, U.; Mandefro, B.; Maimon, A.; Golan-Lev, T.; Narwani, K.; Clark, A.T.; Andrews, P.W.; Benvenisty, N.; et al. Aneuploidy induces profound changes in gene expression, proliferation and tumorigenicity of human pluripotent stem cells. *Nat. Commun.* **2014**, *5*, 4825–4835. [CrossRef] [PubMed]
39. Na, J.; Baker, D.; Zhang, J.; Andrews, P.W.; Barbaric, I. Aneuploidy in pluripotent stem cells and implications for cancerous transformation. *Protein Cell* **2014**, *5*, 569–579. [CrossRef] [PubMed]
40. Werbowetski-Ogilvie, T.E.; Bosse, M.; Stewart, M.; Schnerch, A.; Ramos-Mejia, V.; Rouleau, A.; Wynder, T.; Smith, M.J.; Dingwall, S.; Carter, T.; et al. Characterization of human embryonic stem cells with features of neoplastic progression. *Nat. Biotechnol.* **2009**, *27*, 91–97. [CrossRef] [PubMed]
41. Zhang, M.; Cheng, L.; Jia, Y.; Liu, G.; Li, C.; Song, S.; Bradley, A.; Huang, Y. Aneuploid embryonic stem cells exhibit impaired differentiation and increased neoplastic potential. *EMBO J.* **2016**, *35*, 2285–2300. [CrossRef] [PubMed]
42. Bodnar, A.G.; Ouellette, M.; Frolkis, M.; Holt, S.E.; Chiu, C.P.; Morin, G.B.; Harley, C.B.; Shay, J.W.; Lichtsteiner, S.; Wright, W.E. Extension of life-span by introduction of telomerase into normal human cells. *Science* **1998**, *279*, 349–352. [CrossRef] [PubMed]

43. Potapova, T.A.; Seidel, C.W.; Box, A.C.; Rancati, G.; Li, R. Transcriptome analysis of tetraploid cells identifies Cyclin D2 as a facilitator of adaptation to genome doubling in the presence of p53. *Mol. Biol. Cell* **2016**, *27*, 3065–3084. [CrossRef] [PubMed]
44. Zhang, C.Z.; Spektor, A.; Cornils, H.; Francis, J.M.; Jackson, E.K.; Liu, S.; Meyerson, M.; Pellman, D. Chromothripsis from DNA damage in micronuclei. *Nature* **2015**, *522*, 179–184. [CrossRef] [PubMed]
45. Di Nicolantonio, F.; Arena, S.; Gallicchio, M.; Zecchin, D.; Martini, M.; Flonta, S.E.; Stella, G.M.; Lamba, S.; Cancelliere, C.; Russo, M.; et al. Replacement of normal with mutant alleles in the genome of normal human cells unveils mutation-specific drug responses. *Proc. Natl. Acad. Sci. USA* **2008**, *105*, 20864–20869. [CrossRef] [PubMed]
46. Ly, P.; Eskiocak, U.; Kim, S.B.; Roig, A.I.; Hight, S.K.; Lulla, D.R.; Zou, Y.S.; Batten, K.; Wright, W.E.; Shay, J.W. Characterization of aneuploid populations with trisomy 7 and 20 derived from diploid human colonic epithelial cells. *Neoplasia* **2011**, *13*, 348–357. [CrossRef] [PubMed]
47. Rutledge, S.D.; Douglas, T.A.; Nicholson, J.M.; Vila-Casadesus, M.; Kantzler, C.L.; Wangsa, D.; Barroso-Vilares, M.; Kale, S.D.; Logarinho, E.; Cimini, D. Selective advantage of trisomic human cells cultured in non-standard conditions. *Sci. Rep.* **2016**, *6*, 22828–22839. [CrossRef] [PubMed]
48. Geigl, J.B.; Obenauf, A.C.; Schwarzbraun, T.; Speicher, M.R. Defining 'chromosomal instability'. *Trends Genet.* **2008**, *24*, 64–69. [CrossRef] [PubMed]
49. Gordon, D.J.; Resio, B.; Pellman, D. Causes and consequences of aneuploidy in cancer. *Nat. Rev. Genet.* **2012**, *13*, 189–203. [CrossRef] [PubMed]
50. Giam, M.; Rancati, G. Aneuploidy and chromosomal instability in cancer: A jackpot to chaos. *Cell Div.* **2015**, *10*, 3–14. [CrossRef] [PubMed]
51. Potapova, T.A.; Zhu, J.; Li, R. Aneuploidy and chromosomal instability: A vicious cycle driving cellular evolution and cancer genome chaos. *Cancer Metastasis Rev.* **2013**, *32*, 377–389. [CrossRef] [PubMed]
52. Barnhart, E.L.; Dorer, R.K.; Murray, A.W.; Schuyler, S.C. Reduced Mad2 expression keeps relaxed kinetochores from arresting budding yeast in mitosis. *Mol. Biol. Cell* **2011**, *22*, 2448–2457. [CrossRef] [PubMed]
53. Zhu, J.; Pavelka, N.; Bradford, W.D.; Rancati, G.; Li, R. Karyotypic determinants of chromosome instability in aneuploid budding yeast. *PLoS Genet.* **2012**, *8*, e1002719. [CrossRef] [PubMed]
54. Sheltzer, J.M.; Blank, H.M.; Pfau, S.J.; Tange, Y.; George, B.M.; Humpton, T.J.; Brito, I.L.; Hiraoka, Y.; Niwa, O.; Amon, A. Aneuploidy drives genomic instability in yeast. *Science* **2011**, *333*, 1026–1030. [CrossRef] [PubMed]
55. Nicholson, J.M.; Macedo, J.C.; Mattingly, A.J.; Wangsa, D.; Camps, J.; Lima, V.; Gomes, A.M.; Doria, S.; Ried, T.; Logarinho, E.; et al. Chromosome mis-segregation and cytokinesis failure in trisomic human cells. *Elife* **2015**, *4*, e05068. [CrossRef] [PubMed]
56. Passerini, V.; Ozeri-Galai, E.; de Pagter, M.S.; Donnelly, N.; Schmalbrock, S.; Kloosterman, W.P.; Kerem, B.; Storchova, Z. The presence of extra chromosomes leads to genomic instability. *Nat. Commun.* **2016**, *7*, 10754–10765. [CrossRef] [PubMed]
57. Hansemann, D. Ueber asymmetrische Zelltheilung in Epithelkrebsen und deren biologische Bedeutung. *Virchows Arch.* **1890**, *119*, 299–326. [CrossRef]
58. Ribbert, H. Zur frage der entstehung maligner tumoren. *Naturwissenschaften* **1914**, *2*, 676–679. (In German) [CrossRef]
59. Zasadil, L.M.; Britigan, E.M.; Weaver, B.A. 2n or not 2n: Aneuploidy, polyploidy and chromosomal instability in primary and tumor cells. *Semin. Cell Dev. Biol.* **2013**, *24*, 370–379. [CrossRef] [PubMed]
60. Duesberg, P.; Rasnick, D.; Li, R.; Winters, L.; Rausch, C.; Hehlmann, R. How aneuploidy may cause cancer and genetic instability. *Anticancer Res.* **1999**, *19*, 4887–4906. [PubMed]
61. Zimonjic, D.; Brooks, M.W.; Popescu, N.; Weinberg, R.A.; Hahn, W.C. Derivation of human tumor cells in vitro without widespread genomic instability. *Cancer Res.* **2001**, *61*, 8838–8844. [PubMed]
62. Lengauer, C.; Kinzler, K.W.; Vogelstein, B. Genetic instabilities in human cancers. *Nature* **1998**, *396*, 643–649. [CrossRef] [PubMed]
63. Hanahan, D.; Weinberg, R.A. Hallmarks of cancer: The next generation. *Cell* **2011**, *144*, 646–674. [CrossRef] [PubMed]
64. Iwanaga, Y.; Chi, Y.H.; Miyazato, A.; Sheleg, S.; Haller, K.; Peloponese, J.M., Jr.; Li, Y.; Ward, J.M.; Benezra, R.; Jeang, K.T. Heterozygous deletion of mitotic arrest-deficient protein 1 (MAD1) increases the incidence of tumors in mice. *Cancer Res.* **2007**, *67*, 160–166. [CrossRef] [PubMed]

65. Jeganathan, K.; Malureanu, L.; Baker, D.J.; Abraham, S.C.; Van Deursen, J.M. Bub1 mediates cell death in response to chromosome missegregation and acts to suppress spontaneous tumorigenesis. *J. Cell Biol.* **2007**, *179*, 255–267. [CrossRef] [PubMed]
66. Michel, L.S.; Liberal, V.; Chatterjee, A.; Kirchwegger, R.; Pasche, B.; Gerald, W.; Dobles, M.; Sorger, P.K.; Murty, V.V.; Benezra, R. MAD2 haplo-insufficiency causes premature anaphase and chromosome instability in mammalian cells. *Nature* **2001**, *409*, 355–359. [CrossRef] [PubMed]
67. Ricke, R.M.; Jeganathan, K.B.; Van Deursen, J.M. Bub1 overexpression induces aneuploidy and tumor formation through Aurora B kinase hyperactivation. *J. Cell Biol.* **2011**, *193*, 1049–1064. [CrossRef] [PubMed]
68. Sotillo, R.; Hernando, E.; Diaz-Rodriguez, E.; Teruya-Feldstein, J.; Cordon-Cardo, C.; Lowe, S.W.; Benezra, R. Mad2 overexpression promotes aneuploidy and tumorigenesis in mice. *Cancer Cell* **2007**, *11*, 9–23. [CrossRef] [PubMed]
69. Baker, D.J.; Jeganathan, K.B.; Cameron, J.D.; Thompson, M.; Juneja, S.; Kopecka, A.; Kumar, R.; Jenkins, R.B.; De Groen, P.C.; Roche, P.; et al. BubR1 insufficiency causes early onset of aging-associated phenotypes and infertility in mice. *Nat. Genet.* **2004**, *36*, 744–749. [CrossRef] [PubMed]
70. Dai, W.; Wang, Q.; Liu, T.; Swamy, M.; Fang, Y.; Xie, S.; Mahmood, R.; Yang, Y.M.; Xu, M.; Rao, C.V. Slippage of mitotic arrest and enhanced tumor development in mice with BubR1 haploinsufficiency. *Cancer Res.* **2004**, *64*, 440–445. [CrossRef] [PubMed]
71. Wang, Q.; Liu, T.; Fang, Y.; Xie, S.; Huang, X.; Mahmood, R.; Ramaswamy, G.; Sakamoto, K.M.; Darzynkiewicz, Z.; Xu, M.; et al. BUBR1 deficiency results in abnormal megakaryopoiesis. *Blood* **2004**, *103*, 1278–1285. [CrossRef] [PubMed]
72. Hartman, T.K.; Wengenack, T.M.; Poduslo, J.F.; Van Deursen, J.M. Mutant mice with small amounts of BubR1 display accelerated age-related gliosis. *Neurobiol. Aging* **2007**, *28*, 921–927. [CrossRef] [PubMed]
73. Matsumoto, T.; Baker, D.J.; D'USCIO, L.V.; Mozammel, G.; Katusic, Z.S.; Van Deursen, J.M. Aging-associated vascular phenotype in mutant mice with low levels of BubR1. *Stroke* **2007**, *38*, 1050–1056. [CrossRef] [PubMed]
74. North, B.J.; Rosenberg, M.A.; Jeganathan, K.B.; Hafner, A.V.; Michan, S.; Dai, J.; Baker, D.J.; Cen, Y.; Wu, L.E.; Sauve, A.A.; et al. SIRT2 induces the checkpoint kinase BubR1 to increase lifespan. *EMBO J.* **2014**, *33*, 1438–1453. [CrossRef] [PubMed]
75. Baker, D.J.; Dawlaty, M.M.; Wijshake, T.; Jeganathan, K.B.; Malureanu, L.; Van Ree, J.H.; Crespo-Diaz, R.; Reyes, S.; Seaburg, L.; Shapiro, V.; et al. Increased expression of BubR1 protects against aneuploidy and cancer and extends healthy lifespan. *Nat. Cell Biol.* **2013**, *15*, 96–102. [CrossRef] [PubMed]
76. Hanks, S.; Coleman, K.; Reid, S.; Plaja, A.; Firth, H.; Fitzpatrick, D.; Kidd, A.; Mehes, K.; Nash, R.; Robin, N.; et al. Constitutional aneuploidy and cancer predisposition caused by biallelic mutations in BUB1B. *Nat. Genet.* **2004**, *36*, 1159–1161. [CrossRef] [PubMed]
77. Jacquemont, S.; Boceno, M.; Rival, J.M.; Mechinaud, F.; David, A. High risk of malignancy in mosaic variegated aneuploidy syndrome. *Am. J. Med. Genet.* **2002**, *109*, 17–21. [CrossRef] [PubMed]
78. Matsuura, S.; Matsumoto, Y.; Morishima, K.; Izumi, H.; Matsumoto, H.; Ito, E.; Tsutsui, K.; Kobayashi, J.; Tauchi, H.; Kajiwara, Y.; et al. Monoallelic BUB1B mutations and defective mitotic-spindle checkpoint in seven families with premature chromatid separation (PCS) syndrome. *Am. J. Med. Genet. A* **2006**, *140*, 358–367. [CrossRef] [PubMed]
79. Weaver, B.A.; Silk, A.D.; Montagna, C.; Verdier-Pinard, P.; Cleveland, D.W. Aneuploidy acts both oncogenically and as a tumor suppressor. *Cancer Cell* **2007**, *11*, 25–36. [CrossRef] [PubMed]
80. Silk, A.D.; Zasadil, L.M.; Holland, A.J.; Vitre, B.; Cleveland, D.W.; Weaver, B.A. Chromosome missegregation rate predicts whether aneuploidy will promote or suppress tumors. *Proc. Natl. Acad. Sci. USA* **2013**, *110*, E4134–E4141. [CrossRef] [PubMed]
81. Schiff, P.B.; Fant, J.; Horwitz, S.B. Promotion of microtubule assembly in vitro by taxol. *Nature* **1979**, *277*, 665–667. [CrossRef] [PubMed]
82. Schiff, P.B.; Horwitz, S.B. Taxol stabilizes microtubules in mouse fibroblast cells. *Proc. Natl. Acad. Sci. USA* **1980**, *77*, 1561–1565. [CrossRef] [PubMed]
83. Jordan, M.A.; Toso, R.J.; Thrower, D.; Wilson, L. Mechanism of mitotic block and inhibition of cell proliferation by taxol at low concentrations. *Proc. Natl. Acad. Sci. USA* **1993**, *90*, 9552–9556. [CrossRef] [PubMed]
84. Waters, J.C.; Chen, R.H.; Murray, A.W.; Salmon, E.D. Localization of Mad2 to kinetochores depends on microtubule attachment, not tension. *J. Cell Biol.* **1998**, *141*, 1181–1191. [CrossRef] [PubMed]

85. Komlodi-Pasztor, E.; Sackett, D.; Wilkerson, J.; Fojo, T. Mitosis is not a key target of microtubule agents in patient tumors. *Nat. Rev. Clin. Oncol.* **2011**, *8*, 244–250. [CrossRef] [PubMed]
86. Mitchison, T.J. The proliferation rate paradox in antimitotic chemotherapy. *Mol. Biol. Cell* **2012**, *23*, 1–6. [CrossRef] [PubMed]
87. Zasadil, L.M.; Andersen, K.A.; Yeum, D.; Rocque, G.B.; Wilke, L.G.; Tevaarwerk, A.J.; Raines, R.T.; Burkard, M.E.; Weaver, B.A. Cytotoxicity of paclitaxel in breast cancer is due to chromosome missegregation on multipolar spindles. *Sci. Transl. Med.* **2014**, *6*, 229ra43. [CrossRef] [PubMed]
88. Blasco, M.A. Telomeres and human disease: Ageing, cancer and beyond. *Nat. Rev. Genet.* **2005**, *6*, 611–622. [CrossRef] [PubMed]
89. Frias, C.; Pampalona, J.; Genesca, A.; Tusell, L. Telomere dysfunction and genome instability. *Front. Biosci.* **2012**, *17*, 2181–2196. [CrossRef]
90. Gimenez-Abian, J.F.; Sumara, I.; Hirota, T.; Hauf, S.; Gerlich, D.; De La Torre, C.; Ellenberg, J.; Peters, J.M. Regulation of sister chromatid cohesion between chromosome arms. *Curr. Biol.* **2004**, *14*, 1187–1193. [CrossRef] [PubMed]
91. Gorbsky, G.J. The mitotic spindle checkpoint. *Curr. Biol.* **2001**, *11*, R1001–R1004. [CrossRef]
92. Bompard, G.; Rabeharivelo, G.; Cau, J.; Abrieu, A.; Delsert, C.; Morin, N. P21-activated kinase 4 (PAK4) is required for metaphase spindle positioning and anchoring. *Oncogene* **2013**, *32*, 910–919. [CrossRef] [PubMed]
93. Schvartzman, J.M.; Duijf, P.H.; Sotillo, R.; Coker, C.; Benezra, R. Mad2 is a critical mediator of the chromosome instability observed upon Rb and p53 pathway inhibition. *Cancer Cell* **2011**, *19*, 701–714. [CrossRef] [PubMed]
94. Zhang, P.; Cong, B.; Yuan, H.; Chen, L.; Lv, Y.; Bai, C.; Nan, X.; Shi, S.; Yue, W.; Pei, X. Overexpression of spindlin1 induces metaphase arrest and chromosomal instability. *J. Cell. Physiol.* **2008**, *217*, 400–408. [CrossRef] [PubMed]
95. Manning, A.L.; Yazinski, S.A.; Nicolay, B.; Bryll, A.; Zou, L.; Dyson, N.J. Suppression of genome instability in pRB-deficient cells by enhancement of chromosome cohesion. *Mol. Cell* **2014**, *53*, 993–1004. [CrossRef] [PubMed]
96. Manning, A.L.; Longworth, M.S.; Dyson, N.J. Loss of pRB causes centromere dysfunction and chromosomal instability. *Genes. Dev.* **2010**, *24*, 1364–1376. [CrossRef] [PubMed]
97. Daum, J.R.; Potapova, T.A.; Sivakumar, S.; Daniel, J.J.; Flynn, J.N.; Rankin, S.; Gorbsky, G.J. Cohesion fatigue induces chromatid separation in cells delayed at metaphase. *Curr. Biol.* **2011**, *21*, 1018–1024. [CrossRef] [PubMed]
98. Lara-Gonzalez, P.; Taylor, S.S. Cohesion fatigue explains why pharmacological inhibition of the APC/C induces a spindle checkpoint-dependent mitotic arrest. *PLoS ONE* **2012**, *7*, e49041. [CrossRef] [PubMed]
99. Stevens, D.; Gassmann, R.; Oegema, K.; Desai, A. Uncoordinated loss of chromatid cohesion is a common outcome of extended metaphase arrest. *PLoS ONE* **2011**, *6*, e22969. [CrossRef] [PubMed]
100. Janssen, A.; van der Burg, M.; Szuhai, K.; Kops, G.J.; Medema, R.H. Chromosome segregation errors as a cause of DNA damage and structural chromosome aberrations. *Science* **2011**, *333*, 1895–1898. [CrossRef] [PubMed]
101. Brinkley, B.R.; Zinkowski, R.P.; Mollon, W.L.; Davis, F.M.; Pisegna, M.A.; Pershouse, M.; Rao, P.N. Movement and segregation of kinetochores experimentally detached from mammalian chromosomes. *Nature* **1988**, *336*, 251–254. [CrossRef] [PubMed]
102. O'Connell, C.B.; Loncarek, J.; Hergert, P.; Kourtidis, A.; Conklin, D.S.; Khodjakov, A. The spindle assembly checkpoint is satisfied in the absence of interkinetochore tension during mitosis with unreplicated genomes. *J. Cell Biol.* **2008**, *183*, 29–36. [CrossRef] [PubMed]
103. Thompson, S.L.; Compton, D.A. Chromosome missegregation in human cells arises through specific types of kinetochore-microtubule attachment errors. *Proc. Natl. Acad. Sci. USA* **2011**, *108*, 17974–17978. [CrossRef] [PubMed]
104. Cimini, D.; Cameron, L.A.; Salmon, E.D. Anaphase spindle mechanics prevent mis-segregation of merotelically oriented chromosomes. *Curr. Biol.* **2004**, *14*, 2149–2155. [CrossRef] [PubMed]
105. Daum, J.R.; Wren, J.D.; Daniel, J.J.; Sivakumar, S.; McAvoy, J.N.; Potapova, T.A.; Gorbsky, G.J. Ska3 is required for spindle checkpoint silencing and the maintenance of chromosome cohesion in mitosis. *Curr. Biol.* **2009**, *19*, 1467–1472. [CrossRef] [PubMed]

106. Opyrchal, M.; Salisbury, J.L.; Iankov, I.; Goetz, M.P.; McCubrey, J.; Gambino, M.W.; Malatino, L.; Puccia, G.; Ingle, J.N.; Galanis, E.; et al. Inhibition of Cdk2 kinase activity selectively targets the CD44+/CD24-/Low stem-like subpopulation and restores chemosensitivity of SUM149PT triple-negative breast cancer cells. *Int. J. Oncol.* **2014**, *45*, 1193–1199. [CrossRef] [PubMed]
107. Guerrero, A.A.; Gamero, M.C.; Trachana, V.; Futterer, A.; Pacios-Bras, C.; Diaz-Concha, N.P.; Cigudosa, J.C.; Martinez, A.C.; Van Wely, K.H. Centromere-localized breaks indicate the generation of DNA damage by the mitotic spindle. *Proc. Natl. Acad. Sci. USA* **2010**, *107*, 4159–4164. [CrossRef] [PubMed]
108. Martinez, A.C.; Van Wely, K.H. Centromere fission, not telomere erosion, triggers chromosomal instability in human carcinomas. *Carcinogenesis* **2011**, *32*, 796–803. [CrossRef] [PubMed]
109. Cimini, D.; Howell, B.; Maddox, P.; Khodjakov, A.; Degrassi, F.; Salmon, E.D. Merotelic kinetochore orientation is a major mechanism of aneuploidy in mitotic mammalian tissue cells. *J. Cell Biol.* **2001**, *153*, 517–527. [CrossRef] [PubMed]
110. Weinstein, J.N.; Collisson, E.A.; Mills, G.B.; Shaw, K.R.; Ozenberger, B.A.; Ellrott, K.; Shmulevich, I.; Sander, C.; Stuart, J.M. The cancer genome atlas pan-cancer analysis project. *Nat. Genet.* **2013**, *45*, 1113–1120. [PubMed]
111. Lee, J.K.; Choi, Y.L.; Kwon, M.; Park, P.J. Mechanisms and Consequences of Cancer Genome Instability: Lessons from Genome Sequencing Studies. *Annu. Rev. Pathol.* **2016**, *11*, 283–312. [CrossRef] [PubMed]
112. Carter, S.L.; Eklund, A.C.; Kohane, I.S.; Harris, L.N.; Szallasi, Z. A signature of chromosomal instability inferred from gene expression profiles predicts clinical outcome in multiple human cancers. *Nat. Genet.* **2006**, *38*, 1043–1048. [CrossRef] [PubMed]
113. How, C.; Bruce, J.; So, J.; Pintilie, M.; Haibe-Kains, B.; Hui, A.; Clarke, B.A.; Hedley, D.W.; Hill, R.P.; Milosevic, M.; et al. Chromosomal instability as a prognostic marker in cervical cancer. *BMC Cancer* **2015**, *15*, 361–368. [CrossRef] [PubMed]
114. Birkbak, N.J.; Eklund, A.C.; Li, Q.; McClelland, S.E.; Endesfelder, D.; Tan, P.; Tan, I.B.; Richardson, A.L.; Szallasi, Z.; Swanton, C. Paradoxical relationship between chromosomal instability and survival outcome in cancer. *Cancer Res.* **2011**, *71*, 3447–3452. [CrossRef] [PubMed]
115. Gisselsson, D.; Bjork, J.; Hoglund, M.; Mertens, F.; Dal Cin, P.; Akerman, M.; Mandahl, N. Abnormal nuclear shape in solid tumors reflects mitotic instability. *Am. J. Pathol.* **2001**, *158*, 199–206. [CrossRef]
116. Beroukhim, R.; Mermel, C.H.; Porter, D.; Wei, G.; Raychaudhuri, S.; Donovan, J.; Barretina, J.; Boehm, J.S.; Dobson, J.; Urashima, M.; et al. The landscape of somatic copy-number alteration across human cancers. *Nature* **2010**, *463*, 899–905. [CrossRef] [PubMed]
117. Bignell, G.R.; Greenman, C.D.; Davies, H.; Butler, A.P.; Edkins, S.; Andrews, J.M.; Buck, G.; Chen, L.; Beare, D.; Latimer, C.; et al. Signatures of mutation and selection in the cancer genome. *Nature* **2010**, *463*, 893–898. [CrossRef] [PubMed]
118. Davoli, T.; Xu, A.W.; Mengwasser, K.E.; Sack, L.M.; Yoon, J.C.; Park, P.J.; Elledge, S.J. Cumulative haploinsufficiency and triplosensitivity drive aneuploidy patterns and shape the cancer genome. *Cell* **2013**, *155*, 948–962. [CrossRef] [PubMed]
119. Navin, N.; Krasnitz, A.; Rodgers, L.; Cook, K.; Meth, J.; Kendall, J.; Riggs, M.; Eberling, Y.; Troge, J.; Grubor, V.; et al. Inferring tumor progression from genomic heterogeneity. *Genome Res.* **2010**, *20*, 68–80. [CrossRef] [PubMed]
120. Gerlinger, M.; Rowan, A.J.; Horswell, S.; Larkin, J.; Endesfelder, D.; Gronroos, E.; Martinez, P.; Matthews, N.; Stewart, A.; Tarpey, P.; et al. Intratumor heterogeneity and branched evolution revealed by multiregion sequencing. *N. Engl. J. Med.* **2012**, *366*, 883–892. [CrossRef] [PubMed]
121. Shah, S.P.; Roth, A.; Goya, R.; Oloumi, A.; Ha, G.; Zhao, Y.; Turashvili, G.; Ding, J.; Tse, K.; Haffari, G.; et al. The clonal and mutational evolution spectrum of primary triple-negative breast cancers. *Nature* **2012**, *486*, 395–399. [CrossRef] [PubMed]
122. Navin, N.E. Cancer genomics: One cell at a time. *Genome Biol.* **2014**, *15*, 452–464. [CrossRef] [PubMed]
123. Navin, N.E. The first five years of single-cell cancer genomics and beyond. *Genome Res.* **2015**, *25*, 1499–1507. [CrossRef] [PubMed]
124. Wang, Y.; Waters, J.; Leung, M.L.; Unruh, A.; Roh, W.; Shi, X.; Chen, K.; Scheet, P.; Vattathil, S.; Liang, H.; et al. Clonal evolution in breast cancer revealed by single nucleus genome sequencing. *Nature* **2014**, *512*, 155–160. [CrossRef] [PubMed]

125. Navin, N.; Kendall, J.; Troge, J.; Andrews, P.; Rodgers, L.; McIndoo, J.; Cook, K.; Stepansky, A.; Levy, D.; Esposito, D.; et al. Tumour evolution inferred by single-cell sequencing. *Nature* **2011**, *472*, 90–94. [CrossRef] [PubMed]
126. Chen, G.; Bradford, W.D.; Seidel, C.W.; Li, R. Hsp90 stress potentiates rapid cellular adaptation through induction of aneuploidy. *Nature* **2012**, *482*, 246–250. [CrossRef] [PubMed]
127. Chen, G.; Mulla, W.A.; Kucharavy, A.; Tsai, H.J.; Rubinstein, B.; Conkright, J.; McCroskey, S.; Bradford, W.D.; Weems, L.; Haug, J.S.; et al. Targeting the adaptability of heterogeneous aneuploids. *Cell* **2015**, *160*, 771–784. [CrossRef] [PubMed]
128. Thomson, D.D.; Berman, J.; Brand, A.C. High frame-rate resolution of cell division during Candida albicans filamentation. *Fungal Genet. Biol.* **2016**, *88*, 54–58. [CrossRef] [PubMed]
129. Gerstein, A.C.; Berman, J. Shift and adapt: The costs and benefits of karyotype variations. *Curr. Opin. Microbiol.* **2015**, *26*, 130–136. [CrossRef] [PubMed]
130. Lee, A.J.; Endesfelder, D.; Rowan, A.J.; Walther, A.; Birkbak, N.J.; Futreal, P.A.; Downward, J.; Szallasi, Z.; Tomlinson, I.P.; Howell, M.; et al. Chromosomal instability confers intrinsic multidrug resistance. *Cancer Res.* **2011**, *71*, 1858–1870. [CrossRef] [PubMed]
131. Duesberg, P.; Stindl, R.; Hehlmann, R. Explaining the high mutation rates of cancer cells to drug and multidrug resistance by chromosome reassortments that are catalyzed by aneuploidy. *Proc. Natl. Acad. Sci. USA* **2000**, *97*, 14295–14300. [CrossRef] [PubMed]
132. Swanton, C.; Nicke, B.; Schuett, M.; Eklund, A.C.; Ng, C.; Li, Q.; Hardcastle, T.; Lee, A.; Roy, R.; East, P.; et al. Chromosomal instability determines taxane response. *Proc. Natl. Acad. Sci. USA* **2009**, *106*, 8671–8676. [CrossRef] [PubMed]
133. Guo, J.; Xu, S.; Huang, X.; Li, L.; Zhang, C.; Pan, Q.; Ren, Z.; Zhou, R.; Ren, Y.; Zi, J.; et al. Drug Resistance in colorectal cancer cell lines is partially associated with aneuploidy status in light of profiling gene expression. *J. Proteome Res.* **2016**, *15*, 4047–4059. [CrossRef] [PubMed]
134. Duesberg, P.; Stindl, R.; Hehlmann, R. Origin of multidrug resistance in cells with and without multidrug resistance genes: Chromosome reassortments catalyzed by aneuploidy. *Proc. Natl. Acad. Sci. USA* **2001**, *98*, 11283–11288. [CrossRef] [PubMed]
135. Pavelka, N.; Rancati, G.; Li, R. Dr Jekyll and Mr Hyde: Role of aneuploidy in cellular adaptation and cancer. *Curr. Opin. Cell Biol.* **2010**, *22*, 809–815. [CrossRef] [PubMed]
136. Luzhna, L.; Kathiria, P.; Kovalchuk, O. Micronuclei in genotoxicity assessment: From genetics to epigenetics and beyond. *Front. Genet.* **2013**, *4*, 131–147. [CrossRef] [PubMed]
137. Maciejowski, J.; Li, Y.; Bosco, N.; Campbell, P.J.; De Lange, T. Chromothripsis and Kataegis Induced by Telomere Crisis. *Cell* **2015**, *163*, 1641–1654. [CrossRef] [PubMed]
138. Hoffelder, D.R.; Luo, L.; Burke, N.A.; Watkins, S.C.; Gollin, S.M.; Saunders, W.S. Resolution of anaphase bridges in cancer cells. *Chromosoma* **2004**, *112*, 389–397. [CrossRef] [PubMed]
139. Crasta, K.; Ganem, N.J.; Dagher, R.; Lantermann, A.B.; Ivanova, E.V.; Pan, Y.; Nezi, L.; Protopopov, A.; Chowdhury, D.; Pellman, D. DNA breaks and chromosome pulverization from errors in mitosis. *Nature* **2012**, *482*, 53–58. [CrossRef] [PubMed]
140. Hatch, E.M.; Fischer, A.H.; Deerinck, T.J.; Hetzer, M.W. Catastrophic nuclear envelope collapse in cancer cell micronuclei. *Cell* **2013**, *154*, 47–60. [CrossRef] [PubMed]
141. Stephens, P.J.; Greenman, C.D.; Fu, B.; Yang, F.; Bignell, G.R.; Mudie, L.J.; Pleasance, E.D.; Lau, K.W.; Beare, D.; Stebbings, L.A.; et al. Massive genomic rearrangement acquired in a single catastrophic event during cancer development. *Cell* **2011**, *144*, 27–40. [CrossRef] [PubMed]
142. Heng, H.H.; Stevens, J.B.; Bremer, S.W.; Liu, G.; Abdallah, B.Y.; Ye, C.J. Evolutionary mechanisms and diversity in cancer. *Adv. Cancer Res.* **2011**, *112*, 217–253. [PubMed]
143. Leibowitz, M.L.; Zhang, C.Z.; Pellman, D. Chromothripsis: A new mechanism for rapid karyotype evolution. *Annu. Rev. Genet.* **2015**, *49*, 183–211. [CrossRef] [PubMed]
144. Ly, P.; Teitz, L.S.; Kim, D.H.; Shoshani, O.; Skaletsky, H.; Fachinetti, D.; Page, D.C.; Cleveland, D.W. Selective Y centromere inactivation triggers chromosome shattering in micronuclei and repair by non-homologous end joining. *Nat. Cell Biol.* **2017**, *19*, 68–75. [CrossRef] [PubMed]
145. Nagaoka, S.I.; Hassold, T.J.; Hunt, P.A. Human aneuploidy: Mechanisms and new insights into an age-old problem. *Nat. Rev. Genet.* **2012**, *13*, 493–504. [CrossRef] [PubMed]

146. Sansam, C.L.; Pezza, R.J. Connecting by breaking and repairing: mechanisms of DNA strand exchange in meiotic recombination. *FEBS J.* **2015**, *282*, 2444–2457. [CrossRef] [PubMed]
147. Bomblies, K.; Jones, G.; Franklin, C.; Zickler, D.; Kleckner, N. The challenge of evolving stable polyploidy: Could an increase in "crossover interference distance" play a central role? *Chromosoma* **2016**, *125*, 287–300. [CrossRef] [PubMed]
148. Eaker, S.; Pyle, A.; Cobb, J.; Handel, M.A. Evidence for meiotic spindle checkpoint from analysis of spermatocytes from Robertsonian-chromosome heterozygous mice. *J. Cell Sci.* **2001**, *114*, 2953–2965. [PubMed]
149. Vrooman, L.A.; Nagaoka, S.I.; Hassold, T.J.; Hunt, P.A. Evidence for paternal age-related alterations in meiotic chromosome dynamics in the mouse. *Genetics* **2014**, *196*, 385–396. [CrossRef] [PubMed]
150. Brunet, S.; Pahlavan, G.; Taylor, S.; Maro, B. Functionality of the spindle checkpoint during the first meiotic division of mammalian oocytes. *Reproduction* **2003**, *126*, 443–450. [CrossRef] [PubMed]
151. Homer, H.A.; McDougall, A.; Levasseur, M.; Murdoch, A.P.; Herbert, M. Mad2 is required for inhibiting securin and cyclin B degradation following spindle depolymerisation in meiosis I mouse oocytes. *Reproduction* **2005**, *130*, 829–843. [CrossRef] [PubMed]
152. LeMaire-Adkins, R.; Radke, K.; Hunt, P.A. Lack of checkpoint control at the metaphase/anaphase transition: A mechanism of meiotic nondisjunction in mammalian females. *J. Cell Biol.* **1997**, *139*, 1611–1619. [CrossRef] [PubMed]
153. Hoffmann, S.; Maro, B.; Kubiak, J.Z.; Polanski, Z. A single bivalent efficiently inhibits cyclin B1 degradation and polar body extrusion in mouse oocytes indicating robust SAC during female meiosis I. *PLoS ONE* **2011**, *6*, e27143. [CrossRef] [PubMed]
154. Shao, H.; Li, R.; Ma, C.; Chen, E.; Liu, X.J. Xenopus oocyte meiosis lacks spindle assembly checkpoint control. *J. Cell Biol.* **2013**, *201*, 191–200. [CrossRef] [PubMed]
155. Murray, A.W.; Kirschner, M.W. Cyclin synthesis drives the early embryonic cell cycle. *Nature* **1989**, *339*, 275–280. [CrossRef] [PubMed]
156. Minshull, J.; Sun, H.; Tonks, N.K.; Murray, A.W. A MAP kinase-dependent spindle assembly checkpoint in Xenopus egg extracts. *Cell* **1994**, *79*, 475–486. [CrossRef]
157. Kuliev, A.; Zlatopolsky, Z.; Kirillova, I.; Spivakova, J.; Cieslak Janzen, J. Meiosis errors in over 20,000 oocytes studied in the practice of preimplantation aneuploidy testing. *Reprod. Biomed. Online* **2011**, *22*, 2–8. [CrossRef] [PubMed]
158. Fragouli, E.; Alfarawati, S.; Goodall, N.N.; Sanchez-Garcia, J.F.; Colls, P.; Wells, D. The cytogenetics of polar bodies: Insights into female meiosis and the diagnosis of aneuploidy. *Mol. Hum. Reprod.* **2011**, *17*, 286–295. [CrossRef] [PubMed]
159. Steuerwald, N.M.; Bermudez, M.G.; Wells, D.; Munne, S.; Cohen, J. Maternal age-related differential global expression profiles observed in human oocytes. *Reprod. Biomed. Online* **2007**, *14*, 700–708. [CrossRef]
160. Yun, Y.; Holt, J.E.; Lane, S.I.; McLaughlin, E.A.; Merriman, J.A.; Jones, K.T. Reduced ability to recover from spindle disruption and loss of kinetochore spindle assembly checkpoint proteins in oocytes from aged mice. *Cell Cycle* **2014**, *13*, 1938–1947. [CrossRef] [PubMed]
161. Pan, H.; Ma, P.; Zhu, W.; Schultz, R.M. Age-associated increase in aneuploidy and changes in gene expression in mouse eggs. *Dev. Biol.* **2008**, *316*, 397–407. [CrossRef] [PubMed]
162. Duncan, F.E.; Hornick, J.E.; Lampson, M.A.; Schultz, R.M.; Shea, L.D.; Woodruff, T.K. Chromosome cohesion decreases in human eggs with advanced maternal age. *Aging Cell* **2012**, *11*, 1121–1124. [CrossRef] [PubMed]
163. Yun, Y.; Lane, S.I.; Jones, K.T. Premature dyad separation in meiosis II is the major segregation error with maternal age in mouse oocytes. *Development* **2014**, *141*, 199–208. [CrossRef] [PubMed]
164. Rankin, S.; Dawson, D.S. Recent advances in cohesin biology. *F1000Research* **2016**, *5*. [CrossRef] [PubMed]
165. Revenkova, E.; Herrmann, K.; Adelfalk, C.; Jessberger, R. Oocyte cohesin expression restricted to predictyate stages provides full fertility and prevents aneuploidy. *Curr. Biol.* **2010**, *20*, 1529–1533. [CrossRef] [PubMed]
166. Tachibana-Konwalski, K.; Godwin, J.; Van Der Weyden, L.; Champion, L.; Kudo, N.R.; Adams, D.J.; Nasmyth, K. Rec8-containing cohesin maintains bivalents without turnover during the growing phase of mouse oocytes. *Genes Dev.* **2010**, *24*, 2505–2516. [CrossRef] [PubMed]
167. Chiang, T.; Schultz, R.M.; Lampson, M.A. Age-dependent susceptibility of chromosome cohesion to premature separase activation in mouse oocytes. *Biol. Reprod.* **2011**, *85*, 1279–1283. [CrossRef] [PubMed]

168. Lister, L.M.; Kouznetsova, A.; Hyslop, L.A.; Kalleas, D.; Pace, S.L.; Barel, J.C.; Nathan, A.; Floros, V.; Adelfalk, C.; Watanabe, Y.; et al. Age-related meiotic segregation errors in mammalian oocytes are preceded by depletion of cohesin and Sgo2. *Curr. Biol.* **2010**, *20*, 1511–1521. [CrossRef] [PubMed]
169. Liu, L.; Keefe, D.L. Defective cohesin is associated with age-dependent misaligned chromosomes in oocytes. *Reprod. Biomed. Online* **2008**, *16*, 103–112. [CrossRef]
170. Tsutsumi, M.; Fujiwara, R.; Nishizawa, H.; Ito, M.; Kogo, H.; Inagaki, H.; Ohye, T.; Kato, T.; Fujii, T.; Kurahashi, H. Age-related decrease of meiotic cohesins in human oocytes. *PLoS ONE* **2014**, *9*, e96710. [CrossRef] [PubMed]
171. Ottolini, C.S.; Newnham, L.J.; Capalbo, A.; Natesan, S.A.; Joshi, H.A.; Cimadomo, D.; Griffin, D.K.; Sage, K.; Summers, M.C.; Thornhill, A.R.; et al. Genome-wide maps of recombination and chromosome segregation in human oocytes and embryos show selection for maternal recombination rates. *Nat. Genet.* **2015**, *47*, 727–735. [CrossRef] [PubMed]
172. Ljunger, E.; Cnattingius, S.; Lundin, C.; Anneren, G. Chromosomal anomalies in first-trimester miscarriages. *Acta Obstet. Gynecol. Scand.* **2005**, *84*, 1103–1107. [CrossRef] [PubMed]
173. Morales, C.; Sanchez, A.; Bruguera, J.; Margarit, E.; Borrell, A.; Borobio, V.; Soler, A. Cytogenetic study of spontaneous abortions using semi-direct analysis of chorionic villi samples detects the broadest spectrum of chromosome abnormalities. *Am. J. Med. Genet. A* **2008**, *146A*, 66–70. [CrossRef] [PubMed]
174. Fritz, B.; Hallermann, C.; Olert, J.; Fuchs, B.; Bruns, M.; Aslan, M.; Schmidt, S.; Coerdt, W.; Muntefering, H.; Rehder, H. Cytogenetic analyses of culture failures by comparative genomic hybridisation (CGH)-Re-evaluation of chromosome aberration rates in early spontaneous abortions. *Eur. J. Hum. Genet.* **2001**, *9*, 539–547. [CrossRef] [PubMed]
175. Bittles, A.H.; Bower, C.; Hussain, R.; Glasson, E.J. The four ages of Down syndrome. *Eur. J. Public Health* **2007**, *17*, 221–225. [CrossRef] [PubMed]
176. Sybert, V.P.; McCauley, E. Turner's syndrome. *N. Engl. J. Med.* **2004**, *351*, 1227–1238. [CrossRef] [PubMed]
177. Nguyen, D.K.; Disteche, C.M. Dosage compensation of the active X chromosome in mammals. *Nat. Genet.* **2006**, *38*, 47–53. [CrossRef] [PubMed]
178. Lopes, A.M.; Burgoyne, P.S.; Ojarikre, A.; Bauer, J.; Sargent, C.A.; Amorim, A.; Affara, N.A. Transcriptional changes in response to X chromosome dosage in the mouse: Implications for X inactivation and the molecular basis of Turner Syndrome. *BMC Genom.* **2010**, *11*, 82–94. [CrossRef] [PubMed]
179. Otter, M.; Schrander-Stumpel, C.T.; Curfs, L.M. Triple X syndrome: A review of the literature. *Eur. J. Hum. Genet.* **2010**, *18*, 265–271. [CrossRef] [PubMed]
180. Visootsak, J.; Graham, J.M., Jr. Klinefelter syndrome and other sex chromosomal aneuploidies. *Orphanet J. Rare Dis.* **2006**, *1*, 42–46. [CrossRef] [PubMed]
181. Abramowitz, L.K.; Olivier-Van Stichelen, S.; Hanover, J.A. Chromosome imbalance as a driver of sex disparity in disease. *J. Genom.* **2014**, *2*, 77–88. [CrossRef] [PubMed]
182. Khan, I.; Malinge, S.; Crispino, J. Myeloid leukemia in Down syndrome. *Crit. Rev. Oncog.* **2011**, *16*, 25–36. [CrossRef] [PubMed]
183. Hama, A.; Muramatsu, H.; Makishima, H.; Sugimoto, Y.; Szpurka, H.; Jasek, M.; O'Keefe, C.; Takahashi, Y.; Sakaguchi, H.; Doisaki, S.; et al. Molecular lesions in childhood and adult acute megakaryoblastic leukaemia. *Br. J. Haematol.* **2012**, *156*, 316–325. [CrossRef] [PubMed]
184. Rainis, L.; Toki, T.; Pimanda, J.E.; Rosenthal, E.; Machol, K.; Strehl, S.; Gottgens, B.; Ito, E.; Izraeli, S. The proto-oncogene ERG in megakaryoblastic leukemias. *Cancer Res.* **2005**, *65*, 7596–7602. [PubMed]
185. Stankiewicz, M.J.; Crispino, J.D. ETS2 and ERG promote megakaryopoiesis and synergize with alterations in GATA-1 to immortalize hematopoietic progenitor cells. *Blood* **2009**, *113*, 3337–3347. [CrossRef] [PubMed]
186. Scott, R.H.; Stiller, C.A.; Walker, L.; Rahman, N. Syndromes and constitutional chromosomal abnormalities associated with Wilms tumour. *J. Med. Genet.* **2006**, *43*, 705–715. [CrossRef] [PubMed]
187. Wheatley, D.N. Binucleation in mammalian liver. Studies on the control of cytokinesis in vivo. *Exp. Cell Res.* **1972**, *74*, 455–465. [CrossRef]
188. Eggert, U.S.; Mitchison, T.J.; Field, C.M. Animal cytokinesis: From parts list to mechanisms. *Annu. Rev. Biochem.* **2006**, *75*, 543–566. [CrossRef] [PubMed]
189. Glotzer, M. The molecular requirements for cytokinesis. *Science* **2005**, *307*, 1735–1739. [CrossRef] [PubMed]

190. Steigemann, P.; Wurzenberger, C.; Schmitz, M.H.; Held, M.; Guizetti, J.; Maar, S.; Gerlich, D.W. Aurora B-mediated abscission checkpoint protects against tetraploidization. *Cell* **2009**, *136*, 473–484. [CrossRef] [PubMed]
191. Mullins, J.M.; Biesele, J.J. Terminal phase of cytokinesis in D-98s cells. *J. Cell Biol.* **1977**, *73*, 672–684. [CrossRef] [PubMed]
192. Brito, D.A.; Rieder, C.L. Mitotic checkpoint slippage in humans occurs via cyclin B destruction in the presence of an active checkpoint. *Curr. Biol.* **2006**, *16*, 1194–1200. [CrossRef] [PubMed]
193. Potapova, T.A.; Daum, J.R.; Pittman, B.D.; Hudson, J.R.; Jones, T.N.; Satinover, D.L.; Stukenberg, P.T.; Gorbsky, G.J. The reversibility of mitotic exit in vertebrate cells. *Nature* **2006**, *440*, 954–958. [CrossRef] [PubMed]
194. Schmidt, M.; Budirahardja, Y.; Klompmaker, R.; Medema, R.H. Ablation of the spindle assembly checkpoint by a compound targeting Mps1. *EMBO Rep.* **2005**, *6*, 866–872. [CrossRef] [PubMed]
195. Santaguida, S.; Tighe, A.; D'Alise, A.M.; Taylor, S.S.; Musacchio, A. Dissecting the role of MPS1 in chromosome biorientation and the spindle checkpoint through the small molecule inhibitor reversine. *J. Cell Biol.* **2010**, *190*, 73–87. [CrossRef] [PubMed]
196. Nagata, Y.; Muro, Y.; Todokoro, K. Thrombopoietin-induced polyploidization of bone marrow megakaryocytes is due to a unique regulatory mechanism in late mitosis. *J. Cell Biol.* **1997**, *139*, 449–457. [CrossRef] [PubMed]
197. Vitrat, N.; Cohen-Solal, K.; Pique, C.; Le Couedic, J.P.; Norol, F.; Larsen, A.K.; Katz, A.; Vainchenker, W.; Debili, N. Endomitosis of human megakaryocytes are due to abortive mitosis. *Blood* **1998**, *91*, 3711–3723. [PubMed]
198. Lee, H.O.; Davidson, J.M.; Duronio, R.J. Endoreplication: polyploidy with purpose. *Genes Dev.* **2009**, *23*, 2461–2477. [CrossRef] [PubMed]
199. Zhimulev, I.F.; Belyaeva, E.S.; Semeshin, V.F.; Koryakov, D.E.; Demakov, S.A.; Demakova, O.V.; Pokholkova, G.V.; Andreyeva, E.N. Polytene chromosomes: 70 years of genetic research. *Int. Rev. Cytol.* **2004**, *241*, 203–275. [PubMed]
200. Zybina, E.V.; Zybina, T.G. Polytene chromosomes in mammalian cells. *Int. Rev. Cytol.* **1996**, *165*, 53–119. [PubMed]
201. Hu, D.; Cross, J.C. Development and function of trophoblast giant cells in the rodent placenta. *Int. J. Dev. Biol.* **2010**, *54*, 341–354. [CrossRef] [PubMed]
202. Faggioli, F.; Sacco, M.G.; Susani, L.; Montagna, C.; Vezzoni, P. Cell fusion is a physiological process in mouse liver. *Hepatology* **2008**, *48*, 1655–1664. [CrossRef] [PubMed]
203. Okamura, K.; Asahina, K.; Fujimori, H.; Ozeki, R.; Shimizu-Saito, K.; Tanaka, Y.; Teramoto, K.; Arii, S.; Takase, K.; Kataoka, M.; et al. Generation of hybrid hepatocytes by cell fusion from monkey embryoid body cells in the injured mouse liver. *Histochem. Cell Biol.* **2006**, *125*, 247–257. [CrossRef] [PubMed]
204. Duelli, D.; Lazebnik, Y. Cell-to-cell fusion as a link between viruses and cancer. *Nat. Rev. Cancer* **2007**, *7*, 968–976. [CrossRef] [PubMed]
205. Pawelek, J.M.; Chakraborty, A.K. Fusion of tumour cells with bone marrow-derived cells: A unifying explanation for metastasis. *Nat. Rev. Cancer* **2008**, *8*, 377–386. [CrossRef] [PubMed]
206. Okagaki, L.H.; Nielsen, K. Titan cells confer protection from phagocytosis in Cryptococcus neoformans infections. *Eukaryot. Cell* **2012**, *11*, 820–826. [CrossRef] [PubMed]
207. Harrison, B.D.; Hashemi, J.; Bibi, M.; Pulver, R.; Bavli, D.; Nahmias, Y.; Wellington, M.; Sapiro, G.; Berman, J. A tetraploid intermediate precedes aneuploid formation in yeasts exposed to fluconazole. *PLoS Biol.* **2014**, *12*, e1001815. [CrossRef] [PubMed]
208. Gallardo, M.H.; Bickham, J.W.; Honeycutt, R.L.; Ojeda, R.A.; Kohler, N. Discovery of tetraploidy in a mammal. *Nature* **1999**, *401*, 341. [CrossRef] [PubMed]
209. Otto, S.P. The evolutionary consequences of polyploidy. *Cell* **2007**, *131*, 452–462. [CrossRef] [PubMed]
210. Hellsten, U.; Khokha, M.K.; Grammer, T.C.; Harland, R.M.; Richardson, P.; Rokhsar, D.S. Accelerated gene evolution and subfunctionalization in the pseudotetraploid frog Xenopus laevis. *BMC Biol.* **2007**, *5*, 31–44. [CrossRef] [PubMed]
211. Session, A.M.; Uno, Y.; Kwon, T.; Chapman, J.A.; Toyoda, A.; Takahashi, S.; Fukui, A.; Hikosaka, A.; Suzuki, A.; Kondo, M.; et al. Genome evolution in the allotetraploid frog Xenopus laevis. *Nature* **2016**, *538*, 336–343. [CrossRef] [PubMed]

212. Kaufman, M.H. New insights into triploidy and tetraploidy, from an analysis of model systems for these conditions. *Hum. Reprod.* **1991**, *6*, 8–16. [PubMed]
213. Nakamura, Y.; Takaira, M.; Sato, E.; Kawano, K.; Miyoshi, O.; Niikawa, N. A tetraploid liveborn neonate: Cytogenetic and autopsy findings. *Arch. Pathol. Lab. Med.* **2003**, *127*, 1612–1614. [PubMed]
214. Stefanova, I.; Jenderny, J.; Kaminsky, E.; Mannhardt, A.; Meinecke, P.; Grozdanova, L.; Gillessen-Kaesbach, G. Mosaic and complete tetraploidy in live-born infants: Two new patients and review of the literature. *Clin. Dysmorphol.* **2010**, *19*, 123–127. [CrossRef] [PubMed]
215. Roberts, H.E.; Saxe, D.F.; Muralidharan, K.; Coleman, K.B.; Zacharias, J.F.; Fernhoff, P.M. Unique mosaicism of tetraploidy and trisomy 8: Clinical, cytogenetic, and molecular findings in a live-born infant. *Am. J. Med. Genet.* **1996**, *62*, 243–246. [CrossRef]
216. Soltis, P.S.; Marchant, D.B.; Van de Peer, Y.; Soltis, D.E. Polyploidy and genome evolution in plants. *Curr. Opin. Genet. Dev.* **2015**, *35*, 119–125. [CrossRef] [PubMed]
217. Korthout, H.A.; Caspers, M.P.; Kottenhagen, M.J.; Helmer, Q.; Wang, M. A tormentor in the quest for plant p53-like proteins. *FEBS Lett.* **2002**, *526*, 53–57. [CrossRef]
218. Salman-Minkov, A.; Sabath, N.; Mayrose, I. Whole-genome duplication as a key factor in crop domestication. *Nat. Plants* **2016**, *2*, 16115. [CrossRef] [PubMed]
219. Renny-Byfield, S.; Wendel, J.F. Doubling down on genomes: Polyploidy and crop plants. *Am. J. Bot.* **2014**, *101*, 1711–1725. [CrossRef] [PubMed]
220. Ganem, N.J.; Storchova, Z.; Pellman, D. Tetraploidy, aneuploidy and cancer. *Curr. Opin. Genet. Dev.* **2007**, *17*, 157–162. [CrossRef] [PubMed]
221. Quintyne, N.J.; Reing, J.E.; Hoffelder, D.R.; Gollin, S.M.; Saunders, W.S. Spindle multipolarity is prevented by centrosomal clustering. *Science* **2005**, *307*, 127–129. [CrossRef] [PubMed]
222. Ganem, N.J.; Godinho, S.A.; Pellman, D. A mechanism linking extra centrosomes to chromosomal instability. *Nature* **2009**, *460*, 278–282. [CrossRef] [PubMed]
223. Duncan, A.W.; Hanlon Newell, A.E.; Smith, L.; Wilson, E.M.; Olson, S.B.; Thayer, M.J.; Strom, S.C.; Grompe, M. Frequent aneuploidy among normal human hepatocytes. *Gastroenterology* **2012**, *142*, 25–28. [CrossRef] [PubMed]
224. Storchova, Z.; Breneman, A.; Cande, J.; Dunn, J.; Burbank, K.; O'Toole, E.; Pellman, D. Genome-wide genetic analysis of polyploidy in yeast. *Nature* **2006**, *443*, 541–547. [CrossRef] [PubMed]
225. Carter, S.L.; Cibulskis, K.; Helman, E.; McKenna, A.; Shen, H.; Zack, T.; Laird, P.W.; Onofrio, R.C.; Winckler, W.; Weir, B.A.; et al. Absolute quantification of somatic DNA alterations in human cancer. *Nat. Biotechnol.* **2012**, *30*, 413–421. [CrossRef] [PubMed]
226. Dewhurst, S.M.; McGranahan, N.; Burrell, R.A.; Rowan, A.J.; Gronroos, E.; Endesfelder, D.; Joshi, T.; Mouradov, D.; Gibbs, P.; Ward, R.L.; et al. Tolerance of whole-genome doubling propagates chromosomal instability and accelerates cancer genome evolution. *Cancer Discov.* **2014**, *4*, 175–185. [CrossRef] [PubMed]
227. De Bruin, E.C.; McGranahan, N.; Mitter, R.; Salm, M.; Wedge, D.C.; Yates, L.; Jamal-Hanjani, M.; Shafi, S.; Murugaesu, N.; Rowan, A.J.; et al. Spatial and temporal diversity in genomic instability processes defines lung cancer evolution. *Science* **2014**, *346*, 251–256. [CrossRef] [PubMed]
228. Zack, T.I.; Schumacher, S.E.; Carter, S.L.; Cherniack, A.D.; Saksena, G.; Tabak, B.; Lawrence, M.S.; Zhsng, C.Z.; Wala, J.; Mermel, C.H.; et al. Pan-cancer patterns of somatic copy number alteration. *Nat. Genet.* **2013**, *45*, 1134–1140. [CrossRef] [PubMed]
229. Galipeau, P.C.; Cowan, D.S.; Sanchez, C.A.; Barrett, M.T.; Emond, M.J.; Levine, D.S.; Rabinovitch, P.S.; Reid, B.J. 17p (p53) allelic losses, 4N (G2/tetraploid) populations, and progression to aneuploidy in Barrett's esophagus. *Proc. Natl. Acad. Sci. USA* **1996**, *93*, 7081–7084. [CrossRef] [PubMed]
230. Olaharski, A.J.; Sotelo, R.; Solorza-Luna, G.; Gonsebatt, M.E.; Guzman, P.; Mohar, A.; Eastmond, D.A. Tetraploidy and chromosomal instability are early events during cervical carcinogenesis. *Carcinogenesis* **2006**, *27*, 337–343. [CrossRef] [PubMed]
231. Ornitz, D.M.; Hammer, R.E.; Messing, A.; Palmiter, R.D.; Brinster, R.L. Pancreatic Neoplasia induced by SV40 T-antigen expression in acinar cells of transgenic mice. *Science* **1987**, *238*, 188–193. [CrossRef] [PubMed]
232. Giannoudis, A.; Evans, M.F.; Southern, S.A.; Herrington, C.S. Basal keratinocyte tetrasomy in low-grade squamous intra-epithelial lesions of the cervix is restricted to high and intermediate risk HPV infection but is not type-specific. *Br. J. Cancer* **2000**, *82*, 424–428. [CrossRef] [PubMed]

233. Fujiwara, T.; Bandi, M.; Nitta, M.; Ivanova, E.V.; Bronson, R.T.; Pellman, D. Cytokinesis failure generating tetraploids promotes tumorigenesis in p53-null cells. *Nature* **2005**, *437*, 1043–1047. [CrossRef] [PubMed]
234. Davoli, T.; De Lange, T. Telomere-driven tetraploidization occurs in human cells undergoing crisis and promotes transformation of mouse cells. *Cancer Cell* **2012**, *21*, 765–776. [CrossRef] [PubMed]
235. Castillo, A.; Morse, H.C., 3rd; Godfrey, V.L.; Naeem, R.; Justice, M.J. Overexpression of Eg5 causes genomic instability and tumor formation in mice. *Cancer Res.* **2007**, *67*, 10138–10147. [CrossRef] [PubMed]
236. Nigg, E.A. Origins and consequences of centrosome aberrations in human cancers. *Int. J. Cancer* **2006**, *119*, 2717–2723. [CrossRef] [PubMed]
237. Godinho, S.A.; Kwon, M.; Pellman, D. Centrosomes and cancer: How cancer cells divide with too many centrosomes. *Cancer Metastasis Rev.* **2009**, *28*, 85–98. [CrossRef] [PubMed]
238. Gisselsson, D.; Jin, Y.; Lindgren, D.; Persson, J.; Gisselsson, L.; Hanks, S.; Sehic, D.; Mengelbier, L.H.; Ora, I.; Rahman, N.; et al. Generation of trisomies in cancer cells by multipolar mitosis and incomplete cytokinesis. *Proc. Natl. Acad. Sci. USA* **2010**, *107*, 20489–20493. [CrossRef] [PubMed]
239. Kwon, M.; Godinho, S.A.; Chandhok, N.S.; Ganem, N.J.; Azioune, A.; Thery, M.; Pellman, D. Mechanisms to suppress multipolar divisions in cancer cells with extra centrosomes. *Genes Dev.* **2008**, *22*, 2189–2203. [CrossRef] [PubMed]
240. Galimberti, F.; Thompson, S.L.; Ravi, S.; Compton, D.A.; Dmitrovsky, E. Anaphase catastrophe is a target for cancer therapy. *Clin. Cancer Res.* **2011**, *17*, 1218–1222. [CrossRef] [PubMed]
241. Leber, B.; Maier, B.; Fuchs, F.; Chi, J.; Riffel, P.; Anderhub, S.; Wagner, L.; Ho, A.D.; Salisbury, J.L.; Boutros, M.; et al. Proteins required for centrosome clustering in cancer cells. *Sci. Transl. Med.* **2010**, *2*, 33ra38. [CrossRef] [PubMed]
242. Kuznetsova, A.Y.; Seget, K.; Moeller, G.K.; de Pagter, M.S.; de Roos, J.A.; Durrbaum, M.; Kuffer, C.; Muller, S.; Zaman, G.J.; Kloosterman, W.P.; et al. Chromosomal instability, tolerance of mitotic errors and multidrug resistance are promoted by tetraploidization in human cells. *Cell Cycle* **2015**, *14*, 2810–2820. [CrossRef] [PubMed]
243. Balsas, P.; Galan-Malo, P.; Marzo, I.; Naval, J. Bortezomib resistance in a myeloma cell line is associated to PSMbeta5 overexpression and polyploidy. *Leuk. Res.* **2012**, *36*, 212–218. [CrossRef] [PubMed]
244. Puig, P.E.; Guilly, M.N.; Bouchot, A.; Droin, N.; Cathelin, D.; Bouyer, F.; Favier, L.; Ghiringhelli, F.; Kroemer, G.; Solary, E.; et al. Tumor cells can escape DNA-damaging cisplatin through DNA endoreduplication and reversible polyploidy. *Cell Biol. Int.* **2008**, *32*, 1031–1043. [CrossRef] [PubMed]
245. Zhang, S.; Mercado-Uribe, I.; Xing, Z.; Sun, B.; Kuang, J.; Liu, J. Generation of cancer stem-like cells through the formation of polyploid giant cancer cells. *Oncogene* **2014**, *33*, 116–128. [CrossRef] [PubMed]
246. Sharma, S.; Zeng, J.Y.; Zhuang, C.M.; Zhou, Y.Q.; Yao, H.P.; Hu, X.; Zhang, R.; Wang, M.H. Small-molecule inhibitor BMS-777607 induces breast cancer cell polyploidy with increased resistance to cytotoxic chemotherapy agents. *Mol. Cancer Ther.* **2013**, *12*, 725–736. [CrossRef] [PubMed]
247. Hinchcliffe, E.H.; Day, C.A.; Karanjeet, K.B.; Fadness, S.; Langfald, A.; Vaughan, K.T.; Dong, Z. Chromosome missegregation during anaphase triggers p53 cell cycle arrest through histone H3.3 Ser31 phosphorylation. *Nat. Cell Biol.* **2016**, *18*, 668–675. [CrossRef] [PubMed]
248. Lentini, L.; Piscitello, D.; Veneziano, L.; Di Leonardo, A. Simultaneous reduction of MAD2 and BUBR1 expression induces mitotic spindle alterations associated with p53 dependent cell cycle arrest and death. *Cell Biol. Int.* **2014**, *38*, 933–941. [CrossRef] [PubMed]
249. Sanchez Alvarado, A. Cellular hyperproliferation and cancer as evolutionary variables. *Curr. Biol.* **2012**, *22*, R772–R778. [CrossRef] [PubMed]
250. Leroi, A.M.; Koufopanou, V.; Burt, A. Cancer selection. *Nat. Rev. Cancer* **2003**, *3*, 226–231. [CrossRef] [PubMed]
251. Caulin, A.F.; Maley, C.C. Peto's Paradox: Evolution's prescription for cancer prevention. *Trends Ecol. Evol.* **2011**, *26*, 175–182. [CrossRef] [PubMed]
252. Peto, R. Quantitative implications of the approximate irrelevance of mammalian body size and lifespan to lifelong cancer risk. *Philos. Trans. R. Soc. Lond. B Biol. Sci.* **2015**, *370*. [CrossRef] [PubMed]
253. Sulak, M.; Fong, L.; Mika, K.; Chigurupati, S.; Yon, L.; Mongan, N.P.; Emes, R.D.; Lynch, V.J. TP53 copy number expansion is associated with the evolution of increased body size and an enhanced DNA damage response in elephants. *Elife* **2016**, *5*, e11994. [PubMed]

254. Olivier, M.; Hollstein, M.; Hainaut, P. TP53 mutations in human cancers: Origins, consequences, and clinical use. *Cold Spring Harbor Perspect. Biol.* **2010**, *2*, a001008. [CrossRef] [PubMed]
255. Brown, C.J.; Lain, S.; Verma, C.S.; Fersht, A.R.; Lane, D.P. Awakening guardian angels: Drugging the p53 pathway. *Nat. Rev. Cancer* **2009**, *9*, 862–873. [CrossRef] [PubMed]
256. Vogelstein, B.; Lane, D.; Levine, A.J. Surfing the p53 network. *Nature* **2000**, *408*, 307–310. [CrossRef] [PubMed]
257. Horn, H.F.; Vousden, K.H. Coping with stress: Multiple ways to activate p53. *Oncogene* **2007**, *26*, 1306–1316. [CrossRef] [PubMed]
258. Vousden, K.H.; Lane, D.P. P53 in health and disease. *Nat. Rev. Mol. Cell Biol.* **2007**, *8*, 275–283. [CrossRef] [PubMed]
259. Golomb, L.; Volarevic, S.; Oren, M. P53 and ribosome biogenesis stress: The essentials. *FEBS Lett.* **2014**, *588*, 2571–2579. [CrossRef] [PubMed]
260. Lambrus, B.G.; Uetake, Y.; Clutario, K.M.; Daggubati, V.; Snyder, M.; Sluder, G.; Holland, A.J. P53 protects against genome instability following centriole duplication failure. *J. Cell Biol.* **2015**, *210*, 63–77. [CrossRef] [PubMed]
261. Wong, Y.L.; Anzola, J.V.; Davis, R.L.; Yoon, M.; Motamedi, A.; Kroll, A.; Seo, C.P.; Hsia, J.E.; Kim, S.K.; Mitchell, J.W.; et al. Reversible centriole depletion with an inhibitor of Polo-like kinase 4. *Science* **2015**, *348*, 1155–1160. [CrossRef] [PubMed]
262. Ganem, N.J.; Cornils, H.; Chiu, S.Y.; O'Rourke, K.P.; Arnaud, J.; Yimlamai, D.; Thery, M.; Camargo, F.D.; Pellman, D. Cytokinesis failure triggers hippo tumor suppressor pathway activation. *Cell* **2014**, *158*, 833–848. [CrossRef] [PubMed]
263. Casenghi, M.; Mangiacasale, R.; Tuynder, M.; Caillet-Fauquet, P.; Elhajouji, A.; Lavia, P.; Mousset, S.; Kirsch-Volders, M.; Cundari, E. P53-independent apoptosis and p53-dependent block of DNA rereplication following mitotic spindle inhibition in human cells. *Exp. Cell Res.* **1999**, *250*, 339–350. [CrossRef] [PubMed]
264. Hirano, A.; Kurimura, T. Virally transformed cells and cytochalasin B: I. The effect of cytochalasin B on cytokinesis, karyokinesis and DNA synthesis in cells. *Exp. Cell Res.* **1974**, *89*, 111–120. [CrossRef]
265. Incassati, A.; Patel, D.; McCance, D.J. Induction of tetraploidy through loss of p53 and upregulation of Plk1 by human papillomavirus type-16 E6. *Oncogene* **2006**, *25*, 2444–2451. [CrossRef] [PubMed]
266. Khan, S.H.; Wahl, G.M. P53 and pRb prevent rereplication in response to microtubule inhibitors by mediating a reversible G1 arrest. *Cancer Res.* **1998**, *58*, 396–401. [PubMed]
267. Lanni, J.S.; Jacks, T. Characterization of the p53-dependent postmitotic checkpoint following spindle disruption. *Mol. Cell Biol.* **1998**, *18*, 1055–1064. [CrossRef] [PubMed]
268. Vitale, I.; Senovilla, L.; Jemaa, M.; Michaud, M.; Galluzzi, L.; Kepp, O.; Nanty, L.; Criollo, A.; Rello-Varona, S.; Manic, G.; et al. Multipolar mitosis of tetraploid cells: Inhibition by p53 and dependency on Mos. *EMBO J.* **2010**, *29*, 1272–1284. [CrossRef] [PubMed]
269. Krzywicka-Racka, A.; Sluder, G. Repeated cleavage failure does not establish centrosome amplification in untransformed human cells. *J. Cell Biol.* **2011**, *194*, 199–207. [CrossRef] [PubMed]
270. Crockford, A.; Zalmas, L.P.; Gronroos, E.; Dewhurst, S.M.; McGranahan, N.; Cuomo, M.E.; Encheva, V.; Snijders, A.P.; Begum, J.; Purewal, S.; et al. Cyclin D mediates tolerance of genome-doubling in cancers with functional p53. *Ann. Oncol.* **2016**. [CrossRef] [PubMed]
271. Manning, A.L.; Benes, C.; Dyson, N.J. Whole chromosome instability resulting from the synergistic effects of pRB and p53 inactivation. *Oncogene* **2014**, *33*, 2487–2494. [CrossRef] [PubMed]
272. Rehen, S.K.; Yung, Y.C.; McCreight, M.P.; Kaushal, D.; Yang, A.H.; Almeida, B.S.; Kingsbury, M.A.; Cabral, K.M.; McConnell, M.J.; Anliker, B.; et al. Constitutional aneuploidy in the normal human brain. *J. Neurosci.* **2005**, *25*, 2176–2180. [CrossRef] [PubMed]

MDPI AG
St. Alban-Anlage 66
4052 Basel, Switzerland
Tel. +41 61 683 77 34
Fax +41 61 302 89 18
http://www.mdpi.com

Biology Editorial Office
E-mail: biology@mdpi.com
http://www.mdpi.com/journal/biology

www.ingramcontent.com/pod-product-compliance
Lightning Source LLC
Chambersburg PA
CBHW051712210326
41597CB00032B/5457